A SOCIOLOGY OF SPORT

A SOCIOLOGY OF SPORT

HOWARD L. NIXON II
Towson State University

JAMES H. FREY
University of Nevada, Las Vegas

Wadsworth Publishing Company
I(T)P™ **An International Thomson Publishing Company**

Belmont • Albany • Bonn • Boston • Cincinnati • Detroit • London • Madrid • Melbourne • Mexico City • New York • Paris • San Francisco • Singapore • Tokyo • Toronto • Washington

Sociology Editor: Eve Howard
Editorial Assistant: Julie McDonald
Assistant Editor: Susan Shook
Production: Matrix Productions
Text and Cover Designer: Cynthia Bogue
Print Buyer: Karen Hunt
Photo Research: Image Quest
Permissions Editor: Jeanne Bosschart
Cover Art: *Strike* by Jacob Lawrence, tempura on masonite, 1949; The Howard University Gallery of Art, Washington, D.C.
Compositor: Futura Graphics
Printer: Malloy Lithographing, Inc.

Printed in the United States of America
1 2 3 4 5 6 7 8 9 10—02 01 00 99 98 97 96

For more information, contact Wadsworth Publishing Company:

Wadsworth Publishing Company
10 Davis Drive
Belmont, California 94002, USA

International Thomson Publishing Europe
Berkshire House 168-173
High Holborn
London, WC1V 7AA, England

Thomas Nelson Australia
102 Dodds Street
South Melbourne 3205
Victoria, Australia

Nelson Canada
1120 Birchmount Road
Scarborough, Ontario
Canada M1K 5G4

International Thomson Editores
Campos Eliseos 385, Piso 7
Col. Polanco
11560 México D.F. México

International Thomson Publishing GmbH
Königswinterer Strasse 418
53227 Bonn, Germany

International Thomson Publishing Asia
221 Henderson Road
#05-10 Henderson Building
Singapore 0315

International Thomson Publishing Japan
Hirakawacho Kyowa Building, 3F
2-2-1 Hirakawacho
Chiyoda-ku, Tokyo 102, Japan

Library of Congress Cataloging-in-Publication Data

Nixon, Howard L.
A sociology of sport / Howard L. Nixon, II, James H. Frey.
p. cm.
Includes bibliographical references and index.
ISBN 0-534-24762-8 (acid-free paper)
1. Sports—Sociological aspects. I. Frey, James H. II. Title.
GV706.5.N58 1996 95-15353
306.4'83—dc20 CIP

 *This book is printed on acid-free recycled paper.*

To Carol, Cory, Chris, and Ryan Frey
Sara, Matthew, Luke, and Daniel Nixon
and parents Claire and Chester Nixon, who have all been consistent supporters.

Preface

We are longtime sport sociologists with deep commitments to furthering the growth of sport sociology and sharing its insights and knowledge as widely as possible. Writing this text gave us a chance to fulfill these commitments. Our aim is to introduce readers to sociology through sport in ways that make it easier to understand enduring and significant social patterns, issues, and problems in society. In achieving this aim, we will show how sport is a complex social and cultural phenomenon that bears the distinctive imprint of the society and culture in which it exists. Thus, looking at such issues and problems as violence, racial, gender, and sexual tensions, illicit drug use, burnout, eating disorders, pain and injuries, cheating, strikes, and political and economic boycotts in sport allows us to learn more about broader aspects of society and culture. Because this book provides insights and understandings about society and sport, it is relevant to courses in sociology and other social sciences, physical education, and sport studies. Indeed, the general educational value of studying sport sociology was appreciated enough at the university of one of the authors that the introductory sport sociology course was included as one of the social science choices in the core curriculum of that university.

Preparing an introductory textbook forces authors to look at what other text authors have done to assess the current state of knowledge in the field, and to decide how to expose instructors and students to what is known in ways that inform, stimulate, and challenge. We recognize that we could not be exhaustive in our coverage of sport sociology, but we made every effort to provide broad coverage of the field. We examine the nature of the sociological study of sport and the social development and culture of sport, and we look at the relationship of sport to other social institutions such as the economy, government and politics, family, religion, education, and the mass media. The role of social class, race, and gender in sport and how social processes such as discrimination have created and sustained inequalities in these areas are also analyzed. In addition, we consider how sport is related to socialization and to structures and processes of social deviance. Along with examining various forms of "negative" deviance, such as violence, drug use, cheating, and corruption, we examine forms of "positive deviance," such as potentially self-destructive patterns of playing with pain and injury. We look at the major types of role players at various levels of sport, from youth sports to the major forms of commercialized amateur and professional sports. Whenever possible, we inject a global perspective by including a substantial amount of discussion materials from foreign sources or about sport in cultures and societies outside mainstream North America. The emphasis on global, cross-cultural, and cross-national sources is a distinctive focus of this book. Another important theme is how a masculine ethos influences participation in sport and its participants. In particular, we examine how concerns about masculinity are related to gender relations and issues of sexual orientation in sport.

In an effort to offer more depth in the analysis of interesting issues and problems in sport, "Special Focus" features were created in each chapter. Being instructors ourselves, we also recognized the value of including teaching aids such as chapter summaries, a glossary, and a test bank with this text. Along with its distinctive special features, teaching aids, and broad coverage of the field, this text is marked by our conception of sport sociology. As sociologists, we grounded this book in sociology. We earlier noted our approach to sport sociology as sociology through sport. That is, we suggested how an examination of the sociocultural aspects of sport provides valuable insights into what societies are like and how they operate. In this age of the proliferation of sports and the worldwide reach of the mass media, it is difficult to avoid exposure to some form of sport. Thus, sport is familiar to most people and seriously followed by many. We aim to move readers beyond their general familiarity with sport, their personal knowledge of it, or their fanatical devotion to it—which may have aroused their initial curiosity about a sport sociology course—to a sociological understanding of sport and society. We strongly emphasize the importance of sociological perspectives, facts, and insights for understanding sport and society. Without burdening students with unnecessary jargon, we carefully define sociological ideas and state social research findings that convey what sport sociologists currently know about the sociocultural aspects of sport and the relationship of sport to society. We distinguish sociological perspectives from other ways of seeing sport and society and characterize sport sociology as a social science. We provide instructors who wish to emphasize social research methods with an appendix on this topic. We hope that students finish the book with a more sophisticated sense of the sociological perspective, and of the culture, social structures, and social processes that characterize their own and other societies. In the process, students surely will develop a more sophisticated understanding of sport, which will take them beyond taken-for-granted beliefs, biases, common sense, and intuition.

We recognized that it was essential to provide features in each chapter that emphasize critical thinking. Thus, we developed the "Point-Counterpoint" as a highlighted feature offering pro and con arguments in response to major issues in sport today. This feature can be used for homework assignments and for debates in the classroom. The emphasis on critical thinking has become an explicit component of the liberal arts curriculum of most universities and colleges in recent years, but is sometimes difficult for students unaccustomed to asking questions about sport as athletes or fans to apply this emphasis to sport. We have known students who loved sport and were disturbed or angered when they read or heard criticisms of their long-held and cherished beliefs about sport and society. We do not advance a critical perspective, or what sociologist Peter Berger called a "debunking motif," because we dislike sport or want to tear it down. Both of us have had many gratifying experiences in a variety of roles in sport as well as in sport sociology. Our aim instead is to emphasize the importance of raising questions, testing assumptions, unmasking myths and unfounded biases, and finding what is true about sport and society. We believe that the process of critical, objective, systematic thinking to discover sociological insights and facts about sport and society will surprise, impress, and inform students of sport sociology. Of course, we do not pretend to offer all the answers to questions that might be asked about sport and society or all the solutions to the social issues and problems of sport.

Sport sociology is a relatively young discipline, and much remains to be studied and learned. Thus, readers may find that they discover more questions than answers, more hypotheses than clear-cut research findings, and even at times, more theories than facts about a particular issue, problem, or behavior. We hope that the incompleteness of sport sociology, like all science, and the excitement of sociological discovery will stimulate a continuing interest in understanding the sociology of sport and in sociology in general.

ACKNOWLEDGMENTS

For their contributions to the production of this book, we wish to thank Wadsworth editors Serina Beauparlant, Eve Howard, and Susan Shook; Tricia Schumacher of Wadsworth's Marketing Department; Merrill Peterson of Matrix Productions; and photo researcher Sarah Evertson.

Reviewers of the various drafts of the manuscript, who provided valuable suggestions and insights, include Peter Adler (University of Denver); Todd Crosset (University of Massachusetts); Tom Feucht (Cleveland State University); Donald W. Hastings (University of Tennessee, Knoxville); Gary Sailes (Indiana University, Bloomington); Jerry Slatin (University of Kentucky); Peter Stein (William Patterson College); Charles Tolbert (Louisiana State University); and Beverly Wesley (Moorhead State University).

Our appreciation is also due to UNLV Department of Sociology, especially Veona Hunsinger and administrative assistant Susie La Frentz, and to the Appalachian State University Department of Sociology and Social Work, especially administrative assistant Joyce

Rhymer, for their assistance with the manuscript. Special thanks go to Kristin Seaburg, former student of Howard Nixon and author of the poem "An Old Athlete." We are also grateful to the Internet for making this collaboration possible.

Finally, we would like to express our gratitude to our families for their patience and support.

Brief Contents

Contents

CHAPTER 1

Sociology and Sport

LOOKING DIFFERENTLY AT SPORT

Many American parents follow their children from one recreational sports event to another, some watch their children advance from the peewee recreational leagues to progressively higher levels of competition through the high school and college years, and a few even see their children become professional or Olympic athletes. Athletes with high-level sports experiences may continue their sports participation on a less intense level as a leisure activity, joined by others who never were serious athletes in their childhood, youth, or young adulthood. Young people and adults who have forsaken or avoided the demands of regular sports involvement have the opportunity to follow a multitude of sports on television or radio, in newspapers, and in sports magazines.

Although specific types of sports participation and exposure vary widely among Americans and between Americans and people in other countries, sport of some sort has become deeply embedded in the popular consciousness, cultures, and social fabric of most societies today. Television beams sports coverage of major events such as the American Super Bowl and the Olympic Games around the world, and international sports events draw competitors from nations of all sizes and types. It is difficult indeed to imagine a person exposed to the modern mass media who has not seen, read about, or heard a sports contest.

Paradoxically, sport is not taken seriously as a social issue even though it is highly popular as both activity and spectacle. Or so it appears in the United States, where people speak or write nostalgically about the joys of past sports conquests, yet their eyes often seem to glaze over when sport is mentioned in the same breath as economics, politics, poverty, pollution, racial and gender discrimination, crime, or the quality of education.

The paradox is at least partially or superficially explained by the dual character of sport as fantasy and as social reality. *Because* people want sport to be a set of romantic memories, a diversion, or a distraction from the problems or pressures of their everyday lives and *because* they want it to be fun, they may resist its very real role in modern society. Sport, though, belongs both to the fantasy world of its followers and fans and to everyday life and the social institutions of society.

We argue that whether sport is perceived or experienced as fantasy or as reality, it plays an important role and has real effects on society and people's lives. We use sociological knowledge to reveal and explain important social patterns, effects, and implications of sport in mainstream America and in a variety of other societies and cultures. We use sociological perspectives to understand issues, problems, aspects, and dimensions of sport that may not be readily understood from common sense or experience alone. Sociological perspectives enable us to look differently at sport than we do using common sense and experience, and we see more of sport by looking at it in this way. By the end of this book, both the social importance of sport and the value of seeing sport equally as fantasy and as social reality should be apparent.

THE SOCIOLOGICAL IMAGINATION AND SPORT

Even if we accept the argument that sport ought to be viewed like other social issues, a question remains about the value of using sociology to understand

POINT-COUNTERPOINT

DO COMMON SENSE AND PERSONAL EXPERIENCE TELL US ALL WE NEED OR WANT TO KNOW ABOUT SPORT?

YES

People who play sports, read the sports pages, and watch and listen to sports broadcasts and news have all the information they need to understand sports and their experiences in it. The most important source of knowledge about sport is personal experience because experience is knowledge.

NO

People who only know sports from their personal observations and experiences will not be able to generalize beyond their personal perspective. They are not likely to see the larger social forces and social contexts that influence what happens in sport. Their knowledge is affected by personal biases.

How should this debate be resolved?

INSTRUCTIONS FOR POINT–COUNTERPOINT DEBATES: POINT–COUNTERPOINT debates featured throughout the text evaluate the general validity of sports opinions and experiences through critical thinking and debate. Students should assess pro and con positions on the basis of the best available sociological analysis and evidence drawn from the text, supplementary materials, and personal experiences and observations. Assessments should lead to a pro, con, or mixed resolution that builds on the strong or valid points and refutes the weak or invalid points in the original Point-Counterpoint.

sport. A sociological imagination is necessary, however, if we are to see how sport is embedded in general patterns of social interaction and social organization in society and if we are to explain why sports experiences and the organization of sports institutions take the forms they do. **Sociology** is the scientific study of human interaction and social organization, and its purpose as a social science is to produce knowledge to help us interpret, understand, and predict patterns of human social behavior and solve human social problems.

Mills (1959) described the **sociological imagination** as a form of consciousness or frame of mind. It allows us to understand where individuals are located in society and in history, how their circumstances are like or unlike those of similar others in their society and historical period, and how human experiences are structured and made meaningful by the larger social forces of their society, world, and period. Mills used a distinction between what he called personal troubles and public issues to illustrate how the sociological imagination is applied.

Personal troubles, according to Mills, are individual matters that have to do with an individual's character or personality and the range of his or her immediate social relations. **Public issues** are matters that are beyond an individual's immediate experience, local environment, and personal awareness; they reflect the larger social forces in a society. For example, if one individual out of the hundreds playing professional football in the United States uses steroids, that is a personal trouble. Viewing such drug use in these terms, we are likely to recommend a remedy or punishment that focuses on the flawed character, deficient skills, or limited opportunities of that player. When we recognize that steroid use occurs among a relatively large number of pro football players, we treat it as a public issue and look for explanations and answers that go beyond the qualities or behaviors of individuals (see Yesalis, 1993). Such sociological explanations and answers could be found in the organization of the sport of football, the demands of certain positions in the sport, and cultural norms and values in the sport and larger society that seem to encourage or condone drug use and other illicit practices to achieve success.

Thus, in changing the level or type of analysis from a personal trouble to a public issue, we are able to see a form of behavior as a pattern in society rather than merely a product of the experience of a few isolated individuals. Using a sociological imagination to identify and explain public issues, we can understand that the behavior of individuals is part of, and is affected by, the social arrangements that people make to live together in a society. Such behavior bears the imprint not only of each individual's own thoughts and feelings, but also cultural ideas and attitudes of society and social groups. To continue with our example, although individual football players may use steroids, their decisions to do so are shaped by the pressures of their sport and by the ideas and attitudes that others in their sport and society hold about drug use and the importance of winning. Decisions by many players to use steroids can be seen as a social pattern for this category of people, and the reasons can be found in social and cultural forces that are broader than the awareness and experiences of any individual player.

SPORT AS A MIRROR OF SOCIETY

In his classic analysis of social structure and anomie, Merton (1938) used sport to illustrate how social rules or norms may be ignored, weakened, or purposely violated when people are driven to achieve highly valued cultural goals such as success or winning by the most efficient possible means. Merton developed a conception of **anomie**, or conditions when social norms are weak, inconsistent, or vague. When these conditions exist in sport, people may be tempted to use illegitimate means, such as cheating, to pursue their goal of winning. This kind of deviant behavior parallels the social deviance that can be produced by anomie in settings outside sport. Thus, Merton's analysis shows us how we can use sport as a mirror of society as a whole, a mirror that helps us see and understand what happens in society and why it happens.

If sport can be a mirror of society, then the **sociology *of* sport** actually is **sociology *through* sport**. While we are interested in understanding the social aspects of sport, a major focus of our sociological approach to sport will be on how the study of sport helps us understand the nature of societies; the ways that categories and groups of people organize, pattern, and structure their behavior in societies; the ways people interpret and give meaning to their social interaction and collective experiences; and the major social problems, issues, and processes that affect societies and how they change. We will strongly emphasize sociology through sport even as we try to make sense of the social nature of sport itself.

SPORT AS A SOCIAL PHENOMENON

To understand sport in a sociological framework, it is necessary to describe sport as a social phenomenon. Although sport sociologists lack consensus in agreeing on a single definition, we define **sport** as institutionalized physical competition occurring in a formally organized or corporate structure (Nixon, 1984: 13–15).

Several aspects of this definition are sociologically significant. First, the term **institutionalized** implies that sport has an established structure of patterned and relatively persistent norms, statuses, roles, and social relationships. The **norms** of sport are the formal and informal rules that govern or regulate social interaction. **Statuses** are positions in the organizational structure of sport. Certain rights and duties are associated with statuses, and positions may be defined in more or less specific terms. For example, we can generally refer to the position of coach, and we also can identify positions of head, associate head, and assistant coaches in a specific sports program. The ways that people carry out the behavior expected in particular statuses are **roles**. Thus, people occupy statuses but perform roles. *Achieved statuses* are occupied on the basis of what people have achieved, and *ascribed statuses* are occupied on the basis of characteristics over which individuals have little or no control, such as attributes assigned by birth (e.g., gender, race, and social class background) or stage in the life span (e.g., age). **Social relationships** link individuals, groups, and larger collectivities or social units in enduring patterns of social interaction. The various webs of social relationships that connect individuals, statuses, roles, groups, and larger social units in a community or society are called **social networks**. The second significant element in the definition of sport, physical **competition**, refers to the predominant type of interaction that occurs in sport. That is, sport involves individuals or groups competing against each other in contests mainly of physical skill in which there are winners and losers. The competition in sport is serious or real. The outcome is uncertain and is not prearranged, as it is in professional wrestling.

The third significant element of our definition is that the organizational structure of sport is formal and corporate. The **formal social structure** of sport includes components such as explicit or official rules, official enforcement by regulatory bodies, a hierarchy of positions, and, often, bureaucratic organization. **Bureaucracy** is a type of formal organization emphasizing rational-legal authority linked to positions rather than specific individuals, specialized positions within a clear-cut division of labor, delegation of responsibilities within the hierarchy and division of labor, and rational calculation and efficiency in the pursuit of organizational tasks and goals. Bureaucracies frequently are large in size and relatively complex in their administrative organization (Weber, 1922; Hess, Markson & Stein, 1991: 91–92).

Sport today frequently has a **corporate structure** as well as formal or bureaucratic one (Eitzen & Sage, 1986: 17). Although the corporate organization of sport has many dimensions, the commercial or business dimension is most important. Corporate sport is organized to make money. In this conception of sport, sports contests, participants, and paraphernalia are commodities to produce, market, and sell. The corporate character of sport draws special attention to the economics of sport, but we will see later, in discussions of the political economy of sport, that it also carries a related political aspect. A **political economy perspective** reveals how money and power are interwoven in corporate sport.

Defining sport as we have here distinguishes it from a number of other similar kinds of activities (Nixon, 1984: 13–15). Table 1.1 summarizes the differences among a number of types of physical activities, including sport. The term **athletics** is not included in this table because we use it as a synonym for sport played in school or college settings. Each type of activity defined in Table 1.1 is a noun, which implies, for example, that sport may be "played" even though the activity of play—the noun—is distinct from sport.

Play is a voluntary activity that is nonutilitarian and relatively spontaneous and unstructured. Two children who kick a ball back and forth for no other purpose than to have fun are engaging in physical play. Unrestricted by rules, these children may invent new ways of kicking, reflecting the creativity that is possible in play.

Recreation is a mostly voluntary leisure activity meant to refresh the mind and/or body. As a leisure activity, it is distinguished from work or gainful employment. A person who kicks a ball for a specified number of minutes or yards in a daily exercise routine is engaged in a recreational activity. We can see that this form of kicking is more purposeful, somewhat less voluntary, and somewhat more organized than the example of kicking as play. For these reasons, recreation is closer to sport than play.

Games are even closer to sport than play and recreation are. Kicking a ball becomes a game when specific rules are developed to govern aspects of interaction such as the way the ball can be kicked, who can kick to whom, the boundaries for acceptable kicks, and how the interaction between kickers starts and ends. We might find many groups of children playing the same version of a kickball game in different parts of the country or a community, but they will not be engaged in sport if their games are self-initiated, self-governed, and unrelated to each other. We often refer to sports events or sports contests as games, but games are only part of sport when they occur within a formal organizational or corporate network that links them to other games in the same sports network and regulates them.

Games, then, are relatively rule-bound activities that may be more or less institutionalized, formally organized, and serious. The definition of sport implies that the more institutionalized, formally organized, and serious competitive physical activities are, the more sportlike they are. Thus, when children and youths are formally recruited to play on a team or when they are assigned to teams on the basis of a formal tryout, when they compete in games that are sponsored and regulated by the U.S. Youth Soccer Association and their state association, when they pay money to their local association to purchase uniforms and to subsidize referees and a regional newsletter, and

TABLE 1.1 Sport and Related Physical Activities

Activity	Structure	Regulation	Involvement	Purpose
Play	unstructured and spontaneous	limited internal	voluntary	nonutilitarian fun
Recreation	informal and often routinized	internal	voluntary but may reflect personal commitment	refreshment of mind or body that may be purposeful
Game	often formal and institutionalized	internal or external when part of sport as a contest	usually voluntary but can result from informal social pressure and can be nonvoluntary when part of sport as a contest	fun or serious pursuit of winning
Sport	physical competition that is formal, corporate, institutionalized, and rational-bureaucratic	external	nonvoluntary	serious pursuit of winning and/or money and other material rewards

when they travel to local, regional, and national competitions to vie for awards and honors, these children and youths are involved in the sport of soccer. The coaches and players in this form of soccer may approach their sport at times with a playlike attitude, and the volunteer coaches may view their coaching as a recreational activity. Nevertheless, the activity in which these coaches and players are involved is the *sport* of soccer—because it is a highly institutionalized and formally organized form of soccer.

Saying that sports activities are institutionalized implies that they are a regular and patterned part of social life. The assortment of sports activities that exists in a society can be viewed as components of a sport institution. The social institutions of a society are relatively stable clusters of norms, statuses, roles, and relationships that are organized around important values in the society (Williams, 1970: 37). **Social institutions** respond to enduring needs of people in a society, indicate the nature of problems and approved solutions in relation of the pursuit of enduring needs, and direct the actions of people in particular directions. We could argue about the major or minor institutional status of sport in American and other societies. In fact, Edwards (1973: 90) is among those who have characterized sport in America as a "secular, quasi-religious institution." In his conception, sport as a secular religion functions mainly to disseminate and reinforce values that regulate behavior and goal attainment and indicate approved solutions to secular problems in society.

According to Guttmann (1978), the modern social institution of sport has a number of qualities that distinguish it from earlier forms of sport. These qualities are consistent with the definition of sport proposed in this chapter. They include secularism, a principle of equality of opportunity, specialization of statuses and roles, rational calculation and bureaucracy, quantification of achievement, and the quest for records. Although the type of social phenomena we will call sport in this book clearly meets Guttmann's criteria for modern sport, it should not be assumed that sport, as we have defined it, occurs only in modern Western societies that are generally highly industrialized,

Source: AP/World Wide Photos

FIGURE 1.3 Waiting for the players to return.

bureaucratized, and technologically developed. Sport exists with its modern qualities in Third World and other non-Western nations. The fact that modern sport is found in varied national, cultural, and societal settings around the world merely underlines the point made at the outset of this chapter about the pervasiveness of sport in the global village created by the modern mass media and sports promoters (see McLuhan, 1964).

THE DEVELOPMENT OF SPORT SOCIOLOGY

The recognition that sport is part of society and has enduring social aspects is the fundamental inspiration for a sociology of sport. In fact, the academic field of sport sociology is very young, initially emerging as a subdiscipline of sociology in the 1960s. The first programmatic statement calling for a sociology of sport in North America was written by Kenyon and Loy and appeared in 1965. In 1980, Lüschen noted in the *Annual Review of Sociology* that articles on social aspects of sport had appeared in over a hundred scholarly journals (Lüschen, 1980). Although Lüschen's prediction of the coming acceptance of sport sociology in mainstream sociology has not yet been fulfilled (Frey & Eitzen, 1991), this field has continued to attract scholars, primarily from sociology and physical education, and to spawn new scholarship.

It is possible to raise concerns about the continuing problems of identity and legitimacy for the sociology of sport within mainstream sociology and physical education (Loy, Kenyon & McPherson, 1987; Frey & Eitzen, 1991), but a search of the current sport sociology literature will reveal a fairly large and growing number of scholarly studies on many aspects of sport. Furthermore, one could argue that the theoretical work and research in this field have become increasingly sophisticated as the field has matured. Since 1980, sport sociologists have had their own professional association, the North American Society for the Sociology of Sport (NASSS), with its own journal, the *Sociology of Sport Journal.*

SPORT SOCIOLOGY AND RELATED DISCIPLINES

The social and behavioral aspects of sport can be studied from the perspectives of a number of different academic disciplines, including psychology, history, philosophy, geography, and anthropology as well as sociology. In fact, sport sociologists with physical education backgrounds may be located in broad sports studies programs within departments of physical education, kinesiology, exercise and sport science, or something similar in their university or college, and they may have expertise in one or more other areas of sport studies along with sport sociology (e.g., see Chu, 1982).

Sport sociology differs from physical education, kinesiology, and exercise science because the latter focus on the science of movement and physical activity or performance mainly from physical, physiological, and biomechanical perspectives (e.g., see Brooks, 1981). *Sport psychologists* tend to focus on how factors such as personality attributes, motives, the mind and mental processes, and emotional states are related to the sports performance of individuals (e.g., see Cox, 1985; LeUnes & Nation, 1989). *Sport historians* document and analyze the past of sport, including the individuals, events, groups, and institutions of sport that have preceded contemporary aspects of sport (e.g., see Goldstein, 1989; Roberts & Olson, 1989; Miller, 1990; Rader, 1990). *Social philosophers of sport* study the central values and ultimate meaning or significance of human behavior and social interaction in sport; and they draw attention to ethical, moral, and justice issues (e.g., see Weiss, 1969; Simon, 1985; Lenk, 1986; Morgan & Meier, 1988). Philosophers specializing in logic might also focus on the application of the logic of scientific inquiry to the study of sport as well as on logic and rationality in sport behavior. *Sport geographers* tend to look at the relationship of sport to place, and they examine questions about where sports are most popular, how sports spread from one place to another, and how sport affects and is affected by the landscape where it takes place (e.g., see Rooney, 1974; Bale, 1989).

Anthropologists who study sport from the perspective of cultural anthropology may be most similar to sport sociologists in their focus on patterns of social and cultural behavior in sport. These disciplines tend to be distinguished by the tendency of anthropologists of sport to give relatively more attention to culture rather than society as the basic unit of study and to concentrate relatively more on premodern cultures and social behavior (e.g., see Blanchard & Cheska, 1985). Growing interest among sport sociologists in cultural studies of sport, which will be discussed later in this chapter, has tended to blur the distinction between anthropological and sociological perspectives in the sociology of sport.

Although there are not yet any readily identifiable subdisciplines of sports studies that focus on the political science of sport or the economics of sport, sport sociologists recognize the importance of these dimensions. A number are now analyzing sport from a political economy perspective (e.g., see Sage, 1990). There also are books about sport emphasizing economic (e.g., Hofmann & Greenberg, 1989; Scully, 1989), political (e.g., Espy, 1981; Houlihan, 1991), ideological (e.g., Hoberman, 1984), policy (e.g., Noll, 1974; Johnson & Frey, 1985), religious (e.g., Hoffman, 1992a, 1992b), legal (e.g., Gallner, 1974; Weistart & Lowell, 1979; Freedman, 1987), mass communication (e.g., Wenner, 1989), aesthetic (e.g., Lowe, 1977), and sports medicine (e.g., Dirix, Knuttgen & Tittel, 1988; Bloomfield, Fricker & Fitch, 1992) aspects. The number and variety of perspectives that have been used to study sport reflect its potential as a field. Sport sociology, as the broadest of the social science approaches, allows us to see the richness and complexity of sport.

SOCIOLOGICAL ANALYSIS OF SPORT

The patterned social regularities that characterize societies and social interaction are called *social facts,* and it is the business of sport sociologists—as social scientists—to describe, explain, and predict social facts and how they change. The most basic social facts studied by sport sociologists are elements of social structure such as norms, statuses, roles, and relationships. Sport sociologists also are interested in the elements of **culture**, which include the shared ideas, ideals, and artifacts of a people that they create and sustain for their mutual survival and the perpetuation of their way of life.

In addition, sport sociologists study **social processes**, which are dynamic forces in interaction that can reinforce the existing elements of social structure and culture or transform them into new forms. Social processes operate at the **micro level** of relationships between individuals and in small groups. Social processes also operate at a **macro level**, which refers to a level of analysis that is more abstract and applies to whole social systems or societies and to social networks that tend to be large and complex. Competition is an example of a social process, and it lies at the heart of all sports activities.

In the institutional realm of sport, we also see examples of many other social processes, including social exchange, resource allocation, power, status relations, deviance, control, conflict, leadership, cooperation, accommodation, socialization, and group performance. At the macro level of whole societies, major processes of societal change have created conditions for the emergence of modern sport. These processes have transformed where and how people live and work, what organizations look like, and how people communicate with each other, and they include industrialization, urbanization, bureaucratization, the development of the mass media, and computerization. We will consider how these kinds of social processes have operated in sport and have affected the social organization of sport in a variety of cultural and national settings.

We have said that sport sociology differs from other academic approaches to the study of sport. Sport sociology has a broader focus than other disciplines that examine social or cultural aspects of sport. In addition, the critical and systematic nature of sport sociology analysis distinguishes it from nonacademic and popular analyses of sport and help define sport sociology as a *social science.* By using the term *critical,* we do not mean that sport sociologists try to find fault with sport to debase it or its participants. Instead, as sociologists, sport sociologists employ a *debunking motif* in their analysis (Berger, 1963: 25–53; Nixon, 1984: 11). This means that sport sociologists question commonly accepted or officially stated descriptions or explanations of social life that are not supported by solid evidence. Berger (1963) said that this debunking approach is inherent in sociological consciousness, and it is a quality of sociological thinking that often seems to give sociologists the reputation of being cynical or radical. Sport sociologists share this critical perspective with all other scientists, who do not accept interpretations of social life merely because they are widely believed.

Since proponents of sport, like proponents of other sectors of society, may speak in glowing terms about their sports vocation or avocation without solid grounding in facts, the critical or questioning attitude of sport sociologists about things that advocates take for granted, such as the character-building or health-enhancing effects of sports involvement, may be annoying, upsetting, or threatening (e.g., see Margolis, 1990, versus Nixon, 1990a). The sociological enterprise requires the questioning and testing of traditional, popular, intuitive, common sense, and official assumptions along with the unmasking of social and cultural myths to reveal social facts about society and culture. Sport sociologists await the results of social research before making assertions about social facts.

Social research is part of what makes sport sociology a social science, and it is the means by which sport sociologists seek concrete answers to their questions, hypotheses, and assumptions about sport and society. Sociological analysis of sport is made systematic by a reliance on the careful, logical, and orderly use of established social science procedures for formulating research questions, testing them, and drawing conclusions. Students wishing to gain a fuller appreciation of the nature of social research in sport sociology should turn to the appendix on science and research methods.

Sport sociology as social science is more than a set of research procedures. It involves the testing of precisely stated theoretical ideas through social research. Theoretical ideas make research meaningful, and sport sociologists have used a variety of theoretical ideas and perspectives to analyze sport in society.

TRADITIONAL THEORETICAL DIRECTIONS IN CONTEMPORARY SPORT SOCIOLOGY

Giddens (1991: 12–15) distinguished theoretical questions in sociology from factual, comparative, and developmental questions. Factual questions, also called *empirical* questions, are about how things occur or what happens. Comparative questions relate one social or cultural context within a society to another or draw contrasting examples from different societies or cultures. Developmental questions focus on changes over time. Theoretical questions focus on why things happen, why they are different, and why they have developed. Thus, **theories** propose ideas, assumptions, and arguments to clarify the underlying reasons or explanations for social facts. They involve abstract interpretations that make general statements about the reasons for and meaning of a wide variety of specific and concrete facts about the social world.

Individual social facts, such as higher gradepoint averages of high school athletes than nonathletes, have no intrinsic meaning until they are interpreted in the more abstract and general terms of theories that link these individual facts to aspects of broader social and cultural contexts and social structures, such as relative emphases on cultural values of academic achievement and success in sport and nonsport settings for athletes from various social class backgrounds and the existence of eligibility criteria for athletic participation. Recalling the previous discussion of personal troubles and public issues, we are able to understand the larger meaning of personal troubles when we begin to think of them more theoretically as public issues. A public issues interpretation of problems in society could be seen as a more theoretical than a personal troubles perspective because the former interpretation links individual, personal, or biographical facts to larger and more general social categories, social settings, and historical circumstances.

Most sociology and sport sociology texts identify three major perspectives as the dominant sociological theories of the twentieth century: **structural-functionalism, conflict theory,** and **symbolic interactionism**. Although sport sociology has been criticized for its relatively weak theoretical development (e.g., see Lüschen, 1980; Kenyon, 1986; Coakley, 1987a; MacAloon, 1987; Frey & Eitzen, 1991), the clear imprint of functional, conflict, and interactionist perspectives, as well as the application and development of a number of interesting alternative and new forms of theoretical thinking, can be seen in sociological analyses of sport.

Structural-Functional Analysis

Structural-functionalism and conflict theory generally provide macro-level images of the organization and functioning of societies, whereas symbolic interactionism offers micro-level perspectives of everyday interaction (see Rees & Miracle, 1986; Robertson, 1989: 7–16; Coakley, 1990: ch. 2). The structural-functionalist, or functionalist, perspective in sociology has its roots in the social theories of nineteenth-century sociologists Emile Durkheim and Herbert Spencer. Two of the most influential twentieth-century functionalists have been Talcott Parsons (1951) and Robert Merton (1968). The work of these and other functionalists emphasizes how different parts, structures, or institutions in social systems or societies contribute to the integration, harmony, adaptation, and well-being of the whole system or society. The focus of much structural-functional analysis is large social systems such as societies. **Social systems** are sets of social actors or social units, the patterns of interdependency linking them, and the boundaries that give them a distinct identity.

Social functions are major concepts in structural-functional analysis. They are positive effects or consequences of social structures or behaviors that contribute to the social stability, harmony, or well-being of social systems. According to structural functionalists, social systems, like human organisms, have functional needs that must be met to survive and grow, and social structures can be seen as enduring solutions to these functional issues. In referring to societies and smaller social collectivities in structural-functional terms as social systems, we are implying that they are networks of interrelated structural elements that function in a relatively coherent and coordinated manner to achieve system needs.

To make these somewhat abstract ideas more clear, we can think of the functioning of social institutions in societies (see Robertson, 1989: 54–56). Societies have enduring needs to regulate sexual behavior and to care for children, and the social institution of the family exists to meet these functional needs. Modern societies also have educational institutions to create and transmit cultural knowledge; religious institutions to deal with ultimate questions of meaning and to share and reaffirm common values and a sense of community based on the answers to those questions of meaning; political institutions to distribute power and maintain order; economic institutions for the production, distribution, and exchange of goods and services; legal institutions to maintain social control; military institutions for the expansion or defense of the nation-state; scientific institutions to create more understanding of and control over the human and physical worlds; and medical and health care institutions to treat sick members of society. Although the most obvious social functions of sport institutions are recreation, entertainment, and exercise, sport also functions to enhance the psychosocial adjustment of individuals, to socialize members into their social roles in society, to integrate members of society through shared sports commitments and interests, to produce national and community identification, and to provide avenues of upward social mobility (Stevenson & Nixon, 1972).

In the functionalist view, societies tend to be stable, and their parts tend to be in a state of relative equilibrium or balance. Thus, functionalists see social conflict, deviance, tensions, and pressures as disruptive, causing mechanisms of social control to be activated so that they can be held in check. Merton (1968) referred to elements of social systems with disruptive or negative consequences as *dysfunctions*. He also distinguished between *manifest functions*—which are obvious and intended consequences—and *latent functions*—which are not recognized or intended.

In general, functionalists emphasize legitimate authority, order, stability, and the regulation, control, or elimination of conflict, deviance, and disharmony. Critics see these emphases implying a relatively rigid or conservative view of societies that fails to account for the institutionalization of certain forms of conflict, deviance, and disharmony. The normal tensions and changes in societies that often make social order problematic are ignored by functionalists. Critics also argue that functionalists fail to consider the consequences of structured socioeconomic inequalities. For example, they pay little attention to discrimination, exploitation, and political domination by elites or ruling classes who have a vested interest in maintaining their power and privilege.

To illustrate the application of a functionalist perspective to sport, we can look at the case of a professional football strike. In the functionalist view, a strike could be seen as disruptive to the welfare of the game. In this view, the increased salaries, benefits, and employee rights of movement to other teams sought by the players' union are seen as having destabilizing or dysfunctional effects on the current financial structure and competitive balance of the sport. A strong response from owners is needed to preserve the stability and well-being of the sport. Players who actively support the strike are perceived as troublemakers whose actions need to be controlled by owners. Although the manifest function of the owners' unwillingness to bargain with players might be construed as preserving the stability and well-being of the sport, an important latent function would be to maintain owners' control over players and the purse strings of the sport.

Conflict Theory

An interpretation of player strikes from a conflict theory perspective would contrast sharply with a functionalist interpretation. The image of society conveyed by conflict theory emphasizes processes of competition, conflict, economic exploitation, political oppression, and change as normal parts of social life. The main inspiration for conflict theory was the work of Karl Marx (1977) in the nineteenth century. Marx assumed that history was driven by ongoing social class conflicts that made societies and social order tenuous. In his view, societies were only loosely integrated on the basis of coercion, exploitation, and manipulation of people in the subordinate class by people in the dominant class. From a conflict perspective, the actions of people who are deviant or disruptive are not necessarily threats to the well-being of society, even if they challenge its stability. These actions and the classes who engage in them are analyzed by conflict theorists in terms of how their economic or class interests clash with the interests of the dominant class. Thus, deviant or rebellious behavior could be seen as reactions of victims to exploitation or oppression rather than as threats to a status quo worth preserving.

More contemporary versions of conflict theory are not restricted to economic class conflicts. They might, for example, focus on tensions and conflicts based on cleavages between young and old, manufacturers and consumers, management and labor, women and men, and members of different racial or ethnic groups (Robertson, 1989: 13). In the conflict perspective, the main concern is not with how society benefits from existing social arrangements, but instead with which interests or groups in society benefit most from the unequal distribution of rewards and resources characterizing the status quo.

Critics of conflict theories argue that they show insufficient acknowledgment of the elements of moral and cultural consensus, shared interests, and social cooperation and coordination that characterize and drive most of social life. Some sociologists of sport (e.g., Guttmann, 1978, 1988) also argue that Max Weber's ideas of rationalization and bureaucratization explain more of the evolution of modern sport over the past century than do ideas from Marx and neo-Marxists about the adverse influence of industrial capitalism on the development of sport (see Guttmann, 1988: 181–184).

Looking at the example of the professional football strike through the eyes of a conflict theorist, we see a picture of conflict between the interests of owners, who want to maintain their powerful and privileged position in relation to the players, and the interests of players, who want to gain a share of the control and

Source: Hazel Hankin/Stock Boston

FIGURE 1.2 How long before the playground is a parking lot?

financial rewards of their sport that is more consistent with the value of their contributions to the financial and popular success of their sport. Thus, the strike is a clash of interests between owners and management officials on one side and players on the other. It is not about the well-being of the sport, but about who will benefit most from its revenues and who will exercise the most power over the direction of the sport. The role of striking players, in this perspective, is not that of troublemakers, but instead one of victims of exploitation and oppression who are seeking to elevate their status above the level of being mere commercial commodities or the legal property of owners who use them for their own profit and pleasure.

Symbolic Interactionism

A symbolic interactionist interpretation of a professional football strike would focus on a different level of analysis and different types of questions than functionalist and conflict interpretations. Leonard and Schmitt (1987) used a symbolic interactionist perspective to define the meaning of sport from the viewpoint of the subject or person on the street rather than the sport sociologist. Instead of focusing on structural and institutional aspects of sport—as we have in our definition, their consciously subjective interactionist definition emphasizes the symbolic character of sport. It also draws attention to the subtle, covert, informal, interactive, process, and negotiated aspects of sport as a social activity. In proposing this subjective interpretation of the meaning of sport, Leonard and Schmitt argued for the importance of seeing the similarities between sport and other types of interaction in society. This argument reflects the perspective we have called sociology through sport.

One can trace the origins of a general interactionist approach to Max Weber (1946), who drew attention to the importance of understanding social life from the perspective of the individuals experiencing it (Robertson, 1989: 14). Weber used the term ***verstehen*** to refer to the ability to see the behavior and intentions of others through their eyes. The development of symbolic interactionism can be traced from the early sociology of Charles Horton Cooley (1902) and especially from the work of George Herbert Mead (1934).

Sociologists and social psychologists influenced by Mead, such as Herbert Blumer (1969), have emphasized the importance of focusing on how people interpret their social world through the meanings they derive from interaction with others (see also Lindesmith, Strauss & Denzin, 1991). In symbolic interactionism, interaction or communication through symbolic means—such as signs, gestures, shared rules, and various forms of verbal and nonverbal language—is seen as creating and defining social relationships between social actors as well as transmitting information. Through subjectively taking each other into account, people engaged in interaction are able to influence each other's interpretations of their interaction and respective social worlds and can develop shared meanings of these experiences (Smith, 1987: 236).

Thus, for symbolic interactionists, the subject matter of sociology is what people define as reality through symbolic interaction, and the processes of the construction and reconstruction of meanings in everyday life are a major part of the work of symbolic interactionists. Relevant here is W. I. Thomas's (1927) classic notion of the importance of subjective reality and social actors' **definition of the situation**. When people define situations as real, Thomas asserted, they will be real in their consequences. We act on the basis of what we perceive as reality in our definition of the situation, and when such perceptions or definitions are shared by people who are interacting, interaction can become relatively coordinated, organized, and easier (Hess, Markson & Stein, 1991: 78). The dynamics of interpersonal power relations may have a significant effect on the conception of reality or definition of the situation that is negotiated in social interaction.

A major limitation of the symbolic-interactionist approach is the focus on the microlevel of everyday interaction. This approach neglects larger social institutions and societal processes, which may be significant social facts that influence everyday interaction (Robertson, 1989: 15). Symbolic interactionism—and interpretive sociology in general—also can be seen as a denial of social structure and the possibility of causal explanations and predictions. Symbolic interactionists believe there are multiple social realities in which people perceive the same situation differently. If social rules, roles, and relationships are always subject to different interpretations, it is difficult to see the possibility of patterned social regularities that are embedded in social structures, that are represented in social facts, and that make causal explanations and predictions possible. The hallmark of symbolic interactionism is rich, detailed, and deep description of micro-level processes of interaction. Yet an interactionist approach can be combined with a focus on larger cultural and structural factors to provide insights into more general reasons or explanations for why particular kinds of interaction occur and what they mean.

If a symbolic interactionist were to study a professional football strike, the main focus might be the multiple and covert meanings of the strike for owners, management officials, coaches, or players. A symbolic interactionist studies how people in these different statuses think and talk about the strike, and how the private thoughts, feelings, and actions of these various people correspond to public perceptions as shaped by influential forces such as the mass media. In following the emergence and development of a strike, a symbolic interactionist might look at how management representatives and player union representatives battle to control public perceptions of the strike, and they might carefully document how the definitions of the situation, the language of negotiation, and the negotiators' perceptions of each other and their relationship change as the strike moves toward resolution.

Another dimension of this kind of research could include a focus on how the roles of employee and union member are integrated into football players' interpretations of what it means to them to be professional athletes. In studying mass media coverage of strikes, symbolic interactionists could focus on the labels that the media apply to the various participants in the strike. For example, owners *or* players could be labeled as greedy or unconcerned about the welfare of their sport. Mass media interpretations are likely to be colored by their own or popular sympathy or hostility toward the strike.

The contrasting images of professional sports strikes provided by structural-functionalist, conflict, and symbolic-interactionist perspectives are summarized

TABLE 1.2 Contrasting Theoretical Images of Pro Sports Strikes

Perspective	Image of Strike	Image of Owners	Image of Striking Players
Structural functionalism	dysfunctional or disruptive of the stability of the sport	responsible for stability and well-being of sport; need to exercise strong control for the good of the sport	disruptive troublemakers or rebels
Conflict theory	conflict between economic classes of owners/management and players over who controls the sport and who gets what benefits from it	vested interest in maintaining power and privilege by dominating and exploiting players and controlling the sport for their profit and pleasure	victims of oppression and exploitation of owners; vested interest in fighting owners for a larger share of the control and financial rewards of their sport commensurate with their value and contributions
Symbolic interactionism	continuously unfolding multiple realities that are constructed, negotiated, and redefined by the assorted participants; mass media create images of strikes depending on the amount of their own or popular sympathy or hostility toward it	labeled as targets of ungrateful and greedy players or as greedy exploiters of their players, depending on the popularity of the strike with the media and the public	labeled as victims of greedy owners or as ungrateful and greedy themselves, depending on the popularity of the strike with the media and public

in Table 1.2. Structural-functionalist, conflict, and symbolic-interactionist perspectives provide potentially very different lenses through which to look at sport and other aspects of society. However, the choice of one perspective does not have to be made at the expense of rejecting the others. Each can provide valuable insights. The question of which perspective to believe is a matter that is resolved by research, which provides systematic concrete evidence as the basis for assessing the validity of the hypotheses, assumptions, and conclusions derived from these different theoretical approaches. The perspectives focus on different domains, aspects, or levels of sport. Ideas derived from them can provide complementary, rather than contradictory, ways of understanding the complexity of certain kinds of sports experiences—like strikes.

RECENT THEORETICAL DIRECTIONS IN SPORT SOCIOLOGY

The major theoretical directions that have dominated sport sociology in recent years include ones that are extensions of, or reactions to, the traditional functionalist, conflict, and interactionist-interpretive approaches. There also have emerged in sport sociology theoretical perspectives that take the sociological analysis of sport in new and different directions. Some of these new perspectives emphasize cultural interpretation, others critical analyses, and a number combine both cultural interpretive and critical orientations (Nixon, 1991a).

Cultural interpretive perspectives generally emphasize cultural practices and how they are produced, communicated, and consumed in the context of struggles among various social groups to control cultural symbols and their meanings. It is not difficult to see why such perspectives frequently are classified as critical as well. *Critical perspectives* generally focus on dominant cultural and social structures and practices in society that are tied to capitalism, patriarchy, and racism and tend to contribute to political oppression or economic exploitation of large or specific segments of society.

In a critique of sport sociology that was delivered to a major gathering of sport sociologists, MacAloon (1987) contended that the "real action in sociology" was in the "sociology of culture." He proposed an approach that incorporated major elements of what we are calling here the cultural interpretive and critical perspectives. MacAloon cited the work of the French sociologist Bourdieu (e.g., see Bourdieu, 1977, 1978, 1982, 1988; Bourdieu & Wacquant, 1992) as an exemplar of the direction that should be followed in cultural and critical analyses of sport. In his critical cultural analysis of sport, Bourdieu observed that the increasing professionalization of sports was progressively dispossessing laypeople and relegating them to the role of spectators. Today, professionals perform in front of amateurs. As the practice of sport becomes more dominated by these specialists, spectators become less practically competent in them and increasingly focus on their extrinsic aspects, such as winning and violence, rather than their intrinsic value as athletic contests. Bourdieu suggested the importance of understanding the development of these kinds of practices as outcomes of the rise of spectator sports and the growing gap between professionals and amateurs it has created.

Other examples of the types of analyses MacAloon had in mind for sport sociology are Messner's (1988) examination of the struggles of women in sports realms and relationships heavily influenced by male ideas and ideas of male dominance or superiority; Birrell and Cole's (1990) interpretation of how the mass media's treatment of the transsexuality of tennis player Renee Richards reinforced ideas about the natural physical, and therefore general, inferiority of women in relation to men; and Foley's (1990) study of how small-town high school football embodies important cultural rituals in the community and serves as a place

Source: Dean Abramson/Stock Boston

FIGURE 1.3 Coach challenges authority by questioning the umpire's call.

where people could engage in cultural practices that challenge inequalities of social class, gender, and race. In addition, there is Hoberman's (1984) analysis of the forces that influence interpretations of the meaning of sport by very different political ideologies, ranging from the ideologies of cultural conservatives and fascists to Marxists and neo-Marxists and from the ideology of Nazi Germany to that of the socialist states of the Soviet era and of China before and after Mao.

Although much cultural interpretation of sport is critical, other forms of critical analysis have emerged in recent years in sport sociology (e.g., Sage, 1990a). These recent critical perspectives go beyond the debunking motif (Berger, 1963) that is generally characteristic of sociology, and they do not fit neatly within conventional functionalist, conflict, or interactionist perspectives. As Coakley (1990: 31) has observed, critical theory in sport sociology does not exclusively rely on notions of the needs of social systems or the production needs of a capitalist economy or the ways people interpret the meaning of sport in their everyday lives. Critical theorists understand the interplay between economic and historical forces and interpretations of sport in everyday interaction and between shared values and conflicts of interest.

One of the most prominent critical approaches in contemporary sport sociology and sociology in general is **feminism** (Theberge, 1985; Birrell, 1988; Deem, 1988; Hall, 1988; Messner, 1988; Messner and Sabo, 1990). Feminism generally interprets sport in the context of gender relations or structures that oppress or discriminate against women and reinforce male domination in sport and society. More specifically, feminist sport sociologists have analyzed the persistence of barriers to the growth in female sports participation; the problems encountered by women and feminists in leadership positions in sport; and efforts to exclude women from some sports, limit their participation in others, and minimize or demean their sports capabilities and accomplishments.

A number of common themes can be seen in these critical approaches. *First*, they share the idea that people engage in power relations in struggles over sporting practices. *Second*, they assume that sports are part of structures and relations of class, state, and cultural domination that cause or reinforce the subordination and oppression of disadvantaged social groups in society. *Third*, they imply that sport must be changed because it is tied to structures and relations of subordination and oppression in society or because it has the potential of transforming these patterns of inequality and disadvantage in society. In most critical analyses, sport is seen as contributing to and reinforcing the inequalities of class, race, ethnicity, and gender that stem from capitalism.

There are a variety of other perspectives in sociology, ranging from social network analysis (e.g., Berkowitz, 1982; Nixon, 1993a) to social exchange theory (e.g., Cook, 1987) and poststructuralism and postmodernism (e.g., see Agger, 1991), that could have been given more attention here (see Alexander, 1988). The purpose of this discussion of theories, however, is not to cover all the theoretical perspectives and theorists in sociology and sport sociology. Instead, it means to introduce major theoretical traditions and provide a sense of the diversity and richness of recent sociological thinking about sport.

The numerous theoretical perspectives available to sport sociologists offer diverse and contrasting ways of seeing and understanding sport as a social phenomenon and as part of society. Each perspective provides a different set of glasses for interpreting what sport is like, what it means to participants and members of the larger society, how it is related to the cultural and social structures of the larger society, and how its own cultural and social structures influence those involved in it. Using one pair of glasses does not prevent us from using others as well in this book as we try to understand sociologically interesting aspects of sport and the ways sport mirrors social aspects of the larger society. We assume, however, that sport sociology is a social science, which means that we must understand the essential relationship between social research and theory in the production of knowledge about sport and society.

CROSS-CULTURAL, CROSS-NATIONAL, AND INTERNATIONAL PERSPECTIVES

The importance of perspectives that stretch our vision beyond the United States or its citizens in sport seems self-evident when we recognize the amount of worldwide attention paid to the Olympics, the Super Bowl of American football, and the World Cup of soccer; when we see foreign investment in professional sports played in the United States; when we see increasing attention paid overseas to American sports such as football and baseball; when we see international terrorism played out in a sports arena; when we recognize the impact of an international sports boycott on the system of apartheid in South Africa; and when we see an increasing number of foreign athletes competing in American professional, intercollegiate, and amateur sports and many Americans competing in sports in other countries.

Cross-cultural, cross-national, and international perspectives should expand and deepen our understanding of social and cultural structures and processes in American sport and American society, reveal the connections between the experiences and lives of Americans and people in other nations, develop a social and cultural awareness that extends beyond the boundaries of mainstream America or North America, and see the major issues in American sport in broader perspective. These arguments parallel ones for more emphasis on global perspectives in the study of sociology in general (Ferrante, 1992: 13–14). The **globalization** of sport as a modern phenomenon will be discussed in more detail in ensuing chapters, and the discussion will focus on the process that has created more extensive and complex links among different cultures and nations in the world. The massive changes that have reshaped Central Europe and the former Soviet Union in recent years provide a further justification for looking at sport beyond American borders. Future chapters will consider how the "new world order" emerging from these changes has affected the international landscape of sport (see Frey & Eitzen, 1991: 511–512; Ferrante, 1992: ch. 13; Eitzen & Sage, 1993).

OUTLINE OF THE BOOK

This chapter has focused on the relationship between sociology and sport. In it, we have tried to convey the distinctive nature of a sociological conception of sport, to show the variety of ways that sport sociologists can look at and study sport as social scientists, and to suggest the value of seeing sport from cross-cultural, cross-national, and international perspectives. Ensuing chapters will reveal the richness of sport sociology in more detail.

Sport is placed in historical and cultural context in Chapters 2, 3, and 4. Chapter 2 considers the historical development of modern sport as a social phenomenon, the large-scale social forces that have shaped this development, the political economy of modern corporate sport, and the role of television in modern sports. Chapter 3 examines culture and sport in America, cultural heroes in sport, sport as a national pastime and mythic spectacle, cultural differences in the modern world sport system, the Americanization of global sport, and the place of sport in contemporary American culture. Chapter 4 extends the previous discussion of sport and culture in focusing on religion and sport. It considers sport as religion and quasi-religion, the relationship of sport to religion and morality, the uses of sport by religion, and religion and religious practices in sport.

Sport in society and society in sport are the themes of Chapters 5 and 6. Chapter 5 on socialization, considers how people learn to participate in sport and what they learn in that process. It also examines the consequences of sports participation, especially as sport becomes more serious, for the development of self and for social roles and relationships in the larger society outside sport. Chapter 6 explores how the social deviance and social problems of society penetrate the world of sport. It examines social causes of such deviance and problems in sport, the relationship of overconformity or positive deviance to the Sport Ethic, forms of positive deviance in pain, injury, and risk-taking behavior in sport, and a number of major contemporary social problems in sport, including violence, delinquency, crime, gambling, and drug use.

Sport and educational institutions are the topics of the next two chapters. Chapter 7 looks at major issues and problems in U.S. interscholastic athletics along with the relationship of high school athletics to values, academics, and socialization. It also considers the role of the high school coach, cultural and social tensions in high school athletic participation outside the United States, the role of popular high school sports in the community, and their role in cultural and social assimilation of immigrant minority groups. Chapter 8 examines the world of big-time intercollegiate athletics, with its high pressure, commercialism, social tensions involving women and racial minorities, major social problems, and efforts at reform. It also gives special attention to the role of coaches.

Professional sport and the roles and relations of owners, fans, and athletes in this corporate phenomenon are our next focus of interest. Although the relationship of sport to the economy is a theme that runs throughout the book, Chapter 9 examines that realm of sport where money and profit are legitimately and openly pursued. This open pursuit of money and profit in sport is a paradox, though, that Chapter 9 closely scrutinizes. In addition, Chapter 9 considers the evolution of modern professional sports in the United States, their financial status and future, the political economy of sport and the law, the social consequences of professional sports monopolies, and the global expansion of North American professional sport. Chapter 10 looks at owners, and athletes in professional sports, examining patterns of ownership, labor-management relations, fans, social aspects of the careers of professional athletes and the lives of their families, and professional sports retirement.

No sociology of sport would be complete without an analysis of social classes, social stratification, and race and gender relations in sport. Chapter 11 looks at the kinds of sports that adults actively pursue in their leisure time and the ways social class and stratification shape such adult leisure sports participation. It also considers the leisure sports involvement of people who are disabled and are aging or elderly. Chapter 12 considers race relations in sport by focusing on racial segregation and integration patterns in American sport, the black woman in sport, various forms of residual racism in American sport, and the relationship of sport to the South African apartheid system. Chapter 13 examines gender relations in sport in terms of masculine cultural and social dominance of sport, homophobia in sport, barriers and progress for women in sport, the contemporary sex equity debate, and cross-cultural patterns of women in sport.

Chapter 14 concludes the book with a look at sport in political and global context. Using the Olympics as a case study, the chapter examines government interest in sport, the role of sport in international relations and national development, neocolonialism, the role of sport in the restructuring of nations in the new world order, the political economy of the Olympics, sport and politics in America, and the new world order in sport. Thus, the book ends by reinforcing a central theme made in this first chapter—that sport has a pervasive and growing presence and significance in global affairs.

SUMMARY

Sports devotion is pervasive in society, but paradoxically few people view sport seriously as a social issue. The paradox is partly explained by the dual character of sport as fantasy and as social reality. Sport sociology helps us understand sport from both perspectives as it clarifies and explains sport in society and its effects on the lives of the people who participate in it. The sociological imagination pushes our understanding of sport beyond common sense and immediate personal experiences. It enables us to see sport as a phenomenon that has enduring effects on the lives of large groups or strata of people and that is affected by major social forces in society.

When the social institution of sport is seen as a mirror of other societal institutions and processes, the sociology of sport becomes a means of understanding society by looking at sport. Sport sociology also is a means of understanding the social nature of sport. Both perspectives are important, but the study of sociology through sport is especially emphasized in this book.

Sport can be distinguished from a number of other types of physical activity, including play, recreation, and games. Sport sociology similarly can be distinguished from a variety of related disciplines that examine sport, including sport psychology, sport history, the philosophy of sport, the anthropology of sport, and sport geography. In fact, it is possible to use any of the social or behavioral science perspectives to understand sport in society, but sport sociology is the broadest of these disciplines. The sociological analysis of sport involves the use of sociological concepts, theories, and methods. The major theoretical perspectives in sociology that have been applied to the study of sport are structural functionalism, conflict theory, and symbolic interactionism. Cultural interpretive, critical, and feminist perspectives have also appeared prominently in sport sociology in recent years. In addition, cross-cultural and cross-national perspectives can be used to see sport in a comparative or global context.

CHAPTER 2

The Social Development of Modern Sport

EARLY SPORT

Archaeologists have produced evidence of physical play in prehistoric societies, but signs of organized sport remain elusive, even though the very physical nature of life for hunters and gatherers suggests its possibility (Blanchard & Cheska, 1985, ch. 4). Archaeologists continue to search for clues of sport in Stone Age societies at the same time that sport historians (e.g., Palmer and Howell, 1973: 22) suggest that existing archaeological evidence points to the emergence of sport and games in the early Dynastic period of Sumerian civilization (3000–1500 B.C.). Blanchard and Cheska (1985: 94, 96) document that artistic renditions of sport and games, such as boxing and wrestling, date from approximately 3500 B.C. in Mesopotamia. Clear indications of sport can be seen in the ruins of coliseums and other sporting arenas from the early civilizations of Egypt, India, China, and Mesoamerica as well as later Etruscan, Minoan, Greek, and Roman societies. The range of sporting activities and physical games thought to exist in the period 3500 B.C. through A.D. 500 included wrestling and boxing, acrobatics and tumbling, horse riding, archery, hunting, swimming, ball games of various sorts, stick fighting, bull fighting, marbles, javelin, soccer, running, track and field events, and rudimentary racket games (Blanchard & Cheska, 1985: 98). These games not only provided recreational opportunity, they also became a source of ceremonial, often religious, expression.

The ceremonial, symbolic, and ritualistic aspects of the earliest sports tended to give them a sacred quality associating them with the spiritual or supernatural realm. Early sports participation became an expression of religious worship, designed to appease or please the gods. According to Guttmann (1978: 19–20), the Mayan-Aztec form of an ancient rubber ball game similar to modern soccer was played to challenge the gods of the underworld. The resulting human sacrifice of the losers was done for **sacred** purposes rather than for **secular** purposes associated with the material things and concerns of everyday life in this world. Early forms of organized games and sport in the Western world developed from a combination of religious ritual and physical exercise (Coakley, 1990: 41). In modern sport, in contrast, we do not find the same kind of attachment to transcendent or supernatural forces that was characteristic of ancient sports.

The earliest forms of sport in ancient Greece derived their meaning from mythology and religious beliefs. These sports events and practices were typically associated with festivals that combined prayer, sacrifice, music, dancing, and ritual feasts (Coakley, 1990: 43). The athletes who competed in these early Greek sports came from wealthy families because serious athletic participation required resources for training, travel, and coaching. The sports of ancient Greece included chariot racing, wrestling and boxing, jumping, javelin and discus throwing, foot races, and archery.

Western sport from its very beginnings in Greek society gave less attention to female athletes than to male athletes. The discrimination in sports faced by women in ancient Greece, which included prohibition from the playing fields and stadiums where men played, reflected the patriarchal structure of families and the limited legal rights of women in that society. Yet, as Blanchard and Cheska (1985: 235) have observed, there was a dual ideal of "fertility and femi-

Source: Ninatallah/Art Resource, NY

FIGURE 2.1 Wrestling in Ancient Greece.

ninity" in the Greek city-states. The women of Sparta were obliged to prepare for the physical demands of childbirth, and also to defend the state in case of armed conflict, by routinely participating in strenuous athletic activities. At the same time women had to portray appropriately feminine traits such as dependence and passivity. Thus, at events such as the Haraean Games, girls and women competed in foot races and earned crowns of olive for their successes just as the men did in the Olympic Games.

The construction of permanent buildings and playing fields for the quadrennial Olympics in 550 B.C. occurred as Greek games and sports became more visible, popular, and politically significant. Greek sport became more professionalized and **commercialized** as wealthy patrons and city-states paid for athletes to train for the Olympics and athletes earned monetary prizes for their sports accomplishments. Now Greek athletes competed for the glory of their sponsoring city-state as well as for money. Professional male athletes of the second century B.C. organized guilds similar to modern player unions and used them to bargain for their right to have influence in decisions about scheduling, travel, pensions, and personal amenities (Baker, 1982).

Although they possessed many of the same features as contemporary sport, Greek games and sports before 100 B.C. differed from modern sports in three ways: their grounding in mythology and religion, their lack of complex administrative structures, and their lack of attention to measurement and records. They resembled modern sport, however, in reflecting the dominant structural characteristics of their society, which gave power and privilege disproportionately to wealthy young males (Coakley, 1990: 46).

In the last century B.C. and the first several centuries A.D., Roman sports developed into spectacles pitting gladiators against one another in events such as chariot races, foot races, boxing, and wrestling. Guttmann (1981) noted that in the time of Justinian's rule (A.D. 483–565), chariot racing was more of a political contest than a sport. Teams were supported by different political factions and engaged in fierce and often violent confrontations that in one case led to bloody riots and the destruction of Constantinople in 432 A.D. Roman games and sports were often equally violent, on the field and among spectators. Gladiatorial contests frequently involved armed competitors who were recruited from thieves and murderers as well as Christians, and these contestants engaged in death struggles that were staged for the pleasure of spectators. There also were spectacles pitting animal against animal and men and women against animals.

The spectacles of early Rome were used by political leaders to achieve at least two main purposes: to entertain or distract idle masses and to dispose of "undesirable" members of the population, such as criminals and Christians (Baker, 1982). They also aided in the preparation of obedient military men (Coakley, 1994: 57). Roman leaders were especially inclined to rely on spectacles to try to distract the masses during the period of decline of their empire, roughly from the third to the fifth centuries A.D. Spectators could determine the life or death of the gladiators they watched with thumbs-up or thumbs-down signals (Blanchard and Cheska, 1985: 236). These bloody and perverse spectacles lost popularity as the Roman empire became impoverished and dismantled during the fifth century.

THE EVOLUTION OF MODERN SPORT

The nature of Western sport was changed substantially during the Middle Ages when the traditional large-scale Greek and Roman festivals were replaced by local peasant games and the nobility's tournaments and jousts. During most of the Middle Ages, the Roman Catholic Church endorsed the diverse ball games developed and played by peasants in their local communities, and these games became integrated into community life. The games were often tied to local ceremonial and ritualistic occasions and were accompanied by music, dancing, and religious services. The origins of many contemporary sports, such as soccer, field hockey, football, rugby, bowling, baseball, cricket, and curling, can be traced to the local games of medieval peasants. These antecedents of modern sports lacked most of the structure and formal regulation of contemporary sports. Instead, they were guided by local and regional traditions (Baker, 1982; Coakley, 1990: 47, 50-51).

The tournaments of the feudal aristocracy were motivated by the need to prepare for the physical demands of military service and by the desire for entertainment. These tournaments often simulated violent battlefield confrontations that resulted in many deaths and injuries, the forfeiture of the loser's worldly possessions, and even the imprisonment of losers until proper payments were made. The tournaments became less violent, more festive, and less serious over time, but injuries and occasional deaths still occurred.

As the medieval tournament evolved from a warlike clash with a deadly serious military rationale into a form of entertainment and expression of chivalry, the great jousting tournaments of the Middle Ages reinforced the role of the noble and heroic male and the complementary role of the female as a "patronizing spectator" (Blanchard and Cheska, 1985: 236). Knights provided gallant displays of horsemanship, strength, and other skills for those in attendance. Although women typically were spectators rather than athletes in medieval times, there is evidence of some sports participation by women during this period. "Milkmaids," for example, participated in foot races for prize money at English country fairs (Blanchard & Cheska, 1985: 236).

The specific sports played by the upper classes of feudal times were quite different from those typically pursued by the peasants of this period. With access to equipment and facilities, the nobility was able to develop the precursors of contemporary billiards, shuffleboard, tennis, handball, and jai alai (Coakley, 1990: 51). The upper classes of feudal times tended to ignore the play, games, and sports of the peasantry because they saw these activities performing a cathartic or safety-valve function, funneling the peasants' possible frustrations and discontent into behavior that did not pose a threat to the established structures of power, privilege, and respectability. In the fourteenth and fifteenth centuries, however, wars made political and religious leaders and other members of the dominant strata of feudal society more mindful of the need for military might. During this period, peasants were pushed from their leisure pursuits into activities that prepared them to defend the domains or estates of their feudal lords. The peasants did not willingly give up these pursuits. In some instances their games and sports became means of expressing their opposition to the church and government.

The Protestant Reformation and the emergence of austere, severe, and serious-minded forms of religious expression such as Calvinism and Puritanism directly challenged the legitimacy of sports. Sports were seen as frivolous, profane, useless distractions from religious observance, hard work, family devotion, and expressions of good character that Puritans associated with good, virtuous, godly lives. The restrictive policies produced during this period, roughly between the early 1500s and the late 1600s, were directed mainly at the peasants. The Puritans of sixteenth- and seventeenth-century England especially objected to the drinking and partying that accompanied peasant games and sports and to the practice of holding sports events on Sunday. Objections to the sports pastimes of the upper classes were also raised, but it was difficult for religious opponents of sport to control these activities because wealthy members of society could pursue their horse racing, hunting, tennis, and bowling in the privacy of their estates (Coakley, 1990: 52).

Despite strong opposition from the Puritans in England, sport for all classes persisted through the Reformation era, often in a clandestine or less public forum. King James I of England revived sports in the early 1600s when he challenged English Puritans with the issuing of *The King's Book of Sports*, which asserted that Puritan officials should not discourage lawful leisure activities of the English people. Charles I reissued this book in 1633 and, along with his successors, encouraged the development of English sport. This encouragement led to the emergence of a number of highly organized sports in the late 1600s and the 1700s, including cricket, horse racing, yachting, fencing, golf, and boxing (Baker, 1982; Coakley, 1990: 52–53).

Guttmann has argued that although Puritan ministers in England and America were not "sullenly opposed to any kind of play" (1988: 32), it was nevertheless true that the development of modern English and American sports did not begin in earnest until Puritan magistrates were replaced by their more secular counterparts. The kind of tolerance of sport that Puritan colonial magistrates and ministers displayed in America is illustrated by the case of folk-football, which was officially banned in Boston in 1657. Guttmann points out, however, that this game was tolerated by Puritan leaders if players restrained their aggressive and dangerous practices and if they did not disrupt the peace of more meditative colonists. Shooting matches were actually encouraged because they helped keep the militia prepared to defend the colonial settlers from attacks by hostile Indians. Such encouragement or tolerance by Puritans of sporting activities was tempered by lists of restrictions and by the provision that Sunday was to be devoted to sacred devotion and good works rather the secular indulgence in sport.

During the eighteenth century, modern characteristics such as formal organization, professionalization of participants, and rationalization of procedures emerged in English and North American sports. In the new spirit of the Enlightenment, secular meanings and motivations tended to eclipse sacred values and sports participation was expanded from a restrictive elite participation to mass involvement. Sport was once again seen as legitimate activity, particularly now that it had been freed from the manacles of Puritan theology. Coakley (1990: 53) notes, however, that sport was regarded only as diversion during the Enlightenment period, and sports viewed in this light tend not to become highly organized. This conception of sport was to change during the historical period ushered in by the Industrial Revolution.

THE CIVILIZING PROCESS AND THE DEVELOPMENT OF MODERN SPORT : THE CASE OF BRITISH FOOTBALL

In contrast to Guttmann, Elias (1986) proposes that modern sport is a new phenomenon rather than an extension or revival of sport in earlier times. His specific focus is on a civilizing process that he associates with the establishment of the nation-state and its development of a high degree of stable, centralized state control over means of violence. Pre-modern sports, according to Elias, like the societies in which they were embedded, are much less formally regulated and much more violent than modern sports and societies. A major task in the sociological analysis of the development of modern sport, then, is to explain how it has become more civilized over time.

Studies by Elias and Dunning (1972) and Dunning (1972b) offer an example of this type of analysis applied to the development of modern British football. They show the substantial differences between folk and modern versions of this sport in British society. Elias and Dunning's study reveals that folk football in medieval Britain was governed by informal local rules and was relatively unspecialized; Dunning (1972a: 84–85) observes that the size of playing fields, the number

POINT-COUNTERPOINT

IS SPORT MORE CIVILIZED NOW THAN IN THE PAST?

YES

Sport has become more organized, rationalized, and less violent than in past historical eras, especially before this century. Roles and behavior in sport are much more formally regulated than in earlier eras. The intensity and pervasiveness of violence today in sports such as football may seem substantial, but they do not equal the levels in football at the beginning of this century, which almost caused the banning of the sport. Thus, athletes, coaches, and fans are much more likely now than in the past to conform to established rules and routines and act in a "civilized" manner.

NO

The formal corporate structure of modern sport and the effective marketing and public relations that create favorable images of it for the public mask the "uncivilized" behavior often found in modern sport. The emphasis on violence and the resulting injuries in big-time and professional contact sports such as football and ice hockey are well-documented, as are the violence and injuries in ostensible noncontact sports such as basketball and baseball. We even find sports emphasizing aggression, such as American Gladiators, created for violence-hungry sports fans. British "soccer hooligans" may be the best example of the uncivilized behavior of modern sports fans.

How should this debate be resolved?

of players, and the length of matches varied a great deal. Goals were sometimes 2 or 3 miles apart; games often lasted for many hours and even stretched across days at times; tactics and practices during games were relatively unrestricted; and violence occurred frequently. The violence and limited regulation of folk football reflected the relatively "uncivilized" nature of medieval life in general in Britain. According to Elias and Dunning, these games were a ritualized means of releasing the frustrations that had arisen in relations between groups in the community.

In Dunning's (1972b) study of the development of modern British football, we see the emergence of a sport with more regulation and less violence. Dunning argued, following Elias, that the development of modern football in Britain was tied to a series of interrelated societal changes that made the society, as well as football, more civilized because industrialization and urbanization changed elite private boarding schools in a number of important ways (Dunning, 1972a: 85–86). There was very little order in the English private schools of the late eighteenth and early nineteenth centuries. The masters, who were typically from lower-class backgrounds, could not control their upper-class male students because the students did not like taking orders from social inferiors. As a result, the oldest and more powerful students assumed control of the schools and they would exercise their power over the younger students by means of a system called *prefect-fagging*. This pattern of domination carried through on the football field as well as other aspects of schoolboys' lives and reinforced the violence of traditional football.

Increasing industrialization and urbanization contributed to the reform of the prefect-fagging system by making the English middle class richer and more powerful. By the 1840s the middle class was able to achieve its desired school reforms, including the prefect-fagging system. The efforts of Thomas Arnold, headmaster of the Rugby School from 1828 to 1842, were especially notable (Dunning, 1972b: 141). Arnold coopted the prefects by trying to mold them into a "moral elite" who embodied the ideals of the "Christian gentleman." Thus, instead of being challengers to the masters' authority, the prefects became subordinates who supported their masters' efforts to achieve certain moral and educational objectives.

The reform of the prefect-fagging system by public school masters led them to the realization that sports such as football could be used to achieve character-development and educational goals. On one hand, participation in sports could divert students from other activities that were destructive or deviant, such as destroying property, drinking, and disorderly behavior. On the other, team sports participation might be a means of improving interpersonal relations among the boys and, even more significantly, a means of improving the boys' characters. As they began to conceive of a character-building ideology for sports, the masters became more actively involved as advisers in the organization and playing of school sports. They realized that the wildness of football had to be tamed to make it an activity that could have the positive educational and character-training functions they envisioned. Without directly taking control from their prefects, they encouraged the dominant boys to bring the unruly and violent features of their games and sports under control. Following the lead of the Rugby School, the English elite boarding schools established formal rules and standard procedures for settling disagreements to accomplish these purposes. The institutionalization of this restrictive structure, Dunning contends, served as an important step in the development of modern football as a formally regulated and more civilized sport. Football could now be considered as a respectable activity for young adult "gentlemen."

The new forms of football developed in the elite schools spread beyond the middle and upper classes during the 1870s as the urban working classes formed their own clubs when their work week became shorter. Priests who were trained in the elite schools to believe in the moral and character benefits of sports were important agents in spreading football and rugby to the working classes. As football grew in popularity among varied class and regional groups in the middle of the nineteenth century, interclub matches were organized. The pitting of teams from different social backgrounds and locations and with different orientations to the game created the need for unified rules. Representatives of different factions in the sport met in London in 1863, and formed the Football Association. The rules adopted for this new Association football were not accepted by all who attended the London meeting. A number of clubs who rejected the new rules (such as the prohibition of handling, carrying, and "hacking" or deliberately kicking on opponent's shins) established their own association in 1871, the Rugby Union, to play a different game. In the same year, the first national football competition, the Football Association (FA) Cup, was established. FA Cup football was intensely competitive, involved players from a variety of backgrounds, and made referees and linesmen a regular part of the sport (Dunning, 1972b: 145, 147). By 1883, the initially dominant elite school teams were overtaken by a team of mill workers, and from this point the upper classes no longer had a dominant influence on the development of the sport. The influence of professional clubs, commercialism, spectators, and tensions between middle- and upper-class groups favoring amateurism and working-class groups in favor of professionalism marked the subsequent development of football in Britain.

After a series of disputes, an agreement was reached in 1885 that permitted athletes to accept payments under "stringent conditions." This agreement enabled the Football Association to retain its exclusive control over the sport. The classic characteristics of modern sport around the world—rationalization, specialization, organization—had been set in Britain before the turn of the twentieth century.

We can see from this brief historical survey that the evolution of sport has not been simple, unilinear, or uniform across societies and cultures. As a result of many social forces, however, sport has become a rational, technologically sophisticated, highly organized, and scientifically oriented corporate enterprise. The development of sport as we know it today has been marked by tendencies toward secularism, bureaucratic organization, specialization in a division of labor, and measurement of virtually all aspects of performance, with a specific emphasis on outcome rather than participatory experience. To show what many modern sports have become and how they have become what they are, we will examine the influences of industrialization, urbanization, and bureaucratic and scientific rationalization in more detail.

INDUSTRIALIZATION AND THE DEVELOPMENT OF MODERN SPORT

The Industrial Revolution began in England and Europe during the final decades of the eighteenth century and the early part of the nineteenth century. It spread to North America in the early part of the nineteenth century. Giddens (1991:63) has pointed out that **industrialization** refers to a complex array of technological changes that have affected the economic organization of society and the economic behavior of members of society. The Industrial Revolution included the invention of new machines, new means of producing energy, mass production of goods, and a new reliance on science to improve the manufacture of goods. A major distinguishing characteristic of industrialized societies is that most of their populations are employed in factories or offices rather than in agriculture. In addition, industrialized societies are dominated by urban centers where activities are coordinated by large-scale organizations and bureaucracies, particularly in the economic and governmental sectors. Pre-industrial societies were village-based rural economies dependent upon family agriculture and small-scale craftsmen.

There are many ways that the assorted technological innovations and social changes associated with industrialization during the last three centuries influenced the development of sport. According to Leonard (1988a: 41–45), the technological innovations from industrialization affected sport in three main areas: transportation, facilities and equipment, and communication.

Transportation

As new discoveries such as the steam engine, the smelting of iron for steel production, the internal combustion engine, and the capacity to fly created faster and more efficient means of transportation, it became possible to move sports participants and spectators to increasingly distant locations and thereby expand the geographical boundaries of sports competition. The early growth of intercollegiate and professional sports, for example, was aided by the expansion of railroad travel. In addition, the invention of new forms of transportation created new sports, such as powerboat and auto racing and stunt flying.

In considering the relationship between transportation and urbanization in the development of sport in Canada in the last third of the nineteenth century, Jobling (1976) pointed to the significant impact of railroads. The speed and comfort of railway transportation promoted competition between teams from different cities in many sports and enabled devoted fans to follow their teams on the road. Street railway companies also transported spectators to and from contests within cities.

The emergence of the railroads in the middle of the nineteenth century took away from the influence of the steamboat, which had been the basis of intersectional sports competition earlier in the century. Steamboats remained important throughout the nineteenth century, especially during aquatic regattas. The popularity of rowing and sailing as sports events was largely due to the fact that steamboats made it possible for spectators to get close to the action.

Along with the steamboat and railway systems, the bicycle was an important part of the history of nineteenth-century sport. Jobling observed that the introduction of the "safety bicycle" in the late 1880s led to the organization of many bicycling clubs and also improved the mobility of people who lived in towns, which subsequently increased their participation in other sports (Jobling, 1976). Furthermore, the popularity of the bicycle resulted in the construction of commercialized riding arenas and schools and the emergence of a large number of bicycle manufacturers, retailers, and repair facilities.

Along with effects on the economy, leisure, transportation, and sport, the bicycle had an impact on the attitude of women toward sport and contributed other kinds of changes that had been sought by the women's suffragette movements. Jobling noted that within a few years after the appearance of the bicycle, women began to see fashion alternatives to the awkward long skirts that were easily tangled in the chains, sprockets, and rear wheels of bicycles. After the fashion industry began to manufacture women's clothing more suited to this type of recreation, women became more involved in sports such as tennis, croquet, and golf as they found their participation in these types of athletic activities now more socially acceptable and physically possible (Jobling, 1976: 69).

Equipment and Facilities

The invention of the sewing machine transformed the textile industry in the nineteenth century, and improved production in this industry resulted, among other things, in sports clothing, shoes, balls, and other equipment that were more standardized in form and of better quality. It is not coincidental, for instance, that the modern era of baseball was ushered in by the use of a greatly improved "jack-rabbit" ball in 1920, which was much more lively and traveled much greater distances than the ball of the "deadball era" of approximately 1900 to 1919. With this new ball, there were eight times more home runs (Leonard, 1993: 39).

One of the major changes in the staging of sports events was made possible by the invention of the incandescent light bulb in 1879. The new light bulbs made possible nighttime sports events such as prize fights, horse shows, wrestling matches, basketball games, and walking competitions, and it has been reported that Madison Square Garden in New York City staged night events in the middle of the 1880s (see Leonard, 1993: 41). Baseball was much slower to adopt this invention; the first night game, between Pittsburgh and Cincinnati, did not take place until 1935, and the final baseball stadium to install lights was Wrigley Field in Chicago—in 1988. As the twentieth century progressed, nighttime sports events and the construction of arenas with better sight lines, increasingly comfortable seating, and sophisticated scoreboards made sports attendance considerably more convenient and enjoyable for many sports fans. The recent invention of artificial turf, however, has created controversy because possible gains in the ease and cost of maintenance for stadium owners may be offset by the occurrence of a variety of injuries associated with the turf.

Another example of an invention with a significant impact on sport is the vulcanization of rubber in the 1830s. With this invention, it became possible to manufacture more elastic and more resilient rubber balls in the 1840s and tennis and golf balls later in the century. The related development of the pneumatic tire in the 1880s had a major impact on cycling and harness racing in the 1890s, and subsequently on the growth of auto racing (Eitzen & Sage, 1993).

Communication

The invention of new forms of communication after the Industrial Revolution took place side by side with innovations in transportation. Mass communication media have been critical factors in the commercial success of contemporary sports, but they also had a substantial impact on the popularization of sport earlier in the industrial age. The first American sports magazine appeared in the late 1820s, and seven sports magazines followed in the next decade (McChesney, 1989). These magazines bore the imprint of the well-established sporting magazine industry in Great Britain, which was established in the 1790s. In the 1830s and 1840s, a major surge of industrialization affected the Northeast, and U.S. cities were filled with increasing numbers of migrants from rural areas. The size of the reading public greatly expanded, along with interest in sports. William Trotter Porter's *Spirit of the Times* became the most popular sports magazine in the United States, with its circulation reaching 100,000 by the 1850s. At first Porter covered more respectable sports, such as horse racing, since sport in the early years of his publication was generally regarded as vulgar and disreputable by literate Americans. Porter expanded coverage to boxing, tried to establish cricket as the national sport in the 1840s, and then turned his promotional efforts to the emerging sport of baseball in the 1850s, playing an important role in making the American reading public more familiar with its rules and terminology. McChesney (1989: 51) noted that Frank Queen's *The New York Clipper*, founded in 1853, quickly took the place of *Spirit of the Times* as the top sports weekly. The first genuine American sportswriter, Henry Chadwick, was on the staff of the *Clipper* and became known as the "Father of Baseball" as a result of the attention he gave to the sport.

The U.S. newspaper industry was born in the 1830s and 1840s, but sports coverage in newspapers was relatively limited until the end of the nineteenth century because of sport's disreputable status. The

Source: AP/Wide World Photos

FIGURE 2.2 Lights at Wrigley Field.

New York Herald took the lead in the early years, paying special attention to sports such as prizefighting and horse racing. In the 1880s and 1890s, newspaper circulation greatly increased as a result of technological innovations that reduced printing costs and the soaring populations of cities. As sports coverage by magazines and newspapers became increasingly popular, sport became recognized as a magnet for readers.

By the end of the nineteenth century, newspapers had replaced magazines as the primary medium of sports coverage, and the first distinct sports section appeared in a newspaper in 1895. It was introduced by William Randolph Hearst in his *New York Journal* in the midst of a circulation battle with Joseph Pulitzer's *New York World*. After that, virtually every newspaper began to cover major sports events, such as championship prizefights, horse races, and the World Series in baseball. The invention of the telegraph and the telephone made it possible to provide up-to-date coverage of sports news from around the country. By the 1920s, the sports section was a prominent and popular feature of major daily newspapers. Lever and Wheeler (1984) found in a content analysis of the *Chicago Tribune* sports pages from 1900 to 1975 that the sports section had expanded in proportion to the rest of the newspaper; that coverage of the dominant sports of baseball and football had remained fairly stable; and that there were shifts in coverage from amateur to professional sports, from local or regional to national events, and from individual to team sports. While the printing press, the telegraph, the telephone, magazines, newspapers, and sports reporters contributed substantially to the increasing popularity of sport by the end of the nineteenth century, radio and television became significant factors in the growth of sports in the twentieth century. The emergence of broadcasting added to the nationalization of sport begun earlier by the print media and further embedded sport in the social and cultural fabric of the nation (McChesney, 1989: 59).

Although magazines, newspapers and radio remain important means of conveying sports information to

POINT-COUNTERPOINT

DID LIGHTS RUIN BASEBALL AT WRIGLEY FIELD?

YES

The absence of lights at Wrigley Field represented a last remnant of the glorious past of the "Great American Pastime." The old Wrigley Field reminded fans of tradition and gave them a chance to see baseball played as it was meant to be played, in the daylight—where both players and fans could see the ball clearly and the game could be played most effectively. Each time a piece of the nostalgic past of a sport is removed or destroyed, some of that sport's appeal is lost. The installation of lights at Wrigley Field demonstrates the dominance of commercialism over the integrity of the game in modern sport.

NO

The absence of lights at Wrigley Field was an anachronism. It represented a past that has been surpassed by the much more exciting and accessible modern game of baseball. While having lights at Wrigley has commercial advantages for owners, fans also benefit. Day baseball games prevented employed fans from attending Cubs games or watching or hearing their team play at home. Thus, playing at night makes it easier for Cub fans to follow their team and makes the franchise potentially more popular and profitable. Lights bring the Chicago Cubs into the modern world of sport.

How should this debate be resolved?

POINT-COUNTERPOINT

DOES ARTIFICIAL TURF BELONG IN SPORT?

YES

Although initially expensive, artificial turf ultimately saves money for a team or stadium operator. A field can be used repeatedly for multiple purposes over many years without harming the quality of the playing surface. Maintenance is much easier than with natural grass. The field appearance is always consistent, especially for television.

NO

Artificial turf is dangerous for athletes because it can be very slippery when damp and can cause abrasions. The nature of the surface can make players with joint problems very susceptible to injury. Although short-term maintenance costs may be limited, repairs and replacement are very expensive. Natural grass also looks better.

How should this debate be resolved?

mass audiences, we now tend to think of most major sports at the college, professional, and Olympic levels as televised events. Television has become the vehicle for reaching large sports audiences around the world—and, more importantly, the money this medium generates has become an essential component of the current commercial success of many sports.

URBANIZATION AND THE DEVELOPMENT OF MODERN SPORT

Industrialization pushed people from rural areas into towns and cities. In 1800, less than 20 percent of the British population lived in towns or cities of more than 10,000 people; by 1900, this proportion had increased to 74 percent. This movement of people into towns and cities, called *urbanization*, took place somewhat later in Europe and North America than it did in Britain (Jobling, 1976: 64–65; Giddens, 1991: 676). In 1800, less than 10 percent of the U.S. population lived in communities of more than 2,500 people. Today, more than 75 percent of Americans live in cities (Giddens, 1991: 676). In 1860, only nine American cities had more than 100,000 people, but by 1910, this number had increased to fifty. In 1871, Canada had twenty communities with over 5,000 people, but by 1901 it had sixty-two; twenty-four had more than 10,000 people (Eitzen & Sage, 1993: 43).

We have noted that the growth of the urban industrial working classes contributed to the increased popularity of spectator sports in the nineteenth century. The growing urban areas offered better transportation and access to sports arenas, a higher standard of living, a more affluent class of workers who had more money and more leisure time to spend on sports, and settings that facilitated the organization of teams and leagues (Betts, 1974: 172). Betts observes that horse racing in the second half of the nineteenth century became increasingly dependent on urban spectators and on betting at the tracks in metropolitan areas. He also notes that baseball during this period had become very popular in the towns and cities, and that a variety of commercialized sports, including bicycle racing, boxing, and wrestling, depended on the gate receipts of urban spectators. Furthermore, according to Betts, many city organizations and agencies promoted sports in YMCAs, clubs, and public schools to ease the tensions produced by rapid growth of the cities in the late nineteenth and early twentieth centuries. In a sense the growth of commercialized sports required urbanization because rural areas lacked the transportation facilities, audience, affluence, and other elements of an infrastructure that was required for long-term expansion of sports (Betts, 1974: 173).

BUREAUCRACY, RATIONALITY, AND MODERN SPORT

The assortment of social processes and outcomes associated with industrialization and urbanization have contributed to the commercialization and mass popularity of many sports on national and international levels. Modern sport is more than a commercialized mass phenomenon. It is also rational, technologically sophisticated, highly organized, and oriented to quantifying outcomes in a precise fashion. This distinctive combination of features is a result of the impact of bureaucracy and science, as well as industrialization, on sport.

Bureaucracy in sport can be traced back at least to the Roman empire. Guttmann (1978: 45–47) pointed to the Roman guilds of athletes, which was an imperial organization with elected leaders, detailed rules and regulations, entrance requirements, codes of proper conduct, and material items such as membership certificates. In modern sports, we can find evidence of bureaucracy in the late eighteenth-century English Marlybone Cricket Club. This organization, which became the ultimate authority in cricket, succeeded in standardizing the game in the nineteenth century by establishing precise standards for such things as the weight of the ball, the width of the bat, the distance between wickets, and the dimensions of the wicket. Today, we see highly bureaucratized international federations setting similar standards for most sports. We also see government sports bureaucracies—with the United States an exception—in most industrialized nations today. These government bodies subsidize, regulate, and directly or indirectly control sport of various types.

Bureaucracies, as we saw in Chapter 1, are large-scale social units with highly specialized divisions of labor, elaborate hierarchies of authority, and highly rationalized and formalized activities and regulatory controls meant to make the organization as efficient and effective as possible in the pursuit of its goals (Nixon, 1984: 147). Highly commercialized sports bureaucracies, ranging from the National Football League (NFL), Major League Baseball, the Professional Golf Association (PGA), the National Collegiate Athletic Association (NCAA), and the International Olympic Committee to individual sports franchises or programs in these and other sports realms, are largely oriented to attracting audiences to the sports events they organize, promote, and stage and to deriving financial revenue or profit from their sports productions. These commercialized sports bureaucracies are the major organizational players in modern corporate sport. It is important to recognize that corporate sport includes both overtly professional sports and ostensibly amateur ones, such as highly commercialized forms of intercollegiate athletics and Olympic sports governed by national and international amateur sports federations.

In Weber's classic conception, the development of bureaucratic authority was inevitable because it appears to be the only way to deal with the administrative requirements of large-scale social systems (Giddens, 1991: 348). According to Weber, the more real bureaucracies embody the characteristics he identified—including most prominently hierarchy, written rules, full-time and salaried officials, a separation of work and home life, and a separation of workers from ownership of the means of production—the more successful they would be in achieving their goals. Weber believed that bureaucracies were an effective means of organizing large numbers of people because they ensured that decisions would be made on the basis of general criteria rather than personal or idiosyncratic individual whims or wishes; because they emphasized competence or expertise in carrying out organizational tasks and making decisions; because salaried and full-time positions were assumed to make people more secure and less susceptible to corruption; and because formal and systematic hiring and evaluation procedures were assumed to decrease the influence of personal favor or kinship connections in hiring and other personnel decisions (Giddens, 1991: 350). Weber recognized, though, that actual bureaucracies could be characterized by dull routines that stifled creativity and rigid practices that prevented people from deriving the benefits or services that bureaucracies were supposed to provide. We also can observe patterns of impersonality, buck-passing, inefficient communication, red tape, and excessively complex regulations in contemporary bureaucracies (Hess, Markson & Stein, 1991: 92–93).

According to Guttmann (1978: 47), one of the most important functions of bureaucracy in sport is to establish and enforce universal rules and regulations. To the extent that the organization of sport is consciously designed, ordered, standardized, and routinized by formal rules and regulations, it can also be seen as rational (Guttmann, 1978: 41). Thus, bureaucracy is a mechanism for rationalizing modern sports. The invention and development of basketball is a good example of the influence of rational organization on a sport. Phillips (1993: 65–67) has described how rule changes in basketball reflect its gradual rationalization and bureaucratization to achieve specific goals. Early rule changes, for example, concerning location of the backboards, dribbling, and the nature of the baskets tended to be for the purpose of increasing the enjoyment and fitness of the players. Later rule changes tended to be more oriented to increasing the enjoyment of spectators and the game control exercised by coaches.

Although bureaucracy and rationality are defining characteristics of the organization of modern sport, they are probably not the essential part of how most athletes, coaches, and fans think—or want to think—about sport. They are certainly not what creates the passion that many feel for sport. In fact, one could argue that an important reason for the appeal of sports in highly bureaucratic societies today is that it possesses a basic indeterminacy that is at odds with the predictability that rational bureaucracy is meant to create (Cashmore, 1990: 182). This indeterminacy exists because outcomes cannot be fixed in advance in legitimate sport. When this uncertainty is combined with the colorful action of sports competition, sport becomes exciting for all involved.

In Cashmore's view, sport loses its essential meaning when it loses its indeterminacy and is transformed into a rationally constructed spectacle or drama. (Recall the circuslike quality of Roman sports, in which contests were meant to entertain the spectators more than to challenge the athletic talents of the contestants.) Cashmore would acknowledge, however, that despite the need for indeterminacy in sport, modern sports are distinguished by rational and bureaucratic pressures to organize and operate in such a way that the uncertain is made as certain as possible. There are elements of job specialization, production routines, deadlines, and repetitive training regimens that make sport much like many other bureaucratic work settings in industrial society. Computers are used to assist in the administration of sports organizations and to process data and plan strategies that assure success in sports contests. There are large and complex administrative hierarchies, hierarchies or levels of competition, and thick books of official rules and regulations, even for sports programs for very young children. In addition, rational bureaucratic organization includes the counting and measuring of performance and the keeping of records, which, according to Guttmann (1978: 47), are distinctive characteristics of modern sports.

The combination of rational calculation and determinacy with passion and uncertainty may be paradoxical, but it is not necessarily contradictory. According to Guttmann, modern sports are both an alternative to and a reflection of modern industrial society. They have their roots in the human spirit but take their form from the dominant social structures and processes of our time. In Guttmann's poetic words, sport is the "rationalization of the Romantic" (1978: 89).

SCIENCE, RATIONALITY, AND MODERN SPORT

In the pursuit of victory and ever-higher levels of performance in sport today, we see evidence of both the irrational or romantic and the rational. Rationality here refers to the application of science to sport, and it is applied to enable athletes to become more efficient and productive. With roots in the practices of ancient Greeks, the scientific rationalization of sport has been pushed to increasingly higher levels today through physiology, psychology, sport medicine, and a variety of other sport science disciplines. The former Soviet Union and East Germany are often cited as examples of nations with highly scientific approaches to sport that led to spectacular success in international competition.

When athletes try to push themselves to or beyond their physical or physiological limits, their quest for athletic excellence becomes an irrational and potentially self-destructive form of sacrifice. Where rationality involves a scientific approach to sport, athletes do not merely practice their sport to prepare for competition; they train in a very intense and systematic fashion. In Guttmann's view (1978: 43), training implies a rationalization of the whole sports enterprise and involves a willingness to experiment and to test results constantly. Life becomes a single-minded focus on athletic improvement and success. This scientific orientation is consistent with the modern version of corporate sport in which the emphasis is on outcome (i.e., profit and productivity) rather than on participatory experience.

Hoberman (1992) describes the hundred-year history of "scientific sport" that has tied the development of modern sport to the values of industrial technology. Pervasive scientific experimentation has been performed on athletes, particularly at the highest levels of professional and amateur sport, to turn them essentially into efficient, tireless machines. Exercise physiologists, for example, have been actively engaged in efforts to maximize athletes' stress tolerance and endurance. Industrial psychologists work to establish the link between the mental attitudes and personalities of athletes and the development of strength and energy for athletic competition. The biological influence in sport is seen in the role of performance-enhancing drugs, synthetic hormones, and such practices as blood doping. Sport sociologists and social psychologists of sport also have contributed their scientific expertise to sport through studies of the structural conditions and social dynamics associated with successful team performances.

In addition, sports medicine has emerged to play a critical role in modern sports by providing means of training and treatment to help athletes avoid serious injuries and lost playing time and to help injured athletes to get back into action as quickly as possible. A German sports physician, Wildor Hollmann, stated that Olympic sport has been nothing less than "a gigantic biological experiment carried out on the human organism" over the past hundred years (Hoberman, 1992: 4). A century ago, when scientific interest in physiological and psychological aspects of athletic performance increased sharply, the goal of producing superior athletes was much less important than that of understanding more about the science of the human organism. The increasing importance attached to the measurement of sports performance at the end of the nineteenth century and throughout the twentieth century stimulated efforts to apply science to athletic performance. The initial interest among scientists in studying athletes was not accompanied by efforts to enhance performance because sport lacked

POINT-COUNTERPOINT

HAS SCIENCE IMPROVED SPORT?

YES

Athletes always want to do better. Trying to do better and win are the essence of sports competition. The application of science in a rational approach to sport has enabled athletes to make huge strides in enhancing their athletic performance. Science and rationality enable athletes to improve by maximizing their efficiency and productivity. Sports medicine, an offshoot of the concept of scientific sport, now permits athletes to reduce their chances of injury and recover more quickly from injuries as they push themselves further and harder.

NO

The purpose of rational and scientific approaches to sport is to reduce the uncertainty of the outcome of athletic contests. The essence of sport, however, is competition with an uncertain outcome. Eliminating uncertainty eliminates the challenge that makes sport so exciting and appealing. Scientific experimentation to enhance performance dehumanizes athletes and turns them into machines or experimental guinea pigs. The push to excellence through science and medicine could lead athletes to self-destructive behavior and physical breakdown.

How should this debate be resolved?

scientific legitimacy and the idea of high-performance athletes had not yet captured the popular imagination in the industrializing nations. Today, the application of science to performance enhancement is a very common practice.

The shift from using athletes to further medical research to using medical science to improve athletic performance has occurred during the twentieth century with the parallel development of science and of sport. The theory and practice of boosting the strength and speed of athletes developed among athletes and trainers rather than in the scientific laboratories of a century ago. The goals and ambitions of science and sport intersected in the search for the physical, physiological, and psychological limits of human organisms. During the twentieth century, sport has established a symbiotic relationship with medicine in particular (Hoberman, 1992). Early studies linking the biomedical sciences with performance were conducted using the bicycle as a test device. Cardiovascular changes, muscular development and use patterns, and respiratory stress could be easily assessed on bicycle riders.

The biomedical influence on sport may be most visible today in the widespread use of drugs to enhance or restore performance. For many sports critics, this is one of the most troubling aspects of the medicalization and scientization of modern sports. Hoberman (1992) has pointed out that along with the dangers to the human body posed by many drugs that are used by athletes, drug use also has ethical implications that threaten the integrity of sport itself. The cases of Ben Johnson's drug-assisted triumph and subsequent disgrace in the 1988 Olympics and Diego Maradona's consumption of a stimulant "cocktail" during the 1994 World Cup raise a number of difficult questions about the ethical and practical limits of scientific experimentation in sport. Questions can also be raised about the motivations, ambitions, and character of athletes and coaches who readily use or encourage the use of performance-enhancing drugs in their efforts to win. By broadening our perspective from drugs to the scientific or quasiscientific experimentation it represents, we are led to issues such as the unfair advantage that athletes can gain from drugs and other artificial aids to performance, the abandonment of self-restraint reflected in the use of banned substances, and the drug-induced or scientifically-aided transformation of human athletes into unnatural or artificial creatures similar to the contrived bionic man and bionic woman characters created for television more than a decade ago.

Athletes have been willing to violate ethical standards of sportsmanship and accept medical risks of illicit and potentially dangerous drugs and training techniques to feed their personal ambition. If athletes use these substances in the hopes of improving performance, they are, in effect, volunteering to be subjects in illicit clinical trials (Hoberman, 1992: 18–19). Most of these experiments have involved illegally obtained drugs that athletes have administered to themselves. At times, though, such as in the former East Germany, experiments with drugs such as anabolic steroids have been supervised by highly trained scientists. In the United States, the use of steroids and other drugs has been officially prohibited by sports authorities but implicitly encouraged by coaches, who press athletes to attain levels of strength, size, and aggressiveness that often cannot be attained by conventional or legal training methods or diets (Eitzen & Sage, 1993: 174).

Intentional or unintentional scientific experimentation in sport to improve performance has created what Hoberman (1992) has called **mortal engines**. These developments have a rational cast because they are aimed at making athletes as efficient and productive as possible, an emphasis that parallels efforts to make industrial workers in general more efficient and productive. Hoberman has suggested, though, that the mortal engines produced by scientific sport have been dehumanized by the experimentation and manipulation. In the "scientization" process of being pushed to or beyond their human capabilities, athletes have become less human, more machine than person. This trend of applying science to performance assessment is consistent with other trends that have given sport its corporate character.

THE POLITICAL ECONOMY OF MODERN CORPORATE SPORT

The idea that bureaucratic and scientific rationalization have dehumanized modern sport and athletes reflects a critical sociological interpretation of what elite corporate sport has become and how it affects its participants. Another critical perspective, emphasizing political economy, provides additional insight into the importance of power and money at the corporate level of sport. This perspective emphasizes the influence of corporate capitalism and governmental power on modern sports. As governments have expanded their involvement in social institutions and cultural practices that were traditionally independent of state influence, it has become a major vehicle connecting corporate capitalism and modern sport (Sage 1990a). In the United States, for example, where government has been less obviously involved in sport, it could nevertheless be seen as having encouraged capital accumulation of sport business while using sport as an instrument of national policy. Legislation, court decisions, and administrative decisions of government bodies have provided a favorable economic climate for sports investors and officials to generate healthy profits for their sports enterprises as well as substantial personal financial benefits. Despite limited direct control over sports, the U.S. government has used sports and athletes as a means of conducting foreign policy.

Hart-Nibbrig (1987) has employed a political economy perspective to show how political and economic forces merge in college athletics in the United States. He uses the concept of **corporate athleticism** to describe the influence of corporate business values on the organization of college sport, which has led to an intensification of emphases on recruitment, payment, and training. The financial basis of corporate athleticism, which has fueled much of its development in recent decades, has been television revenue. The commercial growth of college sport has led to efforts by the NCAA to increase its regulatory and financial control over college sport and to legal actions by a number of big-time NCAA member institutions to increase their autonomy and financial opportunities. In addition, reliance on legislative support in state universities and support from business-minded alumni and boosters represents the political dimension of corporate athleticism. Efforts by college presidents to reform corrupt practices in college athletics have been blocked by a coalition of athletic, business, and political interests opposed to any actions that might deemphasize or change drastically the athletic programs they support and control. Much more will be said about college athletics in later chapters, but this discussion of corporate athleticism is meant to draw attention to the importance of corporate economic and political interests in contemporary sports. Nowhere is this trend more evident than in the relation of mass media and sport.

TELEVISION AND MODERN SPORTS

It is impossible to ignore the influence of the mass media, particularly television, on modern sport. No realm of society in the United States or Canada has been as affected by television as sport has. Even as television has left an indelible imprint on the growth of sport, sport also has had a significant influence on television. In other words, sport and the media have a relationship that is mutually beneficial or symbiotic. Commenting on the symbiotic relationship between television and sport, Lobmeyer and Weidinger (1992) have pointed to the entertainment factor as a critical link between the two institutions. The transformation of sport from a participant-oriented practice to a spectator-oriented enterprise made sport attractive to television and led sports officials to appreciate the capacity of television to enhance the commercial development and popular success of their sport. The corporate commercial bond linking television and sport has created the **TV Sport System** which has become especially strong in the United States because this country has provided a very favorable legal climate for profit-oriented American sports entrepreneurs. Unlike nations such as Germany, in which public broadcasting stations dominate the air waves, the United States has permitted the commercial development of the privately owned television industry. The insertion of lucrative commercial advertising in television sports programs has been a key factor in the explosion of the TV Sport System in the United States, making sports profitable for the corporations that control television networks and making television a major source of revenue for sport.

Televised sport began in the United States in 1939 with the broadcast of a college baseball game between Columbia and Princeton. Eitzen and Sage (1993: 277–278) noted that the announcer was in the stands with the spectators, there was a single camera, and it was not able to show the batter and pitcher in the same picture. *The New York Times* reported that players appeared to be "white flies" running across the screen and that the movement of the ball appeared as a "comet-like white pinpoint." The announcer's commentary was the only means for viewers to understand what was happening.

Despite these technical difficulties, interest in television and televised sports grew tremendously after World War II. In 1992, the major networks in the United States broadcast about 1,800 hours of sport, cable stations broadcast about 5,000 hours, and local broadcasts on independent stations added thousands of additional hours of sports coverage. Of the ten most highly rated television programs of all time, five are NFL championships, and over half of the top twenty-five television programs are sports events.

The three major networks, ABC, CBS, and NBC, monopolized television sports until 1980, when ESPN began broadcasting sports on cable systems as the first 24-hour year-round network devoted exclusively to sports. By 1990, ESPN reached more than half of the homes in the United States. Cable networks other than ESPN, such as Prime Cable, TNN, and emerging regional companies, also place a heavy emphasis on sports programming. Another way of watching sports events on television has been to pay a fee in advance. This pay-per-view approach, which is expected to expand in future years, has been used for many years in televising championship boxing events, and it was used by NBC to provide viewers with more extensive access to the 1992 Summer Olympic Games in Barcelona (Eitzen & Sage, 1993: 278–279). Many professional teams are considering maintaining their own pay-per-view networks because of the tremendous profit potential.

The major benefits of television for sport have been revenue and exposure. The steep escalation of rights fees paid by television to sport has been a major factor in the large gains in revenues earned by corporate sports at all levels over the past few decades. It has also been a major reason for the large increases in salaries and tournament prizes during this period. The pattern of these increases is dramatically illustrated by the case of the Olympics. In 1960, the cost for the rights fees was $394,000 for the Summer Games and $50,000 for the Winter Games. By 1980, rights costs were $87 million for the Summer Games (which were boycotted by the United States) and $15.5 million for the Winter Games. The rights costs were $401 million for

Source: Al Tielemans/Duomo

FIGURE 2.3 Dream Team, with Charles Barkley and Magic Johnson, conquers world in 1992 Olympics.

the 1992 Summer Games and $243 million for the 1992 Winter Games. The cost to televise the 1994 Winter Games was set at $300 million. In professional sports, the NFL signed a four-year contract worth $3.6 billion with five broadcast and cable networks; Major League Baseball signed a four-year contract with CBS an ESPN starting in 1990 for $1.48 billion; the NBA signed a contract for $600 million with NBC for coverage between 1990 and 1994. In 1990, CBS signed a $1 billion seven-year contract with the NCAA for the rights to broadcast the men's basketball tournament (Eitzen & Sage, 1993: 279–280, 284).

The value of a contract with a sports league or association is greatest for a television network when it has an exclusive right to cover events sponsored by the league or association. Sports leagues and associations are able to negotiate the most lucrative contracts when their events appeal to large audiences and they act as a monopoly by pooling the sale of their rights fees to a single network. The opportunity for individual sports to pool their television rights for sale was granted to the major professional baseball, basketball, football, and hockey leagues operating in the United States by the 1961 Sports Broadcasting Act and was denied to the NCAA for college football by a Supreme Court decision in 1984 (Nixon, 1984: 186; Eitzen & Sage, 1993: 284). In the case of college football, the University of Georgia and the University of Oklahoma had sued the NCAA to obtain the right to negotiate their own television contracts with networks, cable companies, or independent stations. By taking away the NCAA's exclusive right to negotiate television contracts for its members, the Supreme Court opened the marketplace for individual contracts and expanded coverage of college football. As a result, television now pays less to show more games and there now are fewer viewers of televised college football.

The 1984 court decision did not prevent the College Football Association (CFA), a group of sixty-five institutions with the biggest programs, from pooling its rights fees, or the NCAA from pooling rights fees for its men's basketball tournament. The CFA contract was undermined somewhat when it was challenged by Notre Dame, which ultimately signed its own television contract for football.

Television is important to sport because it provides exposure as well as money. For sports such as golf, tennis, and ski racing, with relatively small audiences, television exposure can be more important than television revenue because national exposure can increase the number of fans, the number of spectators at tour events, and the desire of sponsors to invest in broadcasts of events in these sports. These sports can survive on television with relatively smaller audiences because their viewers tend to be affluent (Nixon, 1984: 188).

Television exposure can be a mixed blessing, and sports officials have become wary of the possibility of overexposure. The history of boxing serves as a warning to sports executives because the heavy exposure of boxing on television between 1946 and 1964 caused its virtual disappearance from the medium for a number of years. Overexposure of boxing on television destroyed interest in attending local club fights, which substantially reduced the number of local clubs, the

SPECIAL FOCUS

Making a TV Spectacle of Sport

In his case study of Canadian Broadcasting Corporation (CBC) coverage of a World Cup downhill ski race, Gruneau (1989) found that the major themes in the CBC broadcast were spectacle, individual performance, human interest, competitive drama, uncertainty, and risk. What links these themes, in his view, is that they all reflect the director's general commitment to entertainment values. The television production process emphasizes these themes through such factors as the positioning of cameras for shots of the action, the focus of cameras on individuals, the language and form of questions during interviews, and the general way in which the story of the event was constructed by commentators.

By making sports programs entertaining in these ways, they can be financially rewarding to television investors and also reinforce a conception of sport that is consistent with capitalist consumer culture. According to Gruneau, this conception of sport suggests that we ought to accept as common sense that sport is an achievement-oriented activity that exists to provide opportunities for the career advancement of individual athletes and for the financial gain of investors. This conception also implies that excellence is achieved through specialization, that enjoyment derives from the development of skills, and that economic reward is necessarily a part of contemporary sports entertainment. Thus, from Gruneau's (1989) perspective, the production process for television sports emphasizes cultural beliefs that legitimize the dominant social, political, and economic structures of society. Thus, televised sport is **mediated** in that any event is not shown as it naturally takes place but is described and interpreted through commentary and the director's scene selection before it is presented to the viewing public. The viewer receives a distorted view (e.g., the only hits in baseball are home runs) of a sports event.

Arguing from a similar critical political-economic position in cultural studies as Gruneau, Real (1989: 256) asserted that the contemporary super media favor "economic profit over social commitment, abstract corporations over actual workers, anticommunism over internationalism, opportunism over principle." Since ownership and control of television and the other major mass media tend to be concentrated in the hands of a relatively small number of wealthy and powerful white men, the images and voices of televised sports could be expected primarily to reflect the interests of this class of people. As a result, we see much emphasis in television sports on themes that appeal to the sports public, make sports programs popular, and help sell the commodities advertised on sports programs. In addition, according to Real (1989: 255), since there are relatively few women and minority-group members in positions of control in the TV Sport System, there is a tendency to resist controversial or critical themes about established status and power arrangements in society concerning such things as gender, race, and ethnicity. In future chapters, we will see the extent to which sport itself reflects the patterns of gender, racial, ethnic, and class stratification of the society in which it exists.

place where the stars of the future were born. With no new stars to attract fans and viewers, television abruptly lost interest.

Mindful of this lesson, other sports, such as professional football, have tried to institute blackout policies to limit television coverage when attendance at live events might be threatened, but Congress intervened to protect the public interest by restricting blackout options for the NFL. In a sense, the NFL had become a victim of its own success, with members of Congress intervening on behalf of a sports public that had become devoted fans of televised pro football and demanded the right to see their favorite teams each week (Nixon, 1984: 191).

Eitzen and Sage (1993) have observed that television plays the role of a broker in linking the sports industry as sellers of sports events with advertisers, who buy commercial time on sports broadcasts, and fans, who consume televised sports programs and the goods and services advertised on these programs. In this sense, then, sport and television are linked by calculated profit-oriented business rationality (Eitzen & Sage, 1993: 279). Thus, we can see why the financial foundation of the sports industry becomes more precarious as it becomes more dependent on television revenue. Television will continue to negotiate lucrative contracts for the sports industry only as long as there are high enough audience ratings to justify the investment. Television executives become less generous in their negotiations with sports executives when ratings for a sport drop and advertising revenues decline. The cases of boxing, college football, and professional baseball illustrate this point.

Television has had effects on sport and the sports public that go beyond money and publicity. As the sports industry has become more dependent on television revenue and exposure, it has also become more willing to adapt to the needs of the medium. Altheide and Snow (1978) believe that the scale of power in the television-sport relationship is tipped in favor of television, and television acts mainly in its interest of making a profit. Although the sports industry is also interested in making money from its relationship to television, television has less commitment to maintaining the integrity of a sport or protecting the interests of athletes or fans than sport does. It schedules and produces televised sports programs for the convenience of the sport's target audience of viewers, and it encourages sports officials to make their events as colorful, exciting, and appealing as possible for the viewing public.

Sports, in turn, have accommodated television in a number of ways: starting times of college basketball games have been pushed late into the night to expand a network's basketball package for an evening; the World Series has become primarily or exclusively an evening affair to maximize its viewers; Major League Baseball has introduced a livelier ball, a bigger strike zone, the designated hitter rule, and a variety of new uniforms to add more excitement and color to the game; the NFL has added "sudden-death" overtimes, longer kickoffs, and a variety of rule changes to protect the quarterback and benefit the offense to make their games more appealing and exciting; professional and college basketball have used shot-clocks and three-point shots to increase action and interest; the slam dunk has been encouraged in basketball to add to the excitement of the game; tennis created the tie-breaker to control the length of matches and added color to the balls and player's clothing to accommodate color television; and, finally, golf has accommodated television by shifting from match to medal play (Nixon, 1984: 189).

While making concessions or accommodations to television can be seen as good business, it can be controversial when the actions taken by the officials of a sport or demanded by television executives adversely affect the quality of play, the comfort of live spectators, or the welfare of the athletes. Media time-outs for advertising spots can change the flow and outcome of a game. Having to talk to television commentators immediately after tough losses can lead to statements that athletes or coaches regret later. Games scheduled late at night, on cold evenings, or early in the morning can pose risks to athletes and cause discomfort for spectators. The experiment with replays in pro football to challenge referees' decisions posed a threat to the credibility of referees. In this last case, NFL owners recognized the threat and ended the experiment, but sports officials often find it difficult to resist the influence or demands of the television medium or its executives because television money and exposure have become so important to the financial structure of their sports.

The creation of special sports events, such as "Celebrity Challenge of the Sexes," "Super Teams," and "Battle of the Network Stars," in which athletes or television and other entertainment personalities compete in contrived games, has been called the "ultimate trivialization of sport" by Eitzen and Sage (1993: 288). Other critics (e.g., Klattell & Marcus, 1988) have contended that these programs are created to advertise the shows on which the competing television performers appear or to sell the products endorsed, advertised, or used by competitors on these programs. Trash sports are said to trivialize "genuine sport" because the events are created solely for commercial and entertainment purposes and because the athletes who participate do not compete in their own sports and the competitors from show business are rarely the exceptional or accomplished athletes we expect to see in showcase sports events.

The impact of televised sport on society can be understood in terms of the influence it has in mediating cultural themes that reflect its interests or the interests of dominant strata of society. In Real's (1989) terms, television is one of the "super media" of contemporary society and as such provides a "clouded mirror" through which we can experience sports events immediately but only indirectly. The filtering process through which television producers decide what we will see and how we will see it gives them the power to shape the nature of our sports experiences as viewers. In addition, the experiences of sport that television constructs for us may influence how we think about the dominant aspects of culture and society.

Although we can readily see major elements of capitalist consumer culture in the ways television mediates sport for its viewers (see the Special Focus discussion), we should not jump to the conclusion that televised sports dupes its audience into uncritical acceptance of the values and practices associated with this culture. Wenner (1989) has proposed that the perceptions that viewers bring with them to sports telecasts will color how they experience televised sports events. He also reminds us, though, that people who watch sports in countries such as the United States are often armed in advance with values, other beliefs, and experiences that tend to be consistent with the themes celebrated by television sports. This finding should not be a surprise since a fundamental argument of this book is that sport mirrors society. Whether or not that mirror is clouded by television, we can expect the pictures and sounds of televised sports to be ones that reflect the major cultural and social patterns of the technologically advanced, corporate, and consumer societies that produced modern sport, television, and the TV Sport System.

Source: Courtesy of World Championship Wrestling, Inc.

FIGURE 2.4 Wrestling as entertainment.

CONCLUSION

Evidence of organized physical activity in some form has been uncovered for the earliest human civilizations. The nature of this activity varied according to environmental conditions, the culture's level of development, and the extent to which this activity was related to other societal institutions such as the military and the economy. It is clear that as societies matured, the nature of sport and games changed from relatively unorganized, individual activity to highly coordinated, rule-driven, team- or club-centered activity with consequences well beyond the physical and normative boundaries of the activity. Just as other societal institutions experienced a transformation that could be attributed to the tendencies toward secularization, rationalization, organization, and specialization, so did physical activity. An evolution took place that saw sport change from playlike to worklike activity that focused on outcome rather than the experience of participation.

Despite debates about the validity of generalizations of the consequences of modern sports, there is little question that analyses of the structure of modern sports must include considerations of bureaucratic and scientific rationalization, quantification, the quest for records, commercialism, television, and corporate organization. We know, too, that modern corporate sports tend to produce spectacles or displays for the entertainment of spectators and fans. Without widespread fan interest and support, television networks, cable companies, superstations, and major independent stations would not spend large sums of money on sport or provide extensive coverage of sports events; and with significantly less television investment and coverage, the financial foundation of commercialized corporate sports would crumble. These and other basic features of the social organization of modern sports have many important consequences. In the coming chapters, we will examine the social sources, implications, and consequences of the social organization of modern sport as a corporate enterprise. We will expand our consideration of the dimensions of modern sport presented in this chapter and will look at the implications of its corporate organization in relation to a variety of levels and types of sports competition in the United States, in other nations, and in international contexts. The next chapter will focus on culture and sport.

SUMMARY

Archeological evidence has revealed the existence of physical play in prehistoric societies and the existence of sporting activities in various parts of the world as early as 3500 B.C. Early sport often embodied ceremonial and sacred qualities. Although women experienced discrimination in early sport, they participated in the Haraean Games in ancient Greece that paralleled the male Olympics. In contrast to the athleticism of the ancient Greek sports, Roman sports developed mainly for two purposes, the entertainment or amusement of spectators and the military preparation of contestants.

Sport has evolved into a distinctively modern enterprise, and, according to some scholars, its development in recent centuries has reflected a larger civilizing process that has characterized the development of many societies. Industrialization, urbanization, and bureaucratization have been among the major social forces driving the modernization of sport. More recently, science and the mass media have left clear and deep imprints on the development of sport. In its modern form, sport is a corporate entity tied to the political and economic structures and processes of the society in which it exists. The most modern forms of sport are intimately tied to television and the money and exposure this communication medium provides.

CHAPTER 3

Culture, The Meaning of Sport, and Fans

DISTINGUISHING CULTURE FROM SOCIETY

Although society and culture are common words in our everyday language, the difference between the two terms may not be immediately obvious. To understand the relationship between culture and sport, we must understand both what a society and a culture are and how they differ from each other. A **society** is a collection of people who share a common territory or space and are at least loosely or indirectly tied together by interdependent networks of interaction and by possession of a common culture. **Culture** is the distinctive way of life that members of a society share. Culture consists of the sum of shared knowledge about what is and what ought to be that is symbolically transmitted across generations as well as the tangible things that people create, acquire from other societies, and share with other members of their society. Culture more specifically consists of knowledge about the skills, facilities, roles, norms, beliefs, goals, and values and of the material products that are associated with living in a particular society (Nixon, 1976: 9; Giddens, 1991: 32–35; Ferrante, 1992: ch. 4).

Culture gives societal members the language they need to engage in meaningful interaction; it gives them the ideas and knowledge they need to think, reason, make judgments, and in general, make sense of their world; and it gives them the material things they need to create new things and accomplish concrete tasks. Culture provides the framework within which the members of a society construct their way of life. For example, if cooperation is a strongly and widely held cultural value in a society, the members of that society will tend to structure their relationships to be cooperative and will have difficulty accepting relationships or activities that are highly competitive or that might lead to conflict. In this kind of society, we would be unlikely to find broad cultural approval of the highly competitive sporting practices of modern societies. Although societies can be viewed in terms of their distinctive cultures, different societies can possess similar cultural characteristics, and a given society may be characterized by differing or contrasting cultural elements in the various groups within its borders.

THE RELATIONSHIP OF SPORT TO CULTURE

Sport and culture are interdependent, with sport bound to society and structured by culture (Lüschen, 1967). The relationship between sport and culture in a society is indicated by the value placed on sport in that society; the value of the social functions of sport, such as entertainment or social integration, in the society; the amount of institutionalization or integration of sport in the society; and the extent that established practices, values, and norms of sport parallel those of the society's dominant culture (Loy, McPherson, and Kenyon, 1978: 300).

Certain forms of sport in certain societies may be elements of high culture, but in modern societies sport is generally part of popular culture. **High culture** generally refers to aesthetic practices such as classical music, ballet, theater, literature, and the fine arts, and it often is assumed to be the special province of the upper or highly educated strata, especially in Western nations (Loy, McPherson, and Kenyon, 1978: 297–298). In these terms, high culture could also include sports such as yacht racing or polo, which are mainly favored by and accessible to the wealthy. As an elite culture, high culture involves only a small portion of society.

Popular culture, in contrast, involves a large segment of society and consists of cultural practices and products designed for mass consumption (Hess, Markson, and Stein, 1991: ch. 22). Like high culture, popular culture is consumed by members of a society during their leisure time as entertainment, and it includes many forms of modern corporate sports as well as pseudosports such as the staged wrestling of performers such as Hulk Hogan, Macho Man, Velvet McIntyre, and Sensational Sherri. When popular culture encompasses the cultural practices and products brought to the masses in a society through the various media of mass communication, it is also **mass culture**. Thus, sport is an element of both popular culture and mass culture in modern societies (Loy, McPherson, and Kenyon, 1978: 298). Looking at sport as popular and mass culture emphasizes the commercial entertainment value of sport and the importance of the mass media in its construction, dissemination, and profitability.

SPORT AND AMERICAN CULTURAL VALUES

The existence and place of a dominant culture in the United States has been hotly debated in recent years in relation to the idea of multiculturalism. Kessler-Harris has written that "at the heart of the recent controversy over multiculturalism lies a concern about what constitutes America" (1992: b3). In fact, U.S. society has elements of cultural unity and cultural diversity. To understand both the shared culture of the United States and its cultural diversity, we must look at culture as a product of a dynamic process of construction and reconstruction. In this process, various social and cultural groups, social classes, political interests, and societal institutions, such as the schools, churches, government, and the mass media, engage in an ongoing dialogue or power struggle about what our national culture and identity include. The recent multiculturalism debate can be seen as part of this continuing process of cultural definition.

American Values and the American Dream

Although the common culture that Americans share at any given time in history must be defined quite broadly, we still can point to certain cultural themes that have had a broad, powerful, and enduring influence in the United States. There is evidence that the themes of individual competition, achievement, success, material comfort, and social advancement or mobility have been widely accepted by Americans as elements of what has been called the "American Dream" (e.g., see Williams, 1970; Rokeach, 1973; Schlozman and Verba, 1979; Nixon, 1984).

The **American Dream** embodies elements of a culture of materialism, which reflects the consumer-oriented economy of advanced industrial capitalism (Hess, Markson, and Stein, 1991: 66–67), and it holds the promise that those who take advantage of opportunities and strive the hardest will be rewarded with the good life of material comfort—and perhaps social prominence and influence as well. The American Dream may not be the culture of all segments of American society, but it probably comes as close as any set of values and beliefs to being the public culture of the dominant classes in the United States. It also represents an image of the United States that has lured many immigrants to its shores because it is a source of hope for people with modest or disadvantaged backgrounds who want a life of greater material comfort.

The American Dream has two faces, though. On one hand, it is an inspiration to the striving and ambitious. On the other hand, it is a form of social control over people who might complain about their life chances, criticize the established structure of their society, or challenge those who have succeeded in their society. The American Dream implies that every American—or, traditionally, every American male—has the privilege of striving for success, but it also implies that every American (male) has the *obligation* to make every effort to be successful. Believers in the American Dream assume that opportunities to achieve and succeed are abundant and that success is earned by individuals who try and failure happens to those who do not try or suffer from some deficiency in character.

Recent evidence indicates that these kinds of individualistic beliefs about success, failure, wealth, and

Source: Focus on Sports

FIGURE 3.1 Fans in the stands; the name of the game.

poverty continue to be widely accepted in U.S. society (Smith and Stone, 1989). To the extent they are, they are likely to act as an ideological force in American society, giving cultural legitimacy to the positions of the successful—and the unsuccessful (Mizruchi, 1964; Nixon, 1984). This framing of the American Dream as a **cultural ideology** means that the American Dream is an argument for the primacy of a particular set of cultural values, beliefs, and practices. As an ideology, the American Dream is supposed to be accepted without question, as a matter of faith. As an ideology, then, it performs a social control function. Believers in the American Dream are relatively unaffected by the facts that opportunities are not as abundant, and success and failure are frequently not as deserved, as the American Dream asserts or implies. As a matter of faith, the American Dream is to be accepted, not questioned.

The American Dream and the Dominant American Sports Creed

Sport is part of the institutional structure of American society. Its culture reflects major American cultural themes, which are embraced by most of the dominant members of the society and are found throughout U.S. society in its social institutions. Thus, we find that the dominant culture of sport in the United States is infused with themes consistent with the American Dream.

In a content analysis of statements about the effects of sport found in newspapers, magazines, and a major athletic journal, Edwards (1973) uncovered a sports belief system that had many similarities to the American Dream. He called this belief system the **Dominant American Sports Creed**, and, like the American Dream, it was seen as an ideology. This sports belief system was meant to convince people of the virtues of sports participation by implicitly or explicitly linking the functions of sport to the kinds of values and beliefs associated with the American Dream.

The major themes of this creed assert that sport builds character, teaches discipline, develops competitiveness, prepares participants to compete in life, enhances physical and mental fitness, and contributes to a belief in (Christian) religion and a patriotic belief in America. We see the ideological character of the

Dominant American Sports Creed in the fact that sports advocates express unqualified faith in its major themes in spite of a lack of clear, consistent, and systematic evidence supporting any of them (Edwards, 1973; Nixon, 1984). The scientific validity of these creedal themes is not what is important, though, in understanding the impact of the Dominant American Sports Creed. People who are committed to sport assert this creed because it legitimizes sport and their commitment to it.

The close connection between the Dominant American Sports Creed and the American Dream is a partial explanation for the tremendous popularity of sport in American society (Eitzen & Sage, 1993: 60). Criticism of sport is not taken seriously because cultural beliefs about the virtues of sport are so pervasive, so deeply entrenched, and so closely tied to dominant American values. Learning to participate in sport is learning about the basic values of American society. Of course, to the extent that powerful people in sport fail to acknowledge the legitimacy of values, beliefs, or practices associated with people or groups outside the mainstream or dominant structures of society, they can create cultural obstacles to the participation of such people and groups. In fact, the barriers faced by women and minorities in sport have included cultural ones, such as discouraging or prejudicial beliefs, as well as structural ones, such as outright discrimination. We will explore these barriers in more depth in later chapters.

Competition, Winning, and Success in Sport

It may seem strange that winning is not explicitly stated as one of the themes of the Dominant American Sports Creed. After all, many coaches at all levels of sport have implicitly or explicitly expressed sentiments similar to the oft-repeated statement by the legendary Green Bay Packer football coach Vince Lombardi that "winning is not everything; it is the only thing" (Nixon, 1984: 21). To assert that winning is the only thing, however, oversimplifies the reasons that many compete in sport, and even misrepresents Lombardi's intention to apply his statement only to the world of professional sports where participants make their living by winning. None of this is meant to minimize the general commitment to winning, but what is important is to look at how an excessive commitment to winning can distort effort and undermine legitimate sports competition.

Some coaches have associated winning with life itself. The late George Allen, who gained fame as coach of the Washington Redskins in the 1970s, said that "the winner is the only individual who is truly alive ... every time you win, you're reborn; when you lose, you die a little" (1973: 76). He believed in outworking his opponents and in aiming for 110 percent effort from his teams. George Allen was a winner, and many players and fans applauded his success, but—as in the case of Vince Lombardi's commitment to winning—we must remember that George Allen and the athletes who played for him were paid for their sports commitment. We must also remember that even in sports realms where competitors are paid for winning, only a small fraction of teams, athletes, and coaches win a large majority of the time, season after season. Thus, the kind of emphasis on winning expressed by George Allen or Vince Lombardi must be assessed in the context of the real prospect that most sports participants will experience some or a great deal of losing during the course of their careers. When winning is defined narrowly as winning the big game or the championship, even fewer can be winners.

In the commercialized realms of American sport, a winning tradition is sold as a product and outstanding athletes are treated as valuable property (Snyder and Spreitzer, 1978: 139–140). In such contexts, sport is more **product** or **outcome oriented** than **process oriented**, and the attention shifts away from the intrinsic value of competing (Vande Berg and Trujillo, 1989). Raising the material and symbolic stakes of competition increases the likelihood that the pleasure of participation will be replaced by the stresses and pressures of having to win and makes it more likely that winning by any means will become more important than playing hard and fairly. Where athletes and coaches *must* win, it should not be surprising to find cheating, substance abuse, and dropouts (see Gilbert, 1988).

The obsessive pursuit of winning in American sport is similar to the single-minded drive to achieve

POINT-COUNTERPOINT

IS WINNING THE MOST IMPORTANT THING IN SPORT?

YES

Sport is structured to be competitive. Its value is diminished when opponents are not totally committed to trying to defeat each other. There is no point in playing if you do not do everything possible to win. You can only be successful if you win; there is no reward in losing. America only cares about its winners. Anyone can be a loser; winning is achieved by the best. Winning proves who is the superior athlete or team.

NO

A great deal can be learned about oneself and life by competing and striving to win in sport. It is the process of striving to win, rather the product of winning or losing, that produces the benefits of sport. It is possible to lose self-respect and the respect of others by winning in deviant or unsportsmanlike ways; it is possible to gain self-respect and the respect of others by losing, after a valiant effort, to a superior or luckier opponent.

How should this debate be resolved?

the American Dream in the larger society (Sage, 1980). The important sociological point here is that competitiveness and striving to win are not natural or universal in human societies. They are behaviors shaped by cultural learning and influenced by social structures (Eitzen & Sage, 1993: 60–65). Important mediators of the culture and rhetoric of winning for the general sports public are sportswriters and sports commentators.

In an analysis of the rise and fall of the pro football Dallas Cowboys, Vande Berg and Trujillo (1989) showed that despite a desire to be associated with winners, sportswriters did not use a single simple interpretation of success as winning in covering the Cowboys. They interpreted the team's success in terms that fit their competitive context and history, initially viewing success as process, shifting to a success-as-product view, and returning to a success-as-process perspective in the post-dynasty period following their years as one of dominant teams in pro football. Coverage of the Cowboys during their most recent dynasty period of the 1990s indicated a return to a product orientation, with quotes about "wasting" a successful regular season if the team did not advance in the playoffs and about not reaching the Super Bowl being a "real disappointment" (Longman, 1993).

Although consistent winning typically creates expectations of future victories, the reality frequently fails to match the expectations. The Dallas Cowboys were viewed as a dynasty during two different periods of their history precisely because they were big winners for so many years, and most recently, repeat winners of the Super Bowl. In most sports now, repeat champions are treated as surprising rather than expected. Research by Mizruchi (1991) supports this view. He studied professional basketball teams between 1947 and 1982 and found, after controlling for the home team advantage and relative team strength, that teams were more likely to lose than win after winning the first game of back-to-back games at the same site. Mizruchi explained his results in somewhat unconventional terms by saying that winning breeds losing in team competition because it decreases the need to win, whereas losing motivates a strong effort to win because it increases the urgency of winning. Mizruchi noted that a survey of players and coaches had overwhelmingly confirmed their *belief* that winning generated more winning. His contrary findings challenged this belief and a possible underlying explanation that winning breeds more confidence, which translates into better performance. On the basis of his research Mizruchi argued, in contrast, that losing increases motivation (i.e., the sense of urgency) and winning decreases it (by causing overconfidence and a tendency to relax).

Although Mizruchi's research suggests that it may be more difficult to sustain winning than the rhetoric of the "culture of winning" in sports claims, it does

SPECIAL FOCUS

Butch Ross, Jim Ryun, and the Burdens of Losing

In settings where only winning counts, athletes and coaches sometimes find that they must bear the burden of defeat long after the game is over. This point is illustrated by the case of Butch Ross (Associated Press, 1992). In 1981, Butch Ross quarterbacked the Shawnee Mission South (Kansas) high school football team in a playoff game against Shawnee Mission West. His team led 24–21 and had possession of the ball with four seconds left in the game. Worried that Shawnee West might regain possession of the ball, his coaches did not have Ross immediately drop to his knee to down the ball. Instead he ran back 15 yards until time expired. At the apparent end of the game, fans went on to the field and Ross, still holding the ball, began to shake the hands of opposing players. One of the Shawnee West players grabbed the ball from him and ran to the end zone. Since Ross had not downed the ball, the referee had not blown his whistle to end the play and the West player's stolen touchdown was counted, giving his team the victory over South.

Butch Ross was understandably upset by the turn of events in his big game. He was so upset that he was unable to return to school for the week following the game. He found on his return that his classmates and local fans were even more upset than he was. Someone had written "Loser" on all 75 of the recipe cards for his home economics class, and at his graduation, some students shouted, "Down the ball," as he received his diploma. Eleven years later, he overheard two men talking about the game and saying, just loudly enough for him to hear as he walked by them, "God, I pity you."

Butch Ross's brother-in-law and business partner said that Butch became a better businessman because of his football experience, pushing him "to strive to succeed more than people realize." At the same time, Ross acknowledged that after being friendly and popular in high school, the effects of reactions to his playoff mistake made him more reclusive and more mistrustful of people.

Butch Ross tried not to have a mistake in a football game adversely affect the rest of his life, but other athletes have suffered greatly from their failure to win and meet others' or their own expectations. Jim Ryun, record-setting American miler in the 1960s, learned how much defeat could sting when he finished second to Kenyan Kip Keino in the 1968 Olympic 1500 meters (Leahy, 1988). Twenty years after his defeat, he said, "Champions aren't supposed to fail. Some people do not let you forget." Running at the 7,300-foot altitude of Mexico City, the frail-looking and asthmatic Ryun pushed himself to maintain the extremely fast pace of the opening lap and ended up unable to keep up with Keino at the end. He was so exhausted by the race that he was unable to move for 20 minutes afterward. Feeling he had run the best race he could at the Mexico City altitude, he found upon his return to the United States that he was treated as a loser for finishing second in the world instead of first. He discovered that American sports fans were greatly disappointed by his loss and treated him with coldness or even scorn. Ultimately he was overwhelmed by the public criticism and pressure, and at the Amateur Athletic Union championship in 1969 he walked off the track at the 660 mark of his race. His confidence shattered, believing that winning was the only kind of success people would accept, he retired after that race. Ryun returned to the track for the 1972 Munich Olympics as a cofavorite, with Kip Keino, for the 1,500 meters, but he tripped in a qualifying race and was unable to recover to qualify for the medal race. This time he was criticized by the press for running a "stupid race." Ryun's career as one of the elite milers in the world ended with that race. Later he participated in Master's Division races for older runners, but the scars from his prior experiences were deeply imprinted on him.

not mean that winning relieves athletes or coaches of pressures or demands to win from the sports public, boosters, or employers. His research shows that when teams do not win, they will feel a great deal of pressure to win. Where there is a great deal of pressure to win, athletes and coaches are subtly or overtly encouraged or pushed to break rules and take excessive risks in their pursuit of winning. The amount of emphasis that is placed on winning versus participation reflects the cultural context in which a sport is learned and played and the structure of the settings in which athletes and teams compete.

Sports Heroes, Success Themes, and Changing American Cultural Values

The way **sports heroes** are portrayed in a society is an important indication of the prevailing dominant values in that society. Heroes are symbolic representations of the dominant social myths and values of a society or subculture (Nixon, 1984: 172–175). In a sense, heroes are fictional characters who are constructed by the mass media and the public from idealized conceptions of real people. They serve as inspiring reminders of what is important in a society or subculture.

Using evidence from selected baseball autobiographies, Haerle (1974) showed how portrayals of American sports heroes changed over this century as the dominant values of American society changed. Autobiographies of Cap Anson (1900) and Bucky Harris (1925) conveyed a picture of heroes embracing the Calvinist **Protestant Ethic**—with its emphasis on hard work, sacrifice, and occupational success as a sign of grace (Weber, 1958)—and idealized conceptions of the American Dream and the Dominant American Sports Creed. According to Haerle, these kinds of values were emphasized during the latter part of the nineteenth century and the first quarter of the twentieth century because this was a period in which the American economy grew rapidly and seemed to provide boundless opportunities for ambitious entrepreneurial capitalists such as Andrew Carnegie to rise above their humble backgrounds to great wealth, power, and prominence.

By the end of the 1940s, the United States had experienced some significant social upheavals, associated with the Great Depression and World War II, that had reshaped American values. There was a general realization that the attainment of success or failure could be affected by forces beyond one's control. This shift in values is revealed in the story of one of the great sports heroes of this era, Joe Dimaggio. His autobiography, *Lucky to be a Yankee*, originally published in 1946, portrays a hero who is aware of the influences of natural talent, fate, and luck in achieving success.

The final historical period Haerle examined was the 1960s, a decade of considerable challenge to established institutions and values. This period followed the economic growth and affluence after World War II, and it was seen as one in which a kind of "self-conscious individualism" or narcissism existed alongside the social protests and rebellion. Haerle used Jim Bouton's *Ball Four*, published in 1971, to represent both the anti-establishment and individualistic values of this era. Since there was much conflict and confusion about values at this time, it is difficult to see Bouton in traditional heroic terms. Insofar as he expressed the changing currents of his time and his views and lifestyle represented a cultural alternative to more traditional values and lifestyles, he could be seen either as an anti-hero or countercultural hero.

Extending Haerle's analysis to the late 1970s and 1980s in the United States, we can see a rejuvenation of conservative values emphasizing business success and material self-aggrandizement. The decade of the 1980s was the Reagan era, and the popular electronic and print media frequently focused on themes of wealth and fame. Stories of the most prominent athletes of this era would focus on contract negotiations, rapidly escalating salaries and tournament prizes, lucrative commercial endorsements, and profitable business ventures as well as on athletic talent and accomplishments. Typical of this period was the battle between agents of members of the basketball "Dream Team" of 1992 and U.S. Olympic officials over displays of manufacturers' labels on their athletic clothing and equipment.

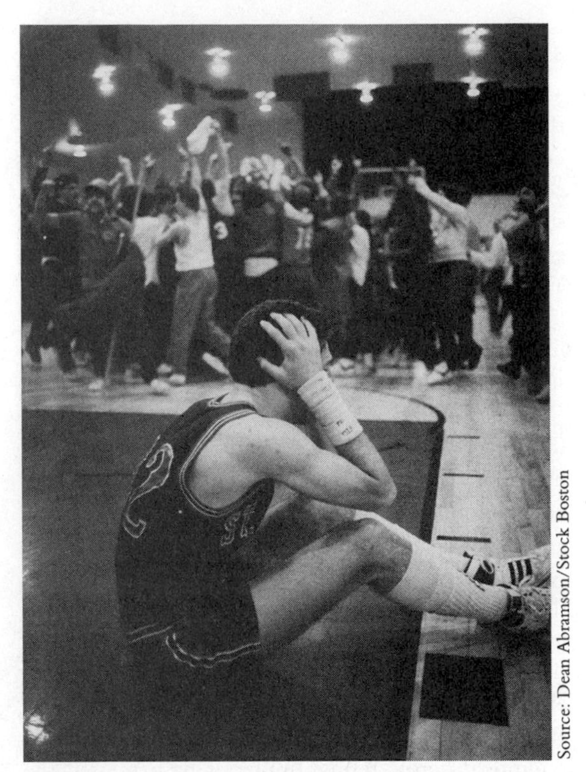
Source: Dean Abramson/Stock Boston

FIGURE 3.2 Winning and losing.

CULTURAL HEROES, SPORT, AND THE QUEST FOR A NATIONAL CULTURE

Haerle's research reminds us that culture has complex and sometimes contradictory elements that change over time. Thus, pointing to one or two heroes of an era can give us a misleading impression of the range of dominant cultural values in a society in that era. Whatever values specific athletes embody, it is likely that certain athletes will be chosen by the mass media to be heroes, at least for a while, because and as long as they can be portrayed as embodiments of widely held values of their time. It does not matter that athletes as real people have flaws that their heroic images lack—as long as the public overlooks these human flaws. A number of commentators have observed, though, that with growing public cynicism about athletes and sports, fueled in part by a more critical generation of sports reporters, athletes have become too humanized or trivialized to be heroes today (Nixon, 1984: 173). It surely has become difficult to sustain traditional images of heroes when we read about prominent athletes involved in gambling, infected with AIDs caused by unrestrained sexual activity, convicted of rape, accused of battering and killing their wives, testing positive for various types of illicit drugs, and engaged in an assortment of other illegal, immoral, or unethical acts.

Since American society and culture are so diverse, it seems difficult or impossible to construct images of particular athletes as "all-American heroes" embodying universal "American values," even if no glaring flaws are evident. A society that sees itself as culturally diverse may be capable only of creating a diverse array of heroes for different subcultures or cultural groups. Sports figures nevertheless may be promoted as such by their sports or the mass media. They may be portrayed as embodiments of *dominant* cultural values or of the values embraced by the society's powerful classes or institutions. In modernizing or otherwise changing societies, which may have some confusion or disagreement about who they are or where they are going, prominent sports figures may serve as cultural heroes who are able to contribute to a sense of national pride or identity. They may be deliberately promoted as heroes by government officials or the mass media to serve such ends.

According to Dyreson (1992), since 1912 the Olympic Games have been a stage for U.S. athletes to perform as "missionaries" of Americanism, with each of their victories meant to convey the message of national as well as athletic superiority. U.S. Olympic successes were heralded as testimony to the "power of [U.S.] democratic ideals, scientific technique, the human energy produced by a national commitment to the strenuous life, and the social mobility embodied in the concept of a 'melting pot'" (Dyreson, 1992: 81). Other nations, many in the orbit of the former Soviet Union, have sought to demonstrate their national superiority through sport in the twentieth century, and they have made their athletic stars willing or unwilling diplomats to serve political ideological aims.

The honor accorded to sports figures as cultural heroes in society reflects both the significance of their athletic prowess in that society and the significance of the values they are seen to embody. Athletes, coaches,

POINT-COUNTERPOINT

DOES THE PUBLIC NEED SPORTS HEROES?

YES

All members of a society, and especially young people, need something to believe in. Heroes embody the elements that a society values most. As idealized media creations, they provide inspiration, motivation, direction, and meaning for people's lives. They remind us what is important. Sports heroes can be a unifying force in society because they make us forget racial, ethnic, and class differences and represent what is good about a realm of society that is known and loved by many people of different backgrounds.

NO

Heroes ultimately are real people, and as such they are bound to disappoint us. As the press probes more and more into the details of star athletes' lives, they uncover more and more that humanizes them and makes them unheroic. Believing in Magic Johnson, Mike Tyson, Michael Jordan, and Pete Rose only makes us cynical as we learn about their faults and flaws. In cases such as that of O. J. Simpson, we see in the most spectacular terms the pain and disillusionment that comes from believing in sports heroes.

How should this debate be resolved?

and other sports figures cannot be heroes unless their sport is highly valued. Thus, we can learn much about the values and culture of a society by understanding not only who its heroes are but also the sports played by those heroes. We also can learn about pervasive cultural themes in a society by considering the sports that are seen as its national pastimes.

NATIONAL PASTIMES AND MYTHIC SPECTACLES OF SPORT

Over the past century, many observers of American sport and society have referred to baseball as the American national pastime. According to Voigt (1976: 82–83), the organizers of the first commercialized baseball league in the United States in 1871 actively promoted baseball's claim to being America's national game. With enthusiastic and tireless promotion, the image of baseball was linked to a dominant American Sports Creed that resonated with the inspirational American Dream. In recent decades, sport historians and sociologists have debated whether baseball has lost its stature as the national pastime in the United States (see Brandmeyer and Alexander, 1987). Some have argued that in the fast-paced cosmopolitan age of urban corporate America, a sport rooted in the pastoral imagery of rural America is passé. One can argue instead, however, that as an established part of American culture and society, with rituals and routines familiar to most Americans, baseball is unlikely soon to disappear from American consciousness. It may have to struggle to retain its financial viability, like other highly commercialized sports in the United States, and it may have to share the national stage with competitors such as football and basketball. But it will remain a popular American pastime, even if it is not *the* national pastime, as long as it continues to evolve to fit and reflect major cultural and social patterns of American life.

Despite the nostalgic feelings kindled by such films as *The Natural, Field of Dreams* and similar media portrayals of baseball, this sport has remained popular and commercially successful at the highest levels because its structure has evolved with society over the past century and because the sports public continues to see in it values that seem appropriate for their lives today. Although it may not have universally recognized heroes for all Americans (see Verducci, 1993), baseball still offers a number of celebrities who capture our fascination or attention as media watchers for at least the moment and some candidates for hero worship among some racial and ethnic subcultural groups.

When sports become popular, their championships, all-star games, and other special events such as hall-of-fame induction ceremonies are widely and often intensely followed by the sports public. Popular and enduring sports and sports events become integrated into the cultural traditions of the societies in which they exist. Some events, such as the Olympics, the Super Bowl of American football, the World Cup of soccer, Wimbledon in tennis, and the World Series of baseball draw audiences that cross national and cultural borders, although events such as the Olympics have also been used to advance nationalistic aims.

In the United States, the Super Bowl is the most popular and lucrative of these megaevents, and Real (1976) has proposed that it is best explained as a contemporary form of **mythic spectacle**. It is a cultural production of the electronic media that functions in the manner of traditional mythic rituals in providing opportunities for collective participation and identification as well as a means of celebrating and reinforcing shared cultural meanings. From Real's perspective, the mediated production of the Super Bowl emphasizes values and meanings concerning the use of territory and property, time, labor, management, physical contact, motivation, infrastructure, packaging, game, and spectacle. More specifically, he states that "North American professional football is an aggressive, strictly regulated team game fought between males who use both violence and technology to gain control of property for the economic gain of individuals within a nationalistic entertainment context" (Real, 1976: 27). According to Real, such emphases are functional in American society because they celebrate in a symbolic way, and therefore reinforce, dominant structural dimensions of the society.

Similar kinds of arguments have been made about social functions of other kinds of popular sports and sports events as cultural productions. For example, Lever (1983) has written about "soccer madness," especially in Brazil, and the paradoxical capacity of this sport to divide and unify. It can reinforce societal cleavages by pitting local teams, with their cultural and social differences, against one another, but it also can integrate people in different communities through a shared devotion to the sport and a common pride in regional and national teams.

The Super Bowl and World Cup are two of the most popular mediated sports events in the world. Real (1989) compared the social organization, popularity, and cultural meaning of these two events. They differ in continuity of action, technological sophistication, amount of scoring, degree of specialization, and how the games are played. They also differ in that American football inspires devotion mainly within the boundaries of the United States, whereas soccer has a passionate following in many nations around the world. Thus, despite its worldwide audience, the Super Bowl is predominantly a national celebration, or, in Real's words, "a high holiday of American civil religion" (1989: 202). The World Cup, in contrast, has a greater geographical scale. These two events are similar, however, in expressing cultural themes that engender partisan loyalties on one hand and draw people together on the other in what amounts to a ritual celebration of transnational consumer culture, commercialism, and rational organization. As television events, both also draw attention to modern science and technology. The embracing of these themes by sports followers in diverse national and cultural settings suggests the breadth of the influence of modern culture and social organization.

THE MODERN WORLD SPORT SYSTEM AND CULTURAL DIFFERENCES

Using an international perspective to compare World Cup soccer with the Super Bowl enables us to see how the appeal of certain sports can cross many national and cultural borders, whereas the appeal of others tends to be associated mainly with a given society, type of society, or culture. Major international events, such as the World Cup, the Olympic Games, and the Davis Cup of tennis feature sports with wide international appeal. As such, these sports events may be seen as carriers of modern culture around the world or as manifestations of a global culture of modernism (Van Bottenburg, 1992). These modern global sports events link athletes of many nationalities in a **world sport system** in which the formal structural characteristics of each sport are basically the same across borders. For example, when athletes and national teams in this world sport system compete against one another, they follow the same rules, are governed by the same international regulatory bodies, face the same penalties and sanctions, play on the same size field, and wear the same type of uniforms; their games involve the same number of players and use the same scoring system.

Although many modern sports can be viewed as global sports or part of a world sports system, even internationally popular sports with the same general type of formal structure in different countries still could be mirrors of important cultural differences between nations, or, when viewed over time, as indicators of cultural changes within nations. Sports are social practices and cultural products that mirror their societies and that can change their forms and meanings as the social and cultural contexts in which they are played change (Van Bottenburg, 1992).

In a classic study, Riesman and Denney (1951) showed how the adoption of rugby football in the United States led to the creation of American football as the original game was modified to fit into the American cultural context. Thus, running, the minimum yardage rule, mass play, and the forward pass were introduced as responses to American spectators' desire for more visible and exciting action. In addition, rules were standardized to permit a common understanding of the game among participants from different places and social backgrounds, and the organization of the game was made more rational to improve efficiency, which was a major focus of industrializing America. As the sport became more distinctively "American," it developed an increasing capacity to become the American mythic spectacle that Real described.

Baseball provides a means of seeing how two nations playing essentially the same sport develop styles of play that reveal important cultural differences between them. When Snyder and Spreitzer (1989) compared Japanese and American baseball, they noted that the Japanese, contrary to American practice, rarely fire managers, and when they do, they go through a ritual that permits the fired manager to save face. Also unlike American baseball, Japanese games can end in a tie, which Snyder and Spreitzer interpreted as a reflection of the emphasis in Japanese culture on process as well as product. We have noted the very strong "winning is everything" product emphasis often found in American sports and society.

A major difference between Japanese and American cultures is the relatively stronger Japanese emphasis on collective goals and team loyalty. In Japan, a star pitcher may be asked to risk a serious or career-ending injury if it is thought he can help the team win a critical game. In the United States, we are likely to find more concern about the individual athlete's welfare. In Japan, individualism and egotism are devalued while selflessness is highly valued, and the willingness of Japanese ball players to sacrifice for the team reflects this cultural fact. The strong team orientation in Japanese baseball parallels the high solidarity among Japanese industrial workers and the high degree of company loyalty frequently observed for Japanese workers in general. Japanese baseball, like Japanese society, also frowns more on emotional displays than Americans do.

Although there are some minor structural differences between baseball in Japan and the United States, the more striking differences are found in the style of play, which reflects the different cultural settings for the sport in the two countries. American players in Japan run afoul of fans when they fail to understand and accept these important culturally derived stylistic differences between the ways Japanese and Americans play what is supposed to be the same sport.

Along with the style of play, the relative popularity of particular sports in a nation can offer clues about the cultural emphases in that nation. For example, we often associate Canada with hockey, Kenya with distance running, Norway with Nordic skiing events, Great Britain and its former colonial empire with cricket and rugby, the former Soviet Union and East European nations with gymnastics and figure skating, China with ping pong, a number of European nations with bicycling and Alpine skiing, soccer with Brazil, baseball with a number of countries in Latin America, and the United States with its own form of football.

The emergence of China as a world power in a number of sports in the past decade, after a lost generation of sporting activity because of the Cultural Revolution, is clear evidence of how sporting practices change in accordance with cultural changes in a society (*Sports Illustrated*, 1988; also see Clumpner & Pendleton, 1978; Knuttgen, Qiwei & Zhongyuan, 1990). Sport as a competitive activity has generally not been highly valued in the Chinese cultural tradition, and Mao especially disapproved of competition in sport and other realms of society. Since Mao, though, China has awakened from its international isolation, entered the world economy, and given more emphasis to competition domestically and internationally. The change in its seriousness about international sports success is a reflection of this major shift in Chinese culture and society.

After looking specifically at the differential popularization of sports in continental Europe, Van Bottenburg (1992) concluded that the major forces shaping sports preferences include the competition among nations, social classes, sexes, and generations. In his view, more social inequality in a society results in more emphasis on the cultural goods favored by the most powerful and prestigious members in that society. In other words, the importance or popularity of cultural goods such as sporting practices tend to reflect the cultural tastes of the dominant groups in a society.

SPORT, SOCIAL DIFFERENCES, AND CULTURAL STRUGGLE

An and Sage (1992) have shown how the development of a sport by dominant groups in a society can make that sport a social and cultural battleground. In their article about the golf boom in South Korea, An and Sage argued that dominant South Korean political and economic groups have promoted the growth of this sport because golf is a means of promoting their own political and economic interests and of enhancing popular support for their cultural values and way of life. With its expensive club memberships and its use of large amounts of land in a country where land is scare and valuable, golf is a sport that meets the needs of the elite. Political and business leaders use the golf course and club for entertainment and making deals. The construction and operation of golf courses can also be a source of substantial profits for large landowners and corporate investors.

Public discourse about the golf boom and frequent coverage of golf in the mass media tend to convey the cultural message that the development of this sport and the lifestyle it reflects are in the national interest. This message has not been universally accepted in South Korea, though. For example, the golf boom has been attacked by farmers, labor organizations, and environmentalists because it removed land from agricultural and industrial uses, contaminated agricultural lands, and polluted drinking water. Although this conflict can be seen in narrow terms as a challenge to the development of a particular sport or leisure activity, it can also be seen in broader cultural terms as a confrontation about the cultural and ideological messages that are conveyed by the use of scarce national resources to accommodate the leisure interests of affluent and powerful members of South Korean society.

In South Africa, rugby symbolizes the great cultural and social divisions between whites and nonwhites (Contreras, 1992). Traditionally, white Afrikaners have been united in their strong passion for this sport and their devotion to the all-white national team, the Springboks. The social transformation of this whitest of South African sports began in 1991 with a merger between the white South African Rugby Board and a nonwhite federation. This merger, however, transformed rugby from a unifying cultural force for white South Africans into a culturally divisive force in the changing South African society. When the Springboks returned to international competition after the world sports boycott against their country was lifted, the players joined a predominantly Afrikaner crowd of 70,000 at the start of the game in shouting "Die Stem," the anthem despised by blacks as a symbol of apartheid. Such behavior by the Springbok players, fans, and officials foreshadowed the challenge faced by the newly integrated government of South Africa in trying to create widespread support among whites for a truly integrated society.

The Montreal Summer Olympics of 1976 provides another example of sport as a source of cultural division, rather than integration, in a nation. The case of the Montreal Olympics is complicated, though. On one hand, these Games were an intercultural festival and hosting them was a source of national pride for many Canadians. On the other hand, they may have fueled French-Canadian separatist sentiments and

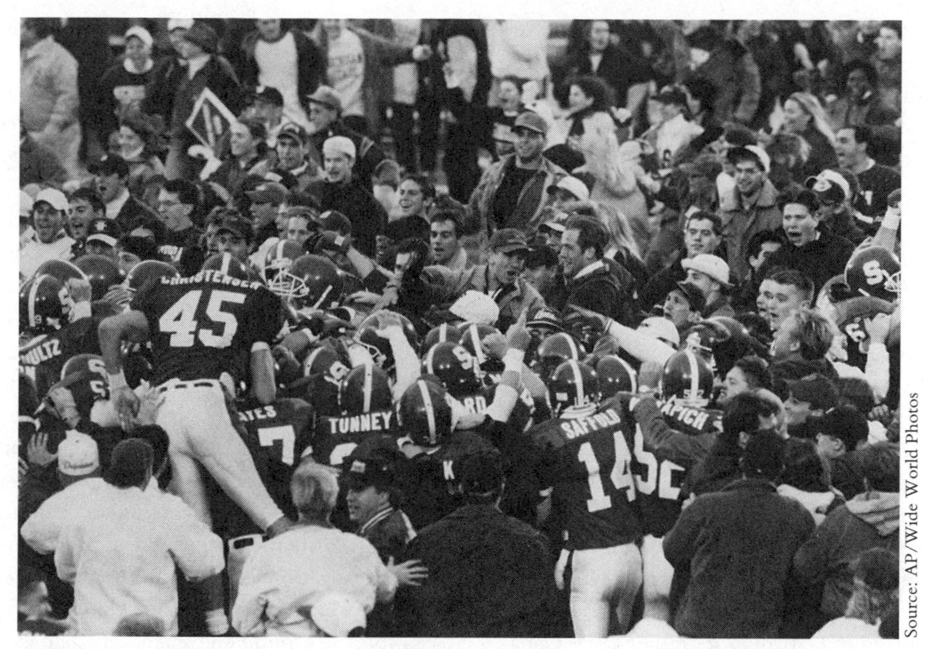

Source: AP/Wide World Photos

FIGURE 3.3 Sport as mass spectacle.

undercut efforts to achieve social equity goals in sports investment in Canada.

All Olympic Games can be viewed as sites of cultural struggles, with different social and cultural groups competing to impose their interpretations of the meaning of the Games (Hargreaves, 1986a; Kidd, 1992). Cultural struggle at the time of the Montreal Olympics was especially intense because the definition of Canada as a unified nation and the purposes of sport were topics of intense debate among Canadians. The assorted troubles of the Montreal Olympic organizers, including construction delays, deaths and scandals, huge deficits, extreme security measures, and hungry athletes, compounded tensions surrounding the staging of these games. A number of the problems of Olympic organization and construction may have been a result of the intersection of arguments between English and French Canadian provincial nationalists and a growing debate about national investment in sport. The pomp and excitement of the games provided only a temporary respite from the "culture wars" and political confrontations in Canada that preceded them.

THE AMERICANIZATION OF GLOBAL SPORT

The cases of golf in South Korea, rugby in South Africa, and the Olympics in Canada demonstrate how sport may be at the center of important cultural struggles about national values, priorities, and even national integration and identity. Sport makes these issues especially salient when foreign sports influence is seen more broadly as a reflection of foreign cultural, social, economic, or political domination. For example, Kidd (1981) wrote about the increasing U.S. domination of commercial sport in Canada, which resulted from increasing amounts of American capital investment. This process has meant that the National Hockey League (NHL) has increasingly become the American Hockey League, with more American owners and more teams in American cities. Especially symbolic of the loss of Canadian control over what was once thought of as a Canadian sport are the increasing influx of American-born players into the NHL and, perhaps most dramatically, the move of Canada's national sports hero, Wayne Gretzky, from Edmonton in Canada to a team in Los Angeles.

Kidd (1991a) has proposed that *Americanization* is a more useful term than *globalization* in interpreting the pattern of development of sport in Canada. By **Americanization**, he meant the commercialization of Canadian sport by American capitalists and state-subsidized American-based cartels, which have saturated the Canadian market with American-focused events, images, and souvenirs. Although this process has been criticized by Canadian nationalists, many Canadians see few cultural differences across the border.

More Canadians have probably played the American national pastime of baseball than the frequently claimed Canadian sports hockey and lacrosse. Canadians have often warmly embraced American coaches and athletes who have competed in Canada, and they have avidly followed Canadian athletes who have competed in the United States. Furthermore, Canadian entrepreneurs and the Canadian government have encouraged the investment of American capitalists and enterprises in Canadian sport. For those Canadian nationalists who have sought to enhance the international stature and interests of Canada by developing its national reputation in sport, American influence over Canadian sport and society has been an irritant and obstacle.

These Canadian nationalists may be more encouraged by the movement of the Canadian Football League (CFL) into U.S. cities, such as Sacramento, Shreveport, Baltimore, and Las Vegas. The CFL seems to have followed the American commercial model in expanding their league to American markets, thereby flipping Americanization on its head with a parallel process of Canadization. In a related process of Japanization, Japanese investors have bought U.S. golf courses, ski resorts, and sports teams, and sports teams and athletes in the United States have bought a variety of athletic products from Japanese manufacturers. As in the case of Americanization in Canada, foreign investment in the United States helps the U.S. economy, but it also increases hostile and anxious feelings toward the foreign investors and their nations (Newman, 1989).

Although Canadization and Japanization, like Americanization, are processes of capitalist commercial development in foreign countries, they are not the same thing as Americanization. Thus, Americanization and the United States are not the only important influences on the global development of modern sport. On the other hand, the popular commercial culture of American-based multinational corporations and mass media has left a strong imprint on nations and sports around the world, shaping the tastes, practices, and images of sports consumers in many nations of the world. According to Kidd, when the American influence is dominant in Canada or other nations, the people in these nations lose their chance to put their distinctive cultural stamp on sporting practices. This could be construed as a condition of **cultural dependency**.

Others have written about Americanization of sports in other nations. For example, Maguire (1990) studied the growing popularity of American football in Great Britain in the context of the more general Americanization of British culture. Maguire argues that the cult of individualism associated with American sport has found cultural acceptance in segments of English society, and that this fact, in combination with what he calls an *enterprise culture,* capitalist development, the influence of multinational corporations, the commercialization of English sport, and the broader context of Anglo-American relations, explained the popularity of American football in Britain in the 1980s. Maguire's analysis emphasizes the role of television in American football's penetration of British popular culture. Thus, for him, Americanization is a reflection of cultural, political, and economic influences. Maguire concludes that American football is unlikely to threaten the popularity of soccer in Britain but may undercut the popularity of rugby because the two games are relatively similar and may appeal to similar audiences.

McKay and Miller (1991) note the parallel Americanization of Australian and British sports. In their view, the recent transformation of amateur to corporate sport in Australia was influenced by domestic and international factors, including its multi-ethnic population, the federal government's policy of multiculturalism and its role in brokering sports funding and sponsorship, Australia's status in world capitalism, and its membership in the Commonwealth of Nations. McKay and Miller recognize the initial importance of American styles of sports production and promotion in

SPECIAL FOCUS

Cultural Resistance in Dominican Baseball

Klein (1991a, 1991b) critically examined Americanization in the Caribbean in his research on the influence of American major league baseball on baseball in the Dominican Republic. He proposes that North American baseball interests operate like multinational corporations in locating inexpensive resources—in this case, talented Dominican baseball players—developing them, and relocating them to the United States for consumption. Especially because Dominicans have developed a strong passion for this American sport, one can find exploitation in the development of the best Dominican ball players for the American major leagues. Yet Klein found forms of cultural resistance to American domination. In the summer, for example, Dominican sports writers and the headlines of their stories gave special attention to the accomplishments of Dominicans in their coverage of major league baseball in the United States and much less to the outcomes of games and the achievements of non-Dominicans. In the winter, the sports pages focused on North American interference in the Dominican baseball league.

Another illustration of cultural resistance was the preference for Dominican over American team hats, which Klein sees as a way of expressing their nationalism. He noted that 78 percent of the 164 Dominican people he surveyed chose a Dominican team's hat, despite the pervasive influence of North American popular culture in their country. He points out that more affluent respondents were more likely than poorer ones to choose a North American hat, indicating that North American culture had a stronger hold on those of higher status, but 70 percent of the upper-class respondents still chose the Dominican hat. This expression of cultural nationalism in regard to their national sport was different from the normal patterns of Dominican consumption of North American cultural goods—including fashions, films, and music. Klein also found some forms of resistance in concrete behavior, such as foot dragging and confrontations over cultural issues between North Americans and Dominicans.

the recent corporate development of Australian sport, but they fail to see American cultural domination. The United States is one of many "players" seen to be affecting the popular culture of Australia. They point to the integration of both amateur and professional sport in Australia into the media industries, advertising agencies, and multinational corporations of the *world* capitalist system.

From a cultural perspective, the meaning of Australian sport can be seen as reflecting what has been called the cultural logic of late capitalism, which strongly bears the imprint of global consumerism and the mass media. Thus, the globalization of contemporary sport may be a modernization process that creates a kind of cultural homogenization of many sports around the world (Wagner, 1990; Guttmann, 1991). This homogenization bears the broader imprint of late capitalism rather than merely Americanism.

Despite the potential for politically and economically more powerful nations, such as the United States, to impose their culture on less powerful nations, these less powerful nations can influence the sports and cultures of the more powerful. The ideas of cultural struggle and resistance imply that cultural influences do not flow only in one direction. For this reason, ideas of Americanization, implying American cultural domination of sporting practices in other nations, are likely to oversimplify the nature of global influences and cultural exchanges affecting how sport develops, is practiced, and derives its cultural meaning in the global sports system. Klein's (1991a, 1991b) study of the Americanization of Dominican baseball shows that even for small nations, American influence over their sporting practices does not necessarily mean total cultural dependency in sport. Klein described a number of forms of **cultural resistance** displayed by the

Dominicans in response to the Americanization of baseball in their nation.

SPORTS FANS AND THE CULTURAL MEANING OF SPORT

In an international context, battles over control of the meaning of sport are means of expressing national pride, class domination, or some form of foreign colonization. For many sports fans, though, sport has a number of other important kinds of meanings. Recent research by Wann and Branscombe (1990) has shown that **BIRGing** (basking-in-reflected-glory) and **CORFing** (cutting-off-reflected-failure) can have identity and self-esteem functions for sports fans.

If by *fan* we mean "fanatic" (see Edwards, 1973: 238), then it should be obvious that for these people sport is an object of special meaning and devotion. The enthusiasm or passion for sport displayed by the most devoted fans is likely to be fed by a reservoir of memories linking them not only to the sport institution but also to meaningful group relationships and personal experiences (Healey, 1991). Their sports devotion also is likely to be fed by feelings of remembrance or nostalgia derived from sport halls of fame, museums, and other depositories of sports memorabilia (Snyder, 1991) and from sports-festival rituals, such as the Olympic flame ceremony, that produce powerful symbolic and nostalgic images (Slowikowski, 1991).

For most modern sports fans, the mass media may be the most important vehicle for enlivening and maintaining sports interest. Real and Mechikoff (1992) characterized **deep fans** in terms of their ritual identification with mass-mediated sports productions and portrayals of athletes and with related commercial advertising. They argue that the essential meaning of such events is not to be found in their commercial, ideological, or technological implications, but in the contest itself. Yet, in strictly utilitarian terms, sports contests have no value. That is, they "make nothing happen" (Real & Mechikoff, 1992: 336). What distinguishes serious sports fans, however, is their belief or illusion that the results of sports contests matter (Smith, 1988).

Mediated sports events may be meaningful for fans only at a symbolic level, but when sports are associated with matters of deep cultural and personal significance, they become very important to fans. In symbolic terms, sports involve matters of striving, struggling, winning, losing, character, masculinity, pride, sacrifice, discipline, achievement, and success. The mass media and advertisers weave these themes into sport in their constructions of sports events as mythic rituals and spectacles to make these themes real and sport meaningful. They create or reinforce the role of the deep fan.

Real and Mechikoff (1992: 337) concluded that "mass-mediated sport today is capable of providing for the deep fan crucial expressive, liminal, cathartic, ideational mechanisms and experiences for representation, celebration, and interpretation of contemporary social life, warts and all." Sport functions in these ways when it serves as a **symbolic refuge** for fans, providing meaningful escape from the dull or unfulfilling routines of everyday life. Contrary to those who have argued for the continuing salience of this symbolic refuge for fans, some sports observers have proposed that sport may be losing its special significance and, perhaps, its capacity to arouse the passions of deep fans.

IS SPORT LOSING ITS SPECIAL PLACE IN AMERICAN CULTURE?

Deford (1985) made the provocative argument that corporate sports in the United States were no longer what he dubbed a "cozy corner." By this he meant that despite the continued love of sports among children and the increasing numbers of people watching sports, society had come to be disillusioned with sports. The disillusionment with sports today, Deford asserted, results from increased media exposure of their flaws and, even more important, from the increased ambiguities and contradictions that have made sports seem more complex and confusing for their public. In short, sports no longer are a cozy corner or symbolic refuge because they have become too much like reality. According to Deford, "Many Americans are still wandering about, vaguely lost, at this strange point where sentiment and cynicism intersect, not quite knowing where they are or what it is they miss" (1985: 61).

In Deford's view, fans have become confused about the basic elements of sport, especially competition,

SPECIAL FOCUS

BIRGing AND CORFing

Wann and Branscombe (1990) conducted a study of BIRGing and CORFing tendencies in regard to sport. BIRGing, basking-in-reflected-glory, is identification with successful others, and has an ego or self-esteem building function. CORFing, cutting-off-reflected-failure, is the effort to distance oneself from unsuccessful others and has an ego or self-esteem protection function. Wann and Branscombe found support among college undergraduates for their hypothesis that the amount of identification with a team moderated the degree to which they displayed BIRGing and CORFing tendencies. Those who highly identified with a team were more likely than those with less team allegiance to display BIRGing. More highly identified fans also were more likely to retain their allegiance when the team lost. That is, they were less likely to CORF than fans with weaker team identification. These results imply that fans differ in the importance they attach to team identification and in the meaning the team's successes and failures have for their identity or self-esteem.

The findings also suggested that fluctuations in attendance are mainly the result of weakly identified fair-weather fans, who showed the strongest tendency to CORF. Their interest wanes especially after poor seasons and jumps especially after good seasons. Highly identified followers are diehard fans, and they will stay with a team through the tough times as well as the successful times. Wann and Branscombe believed that BIRGing and CORFing are induced by the desire to have a positive self identity and that these tendencies may be important mechanisms in the maintenance of self-esteem. Among the significant questions to be answered about BIRGing and CORFing is how diehard fans deal with consistent losing when their strong team loyalty makes it difficult for them to CORF.

achievement, and commitment. Competition has been diluted by too many games, too little consistent effort by players, and too many chances to make the playoffs. Vocationalism has robbed the game of some of its spirit, fun, or playfulness, and encouraged cheating, drug use, and other deviant practices to try to assure success. And the player's loyalty to the team and the owner's loyalty to his players and local fans has been undermined as growth and greed have planted themselves deeper into the structure of modern corporate sports in America. As a consequence of these developments, modern corporate sports now are part of the institutional life of America, just as business, politics, and religion are, but in achieving this status they may have lost the capacity to inspire us as they once did.

Sport as a Symbolic Refuge

Underlying Deford's argument is an idealized conception of sport as a symbolic refuge. This world of sport is a world of cultural ideals, heroes, drama, and excitement—a fantasyland—free from the harsh realities, complexities, confusion, and corruption of everyday life. Fans caught up in lives of ambiguity, frustration, anxiety, uncertainty, and contradictions find a leisure sanctuary of sport appealing or necessary because it is a symbolic refuge or escape from their everyday problems, pressures, or concerns (Nixon, 1984: 208).

The popularity of sport and its acceptance by the public depend in part on the success of athletes, coaches, sports officials, and the mass media in getting the public to believe a set of romanticized myths about the nature and function of sport in society. All societies and formal organizations depend on institutionalized myths to rationalize and legitimize their procedures, policies, and practices. Sports insiders who have personal doubts about the actual embodiment of themes such as the dominant American sports creed in the reality of their sports experiences nevertheless may convey what has been dubbed a "cynical knowledge" (Santomeier, 1979) about sport to legitimize it for sports consumers (and themselves).

Source: National Baseball Library & Archive, Cooperstown, N.Y.

FIGURE 3.4 The Babe: Cultural Hero.

The emphases on character, courage, loyalty, discipline, teamwork, competitive striving, and success inspire Americans because they reflect cherished values of the American Dream. Of course, the American Dream represents a world that is more myth than reality, but because it is an uplifting dream people want to believe it. When people see sport as a clear and simple manifestation of such morally uplifting beliefs, it becomes a symbolic refuge from the realities of everyday life that tarnish the dream. Deford's argument suggests that the growth of bureaucracy, commercialism, and the entertainment ethic in American sports has made fans more cynical and has thereby undermined the legitimation of corporate sports and their capacity to inspire commitment among fans. Scholars have described as "degradation" the process of change in the character of sport that Deford observed (Nixon, 1988a).

The Degradation of American Sport

The **degradation of sport** involves the dilution of victory, vocationalism, and lost commitment and loyalty of which Deford wrote. More fundamentally, however, it is a corruption of athleticism that trivializes superior athletic performance, undermines the integrity and purity of athletic contests, and threatens the welfare of players and even serious fans (Sewart, 1981; Nixon, 1988b). Lasch (1979) proposed that this degradation involves "the intrusion of the market into every corner of the sporting scene." Furthermore, the increasing influence of market principles, which emphasize an entertainment ethic, sports and athletes as commodities, and profits, reduces the capacity of sport to be a symbolic refuge (Sewart, 1981). The degradation of sport may have important elements of what Ritzer (1993) has called the **McDonaldization** of society. That is, sport may be incorporating more and more qualities associated with the fast-food industry, including efficiency, quantification, calculation, predictability, and control, especially through the substitution of nonhuman for human technology. For the most part, of course, these are qualities we earlier associated with the modernization of sport, but they take a more specific and concrete form when we associate them with McDonald's restaurants and other visible symbols of the character of contemporary Americana.

Thus, according to the degradation argument, as sport is degraded by the very processes of bureaucratization, commercialization, and spectacle and commodity production that fueled its growth and mass popularity, it becomes harder to sustain the myths that are an important source of its mystique or special symbolic meaning and popular appeal. Even the Olympics, with its unique qualities as a sports event, must wrestle with the dilemma of trying to maintain the myths and illusions that sustain its mystique and popular appeal in the face of ever-encroaching contradictory realities (Nixon, 1988b). For other types of sports and sports events, maintaining the myths, illusions, and mystique would seem to be harder.

In the degradation argument, sports are degraded by forces that undermine their essential athleticism and spirit of play and cause them to lose their capacity to inspire players or fans. Lasch has asserted that "prudence, caution, and calculation, so prominent in everyday life but so inimical to the spirit of games, come to shape sports as they shape everything else"(1979: 217). Athletes and coaches become more concerned with security and the avoidance of defeat than with the risk, appropriate abandon, and excitement of pure athletic competition.

The popular mystique *and* commercial growth or viability of sports seem to be threatened by cynical or disillusioned fans who no longer find sport a symbolic refuge or cozy corner. Thus, it might be asked: What do we actually know about what American fans believe about sport today? Some of the best available evidence is from a 1982 Miller Lite survey of over 1,300 Americans aged fourteen or older. Relevant results are summarized in Table 3.1.

TABLE 3.1. Miller Light Survey Results

Questionnaire Items	Agreement (%)
Professional athletes are overpaid.	76
Professional athletes are (not) more dedicated to the game than they are to their own gain.[1]	50
Strikes by players decrease my support for a particular sport.	40
Sports events have become too much of a spectacle.	55
The spirit of the game has been hurt by placing too much attention on entertainment and not enough on athletics.	59
Professional sports on television is primarily entertainment.	49
There are too many professional teams.	39
There is just the right amount of sports on television.	59
Athletes are good role models for children.	75
Athletes are often the best role models children can have.	59
Sports writers and sportscasters generally place too much emphasis on athletes' personal lives.	53
Athletes are human like the rest of us and we shouldn't expect them to be of higher moral character than the average citizen.	64

[1]The questionnaire item was stated in the affirmative, with 50 percent disagreeing.

In general—and keeping in mind that the data are more than a decade old—the picture of Americans' views of sport revealed by the Miller Lite survey is not uniform enough to provide definitive support for, or refutation of, Deford's argument or its underlying premises. Nor does it clearly support a "degraded" conception of sport by sports fans. On one hand, there was widespread acceptance of athletes as role models, and people were happy to have as many teams and television sportscasts as they were currently offered. Yet many questioned athletes' high salaries, their dedication to the game, and strikes. The majority of those surveyed also believed that sports events had become too much of a spectacle and the spirit of the game had been hurt by an excessive emphasis on entertainment. Perhaps because they implicitly understood the threat to their symbolic refuge, most felt that athletes' lives were being scrutinized too closely by the press. Although they accepted the idea of athletes as role models, the majority of the respondents believed that athletes were human like the rest of us and need not be morally superior. When age was considered, it was found that on many items, the doubts or concerns adults had about sports were not as strongly felt by young people aged fourteen to twenty-four.

The precise nature of the Miller Lite survey results might differ if the survey were conducted today. Adverse public reactions to player strikes and some souring of the relationship between network television and sports may have made the American sports public somewhat more cynical or disillusioned, or at least somewhat more restrained in its enthusiasm. Yet it remains unclear that increasing bureaucratization, commercialization, and emphasis on entertainment have transformed sports into "just another [consumer] choice," as Deford (1985) argued.

While these forces may be viewed as potential sources of the degradation of sports, they can also be seen as central elements of sports in their modern form. Noting the tendencies toward widening mass media coverage, specialization, rationalization, bureaucratization, and quantification, Guttmann (1988) suggested that serious fans and sports purists today may be ambivalent. This is to be expected, he further suggested, because modern sports are complex, making it difficult to have a simple, unmixed response or totally undiluted passion. Perhaps, like Robert Frost in his view of the world, fans today have a "lover's quarrel" with sport (Guttmann, 1988: 190).

CONCLUSION

The results of the Miller Lite survey suggest that Deford, and more trenchant or radical critics of the structure of American sport in general, may have overestimated the disillusionment, disappointment, sense of degradation, cynicism, and lost interest of the American sports public in recent years. Even though the American sports public today seems to understand that sport is an entertainment spectacle and is willing to accept very human athletes as role models, it may not be ready to give up its sports world as a cozy corner or symbolic refuge. In fact, American sports fans seem to resist the intrusion of reality. For example, they do not want their athletic stars to be scrutinized more closely by the press or become more humanized than they already are. Furthermore, the dismal failure of Ralph Nader to organize sports fans as a consumer movement called FANS (see Nixon, 1984: 226–227) can be seen as a result of an overestimation of the disenchantment of sports fans.

To say that sports fans are not as disenchanted, disillusioned, or cynical as analysts and critics might predict is not to say that these critics have misinterpreted the character of modern American sports or the forces molding that character. However, it appears that these professional observers have overestimated the importance that sports fans ascribe to the larger sports world. That is, critics discuss the degradation of sport in terms of the larger culture, social structures, and social processes of society that affect the meanings, structures, and experiences of sport. However, fans seem more concerned about the meaning of sport as a physical contest. Viewed in such terms, sport is more likely to transcend the ambiguities and contradictions that we see in critically analyzing sport in a larger cultural and social context.

Even if athletes and coaches are somewhat tarnished by reports of their misdeeds by probing investigative reporters or critics, they can be redeemed by their success in the contest. It is not merely that they win. It is that they become connected to the traditions of sport when they succeed, or even lose courageously, in the athletic arena. It may be that as long as fans can interpret their sports experiences in ways that insulate the contests they watch from forces that could degrade them, sport can remain a symbolic refuge for them. The desire to protect the refuge from the intrusion of too much disillusioning reality makes fans more receptive to cynical knowledge or fictions that idealize or romanticize sport—and reinforce the walls of the symbolic refuge. In this way, a self-perpetuating process is sustained.

The capacity of sports fans to resist the intrusion of reality into their refuge can be understood in terms of the subtle and latent symbolic functioning of sport. Because a sports contest can be connected symbolically to traditions larger than the individual event or performers, it can be separated from the corruption of everyday life and even of the athletes, coaches, and other actors in the sports arena. Because it can be seen as separated from everyday life, the sports contest can embody the idealized values of a community or a society and it can inspire in ways that more mundane activities cannot. That is, it can be at least a temporary refuge in an often complicated and stressful world, a world that, in reality, affects sport, too. The tendency for fans not to see, or pay much attention to, the potentially degrading forces of reality can be seen as a reflection of the powerful latent symbolic influence of sport in society.

The degradation of sport trivializes and demystifies it and tarnishes the idealized beliefs that legitimize and popularize it. However, sports fans are not inclined to believe that what they see has been degraded. Many do seem to need a cozy corner, or a place in their lives where events and their meanings seem relatively clear and simple. Thus, they may overlook or minimize the ways that the larger social world of sport mirrors the realities of society, in an effort to find a temporary refuge in sports contests that offer the possibilities of drama, excitement, and enjoyment not readily found elsewhere.

Of course, sport is not a perfect refuge, and it is possible that it will, as Deford warns, cease to be a cozy corner if the modernization or corruption of sport begins to strain the credibility of the cynical knowledge and fictions that sustain its mystique, myths, and illusions. To understand how popular American sports will function for fans in the future, it seems necessary to try to understand more fully the central paradox of modern American sports as symbolic refuge *and* reality.

The symbolic functioning of sport as a source of escape or spiritual uplift may be ephemeral or superficial. As Deford argues, fans may be more confused than they once were, and have more misgivings than they once had, about sports. Yet, as Deford also acknowledges, they continue to watch. Furthermore, as the Miller Lite survey found, they continue to believe that athletes are desirable role models, and they do not seem interested in finding out more about their warts or off-field problems. They want to protect their refuge. Their unwillingness to confront the corrupt realities of sport or allow them to intrude into their symbolic refuge may be a measure of the inadequacy of perceived cultural alternatives to sport as sources of meaning or joyful escape in American society. The inability of sports analysts to account clearly and cogently for the persistence of fans' faith in sport may be a measure of the inadequacy of our efforts to understand the nature of sports fans' commitment and the basically paradoxical meaning of sport in modern society.

SUMMARY

Culture provides the framework within which the members of a society construct their way of life. Sport is a societal institution shaped by the culture of the society. It is an element of popular and mass culture. In the United States, the dominant culture of sport—called the Dominant American Sports Creed—mirrors the dominant culture of the society—called the American Dream. Both function as ideologies, legitimizing the dominant cultural beliefs of the society. A central value of sport not explicitly stated in its creed is winning, which can become an obsession for serious sports participants. Sports heroes are symbolic representations of the dominant social myths and values of a society or subculture. The nature of heroes changes with changes in the culture. The close scrutiny of the human flaws and failings of sports stars today has made it difficult to sustain traditional heroic images of these stars.

Major sports events and popular sports are integrated into the cultural traditions of the societies in which they exist. Certain special sports events, such as the Super Bowl, become mythic spectacles, functioning in the manner of traditional mythic rituals as means of collective identification and celebration. Like heroes, these mythic sports spectacles remind the members of a society what they value. In the modern world, certain widely watched sports events link different nations in a global sports system. In this global system, the sports played in different nations show many similarities, but they show cultural differences, too. Within many nations, there are cultural struggles involving both internal and foreign elements over the types of cultural forms, meanings, and practices associated with major sports. A major issue is the amount of influence exerted by the United States and U.S. corporate interests over the sports of other nations. Cultural resistance to Americanization in nations such as the Dominican Republic means that even small nations are able to resist total cultural domination of their sports by major world sports powers.

Sport can be an important source of identity as well as enjoyment for sports fans, as research on BIRGing and CORFing tendencies has shown. Sport can function as a significant symbolic refuge for fans, but Deford has argued that sport has lost its special place for American sports fans. As it has developed in recent years, sport may have become corrupted or degraded by the processes of modernization. The evidence is mixed, though, as sport continues to be an important element of the culture of many Americans and members of many other nations.

CHAPTER 4

Religion and Sport

RELATIONSHIPS BETWEEN SPORT AND RELIGION

On the surface, sport and religion may appear to have nothing in common. Yet there is much that they share, and some argue today that sport has replaced traditional religion as the basis for consensus and worship in society. Hoffman states: "Sport and religion seem comfortable with each other. Wherever sport is played at high levels, one invariably finds traces of religion. At the same time, wherever one finds the organized church, sport is likely to lurk in the shadows" (1992b: vii). Over time sport has moved from virtual dependence on religion to a less dependent and more complex relationship.

There is some evidence that athletics originated as a religious rite designed to influence the forces of nature in the hope that the land and livestock would be fertile (Sage, 1981: 148). Guttmann (1978) describes how primitive societies utilized running, jumping, wrestling and throwing as integral parts of rituals of fertility, puberty, and worship. Greeks combined sport with religious observances and festivals. Even the Olympic Games were originally a religious festival where the Greek people paid homage to Zeus and looked for evidence of his presence in their sporting activity (Novak, 1976; Guttmann, 1978). In subsequent centuries, various branches of the Christian church controlled sports activity in Western countries because of the concern that the soul should not be compromised by activity that was nonutilitarian and detracted from adherence to religious life. Thus, the church only permitted athletic activities on special occasions or holidays.

Sport was viewed as sinful by Christian denominations who were influential enough in some cases, such as the colonial Puritans in America, to persuade governments to outlaw sports competitions and other frivolous amusements on Sunday. In early colonial America, Puritan traditions dictated against any amusement or otherwise frivolous leisure activity, especially on Sunday. John Wesley's dictum, "He who plays as a boy will play as a man," sums up the status of sport at this time (Hogan, 1967: 121). Play was definitely sinful in the eyes of Methodist, Presbyterian, and other Protestant leaders. Even the mild and pastoral game of croquet was forbidden by some denominations, and the bicycle was seen as evil because its use meant that children would be unsupervised and wives might neglect their family duties (Hogan, 1967).

Puritans did not necessarily restrict all sporting activity during the week, since their English ancestors exhibited a strong heritage of sport and leisure activity (Lucas & Smith, 1978). Advocates of a strict ascetic and devotional lifestyle became more tolerant of sport when they realized that play could be an acceptable and necessary way to offset the drudgery of everyday life and prepare people for God's work.

The industrialization and urbanization of America, along with the secularization of social life, diminished the control of sport by religious institutions and leaders. Sport and recreation activities became a very prominent component of the ministry of many churches. These programs are generally focused on the recruitment of particular populations, such as youth aged sixteen to twenty-five, young families, or senior citizens. Sermons and prayers often rely on sports metaphors to convey religious or church-related

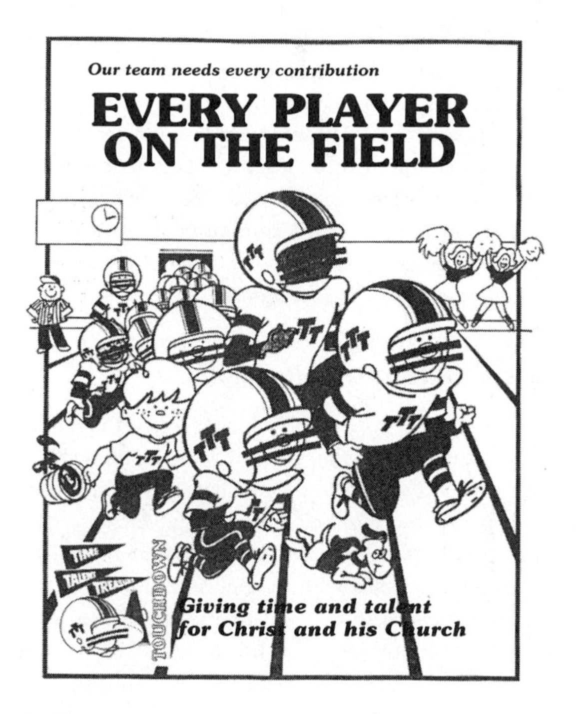

EVERY TEAM NEEDS EVERY PLAYER

God is the one who draws us together as congregations, as church teams. In doing this, the Apostle Paul says, God is careful to place team members together that compliment one another in their abilities to serve. This way, God builds a "body" or "team" which has great potential to serve God's world.

The key to how effective any congregation is in their ministry is wrapped up in how many players are actively participating and how many are sitting on the sidelines. The way each one of us chooses to participate in the life of the church with our time and our talent makes a big difference in the overall performance of our church team.

Time is, in a way, the great equalizer in our total stewardship. We all have the same number of hours in a day to draw from, as we consider what our commitment of time will be to God's work in the church. Try this year to look at your commitment of time to the church as carefully as you do your gifts of talent and money. Consider setting aside a specific number of hours per week that your discipleship will involve being active in the church.

Talent is our opportunity to be unique in our stewardship commitment. Each member of our church team has talent. Each of us has something to contribute in the way of a God given talent which will build up our church.

Every team needs every player! Our church team will be much more effective if all of us consider carefully our opportunity to give of time and talent during the coming year. Let's get all our players on the field!

#2-TD16

Source: Logos Productions

FIGURE 4.1

message, as shown in the church stewardship appeal in Figures 4.1.

Major church-affiliated colleges and universities also use sport to attract students and funding or to draw attention to their religious identity or beliefs. Notre Dame is the best example of how sports success has drawn attention to a church-affiliated institution of higher education. Brigham Young University, Oral Roberts University, and Jerry Falwell's Liberty University are examples of institutions that have used sport as part of their broader evangelical or proselytizing efforts.

Although there are many instances of churches or religions using sport for assorted purposes, a general trend toward secularization has reduced religious influence on sport and society in general. It has been a long time, for example, since Sunday has been restricted to church activity. Religious events, even sacred holidays such as Christmas or Hanukkah, do not receive special treatment when it comes to scheduling sports events. Churches have adjusted their schedules to accommodate the football, basketball, and baseball games regularly scheduled on Sundays.

Religion in America has lost most or all of the hold it once had over sport, but sport has increased its impact on religion. Churches are reluctant to challenge or question the impact of the growing role of corporate sport in American life. Instead of raising questions about ethical and moral issues in sport or about the appropriateness of contemporary athletes and coaches as role models, churches ignore the controversial issues and portray sport and its stars in the positive light of the Dominant American Sports Creed (Edwards, 1973).

Even when we see religious practices in contemporary sports, we cannot be sure that such practices truly are intended to be sacred, spiritual, or religious. In many cases, such as locker room prayers, references to God, and motivational slogans with religious themes, there seems to be more concern about asking God for help in winning or giving thanks for a victory than about glorifying God or expressing religious devotion.

Thus, sport incorporates aspects of religion in its practices, and religion uses sport or sports metaphors to promote its theology or build its following or funding. In general, religion and sport use each other to promote their own self-interest. There are a number of similarities in the ways these two institutions function in society.

THE NATURE AND FUNCTIONS OF RELIGION

Religion is a **cultural universal** that is defined in many ways. In Durkheim's classic sociological definition, **religion** is "a unified system of beliefs and practices relative to sacred things, uniting into a single moral community all those who adhere to those beliefs and practices" (1961: 62). More recently, Coakley defined religion as "a system of inter-related rituals and commonly held beliefs grounded in an established organizational structure and directly linked to some realm of the sacred and supernatural" (1990: 355) Generally religion is defined in terms of some devotion or faith in a supernatural entity or "Ultimate Reality" by which all other aspects of the world are judged (Tillich, 1948).

Durkheim and Coakley also pointed to the connection of religions to the sacred. In contrast to **profane** matters of everyday life, **sacred** things relate to questions of ultimate meaning (Glock & Stark, 1965). Sacred items, such as crosses, vestments, and altars, are separated from daily life, and they inspire awe or reverence from believers. Profane objects are a part of everyday life and do not command any special deference or worship.

What is defined as sacred or profane is, of course, a product of cultural definition. There is nothing inherent in any object that gives it a quality of being sacred. Thus, all religions are characterized by sets of sacred objects worshipped by means of ceremonies and rituals by a group of followers or adherents who believe in the legitimacy or authenticity of these objects and rites. Communities of believers can transform profane objects into sacred ones by attributing to them special symbolic meaning or significance. For example, the ball that sails into the stands to set a new home run record may be given special meaning by fans, and it can acquire special or quasi-sacred status when it is enshrined in the baseball Hall of Fame. Over the years myths and legends will be constructed about the people or circumstances surrounding the record-breaking event (Leonard, 1993: 386). Persons who profess to be religious are often committed to a faith that is beyond the ordinary (Hoffman, 1992a). Thus, when persons exhibit an extraordinary commitment, attraction, or devotion to a sport or sports team, their behavior might be construed as religious.

Social Functions of Religion

Religion has a role to play in the lives of individuals and in the organization of society. Religion serves the needs of individuals and contributes to their sense of well-being in a number of ways (e.g., Eitzen & Sage, 1993: 182–184; Leonard, 1993: 386–388). Religion helps believers deal with the uncertainties of life, their sense of powerlessness, and their fears or frustrations about the finite nature of life on earth by directing their attention to the enduring certainties of the supernatural realm, to an omnipotent God, and to the hope of an eternal life.

Furthermore, the ability to refer to an all-powerful supernatural being to explain away inexplicable or undesirable events can relieve feelings of guilt. Religious faith and a belief in an afterlife also help individuals accept undesirable aspects of their lives, such as alienating working conditions or debilitating family conditions. Thus, religion gives individuals a sense of meaning and significance beyond everyday life.

Religion also contributes to the welfare and stability of societies and communities in a number of ways. Two of the main social functions of religion in society are social control and social integration (Leonard, 1993: 388). Religion acts as a mechanism of social control by encouraging "good" or "moral" behavior that conforms to societal norms and by discouraging "bad" or "immoral" behavior that is deviant. The emphasis of religion on established norms and values related to the Protestant Ethic and the American Dream, legitimizes the cultural basis of the institutional structure of society (Nixon, 1984). The legitimation by religion makes certain values and norms powerful because these are imbued with supernatural and sacred meaning (Hoffman, 1992a: 5).

Religion binds people in a community of common faith and beliefs. Cohesive religious communities offer an important alternative to disconnected, isolated, or depersonalized experiences of people who otherwise lack a sense of belonging or group identification. Religious rituals and ceremonies are important sources of shared meaning.

Marxists have argued that religion is functional for the dominant class rather than for society as a whole. It is used by those in power to create a "false consciousness" in the masses (Eitzen & Sage, 1993: 184). By distracting subordinate and less privileged members of society from the sources of their alienation and disadvantages and turning their hearts and minds toward supernatural things, religion serves, in Marx's term, as an "opiate." It has a narcotic effect on the masses, dulling their inclination to challenge the status quo that keeps them in a subordinate, oppressed, and disadvantaged lifestyle. In addition, religion strengthens the underpinnings of a social structure that benefits the dominant class much more than the masses. In other words, from a Marxist perspective, religion is a tool of ownership to control the workers and to maintain the societal status quo. This function for religion has been transferred to sport with the suggestion that the analogous worship of sports figures and shrines also serves to distract the masses and deter any efforts to change society.

Although religion can be a source of social solidarity, religious differences also can promote division and conflict and even incite wars. Many societies are characterized by longstanding disagreements between religious factions. Ireland, Israel and the Middle East, Eastern Europe, and the Muslim-Hindu conflict in India are only a few examples.

Sport is often seen as a **functional equivalent** of religion, particularly in integrating diverse social and political elements of society, reinforcing dominant norms and values, serving as a mechanism of social control, and diverting anger and emotion from existing social arrangements thus serving as a societal "safety valve." Individuals can find meaning in their lives and feel bonded to a community through their religious and sport affiliations. This does not mean, however, that sport can be defined as religion.

SPORT AS RELIGION

Since sport and religion seem to play similar roles in society and have similar social and cultural characteristics, we might conclude that sport *is* a religion. Major social, cultural, and functional similarities and differences between sport and religion are summarized in Table 4.1. Being able to identify both similarities and differences makes the conclusion that sport is a religion a matter of debate.

A number of aspects of the formal structure of religion can be found in sport, including a core belief system, which may be referred to as theology in religion and ideology in sport; a hierarchy of authority, ruling patriarchy, and high councils; a set of "saints" and "gods" or heroes; a wide assortment of symbols and rituals; regular "pilgrimages" to "shrines" (such as halls of fame, museums, and classic arenas in sports); elements of magic to gain the favor of supernatural forces or to appease them; scribes to report the word; and true believers or fanatics (Edwards, 1973: 260–263; Leonard, 1993: 388–394). Edwards (1973: 262–263) also argues that there is a black-white distinction in the traditions of American sport that parallels the symbolic imagery of the black-white—or evil versus good—distinction in Christian religion.

Coakley (1990: 355) found the following kinds of basic similarities between modern sport and Judeo-Christian religion in his review of the literature: an emphasis on asceticism, perhaps reflected in the "No pain, no gain" philosophy; special times and special events for ritual events; the worship of heroes and belief in legends; procedures and dramatic experiences to attain or demonstrate character or moral development; and well-established institutional structures with clear lines of authority. The values that guide sports participation are strongly related to the traditional values of the Protestant Ethic and the related ideology of the American Dream (Weber, 1958; Nixon, 1984) discussed in the previous chapter on culture. According to Leonard (1993: 399), sports values can be seen as secularized versions of basic Calvinist Protestant values, such as self-discipline and work, initiative and acquisition, and individualism and competition.

TABLE 4.1 Common and Distinct Features of Sport and Religion[1]

	Sport	Religion
Quest for perfection	yes	yes
Asceticism (discipline and self-denial)	yes	yes
Integration of mind, body, and spirit	yes	yes
Intense feelings	yes	yes
Deep devotion and commitment	yes	yes
Established rituals and celebrations	yes	yes
Core beliefs and special language	yes	yes
Bureaucratic organization and hierarchy	yes	yes
Pilgrimages to shrines	yes	yes
Elements of magic	yes	yes
Integration and social control functions	yes	yes
Source of special meaning for individuals	yes	yes
Worship of heroes	yes	yes
Opportunities to demonstrate character	yes	yes
Claims of special virtue	yes	yes
Connection to dominant cultural values	yes	yes
Expression of impulse of freedom	yes	yes
Opportunities for flow	some	yes
Sacred, supernatural, metaphysical focus	no	yes
Emphasis on love	no	yes
Expressive orientation	no	yes
Answers to ultimate questions	no	yes
Resolution of fundamental human dilemmas	no	yes
Consolation for problems and setbacks	no	yes
Secular, material, profane focus	yes	no
Competitive orientation	yes	no
Emphasis on self-promotion	yes	no
Instrumental and goal-oriented focus	yes	no

[1]Based on Edwards (1973); Coakley (1990); Hoffman (1992b); Eitzen and Sage (1993); Leonard (1993).

In the context of the secularization of society, the rise of sport has been seen by some analysts as an indication that it has become a **secular religion** or **civil religion** (e.g., Edwards, 1973: 90; Stein, 1977; Loy, McPherson, and Kenyon, 1978: 301; Leonard, 1993: 400–401). **Secular religion** is worldly activity seen as comparable to religion in its structural features, experiences, rituals, and fervor of its believers. **Civil religion** implies a shared public faith or belief system that elevates secular experiences almost to the level of the religious or sacred (Bellah, 1967). The values, experiences, and rituals associated with sport are very similar to those found in religious ceremonies. Thus, sport can be "religionlike" even though a supernatural being is not a requisite component. Carroll (1986) asserted that sport had become the Western form of meditation, demanding discipline and concentration not unlike what is required in meditation; it serves as a mechanism to distract the individual from the profane and secular experiences of everyday life; and it can produce a flow of coordinated activity that seems to be separate from the individual mind and embedded in a collective sense of purpose and action.

Prebish has made one of the strongest cases for sport as religion by asserting "if sport can bring its advocates to an experience of the ultimate, and this experience is expressed through a formal series of public and private rituals requiring a symbolic language and space deemed sacred by its worshipers, then it is proper and necessary to call sport itself a religion" (1992: 53). According to Prebish (1992), sport has become for many a more appropriate or meaningful expression of personal religiosity than the traditional Western religions, such as Christianity and Judaism. He acknowledges that people will not experience ultimate reality every time they participate in sport, but pointed out that this was true for religious involvement as well. Prebish believes that genuine religious experience is possible in sport. This experience, called a "flow" sensation by some leisure theorists (Csikszentmihalyi, 1975), transports people beyond the sensations of their earthly body to an intense experience of intrinsic contentment that comes with a demanding challenge to physical bodily limits. This experience is generally applied to participants but, to a lesser extent, is also relevant to the spectators at sporting events.

The assertion that sport is a religion has been challenged by many. For example, Chandler (1992) argued that sport does not fulfill the basic functions of either Western and monotheistic religions with written traditions or non-Western and polytheistic religions with oral traditions. Unlike either type of religious tradition, sport offers no answers to ultimate questions of birth, death, and the creation of the universe; it offers no solutions to the basic dilemmas of the human condition; and it offers no consolation for the stresses, strains, frustrations, and disappointments of human experience. In addition, sport cannot be identified as a religion because it has no metaphysical language; it does not articulate any mythical version of the creation of earth and humans; it does not command reverence or awe in the same way religious artifacts and rituals do; and, though it can seemingly be a test to see if "God is on your side," sport does not connect a person with God through faith in the same manner as religion (Carroll, 1986). Sport lacks supernatural or sacred sanction and legitimacy in a secular society.

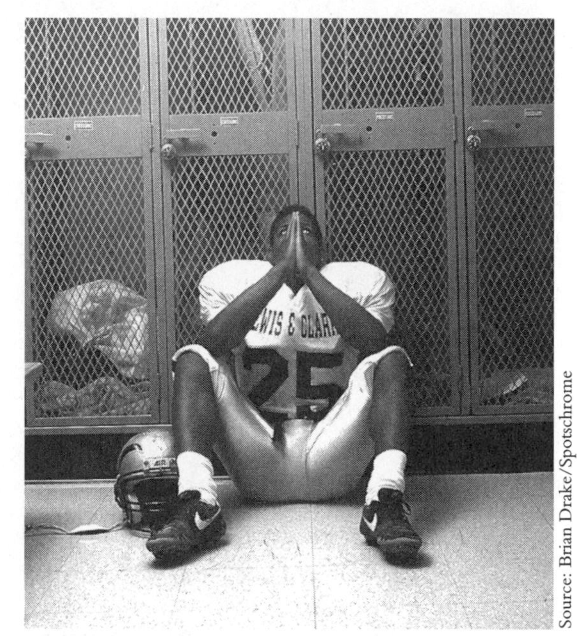

Source: Brian Drake/Spotschrome

FIGURE 4.2 "God, let us win."

Social philosopher Michael Novak (1976) has declared that sport and religion show a number of ideological and organizational similarities to the extent that sport can be designated a civil religion. Sport, Novak says, cannot be a religion in the same sense as the various Protestant denominations, but he sees a religious impulse in sport, which he describes as "an impulse of freedom, respect for ritual limits, a zest for symbolic meaning, and a longing for perfection" (1976: 21). Sport is not mere diversion or simple games; it symbolizes larger struggles that demand more human emotion than the more mundane activities of everyday life.

Major national sporting events, such as the Super Bowl, the World Series, the NBA Finals, and the NCAA Final Four in basketball, can assume the symbolic significance of national religious holidays such as Christmas and Easter. Individuals and families may spend large sums of money to attend these events in person or plan for weeks for these celebrations and clear their schedules to avoid conflict. Major sports events, national holidays, and religious celebrations may all have common symbolic qualities of sacred significance for devoted fans, patriots, and religious believers. Australian philosopher John Carroll describes the connection in his country between the Australian football final and a national holiday as rites "in the powerful sociological sense of an activity full of sacred resonances, with an elaborate mythology, celebrated on a grand scale with strictly prescribed uniforms, rules of practice, and methods of celebration ... providing a common attraction for the population" (1986: 92). These "mythic spectacles" (Real, 1975) are profane events turned into nearly sacred ones by sports marketing people, the mass media, and avid fans.

In general, we can see many parallels between the cultural and structural features of sport and religion along with evidence of various kinds of religious impulses in sport. Furthermore, sport is portrayed by many people as if it were a religion. The case for sport as religion seems tenuous, though, because very little in sport has a direct connection to the sacred or the supernatural.

Classic arenas such as the Boston Garden or ball parks such as Fenway Park, cultlike figures such as John Wooden or Vince Lombardi, or the uniform of a professional sports team may command awe and respect, but they do not carry any significance beyond everyday life and events. The meanings attached to religious events and rituals are much different than those attached to the rituals of sport (Coakley, 1990: 356). The sacred and supernatural dimensions of religion most basically distinguish it from the events, rituals, artifacts and heroic figures of sport.

Even though modern sport lacks a number of qualities of established religions, it nevertheless was seen by Edwards (1973) as a quasireligion. This concept of sport is similar to Bellah's (1967) idea of civil religion. **Quasireligion** implies that sport can inspire a shared public faith in it that gives it religious meaning among its most ardent followers. According to Edwards, sport, like religion, is a source of stability and identity confirmation in an otherwise unpredictable and alienating secular world. Sport is not full-fledged religion because it lacks a genuine deity.

According to Coakley (1990: 356), sport also differs from religion in its emphases on competition, self-promotion, instrumental and goal-oriented activities, and materialistic and profane objects, motivations, and experiences. Sport may provide some hints or clues to our personal life and emotional profile, but it cannot offer final solutions. Religion provides ultimate explanations of events, experiences, and our fate; sport does not. Religion can help us understand uncertainty and grasp the future, but sport cannot do these things. Although sport does not deserve sanctification as a genuine religion (Higgs, 1983), the secularization of religion may have created a religious void for many people who see sport as an alternative to religion.

Sport, Religion, and Morality

Although it has been argued that sport is a civil religion and that sports participation expresses religious impulses, we can question whether participation in sport is *antithetical* to religious commitment. This question is prompted by the frequent stories of immoral or deviant behavior, such as violence, cheating, drug use, rape, and gambling, in sport today. Religion is not the same thing as morality, but we expect religious commitment to guide people in the direction of moral or ethical conduct (see Hoffman, 1992b: part iv).

Veblen (1899) asserted nearly a century ago that sport was inherently immoral and degrading and that it represented either arrested human development or the incapacity of people to reach full spiritual development and maturity. In his view, sport manifested a "predatory temperament" that led to exploitation, excessive aggression, hostility, and the employment of cunning and chicanery among participants (Veblen, 1899; 170). Especially as we look at contact sports today, we see at all levels an intense desire to win that is associated with a willingness to display brutality to accomplish that goal.

The concern for sportsmanship diminishes as sport becomes more serious, organized, commercialized, and corporate. It is in this context that questions arise about the compatibility of genuine religious commitment and serious sports participation. Values and actions such as aggression, revenge, and violence, in direct contrast to religious tenets of compassion and fair play, are openly encouraged in sport. The athletic subculture supports hockey fighting, brushback and bean ball pitches, bench clearing brawls, and spit-drenched baseballs. One pitching coach was said to tell a pitcher in trouble: "If you can cheat, I wouldn't wait one pitch longer" (Shirley, 1984: 16). Fielders trap fly balls and attempt to convince the umpire they were caught on the fly; weak hitting teams have frozen baseballs to lessen the distance they can be hit; linemen deliberately hold opponents; many players have fixed game outcomes for gambling purposes; tennis players curse linesmen; and batters add cork to their bats. A baseball pitcher commented on the brushback pitch: "The brushback pitch is an accepted practice in baseball. It's part of the game; it is not immoral" (Shirley, 1984: 16). Athletes are not likely to see such be-havior as immoral as long as it is deeply embedded in the structure and culture of their sport.

SPORT AS FOLK RELIGION

Mathisen (1992) argues that sport cannot be a civil religion because it does not provide the religious meaning or national self-understanding associated with a civil religion. According to Mathisen, sport instead should be viewed as a folk religion. The concept of **folk religion** is less inclusive than civil religion and implies a more diffuse and subtle religiouslike experience that has an enduring character for partici-

POINT-COUNTERPOINT

DOES RELIGION MAKE ATHLETES MORE MORAL?

YES

Truly religious athletes may play hard and have an intense desire to win, but they understand their religious responsibility (as a devout Christian, Jew, or Muslim) to play by the rules and to avoid intentionally hurting their opponent. Religiously devout athletes have a different value system than other athletes, one that leads them toward moral behavior. They recognize that their obedience to their God takes precedence over secular values and norms of sport and that others will judge them and their religious faith by the righteousness of their actions. A number of star athletes demonstrate their faith by their voluntary involvement in a variety of charitable activities in their community.

NO

Even with all the athletes and coaches in sport today who profess deep religious commitment, we rarely read of open criticism from them about the numerous instances of unethical behavior and deviance in their sport. There is apparent hypocrisy in cases of boxers, football players, and other athletes in contact sports who cause pain and injury to their opponents and then express their faith in God and their thanks to God for helping them win or perform well. Professed religious commitment does not seem to breed compassion for hurt or defeated opponents. There is no research evidence that more devout athletes differ from less devout athletes in the morality of their actions.

How should this debate be resolved?

pants. A folk religion affirms the existing social arrangements and "encourages people to accept popular explanations of American life as having cultural importance along with religious significance" (Mathisen, 1992: 18).

Mathisen asserts that sport as folk religion has replaced a waning civil religion in America. Folk religion is a product of daily life experiences that ultimately serve to integrate diverse peoples and reaffirm values. There is a shared "spirit to life" that emphasizes fair play, humility, sharing, hard work, and other similar behaviors and qualities. This spirit to life guides social relations and becomes the unwritten preferred script for social life. Sport is a visual and behavioral representation of this script and thus represents a kind of religious setting for the reinforcement of its values and beliefs. This script is the product of social practice rather a reproduction of the premises of a particular sacred creed or ideology.

Robert Lipsyte (1975) has used the term *SportsWorld* to refer to the folk religion of sport in the United States. This is an environment in which athletic events and heroes dominate conversations and recreational life and replace religion and nationalistic issues as America's major topics of interest. In the same vein, Novak called sport "the primary lived world of the vast majority of Americans" (1976: 34) Sport is the "American Way of Life," literally dominating the social, political, conversational, and personal lives of many Americans, particularly males.

Sport is like a folk religion in that it promotes a common ideology. It evokes almost cultic rites and rituals. For example, the rites and rituals of Friday night high school football in Texas (Bissinger, 1991) reinforce shared myths about traditions of the community, sport, and the school and evoke a sense of reverence for these traditions and the related events and heroes of the past. Almost all major sporting events, from golf tournaments to major college bowl games, include ceremonies that bring back memories of significant accomplishments by fabled heroes and competitions. The role of sport as a folk religion has been significantly mediated by television, especially since the 1960s (Mathisen, 1992).

Sports commentators on television and in other media use religious language and imagery to describe sport to their audiences. For example, the Super Bowl of professional football has been described as a "religious festival." Price (1984) used this terminology because he saw this event uniting the religious, mythical, political, and sporting elements of social life in America. The major spectacles of sport provide opportunities to remember heroes of the past, who in some cases, such as Vince Lombardi in football or Babe Ruth in baseball, have attained deified status or have been canonized by selection into the Hall of Fame. These events are opportunities to remind participants, viewers, and fans in the stands of the values that undergird the country and sport. According to Albanese (1981), sport can also serve as a *cultural religion* that gives participants clues to their identity and the meaning of the roles they play in everyday life.

USES OF SPORT BY RELIGION

Historically, organizations such as the Young Men's Christian Association (YMCA) and the Young Women's Christian Association (YWCA) have tried to wed sport and religion by using sport to promote the belief that a sound body is important for a strong mind and sin-resistant spirit. A strong body was also seen as a prerequisite for being able to do the work of God, a viewpoint that has been called **muscular Christianity**. This idea, which clashed with the Puritanical aversion to sport and games (Guttmann, 1988), came into prominence in the nineteenth century as a result of the influence of English educators, moral reformers, and writers. It has been an important inspiration for American physical educators, who preached the gospel of character development through physical recreation and sport in schools, colleges, and community and church-affiliated recreation centers (Guttmann, 1988: 73–74).

Games such as volleyball and basketball were invented in the YMCA and later became popular recreational activities in these organizations and in many different types of churches. Gymnasiums were built by churches for community outreach and recruitment purposes. It did not take churches long to see that sport could be a valuable tool to bring new and previously unchurched members, particularly young men and boys, into the faith, to mold personality and character, and to promote church programs to the community (Rader, 1983). This philosophy prevails today as churches continue to sponsor recreational and sport programs as part of their regular ministry.

Private, church-affiliated colleges and universities have a long tradition of using their sports programs to promote the religious ideals associated with the sponsoring faith or denomination. Athletics is treated as a positive influence and a vehicle of public relations as well as an opportunity to promote the attractiveness of particular religious lifestyles. The University of Notre Dame is well known through its football program for its Catholic (and Irish) identity and values. More recently, Catholic institutions at the intercollegiate and interscholastic levels have been very successful in basketball, and have gained substantial publicity from this success (Hoffman, 1992e). Oral Roberts University, Jerry Falwell's Liberty University, Southern Methodist University, Brigham Young University, and others have used athletics to gain attention for their institutions and convey their religious identity. Several of the smaller church-affiliated colleges recently formed their own athletic association, called the National Christian College Athletic Association (NCCAA), with its own administrative structure and national championships.

Church-affiliated institutions have taken advantage of publicity and financial benefits of big-time, commercial, entertainment-oriented college sports. In doing so, these institutions reinforce the corporate and commercial structure of college sports, even though their presidents may express a sincere desire to rid intercollegiate athletics of its obvious excesses and problems. It is not evident from past history that there is much difference between big-time secular and church-affiliated institutions in their desire to see college athletics become less oriented toward winning or money. For big-time athletic programs at all types of universities, a deemphasis means less opportunity to use sport to gain publicity and funds.

The use of sport by religious institutions in the United States has made religion more palatable to a potentially skeptical or uninterested audience because exposure to religion in a sports context gives religion a more secular or socially relevant appearance (Baker, 1975; Flake, 1984; Hoffman, 1992e). In these secular times, linking religion to sport makes the church seem less dogmatic. At the same time, the rhetoric of sport, expressed through the Dominant American Sports Creed (Edwards, 1973), gives sport the appearance of being moral and perhaps even religious. The utilization of religious rhetoric by athletes gives them an opportunity to feel virtuous, moral, or Christian without the need to embrace the spirituality or display the deep commitment of a genuinely religious person. Thus, the secularized religion of contemporary American society is able to benefit from its ties to a sports culture that reinforces basic American religious—or more precisely, Judeo-Christian—beliefs.

Men are often not as comfortable as women with religious practices and church affiliation because these tend to be associated with traditionally feminine ideals of love, caring, and sharing. Involvement in religion in the role of athlete or coach enables men to feel masculine because sport traditionally has been associated with masculine images and values (Flake, 1984). Male athletes and coaches can talk openly about love, brotherhood, and their commitment to Christ because their audience knows that their world of sport involves elements of aggression, toughness, strength, and achievement that identify them as masculine. Thus, the apparently contradictory cultural values and practices of sport and religion, which would seem to create moral or ethical dilemmas for athletes and coaches, may actually complement each other. As a result, many athletes have declared that they are "born-again" Christians, have made drastic changes in their lifestyles, and are ready to tell anyone, or *witness,* of their renewed or newly found faith.

Although the connection of religion to sport at the high school, college, and professional levels may alleviate male doubts about religious involvement, the linking of religion to masculine conceptions of sport reinforces ideas and structures of male dominance in sport, religion, and society. In addition, this kind of relationship of religion to sport reinforces conservative religious ideas about patriarchy, which are thought by Christian fundamentalists in the United States to be biblically based. In this vein, it could be argued that muscular Christianity is a traditionally masculine ideal of Christian or moral character.

Sportianity

Many well-known religious outreach or mission programs, most with a nondenominational fundamentalist or conservative approach, use athletics as a setting to disseminate their religious messages and bring converts into their faith. The term **Sportianity** has been used to refer to the social movement of athletes and coaches who witness for Christ (Deford, 1976, 1979); this movement also has been called "Jocks for Jesus" (Leonard, 1993: 394). According to Hoffman, "as a mode of evangelism, sport has few rivals in modern religious history. Sports fans who wouldn't dream of visiting church to hear the eloquent sermons of a seminarian will listen patiently to an athlete's stammering tribute to God, guts, and glory" (quoted in Eitzen & Sage, 1993: 196).

A number of different traveling athletic teams compete at the national and international levels against local talent in order to witness their faith to audiences at basketball games, wrestling meets, track and field events, and gymnastic competitions. These include Athletes in Action (AIA), Jocks for Jesus, Professional Athletes Outreach, SportsWorld Chaplaincies, Hockey Ministries International, Baseball Chapel, and the Sports Ambassadors. Players testify to the meaning of their faith to spectators and competitors. The fact that they are athletes is used as a means to gain access to the locker room and to potential converts that would not otherwise be available. Using sport to attract converts is not limited to the traditional sports settings of football, baseball and basketball. Fishing champion and born-again Christian Jim Grassi created a unique religious proselytizing opportunity by following a morning of successful fishing in a well-stocked lake with prayer and preaching called "Angling for Souls" (Maxwell, 1992).

The Fellowship of Christian Athletes (FCA) was founded in 1954 as the archetype of Sportianity and the muscular Christianity movement. It has over 4,000 active chapters or "Huddles" on college and high school campuses across the country, having doubled in size in recent years and expanded to include women athletes as well as men. The FCA has a multimillion dollar budget and a staff of nearly a hundred to cater to the spiritual, social, and athletic needs of over 100,000 members. The organization also has chapters for adults who are not currently enrolled in high school or college, and it conducts national and regional conferences each summer as well as coaching clinics and activities such as Bowl Breakfasts, Sunday chapel services for professional athletes, and spiritual rallies that coincide with major sports events, such as the Super Bowl and the NCAA Final Four in basketball. "The FCA applies muscle and action to the Christian faith. It strives to strengthen the moral, mental, and spiritual lives of athletes and coaches in America" (Sage, 1981: 150).

The FCA is an especially good example of the effort in Sportianity to link religion and sport through groups that seek religious converts by taking advantage of the visibility and popularity of sport. High-profile athletes and coaches, such as former Dallas Cowboy mentor Tom Landry, have worked hard on behalf of the FCA. In fact, Landry is often credited by FCA founders with being an important spokesperson in the early days of the fellowship and, perhaps, the reason the FCA survived its early years (Bayless, 1990). Landry's public image as a God- fearing, Bible-carrying coach led him to earn the title of "God's Coach." Skip Bayless, a Dallas sportswriter, described himself in his first interview with Landry as nervous in the presence of a deity:

> But now I could feel my underarms getting wet. And why not? Like Landry, I was raised Methodist. I was conditioned to believe Landry all but sat at the right hand of God the Father. For me, Landry was the most powerful religious figure on earth. (1990: 47)

Athletes and coaches are commonly used to sell religion. The technique is simple: those who are already committed to religion convert the athletes. Since athletes are among the most visible and admired persons in our society, they are used in spreading the gospel to their teammates and others with whom they interact. Tom Landry and college football coach Bill McCartney, sprinter Carl Lewis, former NFL stars Bill Glass and Calvin Hill, and NBA star A. C. Green are examples. Combining the popular appeal they have as celebrities with the metaphors of the sports world, athletes are able to catch and hold the attention of large groups of people (Sage, 1981: 151).

Pro Athletes Outreach organizes speaking engagements for well-known athletes who are willing to profess their faith before audiences across the country. Many of these athletes have had a "born again" experience that they share with their fellow athletes in the locker room or at the speaker's podium. A key element of Sportianity is the use of athletes as evangelists (Deford, 1976). Athletes are viewed as especially effective spokespersons for the values and ideals that make up the Dominant American Sports Creed and the Protestant Ethic. Their status as heroes or icons gives them an advantage in spreading the word or having the ability to reach large and diverse audiences. With such elevated status, athletes can be influential religious or moral emissaries, preaching an ideology or theology of clean living, sacrifice for the team, and hard work, which all directly or indirectly relate to the Protestant Ethic. In addition, athletes can sanctify the somewhat conflicting secular values of the Dominant American Sports Creed and the American Dream, such as team loyalty, conformity, courage, and individual competitive striving and success by associating these precepts with the spiritual messages of Sportianity.

The potentially conflicting values of sports creeds and various theologies suggest the internal struggles that may be experienced by dedicated athletes who are serious about their religion (Hoffman, 1985; Hoffman, 1992b: part iv). Superficially committed religious believers can feel good about their sports participation by accepting the religious tenets of the Dominant American Sports Creed. On the other hand, devout athletes could experience deep ambivalence about athletic role expectations demanding aggressive, self-promoting, dominant, unsportsmanlike, vengeful, or unforgiving behavior. In some cases, athletes feel so much ambivalence or disconnectedness between religion and sport that they quit sports. For example, Nancy Richey quit professional tennis in 1978 because she could not resolve the tension between her competitiveness as an athlete and her newly discovered religious faith. She said, "I had problems relating tennis to Christianity. When I stepped onto a court I felt I was in an isolated area and the Lord was

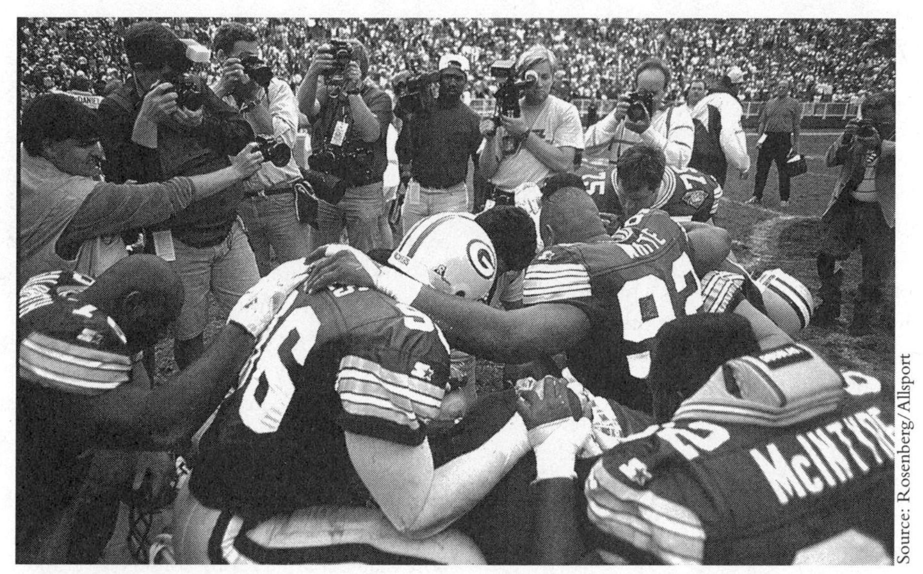

Source: Rosenberg/Allsport

FIGURE 4.3 Reggie White, Green Bay Packers: "Hope God will be on our side."

outside of that [area]. I knew hating my opponent was not a Christian view" (quoted in Hoffman, 1992c: 111–112). After receiving counseling from an evangelist who occasionally was a chaplain for the Washington Redskins, she was able to return to the tennis tour. She learned that tennis was "an act of worship" and that she could worship God by trying to perfect her strokes on the court.

Most religious athletes resolve the tensions be-tween Christian principles and sport before deciding to quit. The Institute for Athletic Performance, an evangelical network, created a doctrine for locker room religion to provide a preemptive resolution of these conflicts between religion and sport (Hoffman, 1992c: 117). It is called "Total-Release-Performance," or T-R-P. Relying on a loose interpretation of biblical passages, this doctrine links God's acceptance of athletic performances to an athlete's love of God and the intensity of his or her athletic participation. Thus, sport gains spiritual meaning when athletes love God and try hard or seek "total release" on the athletic field.

Athletes identify themselves as evangelically minded by marking their jerseys or shoes with crosses or the T-R-P inscription and by shouting "T-R-P" across the line of scrimmage or by whispering it to teammates in their huddle. Glorifying God is done by using as much of one's athletic talent as possible. Total release requires *ascesis,* or the experience of pain and suffering. Ascesis is the essential link between sport and the Christian faith in the T-R-P perspective. Pain and discomfort are the necessary glue for holding together faith, life, and competitive sport (Hoffman, 1992c: 117). Without this suffering, spiritual fulfillment through sport is not possible.

Sport provides highly visible opportunities to showcase religious beliefs, but the impact of these beliefs can be effective only if the player has performed in a remarkable, laudable, and blessed manner that is worthy of God. It is especially significant if the evangelical athlete is a member of a winning team or is successful in individual sports such as golf or tennis. A clear-cut, definitive, or unexpected victory gives the athlete a platform to relate athletic success and success in general to Protestant—or Calvinist—themes such as self-denial, asceticism, hard work, striving, and perseverance. Albanese (1981) referred to this phenomenon as a type of "spiritual reductionism" that is readily applied to sport and to religion. Confusion is eliminated; life is understandable. There are no loose ends when victory is the result and the values that produced this victory are theologically sanctioned by religious expression and behavior.

Born-Again Athletes

Athletes who are born-again Christians often have experienced a traumatic event in their personal lives, such as a near-death accident, a drug overdose, or the death or severe illness of a family member, in which some type of major reassessment of their self-concept and way of life occurred. The conversion to an evangelical version of faith for some is simply a return to the religious experiences of their youth. For others, it represents a dramatic lifestyle change. Athletes also may turn to religion and become "born again" in order to cope with the intense pressures and other negative aspects of their athletic lives, such as loneliness and isolation on the road, performance slumps, and constant scrutiny by the public and the mass media (Flake, 1984). Religion may be seen as a device to augment performance, increase the prospects of victory, or deal with the uncertainties of athletic competition. There have been many athletes who, in the midst of a competition which found them struggling or near defeat, have asked for God's assistance in exchange for a commitment to convert and follow Him forever.

While the hypocrisy or shallowness of such expedient Christian "rebirths" may be readily apparent to some, it is not necessarily so apparent to the athlete who has found a new faith. Such athletes are bound to have this faith sorely tested, however, if their faith remains contingent on whether God continues to help them out of jams and onto the victory platform. In this kind of Christian belief system, God only loves winners.

Many born-again Christian athletes have been involved in an organization called Lay Witnesses for Christ, which professes that athletic success is directly attributable to religious faith. This message is a version of the T-R-P doctrine. Prominent former athletes involved in this organization have included quarterbacks Terry Bradshaw and Roger Staubach, running back Earl Campbell, linemen John Hannah and Reggie White, basketball player Julius Erving, and pitcher Tommy John. Their ministry often reached out to fans as well as other athletes. This group and many other organized evangelical religious organizations have maintained a highly visible presence at the Olympic Games and other elite sporting events.

Over the past three decades, a number of prominent black athletes have been "born again," but not as Christians. The most publicized of these conversions have involved black athletes who have rejected white Western Christianity in favor of the Moslem religion. Two of the highest profile athletes who have made this conversion have been Muhammad Ali (Cassius Clay) in 1964 and Kareem Abdul Jabbar (Lew Alcindor) in 1971. Black Muslim athletes with Islamic names do not provoke a great deal of controversy today, but Jabbar and especially Ali were widely and scathingly criticized for their conversions. They were perceived as expressing anti-American sentiments by espousing such "foreign" religious beliefs. In fact, they were at odds with the basic religious and political thrust of Sportianity, which left little room for religious expression that departed from conservative and fundamentalist Protestant tradition (Deford, 1976).

Management Reactions to Sportianity in the Locker Room

Although religion in the locker room would seem to convey a positive image desired by sports administrators, management officials have reacted with coolness, skepticism, or veiled hostility toward religious expression by athletes on their team. For example, several years ago Seattle Mariners general manager Dick Balderson criticized members of his ball club who held regular chapel services before games (Sullivan, 1986). Balderson was especially critical of these players because he believed that carrying their religious beliefs into the locker room or onto the field interfered with their concentration on baseball. In one case, a pregame service ran overtime and cut 15 minutes from warmup time. Balderson said:

> I think we have too many [players] who think that if we lose, that's the way the Lord meant it to be. But I can't perceive God being on the mound in the ninth inning and saying [a loss] is the way it should be. I perceive Him as being an individual who would beat you any way He can as long as it's within the rules. (quoted in Sullivan, 1986)

A Seattle businessman who was a member of Pro Athletes Outreach dismissed Balderson's criticisms as a "blip on the screen of eternity" (quoted in Sullivan, 1986), but the incident points to the tension that can exist between religion and sport when religion is taken "too seriously." In fact, team officials, agents, and sports marketing personnel may try to downplay the religious activities of genuinely devout athletes because they may limit the commercial or marketing appeal of these players or their teams. In a secular society, secularized or diluted religion in sport may have an appeal, but the more serious forms of Sportianity or genuine religious belief may make most fans uncomfortable or unable to identify with athletes. When former Phoenix Suns star Charlie Scott converted to the Muslim religion in 1972, Suns owner Jerry Colangelo tried to prevent Scott from changing his name to Shaheed Abdul-Aleem. It made Abdul-Aleem feel like he was merely a commodity named "Charlie Scott" that had been purchased by Colangelo. His treatment suggests that there was little sensitivity to his needs as a person or his rights of individual religious expression.

Famous athletes have refused to play on their Sabbath day and on other holy days in their religion. Hank Greenberg and Sandy Koufax are two prominent examples. In the movie *Chariots of Fire,* we saw British Olympian Eric Liddell forced to defend his decision not to compete on Sunday against the entreaties of royalty as well as the British sports establishment.

RELIGION IN SPORT

Religious ritual and symbols permeate American sport with no apparent acknowledgment of the paradoxical or inconsistent relationship between the principles of the Judeo-Christian tradition and the sometimes dubious or deviant motives and norms that guide modern sport. A priest may give an invocation at the start of a race that consists of runners who have taken steroids to augment performance or who have agreed in the locker room on the order of finish. A coach will huddle his football team in prayer prior to the game and in the heat of battle urge his players to "hurt" their opponents or try to circumvent the rules to get the advantage needed to win. Little attention is given to this hypocrisy.

The expression of religious rhetoric in sports settings seems natural and appropriate as long as it provides competitive motivation and does not interfere with concentration on the game. Examples of how coaches and athletes bring religion into sport to aid performance include decorating a locker room with slogans that consecrate the association of religion and sport, posting touchdown prayers, making the sign of the cross before a free throw, or engaging in a pregame prayer or moment of silence. Every Sunday an estimated 4,000 minor and major league players, coaches, trainers, and managers participate in pregame Baseball Chapel sessions like the one that upset the Seattle Mariners' general manager (Rotenberk, 1992). In fact, its pregame religious ritual has become so popular that Baseball Chapel, Inc. has been organized with a formal administrative structure and a newsletter. Some of these players also participate in weekly Bible studies. The most visible and frequent manifestation of religion in athletics is the use of prayer.

Prayer

Athletes, like many of the rest of us, are likely to appeal to a higher authority in stressful situations. They seek reassurance, protection, renewed competitive drive, or a competitive edge (Eitzen & Sage, 1993: 201). We have also seen many cases on television where athletes and coaches thank God for their victory. Marbeto (1967), in one of the few studies of the perceived impact of religion on sport, found that just over 50 percent of the athletes interviewed prayed at some time prior to or during an athletic contest and most of these athletes believed that this practice indirectly affected the outcome of the game in which they were playing.

Prayer often is utilized, usually along with the national anthem, to consecrate events or to dedicate the performance of athletes to the glory of God. Prebish has cited a classic example that was delivered at a World Hockey Association All-Star game in 1976 by Father Edward Rupp:

> Heavenly Father, Divine Goalie, we come before You this evening to seek Your blessing. . . . We are, thanks to You, All-Stars. We pray tonight for Your guidance. Keep us free from actions that would put us in the Sin

POINT-COUNTERPOINT

DOES PRAYER BELONG IN THE LOCKER ROOM?

YES

Prayer contributes to team unity and reminds players of elements of the Protestant Ethic, such as hard work, perseverance, and sacrifice, that are necessary for athletic success. Prayer also gives the players a sense of the competitive edge that comes from knowing that God is watching over them, protecting them from injury, and guiding them toward their goal of victory. Prayer reminds players that competition should take place within the rules and emphasize sportsmanship. Prayer provides inspiration, a sense of purpose, and confidence, which lead to maximum performance.

NO

Prayer in the locker rooms of public school teams violates the constitutional separation of church and state and impinges on the religious freedom of athletes who do not practice religion or who believe that their faith is not represented by the prayer that is offered. Locker room prayer is a manipulative team-building device that has no relevance to the practice of genuine religion. It has little meaning to players when they lose or feel compelled to cheat or hurt their opponent to win. Prayer is part of the ritual to prepare athletes to compete, and it is not meant to lead to genuine religious practice or belief.

How should this debate be resolved?

> Bin of Hell. Inspire us to avoid the pitfalls of our profession. Help us to stay within the blue line of Your commandments and the red line of Your grace. Protect us from being injured by the puck of pride. May we be ever delivered from the high stick of dishonesty. May the wings of Your angels play at the right and left of our teammates. May You always be the divine center of our team, and when our summons comes for eternal retirement to the heavenly grandstand, may we find You ready to give us the everlasting bonus of a permanent seat in Your coliseum. Finally, grant us the courage to skate without tripping, to run without icing, and to score the goal that really counts—the one that makes each of us a winner, a champion, an All-Star in the hectic Hockey Game of Life. (1992: 47)

Coaches have been sanctioned for their efforts to get athletes to take part in religious rituals or convert them to Christianity. Football coach Ray Dempsey of Memphis State University was charged by the American Civil Liberties Union (ACLU) with withholding athletic scholarships and selecting starting players on the basis of players' religious beliefs. Tennessee high school football coaches ignored their state attorney general's ruling that team prayers were illegal because they violated the constitutional separation of church and state. These coaches felt that their players would not be properly prepared if they went on the field without having said the ritualistic pregame team prayer.

The ACLU also pressured the University of Colorado and its outspoken football coach, Bill McCartney, to discontinue religious practices, such as team prayers, team devotionals, and biblical references during practices at this public university. McCartney caused additional controversy by using his public platform as a successful coach to blend his religious views with his fervent opposition to abortion and homosexual rights (Monaghan, 1992). The potential impact of McCartney's beliefs is described by one player:

> Some people felt he played favorites with it. It was the guys who weren't playing who said it: 'This isn't fair, I'm not religious so he's not playing me.' Because of the way he incorporated his religion, I thought it might affect a player's decision to come here. Coach McCartney had his beliefs and he told them to you. He expressed his strong belief in Jesus Christ but that is as far as it went with me. I thought with all the bad rumors about it, it might have an effect on recruiting, that guys would think if you're not a religious guy, you don't fit in here. (Simpson, 1985)

Although a number of church-affiliated institutions of higher education have used sport to further their missions as universities, church affiliation has not insulated them from the problems of intercollegiate athletics (Deford, 1993a). They also have had to recruit nonbelievers of their faith in order to remain competitive on the athletic field. In some cases, such as Southern Methodist University, church affiliation did not prevent a series of deviant practices, which resulted in severe NCAA sanctions. Leaders of these church-affiliated academic institutions generally have had little impact on the patterns of corruption in college athletics. Religion has not been able to change the system. The missionary zeal at church-affiliated universities with big-time athletic programs has tended to focus on converting individual athletes rather than on reforming the college athletic system.

MAGIC AND SUPERSTITION

Prayer can be interpreted as an appeal by athletes and coaches for magical intervention to reduce the uncertainty of game outcomes or to enhance individual or team performance. *Magic* and religion both assume that a supernatural power influences events, but religion is oriented to larger transcendental and personal issues centered on sacred elements of life and ritual. **Magic** belongs to the realm of the profane, focusing on practical or utilitarian answers to everyday problems and situations (Eitzen & Sage, 1993). Gmelch (1971) used his experience in baseball to identify three forms of magic found in sport: *rituals* (aimed at manipulating people, supernatural forces, or nature to produce a desired outcome); *taboos* (referring to avoided activities); and *fetishes* (attributing special powers to objects) (Leonard, 1993: 392). *Superstition* is a belief or action that reflects a desire for magical intervention or that is intended to call forth magical responses.

The use of magic is more likely to occur in situations where the outcome is uncertain or subject to chance (Malinowski, 1948). Magic is called on to prevent the possibility of failure or to help deal with potential danger. Appeals to magic and superstitions are coping mechanisms that help lessen or remove the fear and anxiety associated with the unknown. These practices also offer rationales or excuses for undesirable outcomes or losses.

Superstitious beliefs and practices are employed to gain an advantage over the competition. They are meant to bring the magical intercession of supernatural forces on behalf of a player or team needing help with decisions by referees and umpires or in stressful or uncertain situations, such as being at bat with the bases loaded or facing a fourth down and long yardage late in a close football game. Magic and superstition can serve the athlete in the same way as performance-enhancing drugs, by giving an apparent advantage over the competition when all else—skill, strength, speed, and coaching—seem equal or inferior.

There are many examples of superstitious practices by athletes seeking control over unseen forces that could affect their performance or the outcome of the contest (Leonard, 1993: 393). These practices are usually not random or sporadic; they often take on the status of ritual and are employed in a standard and consistent manner in specific types of contexts. These **rituals**, or behavioral sequences, are *repetitive*, occurring over and over in certain contexts; *stylized* in that the action is formal, not spontaneous; *sequential*, where there is an orderly progression from start to finish; *nonordinary* as distinct from everyday routine activities and clearly set apart from routine; and *potent* or believed to be powerful and capable of controlling supernatural forces or beings (Womack, 1979).

Many players engage in day-of-the-game rituals or behavioral sequences and would never consider varying these rituals except in the most unusual circumstances, such as a lengthy losing streak. These rituals help players focus on the upcoming game or contest and deal with stress and anxiety. They direct individual energy and motivation to team goals, at the same time highlighting individual needs in the face of pressures toward group conformity (Womack, 1979).

The need to control uncertain or unpredictable outcomes appears to be more important to competitors at higher levels of competition. Therefore, it makes sense that superstitious behaviors are more likely to occur at the collegiate and professional levels than at the levels of youth sports or high school. Rituals and other superstitious practices also are more likely to be used in more difficult or uncertain sports roles and sit-

SPECIAL FOCUS

Superstitious Practices in Sport

There have been many interesting and colorful superstitious rituals, taboos, and fetishes in every era in sport, and they reveal the wide variety of magical and superstitious practices that athletes, coaches, and teams have used to deal with uncertainty and anxiety. George Gmelch noted that he engaged in a ritual of eating chicken daily at four o'clock, keeping his eyes closed during the national anthem, and changing his sweatshirt after the fourth inning to maintain a hitting streak. Turk Wendell, Chicago Cubs relief pitcher, brushes his teeth between innings. Ozzie Smith of the St. Louis Cardinals does a back flip at the start of each game. Peter Adler, a well-known sport sociologist, became an academic adviser and counselor to a Division I-AA basketball program because the coach felt that his presence brought the team luck or gave them an advantage other teams did not have. Wade Boggs of the Boston Red Sox and New York Yankees only eats chicken on game days. John Wooden, legendary UCLA basketball coach, looked for hairpins to stick into a tree during game-day walks on campus. To these ritualistic practices, we can add a variety of well-known taboos, such as not mentioning a no-hitter as it unfolds, and fetishes, such as crucifixes, coins, and rabbit's feet. These practices and objects achieve the status of magic or superstition through a process of conditioning in which they become associated with a good or bad performance (Leonard, 1993: 392).

The collective nature of some superstitious practices is illustrated in the following story: early in the 1993 baseball season, the Los Angeles Dodgers brought sod to stadiums with artificial turf in the belief that if the players touched or walked on natural grass turf, they would continue their winning streak that had not included any wins on artificial grass. In every stadium with artificial turf the team ordered sod from a local landscaper to be placed at strategic locations around the visiting team dugout. Each player would walk on the sod when entering or leaving the dugout. The Dodgers continued to play with considerable success for the few weeks that sod was imported on road trips.

Gregory and Petrie (1975) studied the superstitious practices of several Canadian intercollegiate athletic teams and found that respondents utilized over 900 superstitions, which the researchers classified into forty categories. Many of these superstitious practices related to eating, clothing and jewelry fetishes, equipment manipulation, behavior in the dressing room, and game-day routines. The superstitious practices are carried out by individuals, such as a dressing sequence before a game, and others are carried out by entire teams, such as a particular pregame prayer or song.

Buhrmann and Zaugg (1981) found that female athletes were more likely than their male counterparts to engage in superstitious behaviors, such as warmup rituals. They also discovered that the higher the level of sports competition, the more likely participants were to implement superstitious practices. Similar results were obtained in another study of Canadian athletes conducted by Neil, Anderson, and Sheppard (1981). Level of involvement was viewed as more important to explaining the nature and extent of superstitious behavior than gender or type of sport.

uations, such as hitting, pitching, and field goal kicking rather than fielding or blocking (Gmelch, 1971).

Wrigley (1970) pointed out that magic was used in athletic activity to enhance morale, to reduce anxiety and tension, and to compensate for a fear of failure. The inappropriate use or elimination of a routine ritual, such as a song before a game, could also be used as a justification for a loss or particularly bad performance. Since the structure and excitement of sport depend on the uncertainty of the outcome, magic and superstition are a natural outgrowth of the basic nature of sport.

CONCLUSION

The combination of sport and religion has not substantially changed sport or society. Some individual athletes, coaches, and sports administrators have been influenced by a religious impulse to rid sport of corruption or to implant strong ethical principles in the realm of sport. Yet on the structural or institutional level, the association of religion with sport has not seemed to change the direction of sport. Violence, illicit drug use, racism, sexism, various forms of cheating, and other kinds of excesses and deviance remain in the sports.

Athletes and coaches touted as devout Christians and exemplars of muscular Christianity typically have been revealed to be as flawed as their ostensibly less devout and exemplary counterparts in sport (Bayless, 1990; Hoffman, 1992b). Even the omnipresent "Rainbow Man," Rollen Frederick Stewart, was not the saint implied by his display of "John 3:16" placards at numerous nationally televised sporting events. Stewart was recently arrested for kidnapping a hotel maid and making terrorist threats (Malnic, 1993). It often is difficult for evangelists in sport to live out the moral and religious admonitions they profess in the sports arena.

The purposes and meanings of religious rhetoric and displays and the importance and influence of religious beliefs in sport are difficult to assess without more systematic evidence. It appears that the secularization of American society has produced a variety of instrumental and expedient uses of religion by sport and of sport by religion. The mythic spectacles of sport, such as the Super Bowl, have given sport the appearance of a religion, but we must conclude that sport is only a quasireligion—lacking the spiritual and sacred qualities of genuine religions. Its status as a putative form of religion is enhanced by the fact that the secularization of society has diminished the potency and appeal of religious institutions.

SUMMARY

Sport and religion have been intimately intertwined since the very beginnings of organized physical activity. Religion initially dominated sport, but as the world moved in the direction of urbanization, industrialization, and secularization, religion lost its hold on many aspects of social life, including sport. In modern times, sport has a greater impact on religion, particularly in terms of scheduling and program, than the reverse. In fact, the assertion is often made that sport has replaced religion or is the functional equivalent of religion in modern society.

There are many social and cultural similarities between sport and religion in modern society, but sport lacks the spiritual and deistic elements found in religion. Sport may be most accurately described as a civil religion or quasireligion. Today we find religious organizations using sport and sports participants turning to religion in various ways. There are parallels between appeals to religion and appeals to magic and superstition to deal with uncertainty in modern sport. Despite the religionlike qualities of sport and the many professions of religious belief by sports participants, it appears that religion has had little impact on the ethical or moral dilemmas of highly commercialized modern sport.

CHAPTER 5

Socialization and Sport

THE PROCESS OF SOCIALIZATION

Socialization is a lifelong process through which individuals develop and shape their social identities, create a sense of self, and learn how to participate in social roles and relationships (e.g., Nixon, 1976: 9–10; McPherson, 1981; Smith, 1987: ch. 3). Socialization requires internalizing the expectations of others, particularly significant others such as parents, coaches, and good friends, and acting in conformity with these expectations without even thinking about what is being done.

Socialization is a process of learning how to accept and conform to cultural and social constraints embedded in social norms and roles. The conceptions of norms and roles used to guide behavior can be thought of as *social scripts.* Socialization is a process in which we subtly reshape or transform the perceived constraints in scripts by interpreting them somewhat differently and by renegotiating the boundaries of acceptable behavior. Yet since socialization involves the learning of preexisting cultural blueprints and social scripts, it is not necessary for individuals to invent new values, norms, roles, or relationships each time they confront a new person or situation. When cultural blueprints and social scripts are learned, society's members are able to act in ways that make sense and are predictable to others. Social order is possible precisely because people behave generally as they are expected to act and thereby at least loosely conform to the social norms for established roles, relationships, or social networks. In this process of interaction with others, persons learn what they must know in order to be able to survive and function in society (Farley, 1994).

Socialization is usually considered in the context of childhood learning, but it also takes place any time a person enters a new social situation or role or finds it necessary to adjust to a changing cultural, social, or physical environment. It is a continuous process throughout the life cycle in which childhood socialization provides the essential building blocks of our cultural blueprint, social scripts, and self-concept. Adult socialization involves learning specific new roles, values, and self-concept definitions that may be added to or replace existing definitions of role and self acquired during childhood and adolescence (Mortimer and Simmons, 1978). This learning process is called **resocialization**. A related process is **desocialization**, which is the shedding and unlearning of roles. Role and value learning, awareness of self, and establishment or maintenance of one's identity or location in the social structure are dynamic processes in which passage through the life cycle, new experiences, and the reinterpretation of prior experiences can modify a person's roles and self concept (Turner, 1978, 1988). This dynamic process is most effective when it takes place in the context of a **primary group**.

Primary groups, which are characterized by intimate and enduring affective relations, are the most powerful settings for socialization. Examples include families, peer groups, and sports teams. Along with these important **agents of socialization**, the mass media and prominent figures of television, radio, movies, magazines, and newspapers may play an important role in influencing beliefs, behavior, and self concept. Athletes, coaches, and sports groups become powerful socialization agents because people place high value on sport and become strongly attached to the people and groups with which they share sport experiences. For impressionable children and youths who lack effective **role models** or significant others elsewhere, experiences with coaches and teammates in youth sports or school athletic programs can have a substantial impact on their values, attitudes, role conceptions, identities, aspirations, goals, and self-esteem. Thus, it is possible to be socialized *through* sport and learn things that apply to roles, relationships, and experiences outside sport as well as to be socialized *into* sport by being encouraged to participate in sport because of the perceived benefits that participation will provide.

THEORIES OF SOCIALIZATION

There are several theoretical perspectives used to explain socialization. One of the earliest and still prominent is *behaviorist theory.* This is a stimulus-response perspective, which assumes that people repeat behavior that is rewarded or pleasurable and eliminate behavior that is punished or painful. Behaviorists such as B. F. Skinner believed that the mind was not worthy of scientific study because its functioning—for example, in memory and perception—could not be directly observed. He proposed that observable stimuli and behavior were the proper focus of "behavioral scientists." Indeed, in strict behaviorism, there is no focus on self because it is considered to be a product of the mind that cannot be directly observed. *Social learning theory,* integral to study of hero worship and other aspects of socialization through sport, is a derivative of behaviorism.

Social Learning Theory and Role Learning

The social learning perspective uses concepts such as *modeling, imitation,* and *vicarious learning* to explain how we learn new behavior or modify existing behavior (Bandura and Walters, 1963; Bandura, 1969; McPherson, 1981). Modeling begins with exposure to the behavior of **significant others**, whose views, judgments, opinions, and actions shape the socialization of others for whom they are significant. This exposure is followed by the retention, imitation, and reproduction of that behavior (Leonard, 1993). Eventually that behavior (e.g., honesty, courtesy, aggression) is repeated in situations unlike the context in which a particular behavior was originally observed. Thus, behaviors are observed, internalized, and exhibited later in appropriate situations as a reflection of basic role behavior. Behaviors are often not immediately reproduced. Role models can be well known and in immediate proximity to us, or they can be distant or indirect models who exert influence through the mass media or peer group interactions.

Athletes are an example of role models who typically exert their influence from afar, especially in the form of mass media images. The sports public, sports journalists, and coaches in the United States largely believe that athletes are good role models for children, which explains why they believe that behavior such as fighting by athletes is harmful for children to see (Miller Lite, 1983: 62, 101). Sports fans and sports officials in the United States typically would much rather see stories that portray athletes as embodiments of the Dominant American Sports Creed and the American Dream, because in their eyes these stories display athletes as virtuous role models. They also help reinforce the image of sport as a valued part of society. Parents and other adults are likely to encourage sports participation when they believe it will provide experiences that expose children to beliefs and behaviors they value.

Problems in social learning develop when the verbal and nonverbal behavior of significant others or role models sends conflicting messages or expresses conflicting expectations. For example, young athletes become confused when their sports hero uses recreational and performance-enhancing drugs, gambles on game outcomes, brawls on the field, or berates umpires and referees—all such behavior in contrast to messages of control, respect, and fair play. Conflicting messages create confusion about what is appropriate role performance. The portrayal of athletes as heroic role models has been challenged in a dramatic fashion by the trial of football star O. J. Simpson for the murders of his ex-wife and her companion. Part of the issue is to decide if Simpson is a celebrity famous for his athletic feats or a model of the American Sports Creed and the ideology it represents. Similar questions are raised when baseball greats such as Daryl Strawberry, Steve Howe, and Dwight Gooden continually

POINT-COUNTERPOINT

ARE ATHLETES GOOD ROLE MODELS?

YES

Athletes, especially at the professional level, have always been admired as models of virtue and success. Their celebrity status with young people gives them an almost unique opportunity to influence children and youths to stay in school, avoid drugs, refrain from violence, and practice responsible behavior. The involvement of star athletes in a variety of charitable activities in the community serves as an example to young people of the kinds of things that all responsible members of the community should strive to do. In addition and perhaps most important, star athletes demonstrate what can happen if children and youths dedicate themselves to a goal, such as a professional sports career, and achieve it despite staggering odds against such success. Especially with dysfunctional families, athletes are an important substitute for more traditional parental role models.

NO

Athletes are not responsible for the value training of young people in any society. Parents, teachers, and religious and civic leaders are more appropriate role models to socialize the children and youth in their community. Athletes are paid to entertain the public as they strive to perform as well as they can on the athletic field and win. They are not paid to be heroes representing a particular set of values for young people. It is unfair to expect an athlete to be a role model. This expectation places an additional burden on athletes, who already must confront a variety of competitive stresses in their occupation. Furthermore, many athletes are not qualified to serve as role models because they are too young, too immature, too self-centered, or too often involved in deviance. Sports is about physical competition; it is not about displaying model behavior or representing cultural values or ideals.

How should this debate be resolved?

lose battles to drugs and alcohol; Lenny Dykstra continues to play with a prominent chew of tobacco between his gum and cheek even in the face of considerable publicity about tobacco's harmful effects; Albert Belle of the Cleveland Indians is found using an illegal corked bat and is suspended for ten days; professional football players like Steve Enteman and Bryan Bosworth admit to using steroids to enhance their physical features and subsequently improve performance; Wade Boggs admits to marital infidelity and acknowledges that his mistress accompanied him on road trips; and college and professional athletes are involved in violent incidents, sometimes with guns, on and off the field.

Social learning theories can be classified as *functional* in that learning essentially conforms to existing role standards rather than negotiated guidelines. As a result, consensus and solidarity on core values are established and existing societal arrangements are preserved. Functionalists interpret this outcome as beneficial to society, whereas conflict theorists assert that perpetuation of existing relations means inequality and oppression.

Conflict Interpretation of Socialization

If socialization is viewed as a means of maintaining social order and social stability, then conflict theorists would argue that it reinforces structures of power, privilege, and prestige that favor the dominant classes or elites of society. That is, by teaching people how to conform to prevailing or dominant norms and roles in society, socialization reinforces a social order that is unequal and favors the most powerful and the most privileged. It supports existing social practices and ideologies that emphasize loyalty, obedience, control, greater good, and hierarchy. Those in power seek to mold the populace in their own image with behaviors, attitudes, and cultural beliefs that are supportive of the dominant structures and classes. In other words, socialization is seen as a process for dominant classes to exercise superiority or **hegemony** over subordinate classes in society, thereby giving cultural and moral ascendancy to the values, norms, and social practices favored by the dominant classes. Sport becomes a terrain upon which hegemony can be established by socializing participants to accept cultural beliefs, roles, and norms that are consistent with the dominant cultural beliefs, roles, and norms of the larger society. Thus, in socializing participants to accept the Dominant American Sports Creed (Edwards, 1973), sport is implicitly or explicitly developing a commitment to the values that are embedded in the societal ideology of the American Dream (Nixon, 1984). Sports participants may not embrace either set of beliefs, but the power of dominant groups or classes in sport and society often makes it difficult to escape the influence of these ideologies.

One of the socializing mechanisms that accomplishes the reproduction of the class structure is **social channeling**, or preparing children to fit into roles commensurate with their social origins. Therefore, class relations are reproduced from generation to generation. School sports have performed this social channeling function under the rubric of *athleticism* (Sage, 1990a: 197), first in elite nineteenth-century British boys schools (Guttmann, 1988: 73–74) and later in the United States (Nixon, 1984: ch. 3). Although originally disdained by school authorities, sports later became valued for their presumed character building function. Self-confidence, leadership skill, sportsmanship, grace in defeat, experience in competition, self-reliance, and inner control were a few of the valued traits deemed the result of sports participation. These qualities continue to be emphasized in the public schools and youth sports in the United States as elements of the dominant ideology of sport in the United States (Edwards, 1973; Nixon, 1984).

In American public school athletic programs, the emphasis on leadership that characterized athleticism in Britain shifted to conformity, obedience, and respect for authority. Thus, in the United States, athletes learned that good character was associated with doing what you were told. This orientation was functional in early twentieth-century America, which was undergoing rapid industrialization and waves of immigration (Rees and Miracle, 1988). American sport prepared youth to be compliant members of the labor force, accepting authoritarian leadership and fulfilling specialized roles in a division of labor (Ingham and Hardy, 1984). Social acquiescence to coaches became the basic guideline for sports participants. When such acquiescence and the related acceptance of the dominant sports creed occur, socialization tends to reproduce the dominant structural and cultural patterns of society.

Interactionist Interpretation of Socialization

Symbolic interactionist perspectives emphasize the creative capacities of social actors as they construct meanings for themselves and their social situations through social interaction (Lindesmith, Strauss, and Denzin, 1991). The combination of critical and interactionist perspectives emphasizes the tension between human creativity and freedom, called *human agency,* and constraints that are incorporated in the dominant socio-cultural patterns in society (Gruneau, 1983; Nixon, 1990b). Donnelly and Young (1988) proposed that dominant societal values may be transmitted, resisted, or refined through experiences in subcultures. We can either challenge or conform to the constraints embedded in established values, norms, and roles, or we may seek opportunities for individual expression in the context of various structural constraints that convey how we are *supposed* to act.

The sports experience becomes especially meaningful when participants receive feedback from significant others in close, personal relationships, for instance, with family members, friends, teammates, and coaches (Watson, 1976, 1977; Coakley, 1986, 1990). These kinds of relationships provide perspectives, interpretations, and evaluations that help sports participants to understand what their experience means. The process of receiving and exchanging meanings is dynamic and complex because it draws from the varied experiences of those who are interacting and is constrained by potentially different interpretations of what is supposed to be important and what is supposed to happen. This process of meaning construction not only affects role interpretations, it

SPECIAL FOCUS

Play and Games in Mead's Socialization Theory

George Herbert Mead (1934) used play and games in his symbolic interactionist perspective of how the self and role skills develop (Smith, 1987; Kearl & Gordon, 1992). For Mead, the self was a continuing process of interaction between the "I" of spontaneous and largely autonomous actions and the "Me" of self-conceptions from significant others and a generalized other. He noted the importance of the playful imitation of others' roles in early childhood as a means of preparing children for roles later in life. Play provides the opportunity for exploration and discovery, for fantasy, for experimentation in role playing and role taking, for problem solving, for learning coordinated activity in collective tasks, and for learning adult behavior and attitudes. At first limited, simple, and involving only a few significant others, role playing becomes more complex as children get older. Their capacity to take the roles of others becomes more sophisticated as children engage in more complex forms of interaction and social networks.

In moving from playing at roles to the game stage of development, children learn how to deal with a variety of role interactions in a relatively integrated manner. For example, in learning baseball, it is necessary to understand one's position—say, as a shortstop—and its relationship to all the other defensive positions on the field, to the offensive positions of batter and base runner, to the general and specific expectations of coaches, and to the purposes of the opposing teams and the game. Thus, although the rules and roles may seem vague, ambiguous, or incomplete at times to young players, playing baseball teaches them how rules and roles organize interaction. It also teaches young players a general or composite standpoint for viewing the game, representing the *generalized other,* that is different than the specific expectations of individual teammates. This understanding of generalized or community standards and purposes is a major element of the "Me" aspect of self and an important stage in the process of learning how to live in society.

also has the capacity to reshape ideas about self concept and self-esteem. Thus, from this perspective, learning a sport involves learning how the game is played, learning how to play one's part, and learning how good one is thought to be in this sport.

Symbolic interactionists focus on how feedback from others shapes our conceptions of who we are, where we fit in society, and what we are worth. Cooley's concept of the *looking-glass self* refers to the image and evaluation of self that are produced when we interpret how others react to us. In this conception of self, we are the person that we think (significant) others think we are. That is, the self is contained in our ideas about how others think about us, which are our mirror for seeing who we are. Cooley's looking-glass self notion may assume that we are more passive in reacting to others in symbolic interaction than we actually are, but it is important in drawing attention to the influence of others in shaping our self and behavior. For example, we can see from this perspective why a young baseball player who believes that his parents, coach, and teammates think he is good at this sport may get a boost in self-esteem from such perceptions. Of course, in sport, a person has numerous chances to perceive how others think of his or her ability, and research has suggested that sports abilities and interests are correlated with the development of a self-concept and self-esteem (Eitzen & Sage, 1993: 86). Conversely, a youth sports participant who is continually criticized and develops a low sense of self-worth as an athlete will have a difficult time performing well and may eventually decide to drop out. In such cases, an additional outcome may be at least a temporary drop in overall self-esteem, if success in the sport was important to the athlete or his significant others.

McPherson (1986) proposed that interactionist perspectives tend to be most applicable to aspects of socialization that occur at the micro level of society.

At this level, reality is "constructed" or "negotiated" through symbolic exchanges, and the self and roles are formed and transformed in everyday interpersonal interaction. Interactionists tend to see novelty and uncertainties in socialization that are addressed by the active and creative efforts of interacting individuals who are striving to make sense of themselves, their social relationships, and society. At the more abstract or macro levels of society, people learn values, roles, and identities that are defined in terms of membership in groups, organizations, communities, and larger social networks. Cultural and social constraints are likely to have a significant influence on socialization at this level. Social learning theory, role theory, reference group theory, and cognitive-developmental theories offer better explanations of socialization than interactionist approaches. Both micro and macro approaches are useful for a comprehensive understanding of the various dimensions of the relationship between socialization and sport. To understand this relationship fully, we will consider socialization *into* sport and socialization *through* sport.

SOCIALIZATION INTO SPORT

In most countries, especially the United States, there is a prevalent tendency to encourage persons, usually children, to participate in sport. A child's initial involvement will be affected by a variety of social and psychological influences, including the perceived and actual availability of opportunities; the nature of encouragement or discouragement from significant others such as parents, peers, and teachers; direct or indirect exposure to role models; and the child's view of the athletic role as desirable and attainable (Snyder & Spreitzer, 1989; Coakley, 1990). Research suggests that parental encouragement of sports participation increases as the level of sport rises—for example, from intramurals to interscholastic athletics—and becomes more prestigious. At higher levels of sport, athletic ability tends to become a prominent feature of a person's identity. Intramural athletics may perform a cooling-out function for those unable to make varsity athletic teams by helping participants adjust to more limited conceptions of their ability. Intramural participants receive more modest parental athletic encouragement and place less importance on athletic ability in their identity than do varsity sports participants (Snyder & Spreitzer, 1989: 87–88).

Kenyon and McPherson (1973) studied socialization experiences of Olympic athletes and found that almost all participated in sport at a very early age. They were defined as winners by themselves and others during this early involvement. In addition, significant others served as role models, especially for those who entered their sport while in high school. These athletes attended a school where the sport was valued by students, administrators, and others. Early family encouragement was replaced later by the influence of peers, coaches, and teachers (McPherson, 1981). Thus, the source of influence changes with the life cycle.

Social background and status factors such as social class, gender, race, and ethnicity influence major aspects of sport socialization, including the availability of sports opportunities, access to sports facilities, and encouragement to participate. For example, women generally have had fewer sports opportunities and less athletic encouragement than men; lower class members and racial minorities are less likely than upper class and whites to participate in exclusive club sports such as skiing and golf. Furthermore, geographical region and residence in a city, the suburbs, or a rural community can also affect a person's chances of learning particular kinds of sports and receiving athletic encouragement in general (Snyder & Spreitzer, 1989: 88).

The factors leading to initial involvement in a sport have to be supplemented by other influences to keep a person involved or to push that person to higher levels of participation. Stevenson's (1990) study of internationally competitive athletes revealed that these athletes were introduced to a sport by significant others, but that continued participation depended on other factors such as their perception of eventual success in that sport and the existence of a supportive network of fellow athletes and coaches. Commitment developed as the result of the combined influence of "entanglements" or personal relationships (for example, with a parent or brother) that promoted continued involvement in the activity, commitments or felt obligations to meet the expectations of others (such as parents or a coach), reputations or public and self-acknowledgment of achievements in the sport, and the production of a favorable identity with the sports role they were playing. Athletes are likely to choose to continue if feedback is favorable, opportunities are perceived to be available, and performance is seen as successful. In general, Stevenson's research lends support to the symbolic interactionist insight that continued participation depends a great deal on the reaction of others to participation and the interpretation of that reaction by the actor.

In general, the family tends to be the most significant socialization agent influencing a child's entry into sport. But very little is known about how varying family structures, family size, or family authority structures influence sports participation (Eitzen & Sage, 1993). This is particularly important today since so many families are headed by a single parent or represent reconstituted families from multiple marriages. The influence of parents on socialization into sport is important both for boys and girls, especially if the parent participated or continues to participate in physical activity (Greendorfer & Lewko, 1978). The father's role is more important than that of the mother in promoting sports involvement. Fathers tend to place greater demands on their sons than on their daughters to be successful in sport (Sage, 1990).

There are also differences in the socialization into sport by gender. For example, Ryckman and Hamel (1992) discovered that continued participation by adolescent girls in sport was based more on friendship formation and maintenance than on a desire to improve their athletic skills. The researchers also found that girls were more oriented to self testing and participation than to outcomes. This suggests that coaches strongly oriented to winning would have a difficult time motivating some women athletes. Watson (1975) found that boys identified fathers and peers while girls looked to mothers and coaches as significant to their participation in Australian youth sport.

Spreitzer and Snyder's (1976) study of U.S. high school girls' participation in sport found that women athletes were encouraged more than nonathletes by parents, although the parents of athletes were no more interested in sport than the parents of nonathletes. The athletes were also more likely to begin sport participation very early in life. Female athletes received considerable encouragement from their like-sex peers, more so than from boyfriends. Interestingly, the same pattern held for girls participating in music. Parental encouragement was the key factor in explaining participation in both activities. It appears that women are especially dependent on their opportunities in sport and on the influence of significant others (Theberge, 1977).

Many studies have demonstrated that considerable support for athletic participation must exist within the family before a girl will chose to be an athlete (Greendorfer, 1977; Smith, 1979). Men usually need support from a limited number of significant others such as a coach or peers, but women appear to need support from several relationships. Orlick (1973) identified parents as very important to their son's participation in sport. He suggested that parents serve as role models by virtue of their early participation and encouragement of their sons, but parents often respond differently to girls and boys. Boys are encouraged to participate in active pursuits, whereas girls are encouraged to pursue passive activities.

The socialization process differs in a variety of ways for males and females (Ryckman & Hamel, 1992). Women, for example, are more likely than males to organize their sense of self around the establishment and maintenance of personal relationships or affiliation (Gilligan, 1982). Affiliation could become a source of conflict for female athletes as they try to cope with the increasingly competitive and aggressive nature of sport at higher levels.

Very often women receive mixed messages about participation in sport. For example, fathers may encourage participation in sport but at the same time friends will send messages that such participation is "unfeminine." The result is an experience of **role conflict** or stress from differing and conflicting expectations from the roles of athlete and woman. Sage and Loudermilk (1979) reported that 26 percent of the women athletes they studied experienced role conflict. Role conflict is more likely to be experienced by females who participate in sports traditionally defined as male, which usually are team sports such as basketball or softball. Female participants in gymnastics, tennis, or swimming experience less role conflict. More recent research has indicated that high school students tend to believe that female athletic participation is acceptable as long as it does not challenge traditional ideas about appropriate feminine behavior (Kane, 1987). Thus, even though women have had more opportunities to participate in sports, including traditionally male ones, remnants of traditional stereotypes about female-appropriate and female-inappropriate sports seem to have survived (Snyder & Spreitzer, 1989: 191–197).

Variations in socialization experiences can also be attributed to race. Sailes (1985) found that parents tended to have more influence over white athletes than black athletes. Blacks were more likely to be influenced to participate in sport by persons outside the family, such as peers, coaches, and teachers. Blacks also generally had less access than whites to sports facilities and programs. Phillips and Boelter (1985) used the example of the high jump to show that as a result of inferior facilities or equipment, blacks were more limited in the ways they could demonstrate their athletic ability and pursue particular types of sports. Blacks did not use the "Fosbury flop" technique in the high jump, in which they had to land on their upper back or neck, because their schools could not afford inflated or padded mats. Instead, they used the roll or scissors techniques because they could be done with a sawdust landing. Sailes also found that blacks had a greater interest in sport than whites and were more likely to have black than white role models. Whites had role models of both races, but these models were of less importance to them than the black role models were to blacks.

Some research has suggested that birth order affects socialization into sport. First-borns may not have the same sports opportunities as later-borns. Nisbett (1968) found that first borns were more psychologically dependent on adults, had less freedom selecting activities, were more vulnerable to stress, and were less tolerant of pain. In addition, he found that first-borns were less likely than later-borns to participate in the physically riskier and more aggressive sports of hockey, football, and wrestling. It may be that parents are more protective of their first-borns than later-borns and are less likely to encourage participation in sport. Generally, however, there is no relation between birth order and sports participation even if that sport is of

Source: Elizabeth Crews/Stock Boston

FIGURE 5.1 Socialization into sport: teaching sport to kids.

the high-risk variety such as sky diving (Seff, Gecas & Frey, 1993).

Geographical location can dictate what opportunities are available (Rooney & Pillsbury, 1992). Youngsters in Indiana play basketball and those in Texas are more likely to play football. The cold climate of Minnesota is more conducive to hockey participation than the warmer climates of states such as California where hockey is only available in organized clubs with expensive facilities. Californians play water polo.

Although we tend to focus on how parents influence their children's socialization into sport, children may also influence their parents' sports involvement. Most research is unidimensional in focusing on children as the recipients of the socializing influence of their parents. It is possible, however, to apply a reciprocal effects idea to the sport socialization relationship between parents and children. This idea suggests that social influence in this relationship may go in both directions. That is, children may become involved in and learn about sports as a result of their parents' encouragement, and parents may become more involved in a sport or sports in general as a result of interaction with their children. There is research evidence supporting mutual influence between parents and children in sport socialization (Snyder & Purdy, 1982).

Snyder and Spreitzer (1989: 89) have dubbed the influence of children's participation on parental sports interest and knowledge *reverse socialization*. This may be particularly true for mothers whose sons become involved in sports and these mothers become "team moms" who help with refreshments, transportation, keeping team statistics, coaching, or become league officials. Snyder and Purdy (1982) found that reverse socialization is greater for mothers than fathers. Fathers are generally less influenced than mothers by their child's sports participation because they have had more prior sports experience than the women.

Sport socialization research has focused a limited amount of attention on younger children to discover how and why they initially entered sport (Greendorfer & Lewko, 1978; Greendorfer, 1990). This type of research usually targets elite or experienced athletes who are asked to recall the reasons they entered sport. Although this methodological approach may cast doubt on the results, this research generally has suggested that factors such as self perception and goal orientation are very important cognitive factors that combine with social factors such as the influence of significant others to affect how and why a child chooses to participate in sport.

Children who receive more favorable feedback from significant others about their sports participation are more likely to feel good about their participation and to feel that they can master the activity (see Harter, 1978; 1981; Brustad, 1992). The work of Phillips (1987) and Weitzer (1989) showed that parental expectations and evaluations are very crucial to a child's feelings of competence in an activity. For example, it was found that girls were more likely to participate in sport if their parents felt that their daughters had the ability to compete (Weitzer, 1989). Children who did not have a high opinion of their abilities had lower performance records and fewer expectations for future achievement. These patterns were reinforced when parental evaluations of children's competence matched the child's perception of competence.

Criticisms of oversocialized concepts of socialization have emphasized that socialization is not a matter of people as sponges who passively absorb influences from others. In assuming, for example, that children merely respond to influences from significant others or their social contexts, we could reach this conclusion. In fact, though, this approach fails to account for the influence that children may have on their socialization experiences. Hasbrook (1986) found that those youth sports participants who demonstrate considerable interest and skill will receive more parental encouragement than those who are not as skilled. Thus, the encouragement or discouragement of significant others may be mediated by the sports interest and ability displayed by young participants.

Significant others are significant and have influence over others because their relationships with others have the qualities of relationships found in primary groups. That is, relationships with significant others are powerful because they tend to be personal, informal, spontaneous, inclusive, and imbued with strong and intimate feelings (Nixon, 1979: 13–18). It is easy to understand from this perspective why parents, friends, and coaches are significant others. Since children and youth may develop intimate fantasy relationships with their sports heroes through extensive and intense media exposure, we also can understand why heroes or the romanticized images they represent can be significant others and have influence over the sports interests and development of young and aspiring athletes. Research by Lewthwaite and Hasbrook (1989) suggested that positive peer interaction and peer acceptance in sports activities are likely to make these activities more enjoyable for children and encourage them to continue pursuing them. On the other hand, their research also suggested that negative interaction with these significant others in sports would be likely to lead children to avoid these activities in the future.

Lewthwaite and Hasbrook examined peer interaction experiences for participants in formal sport, exercise, and informal games and found that young respondents could distinguish activity type. These respondents reported more negative peer interactions in sport than in exercise and play. Most of this negative treatment came from boys, not other girls, and it tended to emphasize physical skill deficiencies and performance weaknesses relative to other boys. Girls engaged in activities at home that they avoided at school when they were able to participate at home with friends whom they knew liked them and did not care about the level of their playing skills. Children who received criticism for their playing skills and were able to neutralize it were not deterred from future participation. They neutralized criticism in many cases by attributing it to the personality of their critic. Participants who internalized these criticisms, however, tended to discontinue involvement.

Most participants in Lewthwaite and Hasbrook's research said they had fun no matter what the context of the activity was, but anxiety levels were higher in formal sports activities than in exercise and informal games. This finding is consistent with research by Chalip and his colleagues (1984), which showed that the majority of participants they studied had feelings of anxiety during formal sports participation at school but did not feel such anxiety during play or exercise at home. They reported a greater incidence of embarrassment at school in formal physical activity than at home in informal play activity. These patterns of children's affective responses to physical activity are consistent with findings in other research, which generally has shown that children experience more positive feelings in informal physical activities than in formal sports activities, especially during physical education classes (Kirshnit, Ham & Richards, 1989; Kunesh, Hasbrook & Lewthwaite, 1992: 393). Thus, children are encouraged to participate in sport because of the beneficial outcomes of personality and skill development but actual experiences can be so negative that participation is discontinued or the level of participation will change. It is important to evaluate the socialization experience youngsters have while participating in addition to reviewing the influences that impact the initial decision to participate.

SOCIALIZATION THROUGH SPORT

Socialization *into* sport is a process that prepares persons for participation in sports roles, relationships, and networks. Socialization *through* sport is the effect of sports participation on participants' beliefs, attitudes, personality formation, skill development, and self-concept. The lessons learned from sport participation are presumably transferable to other social contexts, thereby helping persons to be able to participate in these contexts in successful fashion.A common belief about sport, which is incorporated into the Dominant American Sports Creed (Edwards, 1973), is that participation in competitive youth sports develops character and instills moral ideals of the culture. These beliefs have been widely accepted even though research evidence does not support this contention (Frey & Eitzen, 1991; Nixon, 1984).

Learning the realities of a culture or the exigencies of the real world is one of the outcomes associated with sports participation. Ritchie and Koller (1964), Bailey (1977), and Webb (1969) all found that the norms, values, roles, and skills learned in the process of playing had relevance for the real world. Lever (1978), for example, found that boys play different types of games than girls. Boys' games tended to be more organized and more age heterogeneous, involve larger groups, and last longer. Girls were less likely to learn the skills needed for occupational success as adults, such as decision-making acumen, competitive success, and participation in collective tasks in a large group, because their games included fewer and more homogeneous individuals who did not routinely engage in competitive activity. Thus, boys had an advantage over girls in preparation for adult life because of the different nature of their childhood games.

There is evidence from DuBois's (1986) research that the value orientations of young soccer players change over a season, but there is no evidence that these values are applied to roles outside sport or that they persist over adolescence and into adulthood. Furthermore, there is no clear or consistent evidence that sports participation is required for people to develop the qualities and learn the behaviors needed for adult occupational success (McPherson, Curtis, & Loy, 1989: 52–53). Questions about generalizable and lasting effects tend to apply to the findings regarding socialization effects of sports participation. Sport is just one avenue for children to express skills and attributes acquired in growth and development (Coakley, 1986, 1990). Despite the claims of the Dominant American Sports Creed and similar cultural assertions about the benefits of sport, there is no consistent pattern of findings showing that sport *generally* builds character, contributes to moral development, develops a competitive or team orientation, makes good citizens, reduces prejudice, develops leaders, enhances social adjustment, affects the self image, or creates valued personality traits (Dubois, 1986; Fine, 1987; Coakley, 1987b; McPherson, Curtis & Loy, 1989; Rees, Howell & Miracle, 1990; Frey & Eitzen, 1991). It may be that sports participation causes one or more of these changes for certain participants, but it is unclear from extant research what conditions of participation produce them or what types of participants are affected by sport in these ways.

Fine (1981; 1987) used ethnographic techniques to study how value themes of effort, sportsmanship, teamwork, and dealing with wins and losses were transmitted and received in Little League baseball. The verbal expressions related to these values have greater impact if they are consistent with the player's self-concept and his or her ability to understand the situation. In some cases, the athlete will verbalize the moral code knowing that this is what the coach wants to hear. Fine proposed that when Little Leaguers were being socialized into the "moral order of Little League baseball," they were also being taught types of rhetoric that could be used to construct moral meanings in social situations in general (1981: 188). Of course, the moral beliefs and rhetoric—or "vocabulary of motives"—they learned reflected the values of their coaches and the other adults who ran Little League. The players did not necessarily internalize the values they learned in baseball into their self-concept, but their experience in Little League taught them how to manipulate the coaches and thereby enhance their status in the coaches' eyes by mouthing the kinds of value statements coaches liked to hear.

In a classic study of socialization into and through sport, Webb (1969) asked third-, sixth-, eighth-, tenth-, and twelfth-graders in public and parochial schools in Michigan to rank the relative importance of playing fairly, playing as well as possible, and winning. He found that winning was consistently ranked last and that playing well replaced playing fairly as paramount in importance as the children got older. In addition, winning became more important with increasing age for males. As these students were exposed to the more competitive and achievement-oriented world of adults as they got older, their value orientations on play and games became, in Webb's terms, more professionalized. Maloney and Petri (1972) and Mantel and Vander Velden (1974) also found that participants in organized competitive youth programs scored higher on Webb's professionalization scale than did participants in a recreational or instructional program. Building on the work of Webb, DuBois (1986) asked participants in a youth soccer program to rank a list of values prior to the start of the season and also at the end of the season. He discovered that children in instructional, noncompetitive programs valued fun, skill development, affiliation, fitness, and ethical behavior over winning.

In general, current research suggests that sport does relatively little to socialize the behavior, values, or personality of participants. If athletes are different from nonathletes, it is probably a result of the selection of particular kinds of people to sport rather than because sport changes them. For example, disciplined and compliant youth may find the structure inherent in athletics to be attractive, whereas nonconventional youth may have been filtered out of sports because of self-selection, eligibility rules, coaching preferences, and opportunity structures (Spreitzer, 1992). In a national study of high school students, Spreitzer (1992) found no association between sports participation and self-esteem but a modest association between participation and educational attainment. This research supported a selectivity hypothesis by identifying a filtering process in which athletics attracts participants who are more advantaged in social background than their nonathlete counterparts and have higher levels of self-esteem, academic achievement, and cognitive ability.

Learning Gender Roles through Sport

If sport has socializing effects, the most evident of its effects concerns the development of gender perceptions. It is difficult, however, to separate the influence of sport from others in the socialization process. What is clear is that sport traditionally conveys strong messages about masculinity and femininity.

POINT-COUNTERPOINT

DOES SPORT BUILD CHARACTER?

YES

Participation in sport is an excellent way for young people to develop social skills, to learn to participate in cooperative group activity, to develop an appreciation of the nature of competition and success and the value of hard work, and to learn how to relate to adult authority outside the home. The social skills and personality traits learned in sport enable athletes to become responsible citizens and successful members of their occupation and community. Sport represents solid moral values of the Dominant American Sports Creed, and those directly exposed to these values are destined to develop good character.

NO

Sport does little to change the social skills and personal qualities and behaviors of athletes. The skills, traits, and behaviors displayed in sport reflect what athletes have learned from powerful socialization agents outside sport. Sport often screens out the troubled and troublesome youths who need help, selects athletes from the middle class or mainstream, and sometimes encourages tendencies, such as violence and drug abuse, in athletes who originally had no inclination to engage in such behavior. The emphasis on winning at all costs teaches athletes to cheat, hurt their opponents, and do whatever is needed to win.

How should this debate be resolved?

According to Messner (1992: 16), modern sport is a *gendered institution* that is constructed by men to reinforce their power over women. Sport has traditionally been a male world in which it was natural to define masculinity in terms of competition, control, aggression, strength, and skill. The structure of sport provided opportunities for boys to affiliate with other boys and adult male coaches or officials without having to establish intimate or emotionally expressive relationships. In other words, boys could interact and associate, but they did not have to be "relational." Messner (1992) argues that boys also internalize a concept of masculinity in sport that is extremely homophobic and sexist. Fine (1987) identifies this phenomenon in his study of Little League baseball. Boys acquire a narrow definition of masculinity at the expense of women, who become objectified by the boys, and of expressive males, who show weakness by displaying their feelings or preferring relational associations rather than ones of power and prestige. As boys get older, their sexist and homophobic feelings become a source of male bonding on sports teams. The bonding males experience in sport is limited and conditional, however.

Bounded relationships based on affiliation but not emotional intimacy characterize the associations of males in sport. The nature of this association is conditioned by the fact that athletes are always in a state of **antagonistic cooperation**. That is, while a strong emphasis is placed on teamwork and cooperation, an underlying aggressive competition takes place among team members for positions, records, attention of the coach, and other indicators of success on the field. This kind of structured relationship allows men to develop a task-oriented bond, but it prevents intimate associations.

Sport is a very important setting for the bonding of men into what Farr (1988) has designated as GOBS or Good Old Boys Sociability groups. GOBS groups accentuate male dominance and male bonding to the exclusion of women. The groups studied by Farr formed a closed male alliance that celebrated masculinity, usually in the context of play or sports activities. These patterns were viewed as simply an extension of childhood play experiences (Farr, 1988: 276). Interestingly enough, within the GOBS subculture, expressions of masculinity were exaggerated compared to similar expressions outside the subculture, and in the company of women or non-GOBS men. Clearly, fellow players are the most important association and extrasport friendships are secondary, especially when they involve women.

Sport also socializes men to believe that "masculine" characteristics should be valued more than "feminine" characteristics (Sabo, 1985). Men learn to accept male dominance as the apparent natural order of things, which reinforces a male-dominated social order. Dunning (1986) observed that rugby was a "male preserve" in which men could mock and vilify women. Men will treat women in this manner when they feel their position or self-image being challenged by the gains in status and power women have made in various realms of society. In Dunning's view, many boys are encouraged as athletes or fans to develop values that define masculinity in terms of competition, violence, supremacy, winning or status aggrandizement, and sexual aggressiveness. In addition, they learn to avoid displaying "feminine" qualities, such as emotional expressiveness, nurturance, and compassion, for fear of being labeled a wimp or worse. In sports environments of hypermasculinity, one of the most devastating criticisms that can be leveled at a male athlete is to be compared to a girl—as in, "You run like a girl" or "You throw like a girl." Men adopt a superior attitude when it comes to women, often treating them like objects to be used, abused, and tossed aside if necessary.

AVERSIVE SOCIALIZATION IN YOUTH SPORT

At the beginning of the twentieth century, sports programs for children and youth in the United States were mainly organized and run by professional educators in the schools (Nixon, 1984: 33). By the 1930s, however, the value of highly organized sports competition for preadolescents had been questioned frequently and compellingly enough to lead American schools to discontinue sports programs in the schools. Boys' clubs, YMCAs, churches, and other volunteer organizations created sports programs for children because many disagreed with educators' evaluation of the values of youth sport. In addition, a number of independent sports programs and leagues for children, such as Little League Baseball and Pop Warner Football, were created at this time. Nonschool youth sports programs currently have approximately 35 million participants in the United States and 2.5 million in Canada (Eitzen & Sage, 1993: 71). Little League baseball is the most popular of these programs worldwide, with 20,000 leagues in over thirty countries and more than 3 million participants each year. The fastest-growing sport over the past decade has been soccer. Millions of other young people participate in YMCA, Junior Olympic, community recreational, and a diverse array of sport-specific programs in North America. Girls now make up 40 percent of the youth sports participants in the United States and about 27 percent of the participants in Canada (Eitzen & Sage, 1993: 71). Most of the participants are concentrated among preadolescents, ages six to thirteen. The amount of participation begins to decline in adolescence, with high attrition rates across all sports during these ages in the United States and in other countries. Kleiber and Roberts (1981) estimated that 80 percent of all youth withdraw from organized sport when they enter adolescence, and this pattern of withdrawal is even more dramatic for females (Duquin, 1978).

Involvement in sport has benefits but can cause problems for participants, such as encouraging deviant behaviors and discouraging participants from wanting to continue. **Aversive or negative socialization** outcomes from sports involvement include an exaggerated or unrealistic sense of self; excessively aggressive behavior in the pursuit of success; and learning cheating, violence, other forms of deviance, and a win-at-all-costs philosophy (e.g., Bend, 1971; McPherson, Curtis &Loy, 1989: 53–54, 55; Snyder & Spreitzer, 1989: 89–95).

In a study of adults and their participation in corporate physical activity programs in Japan and Canada, Yamaguchi (1987) found that those persons who had pleasurable experiences in past involvement would be more likely to participate as adults in sports programs. In other words, childhood experience in sport will impact lifelong participation patterns. Conversely, negative experiences may be so profound that a person's self-worth can be severely devalued, especially in the case of men who depend on sport as a source of social recognition, popularity, and personality formation. Aversive sports experiences make sport less enjoyable and are likely to lead participants to lose interest and drop out (Snyder & Spreitzer, 1989: 89–95).

Dropping Out of Sport

Suffering the public degradation or humiliation of doing poorly in sport can create anxiety and distress for athletes, especially at young ages. A series of self-perceived public failures in sport can damage an athlete's self concept and induce the athlete to quit. Ball (1976) described two processes that screen people out of sport. First, participants are so degraded by reactions to their involvement that they are unable to develop an identity as athletes. Having been regularly yelled at, criticized, or ridiculed by coaches, parents, or teammates, sports participants seek their identity or enjoyment in other activities. Second, participants are "cooled out" by a strategy of downward mobility or assignment to a lower status in sport, such as being traded to a noncontending team, relegated to a substitute role, or assigned to the minor leagues.

Aversive socialization experiences can result in a loss of interest in lifetime participation in sport. This is particularly true in the treatment of substitutes or bench warmers. This badge of failure may become a self-fulfilling prophecy and can leave at least a temporary imprint on the "failed" athlete's general self-concept.

Another factor that leads to withdrawal is the declining experience of success. People tend to continue activities in which they experience success. Thus, when winning is viewed as the primary or exclusive measure of success in sport, as it often is at higher and more professionalized levels, it tends to foster continued participation in sport. Brown (1985) proposed that the cost-to-benefit ratio is perceived as increasing for those who continue to expend considerable energy and gain little in return. In contract, those who continue to experience success enhance their commitment to the role and decrease their interest in other activities. Greendorfer and Hasbrook (1991) applied the social learning paradigm in their study of children in physical activity settings. They found that those children who had negatively valued personal attributes—if, for example, they were short, overweight, or unskilled—received little or no encouragement supporting continued involvement from socializing agents such as peers and coaches. It wasn't long before these children dropped out of sport.

Children lose interest in sport and drop out for a variety of reasons, including an overemphasis on winning and competition; excessive performance expectations from coaches; negative skill evaluations from coaches; punitive coaches; hard physical training; excessive parental pressure to participate and succeed; insufficient playing time; conflicts with other activities; lack of skill improvement; boredom; and lack of fun (Orlick, 1973, 1974; Orlick & Botterill, 1975; 1974; Gould, 1987; Snyder and Spreitzer, 1989: 89–95; Barnett, Smoll & Smith, 1992).

Seefeldt and his colleagues (1989) found in a survey of 14,000 youth sports participants that problems with coaches, their unrealistic expectations for players, and their harsh and unfair treatment of players made up half the list of ten reasons children quit sports. Barnett, Smoll, and Smith (1992) found that children were more likely to stay in sport if their coaches provided positive and reinforcing encouragement as well as technical instruction. They studied this effect by comparing a group of coaches who went through a Coaches' Effectiveness Training (CET) program with a group that did not receive the training. The CET-trained coaches had a much lower rate of attrition, and this relationship persisted when the won-lost record was controlled. Thus, being on a losing team was not a predictor of dropping out in this study. The CET coaches were better liked and were perceived as better teachers. This finding implies that a supportive coach can offset the effects of losing. Such coaches, of course, are unlikely to be those who associate success or fun with winning, and this may be a key to their ability to keep their athletes involved in sport.

Seefeldt and Gould (1980), Scanlan and Lewthwaite (1986), and Wankel and Sefton (1989) found that most boys and girls participated in youth sport to have fun, to improve skills, to be with friends, to improve their physical fitness, and to experience a sense of excitement. The boys and girls discontinued their participation because they became involved in other activities, were no longer interested in competition, did not receive enough playing time, or did not

SPECIAL FOCUS

The Little League Syndrome

Figler (1981) described two forms of dropping out of sport, which he called washout and burnout symptoms of the **Little League syndrome**. He also described a pattern of excessive and dysfunctional competitiveness, which he called the *superstar symptom*. In the *washout symptom,* participants drop out of Little League (and other types of sport) because they have been exposed to substantial criticism and discouragement, and they cannot or do not want to face any more of this treatment. They see others as viewing them as not good enough and may believe this about themselves, too. In the *burnout symptom,* dropping out is a result of too much success too early rather than too little. Success often is accompanied by the pressure of expectations to continue doing well or to do better and better. Relentless pressure and the emotional and physical drain of continually training and competing may drive younger, or older, athletes from their sport. In the superstar symptom, an overdeveloped competitiveness is related to a compulsion to compare oneself with others and to dominate them and an intense need for approval and reassurance. Athletes viewed as superstars do not drop out, but they clearly suffer the adverse psychosocial consequences of their sports involvement. Athletes who engage in highly competitive and demanding sports realms in which only winning is acceptable and in which athletes are revered as long as they win, could become vulnerable to the symptoms of the Little League syndrome, depending on perceptions of their performance, the intensity of their commitment, and their exposure to and internalization of a winning as success philosophy.

like the coach. Wankel and Sefton (1989) studied boys and girls aged seven to fourteen participating in hockey, soccer, and baseball and found that intrinsic factors such as improving skills, personal achievement, and the experience of participation were the most important reasons for being involved in youth sport. Secondary reasons included being on a team and associating with friends. Reasons related to the outcome of sports events, such as winning and getting trophies and other awards, were rated last in priority by both boys and girls regardless of sport. Thus, programs for youth with average talent that emphasize winning and extrinsic awards such as trophies are likely to be at odds with what these young people want from their sports experiences and undermine their enjoyment of them.

Research in youth sport has also shown that feeling better about one's competence and mastery of a sport is correlated with enjoyment of the sport and a greater likelihood of continued participation (Duda, 1987). Anxiety and frustration result when young athletes lack the ability or opportunity to develop a sense of competence or mastery in sports roles. Coaches and parents contribute to this anxiety and frustration when they fail to acknowledge the gap between expectations and ability and when they continue to make demands on performance based on unrealistic expectations of physical ability and emotional maturity. Wankel and Sefton (1989) found that sports were a source of enjoyment or fun for participants when skill development, realistic challenges, associated success with skill development, and a reasonable emphasis on winning characterized the activity. Thus, even though youth sports participants are unlikely to ignore the outcome, it is often not the most important factor explaining their participation, particularly for those who are not competing at the elite levels of sport. Sports can be organized to minimize aversive conditions if they emphasize aspects of the sports experience that are appropriate for the ability and interest levels of participants. One of the most important things to offer participants to keep them motivated is the opportunity to play. A high proportion of youth sports participants would rather play for a losing team than sit on the bench for a winner.

There is a tendency to assume that when a child—or adult, for that matter—withdraws from sport, the

SPECIAL FOCUS

What Kids Want and Do Not Want from Sports

An American Footwear Association study (1991) revealed that the most important reason for sports participation among boys and girls in general was fun *and* that even for the most dedicated athletes winning ranked behind self-improvement and the excitement of competition as reasons for participation. Among boys in general, winning was ranked eighth out of twelve reasons; among girls in general, it ranked last of twelve. This same study also revealed eleven reasons given for discontinuing participation, ranked in order of importance: loss of interest; not having fun; too much time; coach was a poor teacher; excessive pressure or worry; wanting a nonsports activity; being tired of it; needing more study time; coach playing favorites; boredom; and overemphasis on winning (American Footwear Association, 1991: 5).

When asked what changes in sport they would recommend, the boys and girls in this study mentioned making practices fun, providing chances to play more, reducing conflict with social life and studies, and improving coaches as teachers and understanding listeners. The report noted that those who were most interested in improving their skills would be the most likely to continue in sport. Those who were involved because of extreme pressure from outside sport—for example, from parents—generally did not want to practice but were more interested in the sociability aspect of the activity. These latter young people were the ones most likely to drop out. Another group called the "image-conscious socializers," was likely to continue participation because of the rewards of approval from significant others, trophies, recognition of their talent, and looking good as the result of staying in shape. This finding would seem to show that continued participation results from both intrinsic motivational factors such as self-improvements and extrinsic factors such as pressure and rewards from others.

experience will be traumatic and often degrading to his or her self-concept or peer-group status. It is also true that withdrawal may bring a sense of relief and perhaps even rejuvenation because energy is now available to pursue other activities; moreover, it removes a source of pressure and anxiety from the person's life. Coakley (1983) and Curtis and Ennis (1988) suggest that instead of stress and identity problems, athletes feel a sense of rebirth and renewal after making the decision to withdraw from sport.

The trauma of transition is mitigated by several factors. First, withdrawal is made easier if the participant has not defined his or her self-concept primarily in terms of athletics. Second, parents and other significant others ease the transition when they provide social and emotional support for withdrawal rather than degrading criticism. Third, transition is easier when there is an alternative opportunity structure and a readily available alternative social network or community in which to participate.

The national study of 10,000 boys and girls aged ten to eighteen from eleven cities conducted by the American Footwear Association (1991) documented that sports participation and the desire to participate decline sharply and steadily between the ages of ten and eighteen. At age ten, 40 to 45 percent participate or intend to participate in sport outside of school. By age eighteen, participation drops to approximately 25 percent. These rates apply to school and nonschool sports and for both boys and girls.

In general, the most important reason for declining participation in youth sports during the adolescent years probably is the prevalence of elite or **professional models** of sport in youth sports. These models of sports organization emphasize serious high-level competition, winning, playing well, and extrinsic rewards. These elements of sport may be compatible with the interest, motivation, and ability of a relatively small segment of the youth population, but for most they put sport beyond their interest or grasp.

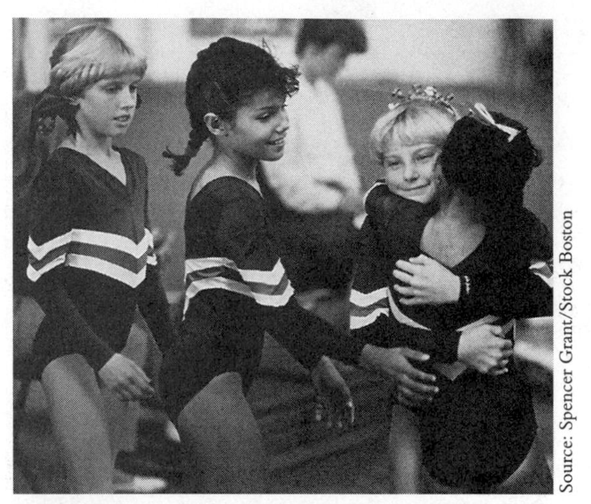

FIGURE 5.2 Getting an early start in sport.

PROFESSIONALIZATION MODEL

According to Coakley (1992), **burnout** results more from the organizational problems of sport than from personal stresses and frustrations. That is, explanations for this type of sports withdrawal should not concentrate on personality factors or individual ability but instead on social structural factors. In Coakley's view, burnout is the result of social relations within sport that prevent the athlete from having any control over his or her experience in sport and from experiencing the diversity of adolescent life that many of their non-athlete peers have the opportunity to experience. Thus, identity development for the athlete, particularly the participant in highly competitive sports, is restrictive and unidimensional.

A professional model of sport for youth has been widely adopted, and the emphasis is on winning or the outcome rather than on the intrinsic value of the participation experience itself. This model generally forces a regimentation and organizational format for youth sport that precludes spontaneous free play, participant decision making, generalized role sharing (such as playing different positions), and experimentation. Watson and Kando (1976) found in their study of Little League baseball that children lack opportunities to devise, interpret, and apply the rules. The Little League is not really their sport; it belongs to the adults who organize and supervise it. All decisions on scheduling, participation, rules, and regulation are made by adults, not children. This is equally true for virtually all forms of contemporary youth sport.

Yablonsky and Brower (1979) and others have contended that youth sport is taking on the form of the professional model with emphases on high-level performance, extrinsic rewards, statistics, and winning. Yablonsky and Brower's (1979) five-year study of the Little League verified that this trend was taking place. Thus, the youth sports experience has come to duplicate in many respects dominant features of life in society for middle-class adults who work in large organizations. Ritzer (1993) has argued that life in Western industrialized societies exposes their members to McDonaldization, which involves bureaucratic standardization, efficiency, rationalization, calculation, predictability, and control. When youth sports programs follow this pattern, they become professionalized **Little Leaguism**, with formal and rational organization and adult control. As such, they deprive young people of playlike experiences that may be developmentally appropriate or even necessary for children and young adolescents.

Participants in such professionalized and bureaucratized sports programs often find that intense competitive pressures interfere with the development of social skills and a secure sense of self-worth. They have few opportunities to experiment, to practice making decisions, to fantasize, or simply to be children or youths. Spontaneous play and informal games are more likely than Little Leaguism to provide these kinds of opportunities and experiences.

This type of organization is why Devereux (1976) argued that the replacement of informal and child-centered play and games by Little Leaguism has resulted in a premature loss of childhood innocence (Nixon, 1984: 31). The irony implied by Devereux's argument is that the preparation for life in a competitive society through sport claimed by the Dominant American Sports Creed (Edwards, 1973) is achieved at the expense of childhood when young people participate in professionalized sports. Devereux's argument is tempered by Coakley's (1990) counterargument that sport is actually much simpler and more clear-cut than other realms of adult society. When we are adults, our relationships, roles, rules, goals, and interpretations of meaning tend to be more complex and ambiguous than the experiences young people have in sport. We can infer from Devereux's notion of Little Leaguism, though, that adult organization and control of youth sports makes them more complex and ambiguous than they would be if young people themselves were in control.

While Coakley pointed to the separation of youth sport from the normal range of everyday roles and relationships, others have emphasized the connection between sport and everyday life for young participants. For example, Lever (1976) argued that the greater involvement of boys than girls in complex games and sport gives boys an advantage in their preparation for adult roles in a competitive and bureaucratic society. Fine (1987) observed that much of the Little League experience could be defined in terms that are normally assigned to describe work. Just as work is involuntary and bounded by formal rules, so is Little League. Little League is serious, goal directed, emotionally intense, and demanding, and it is a setting where injuries are commonplace (Fine, 1987: 43). Little League participation also requires concentration, regimentation, self-control, compliance with authority, and willingness to work hard. Aspects of play still exist, though, particularly in peripheral activities such as dugout horseplay and sideline teasing.

CRITICISMS OF YOUTH SPORT

Many of the criticisms of youth sport are associated with its professionalized organization. We will cite five major ones (Frey, 1980). First, *a number of the claimed benefits of sport, such as character building, have not been consistently documented by research* (Frey & Eitzen, 1991). Fine (1987) concludes that the claims that participation in Little League contributed to moral development and deterred delinquency remained unproven, as did arguments to the contrary. Ogilvie and Tutko (1971) contend that sport does not build character but instead reinforces desirable character traits that participants possess when they enter sport. Sport may *select out* youths with undesirable character traits at the same time that it selects ones with desirable traits. Of course, "desirable" is defined in terms of the dominant values of sport and the values held by coaches.

A second criticism is that *organized sport as it now exists has lost much of its playful character.* Spontaneity, individuality, fantasy, and informality, which are traditional characteristics of peer-centered play, have been replaced by an organizational form that represents efficiency, specialization, hierarchy, external decision making, and a commercial orientation. The current emphasis on all-star teams, championships, travel, and league sanctity make youth sports quite different from play. The focus has moved from participant to organizational welfare. The outcome—including winning games, tournaments, and championships—is more important than the enjoyment of participation. Product has replaced process.

The third criticism of youth sports is that *the adults who run these sports do not have an understanding of the physical, emotional, and social development stages of life and, more specifically, of the development of young people.* These adults expect children to behave like adults on the playing field. For example, children may be expected to display levels of concentration, role and skill mastery, competitiveness, courage, teamwork, and team loyalty that are beyond their developmental capacities (McPherson, 1978: 222). Thus, children are faced with unrealistic expectations from coaches and parents, who expect their young players to act and perform like "little adults."

The fourth criticism of youth sports is that *they can be a significant source of anxiety as well as joy.* Bruns and Tutko (1978) pointed out several sources of anxiety for youth sports participants, such as learning how to deal with success and having to cope with failure.

Failure in sport is usually a public event occurring before peers, parents, coaches, and spectators. There is no way to conceal a muffed ground ball or dropped pass. Thus, failure can be very traumatic, particularly if it is followed by some degree of rejection or denigration. Failure can be very damaging to the developing self-concept of a young person if he or she is not "debriefed" by a concerned coach or parent. Coaches

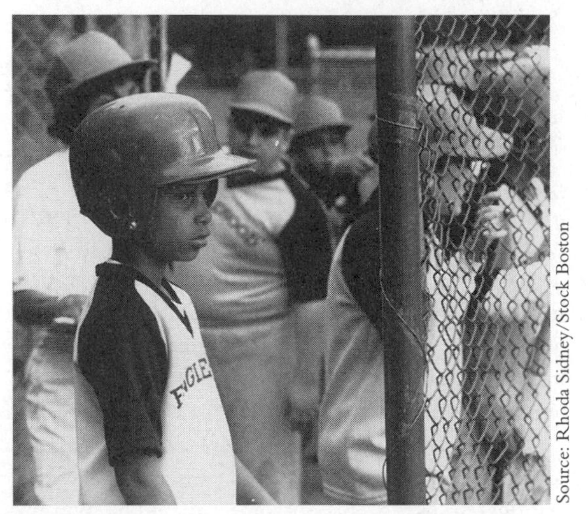

Source: Rhoda Sidney/Stock Boston

FIGURE 5.3 Fear and anxiety are real feelings for many participants in youth sport.

athletic performance is not a life-and-death matter or a reflection of flawed personal character. A strong emphasis on performance and winning, however, creates a *structure of failure* (Messner, 1992: 46). In such circumstances, participants are likely to feel personal responsibility for a loss or error. As a result, they may develop a sense of **conditional self-worth** in which they define and redefine their self-concept each time they compete and win or lose (Schafer, 1975). When "you are only as good as your last game"—which is an implication of the professional model—young athletes are under constant pressure to prove themselves. Such pressure can create tremendous insecurity.

Messner has documented the contingent nature of self in his study of thirty former big-time athletes. Athletes learn that they should never be satisfied with their performance. Messner (1992: 51–52) found that the development of conditional self-worth did not occur in boys and men merely because they had been socialized into sport. It resulted, rather, from the combined effects of the competitive structures and values of sport and the development of masculine identity. According to Messner, in the development of masculine identity young males learn to think of themselves in terms of their reputation in positions rather than in terms of more enduring qualities of the self. Thus, they are attracted to sport and its opportunities to earn distinction from stellar athletic performances. In relying on today's performance to define themselves, though, they are left vulnerable or insecure because they have to prove themselves all over again tomorrow.

This pattern of conditional self-worth in athletes was especially true for lower class males, who have fewer career options than middle- and upper-class males for public demonstrations of their success. The latter see sport more as a steppingstone to success than as the primary basis of success in their lives. The self-worth of more socially and economically advantaged males is less likely to be continually defined by success in sport during youth. For lower-class males and black males from less advantaged backgrounds, sport may be anxiety-producing because it seems so much more important to them than to their more advantaged counterparts to succeed in it.

Another reason for anxiety or fear in youth sports is the possibility of pain and injuries, for which athletes probably will receive little sympathy. "Suck it up" and "sacrifice your body for the team" are part of a prevalent and powerful Sport Ethic in high-level and professionalized sports (Hughes & Coakley, 1991). These slogans and related rhetorical devices and social pressures provide justifications for risking injuries, but they may not be completely effective in allaying the anxieties and fears about pain and injuries among young athletes. In a following chapter, we will discuss in more detail the nature of these types of pressures. The main point here is that the real risk of injury increases as the seriousness of a sport and the intensity of competition increase.

The fifth criticism is that *youth sports do not provide much socio-emotional support for the reserves or substitutes and the bench warmers.* The American Dream and the dominant culture of sport in the United States assert that effort or hard work leads to success. The implication of this belief is that those who are less successful are not trying hard enough and may even be morally suspect for this reason (Nixon, 1984). The irony, of course, is that the reserves may work harder than the starters and stars in their effort to make the starting team or merely to retain their position on the team.

The Substitute

To fate resigned, he sits upon the bench
And leans his chin upon his hands.
He watches every play and vaguely hears
The cheers that thunder from the stands.
Out there his teammates execute the plays
His sweat and toil helped them to learn,
While he, a sub, can only watch and hope
And patiently await his turn.

For weeks and months a lowly place he's had
Gone down in daily, brave defeat,
Submerging self in loyal group desire
That teamwork might be made complete.
For him the drill and drudge of practice hours
And yet no chance of crowd's acclaim.
For him the humble role of substitute,
A satellite of stars who play the game.

The din of cheering rolls o'er his head.
Unknown, the service he performs.
They only see him waiting for his chance,
The chance that often never comes.
Unsung, but still alert to give his best!
Content when thousands laud his mates!
Successful teams were never built, without
The Sub, who hopes and works and waits.

—Anonymous

(Source: H.V.'s *Athletic Anthology*, Chicago H.V. Porter, 1939.)

CONCLUSION

The positive socializing effects of sports participation are the reasons most often cited for encouraging young people to play basketball, swim, or engage in one or more of a myriad of athletic activities while they are growing up. According to the Dominant American Sports Creed (Edwards, 1973), this participation will produce a better citizen who is prepared for life in a competitive society and will act in morally correct and socially appropriate ways. Yet as we noted in our discussion of the Dominant American Sports Creed in an earlier chapter, clear and consistent evidence is lacking to support these beliefs. Indeed, as we will see in future chapters, the sports world may breed social deviance in some cases or may even be, in certain contexts, permeated with deviant practices throughout major aspects of its structure.

In the process of socialization into sport, we have seen that the kinds of sport roles that are learned, the ways they are learned, and their implications for sports participation will vary in many cases according to the dominant values and the structural organization of sport. The type of culture and social organization of a sport or a sports experience is also likely to affect how sport socializes participants. The nature of these socializing effects remains to be documented or clarified in many cases, however.

There is much research that seems to show that engaging in youth sports makes participants different from nonparticipants—for example, by improving their social and psychological adjustment, by enhancing their self-concept, and by producing culturally desirable personality changes. Other evidence appears to show that youth sports participation has no effect or negative effects on these kinds of factors. In fact, it is possible that the socialization effects attributed to youth sport are spurious. That is, youngsters recruited to play already exhibit the personal and behavioral characteristics preferred by coaches and league officials. Those who need sport the most, for developmental and character-building reasons, are denied access or never selected to participate.

The high dropout rates associated with sport suggest that the experience of athletic competition is not as attractive or beneficial as purported. Youth sport participation can create considerable anxiety and stress because the prevalent professionalized organization context forces unrealistic expectations on the participants and makes the experience of competition secondary in importance to the outcome (i.e., winning). Studies show that many participants would rather have an enjoyable or playlike experience rather than worry about being consistently evaluated on performance outcomes. The criticisms of youth sport focus on this contemporary model of youth sport, in which professional character deemphasizes the socioemotional and skill development of participants in favor of an emphasis on elite performance and won-lost records.

The relationship between socialization and sport for children and adolescents can be complex, variable, or difficult to discern from extant research. Much has been claimed about socialization into and through sport, but there is much more to know. Clearly, a substantial amount of what was believed to be true about the socialization impact of sport is more myth than fact. These beliefs are so ingrained in the cultural ideology associated with sport, however, that criticisms are not taken seriously. Even if the negative aspects of sport are revealed and are acknowledged to contain a merit of truth, few would find this sufficient reason to discourage participation.

SUMMARY

Socialization is a lifelong process through which individuals develop and shape their social identities, create a sense of self, and learn how to participate in social roles and relationships. A number of different theoretical perspectives explain socialization processes, including social learning, conflict, and interactionist theories. The relationship of sport to socialization can be examined from each of these different theoretical perspectives. Two basic ways of viewing this relationship is to consider socialization *into* sport and socialization *through* sport. The former perspective focuses on how people learn to participate in sport, and the latter focuses on the effects of sports participation on general psychosocial development and socialization into roles and relationship outside sport. One potentially important consequence of sports participation is the learning of gender identities and roles. Research has attempted to determine to what extent gender identities and roles are affected by various types of sports involvement. The traditionally masculine character of sport is a major focus of much of this research.

Although sports participation in youth offers a number of possible benefits, researchers and observers have given much attention to aversive or negative sports socialization, identifying such phenomena as the Little League syndrome and Little Leaguism. Attention in such studies has focused on the conditions of professionalized youth sports that threaten, stress, and discourage participants, causing them to drop out. Five general criticisms of youth sports are that (1) their claimed benefits have not been documented; (2) they have lost their playful character; (3) those controlling youth sports do not understand the developmental stages and experiences of young people; (4) youth sports create considerable anxiety; and (5) they offer little support or consolation for less talented participants.

CHAPTER 6

Social Deviance, Social Problems, and Sport

THE NATURE OF SOCIAL DEVIANCE AND SOCIAL PROBLEMS

It sometimes seems that more is written or said in sports reporting about drugs, gambling, fines, excessive violence, suspensions, penalties for rules violations, and arrests for violations of the law than about wins, losses, batting and scoring averages, and the pursuit of championships. In fact, a review of the most-written-about baseball topics of the 1980s in newspapers, magazines, and the wire services included drug abuse, on-field rules violations such as doctored bats and balls, and the Pete Rose gambling case among the top seven (*Sports Illustrated*, 1989). We considered earlier how the increasing exposure of fans to these kinds of sports realities may make them more disillusioned about sport and less likely to find a cozy corner or symbolic refuge in the sports world. These realities are interpreted in this chapter as forms of social deviance and social problems.

The difference between social deviance and social problems can be ambiguous because many types of social problems are forms of social deviance. **Social deviance** is behavior that breaks the rules or violates the norms of a group, organization, community, or society. Behavior becomes deviant because people with power or authority treat it as deviant, and these people generally treat certain behavior as deviant because they see it as threatening or disruptive to them, their interests, or the social order. **Social problems** are perceived sources of social disruption, tension, or difficulty in society, and they can be divided into two major types (Eitzen & Zinn, 1994: 10). One type includes forms of social deviance that powerful people or the mass media define as highly undesirable for society, such as violence, drug abuse, and other types of criminal behavior. The second type includes conditions of society that cause widespread or severe psychic or material suffering for some segment of society. This type of social problem includes poverty, hunger, illiteracy, racism, sexism, and assorted environmental problems. It can affect all members of society but usually disproportionately affects members of the lowest strata. In many cases, these types of conditions are problematic *because* they represent social disadvantage and powerlessness. Both types of social problems will be examined in this book, but this chapter will mainly focus on social deviance.

Societal reaction and labeling theorists (e.g., Becker, 1963, 1971; Erikson, 1964; Lemert, 1972) have suggested that behavior becomes meaningful as social deviance or a social problem when influential people perceive and treat it as deviant or problematic. Understanding social deviance and social problems is complicated by the fact that they can be deeply embedded in the social order and even tolerated or encouraged by those with authority. We will consider types of social deviance that are institutionalized in the informal social order of everyday interactions in sport along with types of deviance that are so marked that they arouse strong and widespread public disapproval and are considered serious social problems in sport. We also will consider forms of behavior that are deviant in society but are tolerated, accepted, or even expected and encouraged in sport. For example, the violence that is decried in society may be a major attraction in certain sports such as boxing, football, or hockey. There also are behaviors that are only deviant in sport or a particular realm of sport; for example, the payment of athletes that leads to severe NCAA sanctions for intercollegiate athletic programs has no counterpart in other corporate realms in American society. Indeed, in the realm of professional sports, athletes *expect* to be paid. These complexities of social deviance indicate why sports participants can become confused about what they are supposed to do and why it can be difficult for us to understand the nature of various kinds of social deviance and problems in sport and who or what is responsible for them.

EXPLANATIONS OF SOCIAL DEVIANCE AND SOCIAL PROBLEMS

Different theoretical or conceptual perspectives provide strikingly different pictures of the nature of social deviance and social problems, why they arise, and how they affect society. Some sources of deviance and problems, for example, are interpersonal.

Interpersonal Sources of Social Deviance

Individual athletes and coaches may engage in deviant behavior because they have difficulty meeting role expectations in their social relationships because of role strain or role conflict. **Role strain** occurs when people are faced with conflicting expectations in the same role relationship: for example, a coach expects an athlete to play violently and also display good sportsmanship, or a coach is expected by her athletic director to field an outstanding team without recruiting. **Role conflict** occurs when expectations associated with different roles performed by an individual are incompatible or conflicting: for example, a student-athlete is expected by his instructors to devote as much time as possible to being a serious student and is expected by his coach to devote as much time as possible to being an athlete. Or a college coach might feel pressure from his wife and family to spend more time at home as a husband and father, while he also feels pressure from his athletic director and coaching staff to devote as many hours as possible to his job as coach. Attempts to resolve role strain, role conflict, and other forms of excessive or incompatible role demands can lead to social deviance and social problems when the deviance occurs frequently among those in similar roles in sport.

Social Deviance as Anomie

A structural-functionalist perspective called *anomie theory* proposed by Merton (1957) enables us to see how incompatibilities between culturally approved goals and role opportunities can turn into deviant behavior. In the United States, the goal of success is embedded in the widely accepted American Dream ideology, which we discussed earlier. American males have been encouraged or even have felt obligated to pursue material, occupational, social, or political success. Merton pointed out that there are culturally approved means of striving for such success, such as investing time and energy in training or formal schooling and trying to move up in a potentially prestigious and lucrative career path in business, the professions, or some other legitimate occupation.

Social conformity occurs when individuals follow culturally prescribed avenues to achieve the culturally approved goal of success. Various forms of social deviance occur when individuals adapt to pressures to succeed in ways that involve incompatibilities between cultural means and goals. In Merton's framework (shown in Table 6.1), four types of social deviance derive from these conditions: innovation, ritualism, retreatism, and rebellion. *Innovation* occurs when people strive for success in ways that are deviant from culturally accepted means. For example, an ambitious college basketball coach may cheat in recruiting to be more successful or a competitive weight lifter may turn to steroids to build mass and improve his chances of winning.

TABLE 6.1 Anomie Theory, Conformity, and Deviance

Type of Adaptation	Cultural Goals	Socially Approved Means or Roles
Conformity	accepted	accepted
Innovation	accepted	rejected
Ritualism	rejected	accepted
Retreatism	rejected	rejected
Rebellion	rejected replaced	rejected replaced

Ritualism occurs when people follow the rules but lose interest in or do not care about being successful. Although ritualism seems antithetical to sport as we have defined it, it still is possible to find cases of ritualism in sport. For example, ritualism characterizes the case of college athletes who no longer care about their sport but attend practice and go through the motions of participation merely to hold on to an athletic grant-in-aid so that they can afford to stay in school and graduate. A ritualistic coach may be one who is so preoccupied with formulating sophisticated strategies or having his athletes follow the rules and regimens he has used for the past thirty years that he has lost sight of the skills and capabilities of his current team and what they need to improve and to become more successful.

Retreatists are people who reject both culturally approved goals and the established means of pursuing them. Leonard (1988a: 156–157) cites former college and professional football player Joe Don Looney as an example of a retreatist. After gaining a reputation as a talented but erratic or unpredictable personality, Looney eventually quit football and joined a religious sect in India that renounced worldliness. Coaches who quit because they have lost their desire to win and their interest in recruitment, travel, training sessions, and practices also can be viewed as retreatists.

Rebellion occurs when people reject established conceptions of success and the established means of achieving it and, unlike retreatists, create and pursue alternative goals and means. Stewart Ferguson displayed rebellion as the Arkansas A&M Boll Weevils football coach between 1939 and 1941. What made Ferguson a rebel was his disavowal of an intense commitment to winning and his acceptance of a different approach to coaching and playing college football. In turning over his team to the players and instructing them to play for fun, he encouraged them to make up plays in the course of games that entertained fans and even drew their laughter. In fact, the spontaneous and trick plays created by the Boll Weevils earned them a national reputation as the "Marx Brothers of Football" (Figler & Whitaker, 1991: 73). Although some observers believed they were intentionally trying to lose, Ferguson argued instead that the Boll Weevils did not care about winning. Stewart Ferguson's experiment ended after the 1941 season, and following his service in the military he returned to high school coaching, where his teams once again were more conventional and were consistent winners. This example of rebellion underscores the point that social deviance is not necessarily undesirable.

Conflict and Critical Perspectives

Conflict and critical perspectives focus more attention than anomie theory and other structural-functional perspectives on social deviance, social conflict, and social change as regular features of social relations and societies. Thus, conflict and critical theorists do not assume that cheating, gambling, violence, power struggles, and other social problems and clashes in sport are unusual, nonnormative, or always lead to serious punishment, penalties, or public condemnation from sports authorities. They are more likely to assume that social problems, social deviance, and social tensions are a regular part of sport and society that might be encouraged in some cases by self-serving authority figures.

There are a variety of conflict and critical perspectives (see Nixon, 1991a), but they generally share a view of society that emphasizes the structures of domination, exploitation, and oppression by powerful classes or groups and the ways these classes and groups exercise power to serve their own interests. The case of violence in hockey will help us illustrate how conflict and critical perspectives of social deviance and problems differ from structural-functional or anomie perspectives. We know there are defensive players who have established reputations and careers more on the basis of their brawn and fighting skills than their abilities to skate, pass, check, or shoot according to the rules. These players accumulate penalty minutes and sometimes have to pay fines and serve suspensions for their rough play. The press and National Hockey League officials who criticize this style of play, write or talk about the need to rid professional hockey of such players for the good of the sport. They are regarded as social deviants who must be controlled or banned from the sport. In Merton's terms, such players are innovators because they seem to lack the skills to become professional hockey players by any other means than by flaunting the rules and being bullies.

Source: Courtesy of University of Arkansas-Monticello

FIGURE 6.1 The 'Wandering Weevils of Arkansas A&M'

A different way of looking at hockey violence assumes that violence is encouraged by those who control the sport. Coaches encourage their biggest and strongest players to be violent to intimidate, incite, or neutralize opponents. Owners encourage a violent style of play to draw fans and make money because they believe that violence sells. League officials publicly condemn violence and even impose occasional suspensions or fines for excessive violence, but they also tolerate a regular pattern of violence and chronically violent players because they believe it keeps their league popular with fans. If perceived changes in popular sentiment or new legislative or judicial decisions make violence more costly to the league, the league will more strictly enforce rules against fighting or other forms of violence on the ice, force the "thugs" out of the sport, and encourage coaches to teach different styles of play.

Thus, the amount of violence in hockey is manipulated according to the needs and interests of those in power. From conflict and critical perspectives, hockey violence is only deviant when those who control the sport decide that it must be treated in this way. Those who are treated as deviant are the players, who actually gain relatively little overall from the violence in the game and suffer the costs of dubious reputations, public sanctions, and a constant risk of injury from violent play.

Societal Reaction Labeling Perspectives

In **societal reaction labeling** interpretations, social deviance is a process of interaction between deviants and those with enough status and power to impose deviant labels. Thus, social deviance is a social construction. Societal reaction labeling theorists generally assume, like conflict and critical theorists, that the rules defining deviance and the settings in which such rules and definitions are applied are constructed by the wealthy for poor people, by men for women, and by other dominant strata for subordinate strata or minority groups (Giddens, 1991: 157).

As we saw in the hockey example in the last section, league officials, coaches, and others in managerial or ownership positions in sport have the power to make, enforce, and change the rules according to their interests and needs. For example, the team owners in a professional sports league may reject the bid of a prospective female owner to join the league, saying that even unproven allegations that she has associated with known gamblers could taint the reputation of their league. In constructing a dubious reputation or deviant identity for this prospective owner so that they could preserve the male-only composition of ownership in their league, the owners conveniently obscured the fact that their own numbers included individuals with investments in gambling casinos and racetracks. They created the label they needed to achieve the effect they desired, that is, to keep a female owner out of their sport.

Those with substantial status, power, or wealth may be able to escape the deviant labeling process or minimize its sting when they are labeled and prosecuted as deviants. Thus, as a rich and powerful person in her own right, the prospective female owner in our prior example may be able to fight back and reconstruct her reputation, shedding the deviant label imposed on her. More commonly in sport and elsewhere, we see deviant labels imposed on people and groups with much less power to escape them. We also see that people engaging in the same kind of behavior may not be treated in the same way. A star player may receive a reprimand or be ignored by the coach when he violates the team curfew, but a marginal player may be suspended for the same behavior. Deviance clearly is in the label rather than in the behavior itself or the person.

Deviant labels can be difficult to remove or alter. Athletes sometimes get into trouble with drugs or alcohol and gain a reputation as "troubled," "addicted," or "alcoholic." The public humiliation, suspensions, and other penalties these athletes pay for their drug use can reinforce the image of a substance abuser and make future abuse more likely as they fall into a drug abuse career. At times, it appears that coaches and athletes like the reputations of being crazy or wild, a renegade or outlaw, if it helps their marketability with young consumers, or conversely, if it helps them escape the burdens of being heroes or role models for fans. In many cases, though, deviant labels carry destructive stigmas with them, and those labeled as deviant lack the power to remove the labels or make them more socially desirable.

POSITIVE DEVIANCE AND THE SPORT ETHIC

The concept of **positive deviance** would seem to be self-contradictory because we usually assume that deviance is negative in its effects on the social order or perceived as negative by those responsible for maintaining social order (Sagarin, 1985; Goode, 1990). The idea of positive deviance has been used by sport sociologists, however (e.g., Ewald & Jiobu, 1985; Nixon, 1989; Hughes & Coakley, 1991), to understand cases of conformity that are so intense, extensive, or extreme that they go beyond the conventional boundaries of behavior. They are cases of *over*conformity rather than *counter*conformity, but they are deviant because of their extreme nature.

Hughes and Coakley (1991) examined positive deviance in relation to a code of central values and norms in sport they called the **Sport Ethic**. Related to the Dominant American Sports Creed and the American Dream, it consists of four basic beliefs about what being an athlete means or requires: (1) making sacrifices for one's sport or team; (2) striving for constant improvement and distinction (often reflected by winning); (3) accepting risks and playing through pain; and (4) refusing to accept limits in the pursuit of achievement and success. Athletes who embrace these beliefs and try to fulfill them in extremely intense or extensive ways are positive deviants. They are not deviants because they have rejected the values or broken the rules of sport; they are deviant because they have been exceptionally zealous in their acceptance and conformity.

Hughes and Coakley hypothesized that positive deviance is most likely to occur among athletes with low self-esteem who are highly susceptible to group pressure to sacrifice for the team, whose identity is tied to sport, and who rely heavily or exclusively on sport for social mobility. They also argued that commitment to the Sport Ethic is reinforced or intensified by team dynamics that set team members apart from nonmembers, increase their bonding with one another, and create a sense of superiority over, and disdain for, outsiders who do not have to endure the risks or sacrifices they do. Thus, to the extent sports involvement makes athletes feel important or special, they will seek to embody the standards that distinguish athletic roles from other types in society.

Positive deviance often is not positive in its effects on athletes. Furthermore, extensive patterns of positive deviance in sport reflect significant social problems when viewed from the perspective of the athlete's welfare. Athletes who make extreme commitments and sacrifices to fulfill the Sport Ethic may do so at the expense of other important activities and roles in the family, in education, or in the workplace. They may risk chronic pain or permanent disability when they push their bodies beyond physical limits. They may even engage in forms of more conventionally defined, or negative, deviance, such as cheating or dangerous and illegal drug use, as an extension of their desire to do whatever it takes to improve, win, or achieve distinction.

Positive deviance can result from manipulation by coaches and others who can benefit from athletes' overconformity (Young, 1991). Athletes who make extreme sacrifices and take extreme risks can make their teams and themselves more successful and thereby enrich the investments of owners and sponsors and further the careers of coaches and sports managers. If there is a tendency to promote positive deviants into management and coaching positions in sport (Hughes & Coakley, 1991: 315), it is easy to see why athletes are encouraged to overconform.

Pain, Injury, and Risk Taking as Positive Deviance

A basic tenet of the Sport Ethic is the belief that being an athlete involves risk taking and pain. Implied in this belief is that athletes have to accept the possibility of injuries as part of the game. Since injuries can end an athletic career and cause chronic pain, permanent disability, and even death, one might ask why athletes accept such risks and costs. Exploring this question should provide more insight about positive deviance in sport in general as well as about the seemingly self-destructive behavior of "playing hurt" (see Nixon, 1992a, 1992b, 1993b, 1993c, 1993d).

Baseball manager Sparky Anderson suggested the peculiar acceptance of pain in sport when he said, "Pain don't hurt" (quoted in Morris, 1991: 194). Although seemingly self-contradictory, this statement captures the reality of serious sports, in which athletes routinely face pain and injuries—and ignore or minimize them—as they "play hurt." A number of sociological observers of sport (e.g., Kotarba, 1983; Sabo, 1986; Stebbins, 1987; Curry & Strauss, 1988; Messner, 1990, 1992; Nixon, 1993a) have pointed to the forces in sport that *normalize pain and injuries* and encourage or pressure athletes both to accept and minimize them as a routine part of sport. Along with subtle or obvious pressure from coaches, trainers, or management officials, messages from the mass media that promote a **culture of risk, pain, and injury** are likely to influence athletes to accept playing hurt.

One study (Nixon, 1993a) presented results from a content analysis of over a decade of *Sports Illustrated* articles showing a picture of the male corporate sports world in which athletes are regularly exposed to role constraints, inducements, general cultural values, and processes of institutional or organizational rationalization and athletic socialization that convey the message that they ought to accept the risks, pain, and injuries. When influential media such as *Sports Illustrated* present words and pictures of star athletes and prominent coaches that normalize, rationalize, or legitimize playing hurt, the culture of risk, pain, and injury is likely to become more compelling for aspiring and experienced athletes and coaches.

A survey of male and female athletes at a medium-sized southeastern comprehensive university competing at the NCAA Division I level indicated how much risk, pain, and injury existed in college sports and how athletes responded to these conditions (Nixon, 1993b, 1993c). The nearly 200 respondents represented all eighteen of the men's and women's varsity sports at their institution and were approximately 45 percent of all varsity athletes on campus. Over 150 of the respondents reported having had significant injuries, and nearly all of these 150 athletes also reported having played hurt. In addition, over 45 percent said they had lingering effects of their injuries. About half said they had felt some influence from significant others to play hurt. The most widely accepted belief about risk, pain, and injury (over 90 percent agreement) was: "Being an athlete means that you have to be willing to accept risks." A great deal of strong support was also expressed in response to statements about the difficulty athletes have in quitting, even after serious injuries; the need for athletes to push themselves to their physical limits; the belief that every athlete should expect to have to play with an injury or pain sometime; and the popular slogan "No pain, no gain." Overall, a majority of the athletes expressed strong or reserved agreement, with nearly twenty of thirty-one items suggesting a willingness to play hurt.

POINT-COUNTERPOINT

ARE TAKING RISKS AND PLAYING HURT RATIONAL ACTIONS FOR ATHLETES?

YES

Life involves many risks and possibilities of getting hurt. In sport, playing hurt and risk taking are calculated and necessary. Coaches, sports medicine personnel, and sports administrators take every precaution to protect athletes from getting seriously hurt. Part of the joy of sport, however, is its physical nature, including physical contact and physical risk. If we eliminate all risk from sport, sport could not exist. The chance of pain and injuries is part of the game. Athletes know this but believe that the benefits outweigh the risks and possible costs.

NO

Playing hurt and taking risks in sport are not rational for athletes because they are discouraged by coaches and sports administrators from thinking about the long-term consequences of these actions. Athletes are often socialized to accept playing hurt and taking risks as part of the game. Or they play with injuries or pain as a result of social pressure from coaches, teammates, family, friends, or fans. There is no rational decision making involved in such behavior.

How should this debate be resolved?

This survey revealed that college athletes' responses to their pain and injuries are influenced by how significant others in their sports networks, such as coaches, teammates, trainers, and physicians, interacted with them. The two factors that most strongly influenced athletes to talk to medical experts about their pain and injuries were sympathy and caring from trainers and doctors. Furthermore, athletes were more likely to talk and turn for help to coaches when they were seen as sympathetic and caring. On the other hand, athletes tended to avoid or hide their injuries from authorities, such as coaches, trainers, and physicians, when they were perceived as likely to pressure them to play hurt. Perhaps surprisingly, in view of traditional gender stereotypes, there were relatively few differences between males and females with respect to how the athletes perceived or responded to pain and injury.

The coaches of these surveyed athletes were not obviously or intentionally abusive or exploitative. A survey of the coaches (Nixon, 1992b) revealed that most stated that they cared about the welfare of their athletes and said that athletes should trust them and other athletic officials to protect them from injuries. Nevertheless, it was true that the coaches' beliefs about risk, pain, and injury generally paralleled those of their athletes. A majority of the coaches agreed with nearly two-thirds of the statements consistent with the culture of risk, pain, and injury.

Despite the stated intentions of coaches, sports medicine experts, and other sports officials to take care of athletes, there may be an implicit conspiracy of teammates, other athletes, coaches, fans, and the sports media to induce athletes to take risks and play with pain and injuries when they are in that ambiguous or gray area between disabling injuries (that physically prevent athletes from competing) and nondisabling injuries (with which athletes can continue to play but perhaps at a reduced level of effectiveness). Whatever their actual motivation, coaches who pressure their athletes to play hurt and try to influence team trainers and physicians to allow athletes to play hurt can turn to the Sport Ethic as their justification.

Stebbins's (1987) study of Canadian professional football players suggests that professional athletes become skeptical about the motives of coaches and trainers and sense some manipulation or exploitation. Stebbins found that under these circumstances, self-preservation replaces team sacrifice in regard to pain and injuries. Yet for a complex combination of reasons these athletes still choose to play hurt. Messner's (1990, 1992) interviews with male former athletes at different sports levels suggested to him that internalized ideas about stereotypical masculinity join with external influences to cause athletes to "choose" to play hurt, despite their suspicions or worries. If athletes turn only to others in their sports network for advice or support, and the Sport Ethic is the glue that holds their sports network together, it is not surprising that athletes ultimately decide to play with pain and injuries, despite personal misgivings.

It is impossible to eliminate the risks of pain and injury from high-level contact sports because these sports are structured to produce pain and injuries. Unless athletes are willing and able to exercise power on their own behalf to protect themselves from influences that place their welfare behind the welfare of the team or organization, they will be vulnerable to considerable direct and indirect pressure to take excessive risks with their bodies and health. A series of injuries to star athletes in a popular sport is likely to attract extensive media attention, and this media attention may induce coaches and officials in a sport to think about more protective rules and better and safer equipment and facilities. Otherwise, the protection of players is most likely to come from the law, the courts, or, in the case of professional leagues, provisions of their Standard Player Contract (see Stebbins, 1987: 126; Nixon, 1993b).

MAJOR PROBLEMS OF SOCIAL DEVIANCE IN SPORT

Although social problems generally reflect or arise from deviant practices, social deviance is not necessarily viewed as a social problem. In fact, **ordinary deviance** is so commonplace that it becomes an expected part of sport or sports contests. This kind of deviance may lead to penalties but does not prompt ethical questions or cause a moral outcry. At worst, coaches, teammates, and fans may be unhappy or disappointed if such behavior, such as a holding penalty in the middle of a key touchdown drive, adversely affects a team's chances to win. In some cases, this ordinary deviance may be an unusual practice, such as wearing a mask or a helmet in hockey before they became accepted, which initially causes amusement, scorn, or criticism within a sport but eventually becomes a typical or normative practice as it becomes more widely accepted.

The social deviance in sport that is related to social problems is different from ordinary deviance. It is the type of deviance that is so morally questionable, so damaging in its perceived consequences, so recurrent or pervasive, or so threatening to authorities or powerful interests that it arouses strong and widespread public disapproval and becomes an issue for powerful people in sport or society. When deviance becomes a social problem, efforts are made to exercise social control. This *problematic deviance* and other problematic social conditions in sport are the focus of the ensuing sections about major social problems in sport.

Player Violence

It is easy to confuse physical contact in sport with violence, and the former surely can lead to the latter. Contact and violence are not necessarily the same thing, though. **Violence** can be distinguished from other kinds of physical contact by the intent to harm another person. Types of player violence can vary in intensity, seriousness, and legal status. For example, Smith (1986) distinguishes among body contact, borderline violence, quasicriminal violence, and criminal violence. *Body contact* is physical contact that occurs within the rules of a sport, such as tackles and blocks in football, bodychecking in hockey, and punching and jabbing in boxing. Even though body contact can produce serious injuries and death, causing many sports observers to view it as a social problem, it is legal. Thus, body contact is distinguished by whether or not it falls within the rules or is normative rather than by how brutal its consequences may be. The confusing nature of such contact is illustrated by the case of the hit by Jack Tatum that broke the neck of wide receiver Darryl Stingley and left him permanently paralyzed. Jack Tatum had taken pride in the vicious hitting that had earned him the nickname "the Assassin" (see Tatum, 1972) and was honored and applauded for his style of play. After Tatum injured Stingley, critics within his sport called him part of the "criminal element" in the NFL. Tatum responded that the injury was a "terrible accident," but that it was a "routine

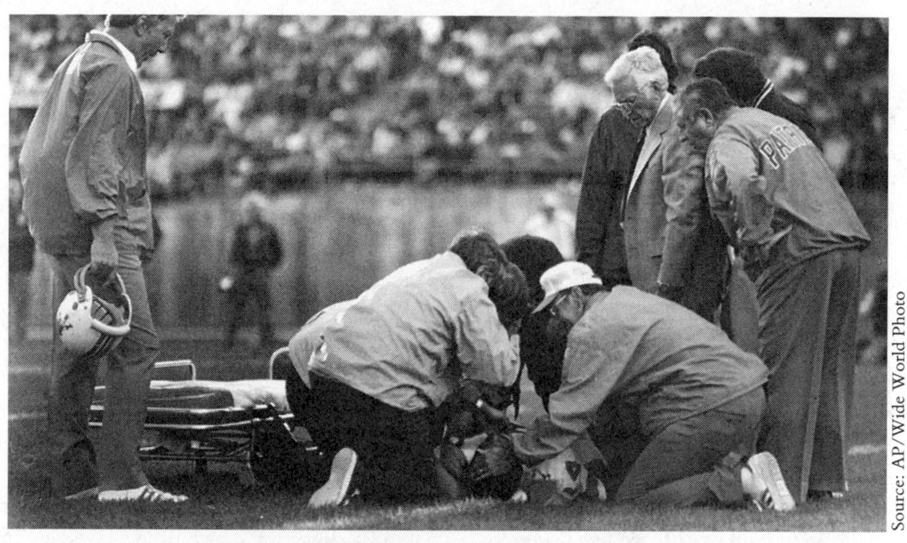

Source: AP/Wide World Photo

FIGURE 6.2 Darryl Stingley is paralyzed by a violent hit.

play" within the rules (Messner, 1990: 210). It was, in Smith's terms, body contact.

Borderline violence includes assaults that occur routinely in a sport, even though they are formally prohibited by the rules. They tend to be accepted or tolerated by sports officials, coaches, players as part of the game, but occasionally or frequently they lead to penalty calls by referees, umpires, or judges. Social control seldom goes beyond the boundaries of a specific sports contest, and the most severe penalties are likely to be suspension from the game or a subsequent fine. This is ordinary deviance, which mainly upsets victimized athletes, teams, or fans when it is unnoticed by contest officials or the penalty that is imposed is not perceived to be severe enough. It includes tripping or high tackling in soccer, brushback pitches in baseball, pushing or elbowing in basketball, and bumping in track, road racing, auto racing, and horse racing. Although borderline violence could result in serious injuries or affect the outcome of a sports contest, it is generally not seen by sports administrators, coaches, players, or fans as a social problem because it is so embedded in the fabric of the sport.

Quasicriminal violence generally prompts more anger or outrage than borderline violence. This is assaultive behavior that violates both the informal code of player conduct and the formal rules of the sport. It involves very dangerous attacks that potentially or actually lead to serious injuries. Penalties generally go beyond the contest and may range from suspensions for several games to lifetime bans. A highly publicized example includes San Francisco Giant pitcher Juan Marichal's attack on Los Angeles Dodger catcher John Roseboro during a baseball game in 1965. Marichal knocked Roseboro down with a bat after the two had engaged in a heated argument. Roseboro was seriously hurt, Marichal was fined and suspended by the league, and Roseboro filed a civil suit for damages against Marichal and the Giants (Kuhlman, 1975; Smith, 1986: 226).

Although quasicriminal violence produces public outrage and calls for criminal prosecution, it differs from criminal violence because it is viewed essentially as a sport problem and does not result in criminal prosecution by the legal system. *Criminal violence*, on the other hand, is treated from the outset by all responsible people in and out of sport as a matter so severe that it requires legal intervention by the criminal justice system. Criminal violence may involve death, and frequently occurs immediately before or after a sports event. In some cases, sports figures engage in criminal violence that has nothing to do with sport itself, as in the cases of athletes who beat or kill their wives and members of sports teams who participate in a gang rape at a postgame party. In other cases, criminal violence occurs outside the sports arena but is prompted by the competition inside it. For example, a shouting or shoving match between opposing coaches or players at the end of a game leads to punches after the game. A third kind of sports-related criminal violence occurs in the sports arena. Legal authorities historically have been reluctant to initiate criminal prosecutions even in serious cases of this latter sort. An exception is when Dino Ciccarelli, right winger for the Minnesota North Stars of the NHL, punched Luke Richardson of Toronto after hitting him over the head with his stick during a game in January 1988. Ciccarellil was convicted and sentenced to a day in jail. This kind of case blurs the line between sport and society. Applying the laws of society to sport calls into question the legality of a number of sports that allow or require forms of body contact that are not normally permitted in society. In effect, violent sports have been permitted to exist because the laws of society have been suspended, allowing these sports to set their own rules and police themselves.

Violence in sport may be permitted or accepted in a society because it is a part of the violent social and cultural life in that society generally. In the United States, violence is part of a cultural tradition that includes wars and rates of homicide and other violent crimes that are very high compared to other industrialized nations. This culture of American violence has been spawned by easy access to firearms, the pervasive influence of the frontier tradition, and the existence of subcultures of violence in urban America (Giddens, 1991: 163).

Research has shown a correlation between the popularity of violent sports in a society and its militarism. More specifically, Sipes (1973) found that warlike societies are much more likely than nonwarlike ones to have combative sports and that the popularity of combative sports (including boxing, hockey, and football) in the United States in particular has tended to rise during wartime. Other research, by Keefer, Goldstein, and Kasiarz (1983) has shown a correlation between national military involvement and the popularity of contact sports in nations in the Olympics.

Violence in sport is learned. Vaz (1972, 1982) reported that young hockey players learn to fight in minor hockey programs by watching the aggressive behavior of professional hockey players and being encouraged to imitate them. Quasicriminal violence and even criminal violence may arise when athletes accustomed to engaging in or seeing violence are caught in the stressful or frustrating dynamics of specific game and social situations that push them to more extreme forms of violence. By the same token, these athletes may have so thoroughly internalized violence as part of their personality or sense of manhood that violent behavior is seen as an acceptable, normal, or natural way of handling conflict and situations or interactions perceived as unfair or threatening (Coakley, 1990: 147–148). Indeed, "enforcers," "goons," and "policemen" in sports ranging from football and hockey to basketball earn their living by being intentionally violent to intimidate opponents. These athletes commonly engage in borderline violence (Eitzen & Sage, 1993: 160) and may account for a disproportionate amount of the violence in their sports. An estimated 2 percent of the players in the National Hockey League receive more than 20 percent of the penalty minutes (McPherson, Curtis & Loy, 1989: 270).

In learning violence as part of the game, athletes also learn that competitive advantages can be gained from being violent. A study of penalties assessed over four years of play in the National Hockey League revealed a positive correlation between the incidence of aggressive penalties (including slashing, spearing, highsticking, boarding, charging, and fighting) early in a game and winning the game (Widmeyer & Birch, 1984).

Little systematic evidence exists about violence in women's sports, but anecdotes indicate the occurrence of quasicriminal violence involving players and coaches in women's sports such as basketball and hockey (McPherson, Curtis & Loy, 1989: 270). For example, women athletes and coaches have thrown balls at referees, punched and pushed them, and speared them with a hockey stick, and they have punched, pushed, and engaged in brawls with each other. The overall level of violence in women's sports in the United States may be much less than the level in men's sports because American women have traditionally been discouraged from aggressive behavior in general and from contact sports in particular. If changes in gender socialization and role opportunities for women blur stereotypical distinctions between male and female behavior, if athletic socialization emphasizes more aggressive behavior and sports for women as well as men, and if competitive advantages are to be gained from more violent behavior for women in sport, the apparent violence gap between men's and women's sport could decrease (Coakley, 1990: 151–152).

Although fans often say they do not like fighting or violence in sport (Miller Lite, 1983), many watch and attend sports events where violence, collisions, and intense contact are a normal part of the game. For a number of fans, violence has a special appeal. For example, a hockey fan from Illinois started a newsletter in 1992 called *Beaver's Mixin' It Up* devoted to "the physical aspect of the NHL" (O'Brien, 1993). Its editor opposed the elimination of fighting in the NHL because it would ruin the sport. Every three weeks the newsletter provided its 5,000 subscribers with editorials, statistics, and stories about the fights, penalties, and tough guys of the NHL.

Fan Violence

Fans may subtly or overtly encourage and enjoy player violence (Eitzen & Sage, 1993: 161). They also engage in sports-related violence themselves, ranging from throwing objects at athletes, coaches, and referees to destroying property in the sports arena, rioting in the streets in ostensible victory demonstrations, and fighting with other spectators. In some cases, disappointed, vindictive, obsessed, or misguided fans and supporters assault or kill athletes. Examples include the tragic killing of Columbian soccer star Andres Escobar in his homeland after his accidental goal for the United States in the 1994 World Cup and the vicious assaults on figure skater Nancy Kerrigan and tennis player Monica Seles to give a competitive advantage to their rivals. In some sports, fan violence is so common and serious that it represents a troubling social problem for society. Perhaps the best example of this type of fan violence is **soccer hooliganism** (Taylor, 1972; Maguire, 1986; Dunning, 1990; Roversi, 1991).

The murder of Andres Escobar demonstrates the intense feelings—of passion and hostility—that soccer can generate around the world. This sport has provoked many extreme cases of fan violence. In one, a questionable call by a referee in a game between Peru and Argentina in 1964 caused a violent confrontation among spectators that led to 293 deaths (Nixon, 1984: 220–221). In another, a World Cup game between El Salvador and Honduras in 1969 caused these nations to sever political and economic relations and pit their armies against each other in what has been called the "soccer war" (Eitzen & Sage, 1993: 165).

Although Great Britain has not gone to war over soccer, British soccer fans may be most notorious around the world for their hooliganism. Violent fan behavior has been associated with British soccer since before World War I, and in 1984 fans engaged in perhaps their most destructive rioting before a European Championship Cup match between British and English teams in Heysel Stadium in Brussels, Belgium. The riot resulted in the deaths of nearly forty people and injuries to more than 400 others (Gammon, 1985; Clifton, 1987). The fact that such fan violence is commonplace in soccer may explain why the game was played before all the dead bodies could be removed from the stadium (Eitzen & Sage, 1993: 165). The violent behavior of British fans has been so much feared by other European nations that at least three English teams—Leeds United, Manchester United, and the Glasgow Rangers—have been prohibited at certain times from competing against European teams because of the destructiveness of their fans (Leonard, 1988: 455).

All the forms of British soccer hooliganism that occur today—including invasions of the pitch, attacks on players and referees, property destruction, and fights with opposing fans—have occurred throughout the history of the professional game since its start in the 1870s. The relative frequency of different types of violence has varied, however, from period to period. Attacks on players and referees predominated before World War I, while fights between opposing fan groups are predominant today.

According to Dunning (1991), the recent pattern of clashes between fans is related to the fact that since the 1960s sections of soccer stadia, of the streets around them, and of specific areas in the towns and cities where professional soccer is played have become contested terrain for working class fans and the police.

Source: UPI/Bettmann Newsphotos

FIGURE 6.3 Soccer violence in the stands.

Soccer contests provide occasion for playing out violent masculinity rituals in which rival gang members battle each other to gain control of territory and assert their physical dominance. These types of fans see the aggressive professional game as symbolic of their culture, and amidst the large crowds attending soccer contests, they are often able to engage in their own symbolic and real violence without detection or arrest. Tough law-and-order responses by the police so far have been largely ineffective, serving only to solidify the gangs and intensify hostilities (Dunning, 1991). Unlike societies where sports-related violence is part of a general pattern of social violence, soccer violence in Britain seems to be an aberration. Soccer hooliganism stands in marked contrast to the generally low rate of violent crime and the civilized image of British society.

Fan violence is not restricted to professional soccer, of course. Instances of disorderly and violent fan behavior are associated with a variety of other sports, including rugby, football, hockey, horse racing, box lacrosse, auto racing, basketball, and baseball, from youth leagues and high school to the professional levels (Nixon, 1984; Leonard, 1988a; Dunning, 1990; Eitzen & Sage, 1993). In addition, a recently escalating pattern of violent celebrations of hometown victories in championships has occurred in North America, with three especially violent riots—in Dallas after the Cowboys' Super Bowl victory, in Montreal after the Canadiens won the Stanley Cup, and in Chicago after the Bulls won their third consecutive NBA championship—in the first six months of 1993 (Johnson, 1993). There is no single or simple explanation for these various kinds of collective violence, but the violence may be a reflection of growing frustration among disadvantaged inner-city residents, a decline in the effectiveness of social control agencies, and an increasing willingness to use violence to express frustrations and other emotions. Violent demonstrations are related to sport because sports arouse great passion and because sports contests often are stages for highly aggressive and violent clashes.

A continuing debate about the cathartic function of violent sport for fans focuses on whether watching violent sports serves the function of siphoning off hostile impulses that would be vented in more violent and less easily controlled ways elsewhere in society (Leonard, 1988a: 170–176; Dunning, 1990). To date, research seems to suggest that fan violence may be stimulated by a variety of game or situational factors, such as player violence, the dynamics of crowd interaction during a sports contest, the consumption of alcohol and other drugs, the history of past rivalries between the opposing teams, the perceived fairness and competence of referees, the effectiveness of crowd control, the importance of winning to fans, the social, economic, and political context of the event, and the social backgrounds of fans and their exposure to violence in their everyday lives and culture (Coakley, 1990: 154–158; Eitzen & Sage, 1993: 166–167).

POINT-COUNTERPOINT

DOES TELEVISED SPORTS VIOLENCE MAKE IT HARDER FOR US TO SEE VIOLENCE AS REAL?

YES

We see so much sports violence on television that we think of it as ordinary deviance. When we see it on television, we also think of it as entertainment and not real. When sports figures are elevated to heroes by the media for their highly aggressive play or for their success in a violent sport, we have difficulty condemning them for their violent behavior on or off the field. Thus, when O. J. Simpson was being pursued by police as a suspect in the murder of his former wife and her friend, fans lined the freeway to cheer him and expressed their sympathy following his arrest. His previously acknowledged domestic abuse and his alleged murders were no more real to them than the violence they saw so often in the sports they watched.

NO

Although televised violence in sport and violence in mass culture in general are pervasive, we are able to recognize the difference between the kind manufactured for our entertainment and real violence. Violence also is pervasive in the everyday lives of ordinary people. They suffer its real effects when they or someone they know is hurt by it, and they fear becoming victims of it. The strange initial public reaction to O. J. Simpson following allegations of murder was the result of the romanticized or sanitized media image of him and a failure to believe he could have acted so brutally. The reality of the killings horrified them, but suspicion that their sports hero was a killer mystified them and seemed unreal.

How should this debate be resolved?

Whatever the specific causes of sports violence among players or fans, research by Semyonov and Farbstein (1989) indicates that it is neither random nor sporadic. They studied nearly 300 soccer teams in Israel and found that violence was systematically related to structural factors associated with the urban ecology and ecology of the sport in which the teams operated. They also found that player violence affected spectator violence but that there was no effect of spectator violence on player violence. In their analysis of the urban ecology of violence, Semyonov and Farbstein found that teams representing communities of subordinate ethnic minorities were more violent than others and that teams representing large urban centers were relatively less violent than others. In regard to the ecology of the sport, they found that teams competing in higher level (professional) divisions and teams at either the bottom or top of high-level divisions were more violent.

It appeared that professionalization and the level of competition promoted violence in soccer. Whereas player violence was not strongly related to the urban ecology, spectator violence was significantly affected by the type of community. Semyonov and Farbstein speculated that player violence was unaffected by where the teams played because socialization into soccer is relatively similar across Israel. Players probably use each other as role models for violence, but they seem unlikely to model the violent behavior of spectators.

Crime and Sport

In the Miller Lite (1983) survey of the American sports public, three-quarters of the respondents said that athletes were good role models for children, and 59 percent said that athletes are often the *best* role models children can have. Among more serious sports fans, 84 percent felt that athletes were good role models. Evidence about the delinquency of athletes of high school age and younger would seem to reinforce these beliefs, with most studies indicating that athletes of these ages, especially those from lower-class backgrounds, are less likely than nonathletes from similar social backgrounds to be involved in delinquent behavior in general and specific criminal acts such as vandalism, robbery, and assault (e.g., Schafer, 1969; Donnelly, 1981; Segrave, 1983; Hastad, Segrave, Pangravi & Petersen, 1984; Segrave & Hastad, 1984; McPherson, Curtis & Loy, 1989: 270–271). Recent reports of the criminal behavior of college and professional athletes offer a different picture of the relationship between sport and deviant behavior.

In recent years, we have read about prominent college and professional athletes, sports agents, coaches, and owners accused or convicted of a growing list of felonies and misdemeanors, including racketeering, mail and wire fraud, extortion, murder, attempted murder, rape, assault and battery, robbery, kidnapping, possession and sale of cocaine, credit card theft, forgery and fraud, criminal trespass, misdemeanor menacing, brawling, and disorderly conduct (e.g., Leonard, 1988a: 163–164; Selcraig, 1988; Givens, 1989; Kirshenbaum, 1989; Reilly, 1989; Telander & Sullivan, 1989; Wulf, 1989; Lederman, 1990; Murphy, 1991; Blum, 1992). One popular American sports magazine (*Sport*, 1987) published a "police blotter" with the names of fifty sports figures who had been convicted of crimes in 1987 (of more than 180 who had been accused that year); another in 1989 (Kirshenbaum, 1989) called the lawlessness of college athletes "An American Disgrace"; and the *Chronicle of Higher Education* published feature articles in 1990 (Lederman) and 1992 (Blum) about the criminal behavior of college athletes. The rape trial and conviction of former world heavyweight boxing champion Mike Tyson captured national headlines. The list of sports touched by these criminal activities is long and varied, including football, boxing, baseball, basketball, hockey, soccer, track and field, power boat and auto racing, and weight lifting.

It is difficult to estimate the precise percentage of athletes at various levels of sport who engage in criminal behavior. In a survey of 350 colleges it conducted in 1986, the *Philadelphia Daily News* discovered that athletes had been implicated in at least sixty-one sexual assaults between 1983 and 1985. It estimated that football and basketball players were 38 percent more likely to be accused of such crimes than the average male college student (Kirshenbaum, 1989: 17). Some have contended that the apparently rising crime rate among college and professional athletes may be a mirror of increased crime in the larger society in the United States. But that explanation does not explain the possibly higher rates of crime for major-sport athletes than for their nonathlete counterparts on the campus or in society. It also does not explain why college and professional athletes in at least certain sports may be more deviant than nonathletes, while athletes of high school age and younger seem to be less deviant.

A number of different factors have been proposed in the mass media to explain the criminal behavior of college and professional athletes. Among them are the poor role models provided by coaches who cheat or encourage their athletes to cheat, lenient or special treatment of athletes who are not held responsible or are excused for their misdeeds on and off the athletic field, general pampering of athletes who receive extravagant attention for their athletic deeds, the recruitment into college athletics of an increasing number of talented athletes who are academically and socially unprepared for college or who have suspect character, the envy created when athletes from poor backgrounds go to institutions with a large number of affluent students who have a lot of spending money and material possessions, the major change between the strict controls imposed on athletes' behavior during the season and the freedom of the off season, and the cultural isolation of many racial and ethnic minority athletes (Kirshenbaum, 1989; Reilly, 1989; Telander & Sullivan, 1989; Wulf, 1989; Murphy, 1991).

It might be argued, too, that the team bonding that leads to a sense of superiority, disdain for outsiders, and positive deviance also might lead to risk taking and rule breaking outside sport (Hughes & Coakley, 1991). Evidence (Lederman, 1990) that male team sport athletes and fraternity members have higher rates of sexual assault than other college students is consistent with this bonding argument. It may be, as feminists have argued, that the group dynamics of male sports teams encourages sexist attitudes and contributes to a "rape culture" that influences athletes to engage in sexual aggression (Messner, 1992: 101). According to Neimark (1991), those who study gang rape are now beginning to believe that highly physical, or brutal, team sports make athletes more prone to engage in brutal group acts outside the athletic arena, such as gang rape. She noted that tennis players, swimmers, and other individual sport athletes are very unlikely to participate in gang rape, but members of football, basketball, and hockey teams are most likely to engage in this behavior. Football, basketball, and hockey players and members of other contact sport teams may be exposed to a rape culture and may translate their desire for sexual aggression into individual or group rape when they believe their privileged status as athletes entitles them to sex with women. When they feel entitled, they may not even see their sexual aggression as rape and may brag about it as evidence of their masculinity (Neimark, 1991).

An expert on campus crime, Michael Clay Smith (quoted in Lederman, 1990), noted that criminals tend to be less educated and to have lower levels of academic achievement than noncriminals. Thus, when coaches recruit athletes with deficient academic backgrounds for their institutions or professional teams draft poorly educated athletes and these programs or teams provide little social support for them, we might expect to see at least some of these athletes get into trouble with the law. Smith pointed out that minority athletes in particular are often placed in environments on predominantly white campuses that may be alien or hostile to them. If they are having difficulty with their studies or relating to other students, they could lash out with criminal behavior.

We do not have precise data showing the rates of criminal activity for college and professional athletes, and we do not know exactly why some athletes get into trouble with the law while others with similar backgrounds do not. We also lack systematic evidence about whether sport can be an effective means of rehabilitation for people who have committed crimes and are in correctional institutions. Even if it does not prevent recidivism or help redirect the lives of prison inmates, some prison officials and correctional recreation personnel believe that sport has a therapeutic effect, constructively occupies inmates, improves their morale, and gives some of them a sense of hope about their lives, (Telander, 1988).

Gambling

Sometimes highly institutionalized and popular activities in a society are associated with serious forms of social deviance. Examples in sport are the association of gambling with the social problems of point shaving (intentional reduction in scoring by players), the fixing of sports contests, and compulsive gambling. The popularity of sports gambling in the United States is well documented, and estimates of the amount of money illegally bet on team sports has ranged from $20 billion to $100 billion per year (Frey, 1985; *Sports Illustrated*, 1986; McPherson, Curtis & Loy, 1989: 139; Wilstein, 1993). With a definition of sports gambling that excluded horse racing, dog racing, and jai alai, Smith (1990) reported estimates of the gross revenue retained by operators from illegal sportsbooks and sportscards in the United States to be $2.4 billion, based on a total illegal sports betting volume of $48 billion, and a total illegal sports gambling volume of $1 billion per year in Canada.

Legal sports betting in Nevada has been the fastest growing form of gambling in the United States in recent years. The betting action in Nevada sportsbooks primarily involves professional and college football (39%); professional and college basketball (34.5%); baseball (23%); and hockey (2%). Wagers on boxing, golf, and tennis make up the remaining 1.5% (Humber, 1988). According to Smith, these rankings are similar to the action in illegal bookmaking, with a major difference being that professional football draws a larger share of the action in illegal betting. These rankings are also consistent with the feelings about betting on different sports expressed in a recent poll of American males and females aged 12 and over (Wilstein, 1993). This poll also showed that females had much weaker positive feelings and much stronger negative feelings than males about gambling on all of these sports.

In Canada, the most popular sport for bettors is NFL football, followed by U.S. college football, baseball, basketball, hockey, and Canadian professional football. Canadian bettors may be less interested in their indigenous sports than American sports because, in the case of hockey, bettors may prefer pools to team-versus-team bets, and in CFL football it may be difficult to establish realistic betting lines because of the lack of parity in the league (Smith, 1990).

Sports gambling has grown a great deal over the past two decades for a variety of reasons. Smith (1990)

POINT-COUNTERPOINT

WOULD THE LEGALIZATION OF ALL U.S. SPORTS GAMBLING BE HARMFUL FOR SPORT AND THE SOCIETY?

YES

Sports gambling is associated with organized crime. Legalizing all sports gambling would provide criminal syndicates with more opportunity to operate freely, and it would encourage them to exert more influence over the outcomes in sports of all kinds in the United States. Those who report betting lines, odds makers, and others who assist gamblers would gain greater prominence and would distract more attention from the enjoyment of sport for its own sake. More people would be interested in final scores than in how the game was played. Children would grow up with a jaundiced view of the meaning of sports. Legalization would create more gambling addicts because it would loosen the controls that now restrain potential addicts.

NO

Sports gambling already is widespread. It is legal in certain sports in the United States such as horse racing, and it is legal in many countries and in states such as Nevada. It is not different than lotteries, which exist in many American states. Gambling has elements of risk taking, challenge, and excitement, which are part of the American cultural heritage. Gambling teaches people how to take risks and how to recover from losses, which are important aspects of the capitalist system of America. Gambling is a victimless crime and, when decriminalized, a harmless form of recreation. It will be easier to regulate when it is legalized under government control, and it can generate tax revenue if it is legalized.

How should this debate be resolved?

proposed that these reasons include a gradual weakening of community attitudes against victimless crimes such as gambling; legalization of new forms of gambling, such as lotteries, in a number of states; the increased number and geographical spread of professional sports teams; and increased media coverage of sports. Since most sports gambling involves friends and acquaintances and relatively small sums of money (Smith, 1990), a number of experts on gambling have proposed that it can serve as a positive form of recreation, containing elements of risk taking, challenge, and excitement that can provide relaxation, fun, and escape (Putnam, 1986). Frey (quoted in Putnam, 1986) contended that fathers may encourage their sons to play poker at a young age because it can teach them how to handle money and how to deal with risk, challenge, and competition, which are all important aspects of the success striving that is central to the pursuit of the American Dream.

Scholars, journalists, politicians, and sports officials have engaged in an ongoing debate about the legalization of sports gambling. They have debated the relative benefits and costs of legalized sports gambling for individuals, sport, and society (*Sports Illustrated*, 1986; Smith, 1990). Practical, political, and moral perspectives shape whether or not one sees sports gambling or its effects as a social problem. Recent poll results (Wilstein, 1993) showed that 55 percent of American males and 44 percent of American females believed that since people were likely to bet on sports whether or not betting was legal, sports gambling should be legalized and taxed. Sports officials have expressed a very different view. A study of 214 coaches and 127 athletic directors at NCAA institutions revealed that they almost unanimously agreed that legalization of sports gambling would be harmful for college athletics (Frey, 1984). They opposed the circulation of point spreads (estimates of the number of points by which one team will defeat another), which would emphasize by how much a team won or lost rather than whether they won or lost. They also feared that legalized gambling would increase the contact of their players with gamblers and increase the incidence of fixed games. While strongly opposed to sports gambling, almost all (93%) of the coaches and athletic directors said that gambling was not a problem on their campus, and thus few of their campuses allocated money or staff for a gambling awareness program.

Gamblers have most tainted the integrity of sports when they have paid players to manipulate or fix the outcome of games within the bookies' pregame point spreads. College basketball has had a series of fixing scandals. The first major one surfaced in the early 1950s (Straw, 1986; Guttmann, 1988: 75–76). A grand jury convened to investigate the point shaving and related gambling network discovered that the scandal involved at least eighty-six games and thirty-two players from seven colleges, including CCNY, Long Island University, Manhattan College, New York University, Bradley, Toledo, and the powerhouse of that era, the University of Kentucky. Subsequent well-publicized fixing scandals occurred in college basketball in 1961 (Guttmann, 1988: 76) and in the 1980s. The scandal at Tulane in 1985 led its president to do what the University of San Francisco had done a few years earlier in a similar situation—eliminate the entire basketball program for a period of time.

Fixing has not been restricted to college basketball, as the infamous Black Sox scandal in baseball (Voigt, 1976:ch. 5) and evidence of "bagged" horse races (Lesieur, 1977:35), dubious knockouts in boxing (Nixon, 1976: 27), and fixed jai alai matches (Kirshenbaum, 1981) have shown. Its history in the United States stretches back to the first documented fix—of a horse race—in 1674 (Shecter, 1969). In fact, many popular sports are vulnerable to fixes by gamblers, but college basketball may be especially vulnerable because bribing one or two players can make a relatively larger difference in basketball than in other sports; because high scores and quick turnovers make it more difficult to detect obvious manipulation of basketball scores; because the high scores enable players to beat the point spread without losing the game; because unlike their professional counterparts the numerous talented college basketball players from economically deprived backgrounds may be tempted by the thousands of dollars gamblers offer; and because highly recruited players may have difficulty distinguishing the illegal inducements they are offered when they are recruited from bribes they are offered to fix games (Guttmann, 1988: 79–80).

Fixing threatens a sport because if it does not have reliably honest outcomes a sport risks losing its fans and probably also the coverage and financial investment from the mass media that enable it to remain popular and profitable if a fix is uncovered. A sport with fixed games also loses its appeal to gamblers, and gamblers are among the most avid followers of sports with betting lines. These people may be less interested in the sport itself than in following their bet, but for the mass media and sports league they add numbers to their audience, and these numbers mean dollars.

The increasing popularity of sport among gamblers and the reliance of corporate sports on the mass media help explain why it is so difficult to disentangle a number of sports from gambling connections. We are reminded by the Pete Rose gambling case that sports officials are very harsh in their treatment of stars who gamble on their sport, especially when there are ties to organized crime (e.g., Murphy, 1989; Neff, 1989; Lieber & Neff, 1989). Major League Baseball, like other professional team sports, prohibits its players, umpires, club and league officials, and employees from betting on games in their sport, and they generally impose harsher penalties for bets on games in which bettors are directly involved (Lieber & Neff, 1989: 13). Despite such prohibitions and suspensions of convicted bettors within their sports, sports leagues have been reluctant to pressure the mass media to stop publicizing point spreads and other information that benefit gamblers, such as odds, betting lines, weekly injury lists, weather reports, and inside information from sports experts (Kaplan, 1986; *Sports Illustrated*, 1986).

Sometimes the officials of a sport or sports league take a stand against gambling. For example, the NCAA has stated in its negotiations with television networks that it wanted touts who report the odds excluded from college sports telecasts, and it has had a policy of withholding credentials for the Final Four basketball tournament from publications that provided an excessive amount of gambling information (*Sports Illustrated*, 1986). The hold of gambling on sport nevertheless seems deeply entrenched and secure as long

SPECIAL FOCUS

The Pathology of Compulsive Gambling

A by-product of gambling that may affect the integrity of sports is pathological or compulsive gambling by people in sport (Lesieur, 1977). Lesieur (1987) has contended that like alcoholism, drug addiction, racism, and violence, pathological gambling is a social problem in sport. The pathology of a gambling addiction can afflict fans, athletes, coaches, and others in sport. This addiction is pathological because it can lead to suicide attempts and substance abuse, disrupt families, occupations, and social relations, and push addicts into crimes as a way of trying to escape their deepening gambling debts. Lesieur noted that pathological gambling could be construed as positively functional for corporate sports because addicted gamblers were avid sports fans who spent their money on sports in various ways to support their addiction and because they often worked in the gambling industry in roles such as bookmaker, providing services to sports fans who were nonaddicted gamblers. As a pathology or social problem, however, compulsive gambling can be very dysfunctional for gamblers and for sport. It is especially dysfunctional for sport when the addicts are prominent athletes or coaches who are made more vulnerable by their addiction to fixers and others intent on corrupting them or their sport for their own profit.

Even when such athletes and coaches do not throw games or bet against their team, the negative publicity they receive for their addiction, the big debts they accrue from gambling, their contacts with crime figures, and their desperate and sometimes illegal efforts to obtain money to feed their addiction or pay their debts can tarnish the image of their sport as well as wreck their own reputations and careers. The personal devastation that can result from this addiction cannot be better illustrated than in the case of Pete Rose, who was transformed in the eyes of many from a hero to a criminal and who lost his chance for a nearly certain first-ballot election into the baseball Hall of Fame (Lieber & Neff, 1989).

as sport is appealing to gamblers, and it remains tied to the mass media that provide essential information to fuel the legal and illegal gambling industries. The glue that holds together the mass media–sport–gambling alliance is money—from audience shares and circulation for the media business; from viewers, listeners, readers, and spectators for the sports business; and from ever-hopeful bettors for the gambling business.

Drugs and Sport

Gambling addictions and associations with gamblers have ruined many promising and prominent sports careers. In the past two decades, though, many more sports careers have been ruined by drugs than by gambling. Both gambling and drug use often involve illegal activity that could result in addictions. Among athletes, the development of drug dependence may be a more insidious process than the development of a gambling addiction because athletes may be drawn to the drugs as a means of directly helping their athletic performance or relieving the physical or emotional stresses produced by athletic effort.

Athletes use restorative drugs, such as painkillers, tranquilizers, barbiturates anti-inflammants, enzymes, and muscle relaxers, to help them overcome the adverse effects of illness, injury, pain, anxiety, or other such physical or emotional conditions on their athletic performance. They also use drugs such as alcohol, marijuana, and cocaine for recreational purposes off the field. Furthermore, they use additive or performance-enhancing drugs, such as anabolic steroids, amphetamines, and methamphetamine or "speed," to stimulate their athletic performance. Although the excessive, unsupervised, or otherwise inappropriate use of restorative and recreational drugs can be illegal and cause serious problems for athletes, their use of additive drugs is especially troublesome for sports officials because athletes may see it as an extension of the Sport Ethic.

Over two decades ago, sports journalists were calling the increasing use of drugs by athletes "a significant menace to sport, one that the athletic establishment is assiduously trying to ignore" (Gilbert, 1969, July 7: 30). More recently, it has been impossible for sport to ignore its "drug problem." The coupling of easy access to drugs of various types in society and in sport with tremendous pressures on athletes to win and improve their performance has made excessive, illicit, and addictive drug use by athletes a widespread and deeply entrenched problem in sport.

In the 1960s and 1970s, amphetamines and speed reportedly were widely used by athletes in certain sports in the United States, such as NFL football (Meggyesy, 1970; Oliver, 1971). According to one study, these drugs were supplied to NFL players by team trainers and doctors in many instances to help them get a competitive edge on opponents (Messner, 1992: 77–78). Over the past decade, the additive drugs causing the greatest controversy have been anabolic steroids. Athletes in sports such as football, weight lifting, and track and field have taken steroids to add body mass, strength, and aggressiveness so that they can perform better (Messner, 1992: 78). In using these drugs, they knowingly or unknowingly risk heart attacks, sterility, ulcers, liver tumors, and psychological and emotional instability. Fuller and La Fountain (1987) interviewed a self-selected sample of fifty weight lifters, football players, wrestlers, and bodybuilders who admitted steroid use to learn how they rationalized such risks to their health as well as the unfair advantage gained from such drug use and the illegal nature of this activity. The athletes ranged in age from fifteen to forty, with an average age of nineteen. These athletes saw their drug use as a victimless crime that would not harm anyone else. Most said they would consider discontinuing their use of steroids if they knew their opponents did not use drugs, but they believed that being a serious competitor at the national and international levels in their sports, especially against athletes from Eastern Europe and the former Soviet bloc, required steroids. The fear that opponents used steroids was an important part of the weight lifting and football subcultures, motivating athletes in these sports to use these drugs. In this subculture the idea that success required steroids was powerful, even if it was not necessarily true.

The athletes were able to sidestep health concerns mainly because they were poorly informed about the actual effects of steroids on their bodies. They tended to be uninterested in learning more about scientific evidence about steroids, and they typically looked at their bodies as mechanical devices to help them achieve their performance goals in sport. The information that guided their steroid use tended to be drawn from more experienced athletes at the gyms where they worked out. One of the ways they confronted the evidence or arguments they heard about steroid use was to question the knowledge, competence, and motives of steroid critics.

At the international level, the various forms of dangerous, dubious, and illegal drug use have been collectively defined as **doping**. First defined and prohibited by the International Olympic Committee in 1967, the term refers to "the use of substances alien or unnatural to the human body with the aim of obtaining an artificial or unfair increase of performance in competition" (Coakley, 1990: 129). In recent years, doping has referred generally to drug use and specifically to the use of steroids. According to Hoberman (1992: 103), doping represents an "ideology of uninhibited performance" and for this reason has had a strong appeal to elite athletes.

Athletes can justify their additive drug use in terms of the Sport Ethic, but among sports bureaucrats and in the mass media and the mind of the public, doping with steroids and similar substances is illegitimate. In the battle pitting drug-using athletes against sports authorities and the press, the athletes have lost. Athletes are now subjected to drug testing and can be suspended or banned for life from their sport if their testing reveals banned substances in their body (Coakley, 1990: 129–135). Media antidrug campaigns have left little doubt about the ill effects or illegitimacy of doping (Davis & Delano, 1992). Survey results (Miller Lite, 1983) have shown that a large majority of Americans favor drug testing of professional athletes and that 89 percent of the general American public, 87 percent of sports journalists, 92 percent of coaches, and 92 percent of sports physicians have said that the use of drugs to increase body weight or strength should never be allowed in sports.

The antidrug messages of sports bureaucrats and the mass media make a strong argument against doping. Yet as critical observers (e.g., Davis & Delano, 1992; Hoberman, 1992) have pointed out, the line between the use of currently banned drugs and other forms of performance enhancement in sport is not perfectly clear. In view of the increasing reliance on new technologies and an array of experts on exercise science, sport nutrition, sport psychology, and even sport pharmacology at many levels of sport, Davis and Delano (1992) asked why certain techniques and practices are treated as natural and legitimate whereas others are condemned and forbidden as artificial or unnatural. Hoberman (1992) pointed to the ingestion of substances by athletes dating back to ancient Greek times and occurring across different cultures, and suggested the difficulty of distinguishing between natural and artificial substances in these varied historical and cultural contexts.

The tragic case of Birgit Dressel (see the Special Focus discussion) graphically demonstrates the costs of doping and the underlying motivations tied to the Sport Ethic and personal ambition. Other athletes have died from doping, but the case of Ben Johnson may have captured even more attention than those cases because Johnson performed under the international spotlight in a glamor event in which he had surpassed the most gifted and accomplished American sprinter of the era, Carl Lewis. Acclaim for Johnson's achievement turned to dishonor when it was discovered that he had used steroids (Johnson & Moore, 1988). Although his physician, coach, trainer, and managers directly or indirectly collaborated in his doping, Ben Johnson bore the responsibility for it, which one Canadian television announcer described as "a tragedy for the country and the Canadian Olympic movement" (Eitzen & Sage, 1993: 168).

Sports officials and coaches who decry doping may contribute to it by pushing qualifying standards to increasingly higher levels. In the intensely competitive world of elite athletes, many share the awareness that higher levels of performance cannot be achieved without doping (Hoberman, 1992: 248–249). Thus, athletes engage in doping at least partially because they are involved in a world that is structured by an "ideology of uninhibited performance" and the use of modern science and technology to encourage it.

Along with illicit additive drug use, recreational drug use by athletes has been a social problem in sport in recent years. A study of a professional hockey team revealed that almost every member of the team engaged in some type of "juicing, burning, or tooting" (alcohol, marijuana, or cocaine use) (Gallmeier, 1988). There have been a number of highly publicized cases of cocaine use by professional athletes (e.g., Chass & Goodwin, 1986), but alcohol very likely has been the drug most abused by athletes at all levels of sport.

A 1990 *USA Today* survey of nearly 800 high school coaches revealed that 88 percent thought alcohol was the drug that posed the greatest threat to athletes on their team, whereas 6 percent said cocaine or crack, 3 percent said marijuana, and 1 percent said steroids (Eitzen & Sage, 1993: 169). These results may reflect general patterns of concern and actual drug use across many levels and types of sports, but they do not mean that the use of recreational drugs other than alcohol is restricted to a few well-publicized cases. A 1984 study revealed, to the contrary, that 17 percent of American college athletes admitted to using cocaine in the previous year, and a second study conducted in 1986 indicated that 7 percent of the elite female athletes surveyed said they used cocaine, with approximately half saying they used it before or during competition (Mottram, 1988; Wadler & Hainline, 1989; Eitzen & Sage, 1993: 169). With its stimulant effects, this latter use of cocaine could be viewed as a form of doping. Although alcohol, cocaine, crack, and marijuana have differing effects (see Voy & Deeter, 1991)—with cocaine, for instance, producing a high and alcohol acting as a depressant—a number of different drugs could provide at least temporary relief from what one drug expert called "the pressure of having to perform in the public eye relentlessly" (quoted in Chass & Goodwin, 1986: 307).

The most spectacular and tragic drug-related cases in sport, such as the steroid-related fall of Ben Johnson and the cocaine-induced deaths of college basketball star Len Bias and NFL player Don Rogers, at least briefly cause a flurry of news coverage and widespread

SPECIAL FOCUS

Doping in German Sports

The East German Olympic teams achieved success that was disproportionate to the size of the nation (Tedeschi, 1991). Recognized for their excellent coaching and sophisticated athlete selection and training techniques, the international sports success of the former German Democratic Republic came into disrepute after the acknowledgment that the use of performance-enhancing drugs was a state policy. The East German doping program for athletes was secret and centrally administered. Athletes were expected to take their drugs and remain silent about it. Some East German athletes reportedly complained that teammates were receiving stronger drugs than they were and thus had an unfair competitive advantage (Hoberman, 1992: 222). These athletes were given steroids, hormones, and other substances, and East German scientists were committed to developing a variety of new biochemical and technological means to boost athletic performance.

Scientists and physicians in the former East Germany rationalized their doping program as serving therapeutic as well as performance-enhancing purposes, and some discounted the dangers. The chief of medical research at the Research Institute for Physical Culture and Sport in the GDR claimed in a 1990 interview that steroids were used in his country "out of concern for the condition of the athlete" who was drained by the heavy and constant demands of high-intensity training (quoted in Hoberman, 1992: 223). He and other physicians also argued that their involvement in the doping program was influenced by pressure from "medal-hungry sports officials" in their country (Hoberman, 1992: 224). In fact, though, the doping program represented collusion between the athletic and medical establishments of the German Democratic Republic, and a number of prominent athletes who had benefited from this application of sport science, such as swimmer Kristin Otto and javelin throwers Ruth Fuchs and Petra Felke, were unapologetic or even defiant in asserting the legitimacy of their doping as a reflection of their state ideology of uninhibited performance and Sport Ethic (Hoberman, 1992: 227).

Although revelations about the GDR's state-sponsored doping program in 1990–1991 created a substantial amount of media coverage of this issue, East Germany clearly was not alone in these practices. Indeed, it may be that the difference between doping in capitalist and former socialist countries such the GDR was that it was done much less systematically and competently in the capitalist countries. After all, as an East German sports medicine official observed, "There was no Dressel case in our country, because we didn't leave the field open to medical charlatans" (quoted in Hoberman, 1992: 227). Birgit Dressel was a West German heptathlete who died in 1987 at the age of twenty-six after a career of massive doping that included over 400 injections of scores of substances by her primary sports physician and by self-administration; Dressel was probably encouraged and monitored by her trainer-lover, who shared her ambition to become a successful international sports star (Hoberman, 1992: 1-2). The Dressel case created a furor for the West German sports establishment, but the boundaries of doping have stretched well beyond Germany (Coakley, 1990: 129–131; Eitzen & Sage, 1993: 167–168; Noden, 1993). For example, one unofficial poll done by a member of the U.S. track team in 1972 indicated that 68 percent of track and field athletes preparing for the 1972 Munich Olympics used some form of anabolic steroids (Todd, 1987).

public concern. Yet research (Donohew, Helm & Haas, 1989) suggests that exposure to a steady stream of stories about athletes' failing drug tests and entering drug rehabilitation clinics may lessen public interest in drugs in sport or the media's perception that such stories are newsworthy. Like other kinds of deviance in sport, most forms of drug use by athletes may now be seen as "ordinary deviance." The fact that Americans live in a society where drug use of various kinds is widespread makes it easier to expect, accept, or ignore

POINT-COUNTERPOINT

SHOULD ATHLETES BE PERMITTED TO USE PERFORMANCE-ENHANCING DRUGS?

YES

The point of sports is to compete and win, and this point is reinforced by the Sport Ethic. With modern science and technology constantly raising the levels of sports performance and with the rewards from winning increasingly lucrative for athletes and coaches, it is necessary to do whatever is possible to win. Doping is a problem because certain athletes have access to performance-enhancing drugs and others do not. If everyone could use these drugs under the supervision of sport scientists and medical personnel, no one would have an unfair advantage. Proper supervision would also minimize the likelihood of adverse effects from additive drug use.

NO

The use of alien or unnatural substances to enhance performance detracts from the meaning of sport. Doping pits pharmacologists, sport scientists, and physicians against one another, and the pure effort of athletes, which is the essence of sport, is lost. Even experts cannot be sure when additive drug use will hurt individual athletes, and thus "permissible doping" inevitably will cause some cases of serious side effects and may lead to long-term debilitation or death. Although it has been difficult to rid society of illicit drug use, sport should not permit or tolerate drug use because it is difficult to control. Effective drug testing and enforcement can stop the use of banned substances.

How should this debate be resolved?

drug use in sports, especially among those who are more cynical—or realistic—about modern sports and athletes.

CONCLUSION

The Dominant American Sports Creed emphasizes character building and other positive functions of sport, and it focuses more on conformity and obedience than on social deviance (Edwards, 1973). Nevertheless, social deviance and social problems are tightly interwoven in the institutional fabric of corporate sports of all kinds and at different levels. Athletes are exposed to powerful social conditions and people in sports that lead them into experiences of violence, crime, gambling, fixing, and drug use on and off the athletic field. In some cases, athletes' dedication to the Sport Ethic, which is strongly encouraged and supported by coaches, sports administrators, the mass media, and fans, can become excessive and turn into a positive deviance that is self-destructive, harms others, or threatens the integrity of their sport.

When athletes become too violent and seriously injure an opponent or become too dependent on drugs and get caught, positive deviance becomes mere deviance. In such cases, applause is likely to be transformed into disapproval, condemnation, or disgrace, and athletes often are left confused, defensive, and alone. They were only doing what they were expected to do or had to do and what everyone else did, they would argue, but this reversal of fortunes reveals the complexity of social deviance.

Deviance is not necessarily what "bad" people or people with bad intentions do; it is typically not defined by the act itself; it is not defined the same way by others all the time; and it is not always easily distinguished from conformity. Positive deviance results from doing too much of the "right" thing. In some situations, people may be expected to do what is officially considered deviant or a violation of the rules, and when they are caught, the people who told them to engage in the deviance—for instance, "taking out" the star receiver or quarterback of the other team or using steroids to "bulk up"—disavow their association with the deviant behavior and proclaim their belief in more virtuous behavior or athletes. When this happens, it reveals another important aspect of the process of social deviance in sport. That is, the definition of what and who is deviant is shaped by powerful people, and in battles over such social definitions athletes usually find themselves vulnerable or powerless in the face of the power structure. The story of Ben Johnson clearly illustrates this point.

SUMMARY

Social deviance and social problems are realities of sport that fans often would like to ignore because they taint the image of sport as a symbolic refuge. Although social problems include conditions causing suffering in society as well as social deviance, problems of social deviance are the main focus of this chapter. Both social deviance and social problems are culturally defined and can be understood from a variety of sociological perspectives, including interpersonal, role, anomie, conflict, critical, and societal reaction labeling approaches. Positive deviance occurs when people conform so intensely, extensively, or extremely that their behavior goes beyond conventional boundaries. Athletes and coaches who embrace the Sport Ethic too zealously are inclined toward positive deviance. Although positive deviance superficially appears to be virtuous because it is inspired by institutionalized norms and values such as the Sport Ethic, it can have destructive effects, as in excessive risk taking and dubious efforts to play with injuries and pain.

Sometimes norm violations in sport are ordinary deviance and are not treated as social problems. This type of deviance does not prompt ethical questions or cause a moral outcry. There are, however, a number of major problems of deviance in sport that raise serious ethical questions, create great moral concerns, and arouse strong official and public disapproval. These problems include player violence, fan violence, criminal behavior, gambling, and illicit and dangerous drug use.

CHAPTER 

High School Athletics

FRIDAY NIGHT LIGHTS

Athletes, coaches, parents, and fans can become highly passionate about high school athletics in the United States. The message of the 1971 film *The Last Picture Show* is that high school sports are the biggest show a small town has to offer, and star high school athletes and coaches are among their biggest heroes. A more recent account of the powerful impact of high school sports on small-town America is *Friday Night Lights,* by H. G. Bissinger (1990). Bissinger narrates how high school football transformed a hot, dusty, depressed West Texas oil town into a place of excitement, hopes, and dreams on Friday nights, when as many as 20,000 fans packed the Odessa football stadium to watch the Permian Panthers seek victory and glory. His story reveals how high school sport can capture the life of a community, how it can arouse devotion and hostility, how it can provide a means for adults to live vicariously through young men's accomplishments in a game, and how it can reflect a community's social aspirations and tensions.

In this chapter, we examine in more detail what interscholastic athletics means to those who participate in it and support it. We also consider the effects of high school athletic participation and athletic programs on athletes and the school, the roles and influence of high school coaches, and the relationship of interscholastic athletics to the cultural and social order of the community. In addition, we look at how athletics in American secondary schools compares with school sports in some other countries.

MEDIA IMAGES OF INTERSCHOLASTIC SPORTS IN THE UNITED STATES

To gain a better understanding of the character, significance, and impact of interscholastic athletics in the United States, let us consider a range of images of American high school sport and athletes conveyed by newspaper and magazine headlines:

- "Officials Study Violence in High School Hockey" (*Burlington Free Press*, November 10, 1981). Approximately 200 coaches, administrators, and hockey officials in Vermont attended a conference to learn more about violence on and off the ice in high school hockey.
- "Losing Coaches Threatened" (*Burlington Free Press*, October 14, 1983). Coaches of a high school football team with a losing record in Baytown, Texas, near Houston, received threatening phone calls, including at least one death threat.
- "Attacks on Sports Officials Spur Bill for Tougher Penalties" (*Charlotte Observer*, April 2, 1993). The North Carolina legislature was expected to debate a bill to create stiffer penalties for attacking sports officials because threats and violence seemed to be increasing against umpires and referees.
- "Mixed Messages: California's 'Spur Posse' Scandal Underscores the Varying Signals Society Sends Teens About Sex" (*Newsweek*, April 12, 1993). This story is about a clique called the Spur Posse, which included many top athletes, at

Lakewood High School in California, that kept count of the number of young girls with whom members had engaged in sex. Some clique members bragged of more than sixty such "scores," and it appeared that these young men were more concerned about their number of scores than the sex itself and that at least some of them saw nothing wrong with what they did.

- In a related article ("A Town's Divided Loyalties," *Newsweek*, April 12, 1993), it was reported that eight members of the Spur Posse had been arrested and were to be charged with a total of seventeen felony counts of "lewd conduct, unlawful intercourse and rape." Their victims allegedly were seven girls aged ten to sixteen. The local district attorney ultimately decided to press charges only against one minor, who was charged with having sex with a ten-year-old girl. Public opinion in Lakewood seemed to be mostly against the clique members, but a few parents defended the boys.
- "'No Pass, No Play' Showdown Draws Nearer in Texas" (*Burlington Free Press*, May 24, 1985). The state of Texas had to confront at least thirty lawsuits from angry parents and their sons and daughters who had been put out of action as a result of a new rule in Texas prohibiting high school athletes flunking one course from participating in extracurricular activities for six weeks. The legal battles ultimately reached the Texas Supreme Court.
- "Why Johnny Can't Play" (*Sports Illustrated*, September 23, 1991). This article considers financial problems facing many high school athletic programs. School administrators, with diminishing funds, have been forced to make difficult budgetary decisions between the competing needs and demands of academic and nonacademic programs and personnel.
- "Stackhouse Decision Will Have Impact Around State: Effects from Stackhouse's Move to Oak Hill Go Beyond Basketball" (*Charlotte Observer*, August 30, 1992). The decision by high school basketball star Jerry Stackhouse to transfer from Kinston High in North Carolina to Oak Hill Academy in Mouth of Wilson, Virginia, for his senior year was expected to cost the North Carolina High School Athletic Association at least $100,000 in gate receipts in the 1992–93 basketball season, according to the executive director of the association. The executive director stated that high school athletics was a business and that superstar players "generated publicity, crowds, and increased revenue." Stackhouse lamented the comment by an athletic director in his conference that they had lost "a great commodity." The young basketball star preferred being viewed "as a basketball player and even more as a person."
- "Miles from Nowhere" (*Sports Illustrated*, February 9, 1987). Virginia's small and remote Oak Hill Academy was a coed boarding school that mainly educated troubled teenagers in a strict Baptist environment. Faced with shrinking enrollment and financial resources, in 1976 it began recruiting star inner-city high school basketball players with scholarship offers, and its fortunes changed. Oak Mill Academy's teams and players gained national reputations, its student enrollment more than doubled, and its operating budget went from $200,000 before big-time basketball to $1.7 million in 1987. The attraction of the school has been enhanced by the declining quality of many public schools, especially in poor communities; the various temptations and destructive forces to which many urban youths are regularly exposed; and the increasingly stringent academic eligibility requirements in college basketball. Oak Hill Academy has tried to prepare its students "to go to college and stay there."
- "Injuries Incurred by High School Athletes Lead to More Lawsuits Against Coaches" (*Wall Street Journal*, September 23, 1992). Until recently, grounds for these high school sports injury lawsuits generally focused on inadequate supervision or defective equipment. Today, parents and athletes are more inclined to sue coaches over sports injuries, and lawyers are finding increasingly novel ways to apply liability law to high

school sports. Suits now may involve such things as rule violations by the opposing team and alleged violations of athletes' constitutional rights.

MAJOR ISSUES AND PROBLEMS IN AMERICAN INTERSCHOLASTIC ATHLETICS

Threats and assaults against players, coaches, and referees; a sex scandal involving prominent athletes; academic pressures; financial woes and athletic program cutbacks; big business; celebrity status and the accompanying pressure for athletes; the use of athletics for publicity, student recruitment, and the generation of revenue; athletes as commodities; injuries; and litigation. The broad sweep of images and issues represented by this admittedly unscientific sampling of headlines and stories suggests the opportunities, problems, complexities, and prominence of high school athletics in the United States.

A panel of six high school athletic directors representing schools across the United States echoed the sentiments, concerns, and issues raised by this sampling of headlines (NIAAA, 1991). These athletic directors identified the following issues as problems for the 1990s:

- Funding and the need for coaches to raise money for their programs
- Recruiting and retaining qualified coaches and referees
- Controlling violence
- Maintaining sportsmanship on the field and in the stands
- Convincing the public of the academic merits of athletics
- Growing specialization and professionalization of athletes
- Abuse of illegal or dangerous substances such as steroids and alcohol
- Providing insurance coverage and protecting against legal liability for injuries
- Intense competition to be a member of the cheerleading squad
- Lack of female role models in coaching for female athletes

Hill and Simons (1989) concentrated on the issue of sport specialization in their study of 152 high school athletic directors in Illinois. **Sport specialization** is the restriction of athletic participation to a single sport, with training, practice, and/or competition throughout the year. The athletic directors saw a pattern of increasing specialization in high school athletics over the 1980s and into the 1990s. The factor perceived by the most athletic directors (80%) as a cause of this trend is pressure from coaches. The coaches identified eight other contributing factors. Ranked in descending importance, they are high parental expectations (77.3%); athletes' desire to participate in state championships (71.8%); encouragement from college recruiters (71.3%); societal trend toward specialization (66.4%); community emphasis on a particular sport (55.6%); availability of summer camps (55%); athletes' desire for pro athletic careers (50.7%); and stiff competition for varsity positions (49.3%).

Whereas most athletic directors responded that sport specialization enabled athletes to enhance their skills in a sport, made teams more skillful, and increased the likelihood of earning a grant-in-aid to play at the college level, 55.6 percent also said that it increased the possibility of athletic burnout. The majority of athletic directors also indicated that sport specialization had negative effects on the total athletic program of the school, on relations with school administrators, and on relations with other coaches. Slightly more than half said that their schools encouraged parents of athletes to think about diverse sports participation for their children.

Sport specialization in high school is part of a larger pattern of rationalization that has emphasized external regulation and control, efficiency, and cost-accounting and has characterized the modern development of sport at all levels (Guttmann, 1978). Specialization and rationalization reflect the increasing seriousness with which sport is viewed, at least in the United States, by youths, their parents, and the adults who organize and control American sports. Increasing seriousness has been associated with a strong emphasis on winning and the constant need to improve, which are basic elements of the professionalization of attitudes that Webb (1969) identified in childhood and adolescent play and games over twenty-five years ago.

The existence of major issues and problems in high school athletics does not mean that all high school athletic programs are conflict ridden or misguided and

POINT-COUNTERPOINT

IS SPECIALIZATION A POSITIVE TREND FOR ATHLETES AND COACHES IN HIGH SCHOOL ATHLETICS?

YES

Sports have become very competitive, making it increasingly difficult to make the team in high school and college athletics. Coaches know they can count on the commitment of athletes who are devoting all their serious athletic attention to one sport; they can expect more from these athletes; and they can develop their team play to high levels of proficiency when they have athletes totally committed to their sport. High school athletes who specialize in one sport have a better chance than those who do not of developing their full athletic potential in a sport, of earning a college athletic grant-in-aid, of playing at the Division I college level, and of becoming a professional athlete.

NO

The purpose of high athletics is not to produce professional athletes. High school athletic participation is an extracurricular activity that is supposed to be fun and have educational benefits. Adolescents develop more from having a range of social and athletic experiences than from concentrating on a single activity. Specialization reflects increasing seriousness and competitiveness, which can cause burnout, an unhealthy obsession with winning, or an unrealistic expectation of high-level athletic success. Less talented athletes who specialize in one sport are left with limited athletic alternatives when they fail to make the varsity team in their specialty sport.

How should this debate be resolved?

lack redeeming qualities. In fact, high school sports are like the community and society in which they are embedded, and they embody the diverse and sometimes conflicting patterns of social life in their community and society. Thus, American high school athletics produces hopes and dreams as well as discouragement, excitement as well as tensions and anxieties, honor along with shame, promise along with problems, and soaring accomplishments as well as crushing disappointments.

INTERSCHOLASTIC ATHLETICS AND AMERICAN EDUCATIONAL VALUES

The unusual amount of emphasis on interscholastic athletics in the United States may be the result, in part, of the tendency to value intellectual development, academics, and education less in the United States than in other countries (Ferrante, 1992: 362–364). Among the reasons given by experts for these cross-national differences in educational orientation is the tendency of American schools to have multiple and sometimes conflicting curricular and social goals and functions, including pursuit of excellence, attainment of social justice, enhancement of psychosocial development, transmission of culture, and training for adult roles. The dilution of a strictly intellectual or academic orientation, reinforced by the highly pragmatic view of education held by Americans, has made it easier for educators to justify the pursuit of nonacademic programs, such as athletics, in the American high school.

With waves of new immigrants at the end of the last century and the beginning of the present one, the socialization or enculturation of these Americans was perceived to be an important function of the high school and a useful function of high school sports. The development of sports programs in American public high schools generally followed the lead of the New York Public School Athletic League, which was formed in 1903 (Nixon, 1984: 71–77). Proponents of these early programs were similar to contemporary proponents in supporting socialization or character-building functions of interscholastic sports participation. That is, by participating in school sports, students ostensibly could learn values or skills that could lead to better adjustment or success as adults in society. The entry of the United States into World War I provided another rationale for school sports and physical education because they served as means of improving the physical fitness of the young men who would be called into military service (Nixon, 1984: 74).

When industrialization required factory workers who were reliable, disciplined, obedient, healthy, and hard working, the Dominant American Sports Creed again provided a rationale for supporting school athletic programs. With the emergence of a corporate economy following industrialization, participation in rationalized corporate interscholastic sports programs could be seen as appropriate preparation for participation in the corporate world, even when traditional myths continue to be used to legitimize these sports programs. In fact, if high school sports participation actually teaches the importance of rationality, specialization, consumption, and other modern values, it could prepare participants for life in a world of corporate capitalism. According to Miracle and Rees (1994), though, there has been recurring skepticism about the educational value of interscholastic athletics among observers of public education in the United States.

Such skepticism is fueled by the problems, values, and emphases in contemporary interscholastic athletics. Questions are raised about public investment in interscholastic programs that seem to benefit relatively few students at a time when budgets are tight (Swift, 1991a) and when the United States seems to be slipping in its economic and educational status among nations (Ferrante, 1992). Furthermore, as Miracle and Rees (1994) have argued, it is hard to see how the traditional claims of the Dominant American Sports Creed justify high school athletics today when students need to learn how to live in a world of high technology and a global marketplace.

Contemporary American high school students need a rigorous curriculum that is attuned to the demands of the modern workplace and society. If high schools are to prepare students for higher education and employment, they need to produce students who are effective communicators; have a basic grasp of mathematics, science, and the new technologies; have sound bodies as well as sound minds; and are capable of learning what they need to know to function in a changing society. This kind of mission places great demands on high schools which often seem to be struggling to meet much simpler educational objectives (Apple, 1983; Hess, Markson & Stein, 1991: ch. 15; Ferrante, 1992: ch. 11), and it is unrealistic to expect interscholastic athletic programs to meet educational objectives that public high schools in general cannot. If, however, interscholastic athletic participation boosts academic performance and aspirations and indirectly leads athletes to additional education, it can be justified on the same practical grounds that education generally has been justified in secondary schools.

INTERSCHOLASTIC ATHLETICS, ACADEMICS, AND EDUCATIONAL ASPIRATIONS

The stereotypical image of the dumb jock is a persisting cartoonlike picture of the athlete as student. It is reinforced by public complaints about policies such as "No pass, no play," which imply that athletes are not able to make the grade academically. In fact, the academic performance of high school athletes is a well-researched topic in sport sociology (see summaries in Marsh, 1993; Phillips, 1993: 119–132). In general, athletic participation does *not* lead to poorer academic performance or lower academic aspirations. For athletes from certain backgrounds, being involved in high school sports is associated with better academic performance and higher aspirations than characterize their counterparts from the same background who do not participate in sports.

In the most extensive and methodologically sophisticated recent study of the effects of high school athletic participation, Marsh (1993) analyzed data from a nationally representative "High School and Beyond" (HSB) sample in the United States that were collected between 1980 and 1984. Focusing on changes occurring in the last two years of high school for over 10,000 students, Marsh found positive effects on social self-concept, academic self-concept, educational aspirations, attending a university, school attendance, taking academic and honors courses, time spent on homework, parental involvement with schoolwork, and parental educational aspirations. Although none of these effects was very strong, all held up after controlling for possible differences based on gender; academic

SPECIAL FOCUS

Reinterpreting *The Adolescent Society*

Marsh's research casts doubt on Coleman's (1961) interpretation of the results of his much-cited and much-replicated (e.g., Eitzen, 1976; MacKillop & Snyder, 1987; Snyder & Spreitzer, 1989: 157–159) study, *The Adolescent Society,* which showed that male adolescents placed much more value on sports success than academic excellence. Coleman studied the attitudes and values of students in eleven public and private high schools in the American Midwest in the late 1950s. One part of his questionnaire focused on whether boys wanted to be remembered as a brilliant student, an athletic star, or most popular. Girls were asked whether they wanted to be remembered as a brilliant student, a leader in activities, or most popular. (The fact that Coleman did not consider the value of sports participation among females reflects the limited opportunities for females in sport at the time of his research.) When they were asked about their parents' attitudes, more students in his sample said that their parents would be proud of them for being on the basketball team or cheerleading squad than for being selected as an assistant by a science teacher.

Coleman believed that the greater importance attached to athletics than academics meant that a commitment to academic excellence was being undermined by a strong interest in athletics, at least among male members of the "leading crowd" in the high school. He believed that parents reinforced the values of the adolescent subculture, not necessarily because they shared their children's values but because they wanted their children to be successful and gain the respect of their peers (Coleman, 1961: 34). In the simplest interpretation of Coleman's research and analysis, students could invest in either athletics or academics, and since they tended to place more value on the former, academics had to suffer.

An interesting variation of Coleman's analysis was proposed by Start (1966). He suggested that in England, where academic achievement was valued more than athletics in school, less successful students could use sport as a compensation for poor academic performance. According to Phillips (1993: 121), though, Start (1966, 1967), other British researchers (McIntosh, 1966), and their colleagues in the United States (Rehberg & Schafer, 1968; Schafer & Armer, 1968; Schafer & Rehberg, 1970; Buhrmann, 1972; Schafer & Armer, 1972; Spreitzer & Pugh, 1973; Picou & Curry, 1974; Snyder & Spreitzer, 1989: 159–165) began to discover that participation and success in athletics *and* academics were not mutually exclusive alternatives. Some research (e.g., Snyder & Spreitzer, 1977) even focused on female sports participants and found positive educational consequences of participation. Thus, one pursuit did not seem to detract from the other. In fact, as we have suggested, good athletes often are good students and participation in athletics may lead to improved academic performance and higher educational aspirations, especially among those whose less advantaged backgrounds seem to discourage academic interests and higher educational aspirations.

ability (measured by standardized test scores); socioeconomic status; the academic, social, and athletic climate of the school and the student's peer group; and school size.

Marsh saw a clear pattern in his results. Interscholastic athletic participation increased the ties linking students to the school and school values and made them more academically committed in a variety of ways. Although Marsh acknowledged some methodological problems in his research—such as the difficulty of determining the precise causal influence of sports participation and a reliance on simple measures of sports participation based on whether or not a subject was a participant and had a leader/officer role—it offers strong fuel for those who support interscholastic athletics as a beneficial activity for a variety of students in diverse educational settings (see also Swift, 1991a: 64).

In trying to understand why Marsh and others have found assorted positive outcomes of high school athletic participation, it is necessary to sort out a number of possible mediating factors between participation and the outcome variables. Phillips (1993: 124–128) proposed that the positive consequences of interscholastic athletic participation are largely mediated by positive peer affiliations that result from being in sport. This explanation fits data showing that the positive impact of athletic participation was most pronounced among students who tended to be from backgrounds that did not encourage success in school or who attended schools that offered few rewards for athletic participation.

Marsh challenged the peer influence interpretation. His results showed that academic self-concept and educational aspirations were more important mediating factors than were the influence of peers and other kinds of significant others, such as parents. In his view, participation in sport improved how individuals saw themselves as students and inflated their educational aspirations. Academic self-concept and educational aspirations, in turn, had a number of positive psychosocial and educational consequences that increased identification with the school and educational values.

Note that the positive outcomes of high school athletic participation found by Marsh generally did not significantly differ according to gender, race, or socioeconomic status. Another study using the HSB data, conducted by Sabo, Melnick, and Vanfossen (1993), found that high school athletic participation most affected the postsecondary status attainment of white males, and, to a lesser extent, suburban white females and rural Hispanic females. Interscholastic athletic participation had almost no effect on the college attendance or educational expectations of black males and females. In addition, it was generally not related to postsecondary occupational status and aspirations. The researchers concluded that the effects of athletic participation on mobility for women and minorities were complex and that they could not find specific processes or factors that enhanced or retarded mobility for female and minority sports participants.

In a related study using HSB data, Melnick, Vanfossen, and Sabo (1992) looked at educational effects of participation in high school sports on blacks and Hispanics. They found that athletic participation had a positive effect on senior-year academic grades for suburban black males and rural Hispanic females; had a positive effect on reading, vocabulary, and mathematics achievement test scores for urban black males and rural Hispanic females; resulted in lower dropout rates for rural black males, suburban Hispanic males, and rural Hispanic females; and resulted in a decline in educational expectations for rural Hispanic males. The relatively small number of positive educational effects on minority youths perplexed Phillips (1993: 130) because athletic participation generally has been thought to have the greatest educational benefits for students from less advantaged backgrounds. Until new research addresses this issue, though, we can only speculate about the reasons for the limited educational boost that minority students receive from sports participation.

Although numerous positive educational outcomes from high school athletic participation have been found, sweeping generalizations of such positive effects would oversimplify the state of current knowledge about such outcomes. We still need to know more about what aspects of high school athletic participation directly or indirectly lead to better academic performance, higher educational expectations, and more identification with the school and its values. We need to know how sports-related variables such as the type of sport and coach, sports status, and sports commitment affect the education and aspirations of high school athletes from different social backgrounds.

INTERSCHOLASTIC ATHLETICS, PSYCHOSOCIAL BENEFITS, AND SOCIALIZATION

We can understand why participation in high school sports might have positive psychosocial developmental effects for males, since being an athlete continues to be a source of high status in many types of high schools in the United States. The more limited focus on sport among females than males would suggest less psychosocial benefits for females, and some evidence supports this view (e.g., Feltz, 1979; Thirer & Wright, 1985). However, a study of Canadian female high school students (Buhrmann & Jarvis, 1971; Buhrmann & Bratton, 1977) as well as the High School and Beyond study in the United States (Melnick, Vanfossen & Sabo, 1988, 1992) show that females derive at least modest benefits of perceived or actual popularity among peers from sports participation. The Canadian results are especially interesting because Canadian high schools, unlike their American counterparts, have typically placed more emphasis on academics than athletics.

Melnick, Vanfossen, and Sabo (1988) found a moderate effect of athletic participation on females' involvement in other extracurricular activities, but no evidence of differences between athletes and nonathletes in their gender-role attitudes, self-esteem, or sociability. In focusing on minority students (Melnick, Vanfossen & Sabo, 1992), they found that black female athletes in urban and suburban schools and Hispanic female athletes in rural schools had higher self-perceived popularity ratings than their nonathlete peers. They also discovered that black female athletes in suburban and rural schools were involved in more extracurricular and community activities than their nonathlete counterparts. In considering the effects of athletic participation on minority males, the researchers found that black and Hispanic males in urban and suburban schools saw themselves as more popular when they were in sports and that black males in suburban schools and Hispanic males in rural schools were involved in more extracurricular and community activities when they were in athletics.

The findings on gender-role attitudes, self-esteem, and sociability from Melnick, Vanfossen, and Sabo's (1988) research raise questions about the character-building claims often used to legitimize sport in the high school and in other realms. Rees, Howell, and Miracle (1990) used data from the nationally representative five-stage Youth in Transition panel study to address the character-building issue specifically in regard to high school athletics in the United States. The initial sample was drawn from male sophomores in 1966, and followups were conducted in 1968, 1969, 1970, and 1974. The researchers found some character-building effects (e.g., on self-esteem, attitudes toward school, educational plans, and occupational goals) as well as some effects indicating a devaluing of honesty and social responsibility, aggressiveness, irritability, self-control, and independence. Rees, Howell, and Miracle concluded, however, that interscholastic athletic participation *generally* had limited beneficial *or* harmful socialization effects. Their results are similar to the general pool of socialization findings, which fail to show a clear, consistent pattern. Although Rees, Howell, and Miracle offered no basis for using a broad character-building argument to justify high school athletics, their results at least reinforce the idea that participation in high school athletics may have some positive effects on identification with the school and educational aspirations.

Source: Spencer Grant/Stock Boston

FIGURE 7.1 High school sport can be the basis for popularity.

THE HIGH SCHOOL COACH

In considering how athletes are influenced by their participation in interscholastic athletics, attention inevitably must turn to the coach. Although Marsh's results minimized the mediating influence of the coach on the education and psychosocial development of athletes, there seems little doubt that the coach can exercise a great deal of power and authority in the athletic arena. In view of the central role of the coach in high school sports, it is surprising that little systematic sociological research or analysis has focused on this role.

Sage (1987) has pointed out that each year high schools sponsor more than 190,000 athletic teams in-

POINT-COUNTERPOINT

DO THE BENEFITS OF ATHLETIC PARTICIPATION JUSTIFY PUBLIC INVESTMENT IN HIGH SCHOOL SPORTS PROGRAMS?

YES

Athletes directly benefit from exposure in high school athletics to middle-class achievement values and the Dominant American Sports Creed by gaining motivation and direction. They learn the importance of doing well in school, even if it is only to remain eligible to participate in athletics. They earn status recognition from peers, which boosts their self-esteem. They learn a number of important lessons of life from competing, winning, and losing in sports. They identify more with their school as athletes, which encourages them to continue their formal education beyond high school. They are also more physically fit as a result of athletic participation.

NO

Athletics has a limited number of educational and psychosocial benefits, which are derived mainly by white middle-class males. The advantages from athletic participation can be derived from other extracurricular activities having more educational benefits, such as academic clubs and the debate team. The high status accorded male athletes reflects a distortion of adolescent values. Athletes who are marginal students, devote all their attention to athletics, and aspire to compete in college do not develop the academic skills and knowledge needed to become successful college students. Varsity athletes are more likely than nonathletes to suffer serious injuries.

How should this debate be resolved?

volving over 6 million youths and that most of these teams are coached by men and women with the dual role of teacher-coach. It has been estimated that one-third to one-half of the teaching staff of every high school coaches one or more sports teams (*National Federation Handbook,* 1984). Even though high school coaching typically is an extracurricular assignment, it actually constitutes a second full-time job for several months of the year for teacher-coaches. In some sports, such as basketball, high school coaching is becoming almost a year-round job. A study of 235 girls' and boys' head high school basketball coaches in six western states (Capel, Sisley & Desertrain, 1987) found a low or medium level of burnout. Burnout can cause people to reduce their role commitment or drop out of their role. Burnout occurs when role demands and stresses exceed a person's endurance and ability to cope, and it can take various forms, including emotional exhaustion, depersonalized treatment of other people, and a sense of diminished personal accomplishment. Role conflict was the most powerful predictor of both burnout frequency and burnout intensity.

Sage (1987) documented the multiple role demands faced by high school coaches in a field study of fifty teacher-coaches in six high schools. He found that when coaching was added to teaching, there was pervasive role stress from an overloading of responsibilities and incompatible expectations or demands from the two roles. Capel, Sisley, and Desertrain (1987) proposed that this role overload was exacerbated by extending coaching responsibilities to almost a year-round job. According to Sage, the conflict between coaching and teaching often meant having to decide role priorities and giving disproportionately more time to one role at the expense of the other. This role conflict was made even more stressful by conflicts with family roles. In Sage's study, the major reason for dropping out of coaching was not being able to spend enough time with their children.

Sage found that approximately one-third of the coaches he interviewed planned to leave coaching within a few years. Others said they planned to leave education entirely at some unspecified time in the future. Prior research showed that high school coaches tended to be young, with an average age of thirty-two (Sage, 1974). The coaches in his more recent study (Sage, 1987) averaged thirty-four years of age. It may be that with growing family pressure and the accumulating role conflicts, stresses, and strains, it is increasingly difficult to fulfill the dual roles of teacher and coach. Although teacher-coaches often prefer to be coaches (Segrave, 1981), their job security and income are more likely to be tied to teaching. Faced with growing pressures from their dual roles and forced to make a choice between them, they are likely to choose teaching or, perhaps, to leave education entirely.

The behavior of coaches is less likely to be shaped by personality than by the distinctive pressures associated with the world of high school athletics and the role relations associated with that world. Public visibility and accountability, unpredictable outcomes, and easy measurement of success by wins and losses create special and substantial pressures for coaches not usually found in other professions (Coakley, 1990: 165). Although such pressures tend to increase up the levels of sport, high school coaches of major sports teams with rabid local fans can feel a great deal of pressure. According to Coakley (1990: 169), in the face of the role strain associated with the coaching role, coaches may seek public support for their methods, try to gain as much control as possible over their programs and the people involved in or with influence over these programs, and act as expediently they can to make the most of their opportunities. Such behavior may make coaches appear to have authoritarian, manipulative, or control-driven personalities, but these traits can also be seen as arising from a combination of the role pressures and environmental conditions associated with coaching. Coaches learn how to adapt to their role mainly through their contacts with other coaches in the coaching networks or subcultures..

Coaches must develop intense and extensive relationships with their athletes to mold the latter's character. Research has not yet demonstrated that high school coaches typically are character builders. On the other hand, research (Sage, 1972) has shown that when coaches take the time to give personal advice to their athletes about critical decisions, such as college and occupational plans, and the athletes have no other source of advice, coaches can have substantial influence (Snyder, 1972). Players from less advantaged backgrounds are more likely than those from more advantaged backgrounds to say that the coach was influential in their educational and occupational planning. This finding suggests that part of being socially and economically disadvantaged is having relatively few reference persons or sources of advice for major life decisions.

Coaches may not give the same amount of personal advice to all their players. In a study of 300 basketball players at 270 Ohio high schools, Snyder (1975) found that starting players were more likely than substitutes to see their coach as an important influence in their lives. Among seniors, starters and especially stars were the most likely to plan to attend college, and they were more likely than substitutes to receive advice about college attendance from their coach. Snyder proposed that coaches should not be viewed as dispensing advice in a purposely biased manner for their own aggrandizement. He suggested, however, that since coaches depend more on starters and stars than on substitutes for their success and gain more recognition by their association with better players, it is understandable that they might be more involved with the personal lives of these better players. In addition, since coaches are consulted by college recruiters about their better prospects, coaches are likely to focus more attention on these players.

HIGH SCHOOL FOOTBALL AND LESSONS OF DEMOCRACY

Coaches can influence their athletes in subtle ways. One way is in terms of the kind of authority they exercise and the kinds of educational lessons they provide through sport. The story of George Davis reveals interesting insights about public attitudes toward coaches who try to teach democratic responsibility to their players (Amdur, 1971; Nixon, 1984: 82–84). Davis was a very successful high school football coach in California. Between 1960 and 1964, his teams at St. Helena High School established a California state record with forty-five consecutive victories. George Davis cared about winning, but he was also commit-

POINT-COUNTERPOINT

DO HIGH SCHOOL ATHLETES BENEFIT FROM PARTICIPATION IN CHOOSING THE STARTING LINEUP AND SUBSTITUTES?

YES

High school athletes benefit from democratic participation in team decisions because it teaches them about the responsibilities of citizenship in a democratic society. Democratic participation is an educational experience through which they learn to make careful and objective decisions for the good of the entire group. Allowing athletes to vote for starters and choose substitutes teaches them that the entire team shares responsibility for the success or failure of the team. When everyone on the team becomes involved in team decisions, team cohesion and commitment increase, which can make the team more successful in competition.

NO

Most high school athletes are not mature and responsible enough to make decisions about important matters such as the starting lineup and substitutions. Their decisions are strongly influenced by peer pressure and are biased by friendships and interpersonal hostilities on the team. Coaches are paid to make major team decisions. Coaches who allow their teams to make these decisions for them are abdicating responsibility. Athletes do not know who deserves to start and who should be substituted at crucial times in a game. Only the coach has the expertise and objectivity to make these kinds of decisions for the good of the team.

How should this debate be resolved?

ted to teaching his players the importance of hard work and sharing responsibility for team success. He allowed his players to vote for the starting lineup, decide what positions they should play, and discuss their roles in the heat of competition. This amount of freedom for his players was unusual or unorthodox enough to create media attention and hostile reactions from parents and fans.

Other coaches have given their players the chance to exercise some authority (e.g., Easterwood, 1979), but George Davis's case probably has been the most publicized, in part because he had an enduring commitment to this approach and in part because he was a big winner. George Davis saw himself as a genuine teacher-coach on the football field. He wanted to prepare his players to become citizens in a democracy, and for this reason he tried to teach them about individual and group responsibility, leadership, discipline, and shared decision making as well as about how to perform on the football field. He reasoned, too, that more motivation to perform well would come from the peer pressure of teammates on a cohesive team than from an autocratic coach.

The fact that some of his players reacted with uncertainty and reluctance, especially right after George Davis moved to a different school and began coaching a new group of young men, may be easier to understand than the resentment and opposition displayed by parents and other community members in response to his methods. For the players, the approach puts responsibility where it typically had not been placed before, and the players needed to gain confidence in themselves before accepting this responsibility and their coach's system. For many adults, though, George Davis seemed to be abdicating responsibility as he turned over authority to his players. The adult doubts are ironic because George Davis was a genuine educator who tried to translate the American Dream and the Dominant American Sport Creed into reality for his athletes. The fact that his teams won makes resistance to his approach even more curious. It seems to reflect either a basic misunderstanding of the educational requirements of **democratic citizenship** and the meaning of the American Dream or a traditionalist paternalistic rejection of the idea that adolescents can and should exercise authority.

The simple act of participating in decision making that George Davis allowed his players may not appear very radical on the surface, but it is a major departure from conventional practice in high school football and many other sports. The role pressures and environment of interscholastic athletics are more likely to push coaches to increase control than to practice democracy (Edwards, 1973: 135–141). Yet George Davis's willingness to give up some of his control to his players enabled him to teach a basic civics lesson about participatory democracy, and in doing so he did not have to give up winning. The primary cost of his experiment was the backlash from community members and some athletes who apparently did not believe that adolescent athletes could handle the responsibility of making team decisions or, perhaps, that democracy belonged in sport. This backlash raises questions about the extent to which high school sports fans care about the incorporation of educational purposes into interscholastic athletic programs in their community.

A COMPARATIVE FOCUS ON INTERSCHOLASTIC ATHLETICS

Public criticism of George Davis is one indicator of the seriousness with which many people view interscholastic athletics in the United States. We have noted that an unusual amount of emphasis is placed on sport in American high schools and that in countries such as England (Start, 1966, 1967) and Canada (Friesen, 1976) athletics typically rank behind academics in public secondary schools. Each year in the United States, over 5 million males and females participate in thirty different high school sports (Sage, 1990b: 59). This level of involvement is unusually high when compared to other countries. European adolescents often participate in sport through community-based athletic clubs; Canada has both community athletic clubs and school athletic programs, but hockey, the most popular sport for male adolescents, is available only through community programs in most parts of Canada. Female sports programs and less popular male sports programs are mainly available in the school for Canadian adolescents (Eitzen & Sage, 1993: 102).

It is difficult to make sweeping cross-national generalizations about sport in the school because nations differ in their organization and financial support of education and sport, especially in developing countries (e.g., Riordan, 1978, 1993b; Broom, Clumpner, Pendleton & Pooley, 1988; Heinemann, 1993). In addition, there have been few studies comparing interscholastic sports programs and participants across nations. In one study of this type, Hardman, Krotee, and Chrissanthopoules (1988) surveyed students and teachers in English, Greek, and American high schools. Their research focused on attitudes and beliefs about identification with school values among athletes and nonathletes, perceptions of interscholastic competition, and characteristics of the programs.

Some interesting cross-cultural similarities and differences emerged from this study. For example, the amount of agreement with the statement that gaining prestige for the school is the main reason for interschool sports competition was highest for American students, next highest for English students, and lowest for Greek students, with very few Greek students registering agreement. The researchers interpreted these results as implying that Greek and English students had a wider appreciation of the value of interschool competition. Another finding was that English and American students both generally agreed with the statement that a school is judged by the performances of its sports teams, whereas Greek students strongly disagreed. Students in all three countries tended to agree that success in sports was conducive to more community awareness of the school.

The English and American teachers generally thought that athletes had more favorable attitudes toward school than nonathletes, but the Greek teachers held a contrary view. A majority of teachers from the English and American schools believed that success in interscholastic athletics was overemphasized. Nearly all teachers in all three countries favored broad sports experiences rather than specialization in one sport. Greek teachers expressed the strongest agreement with the idea that weaker athletes ought to be encouraged to join athletic teams, and American teachers expressed the strongest disagreement with this idea.

American students and teacher-coaches spent the most time involved in athletics, English students and teacher-coaches spent less time than Americans in school sports, and Greeks spent the least time. There was widespread agreement in all three countries that sports competition increased social contacts, and large proportions of student respondents in all three countries indicated that being selected as a member of a school team earned parental approval. Teachers generally believed that parental support had a positive effect on student athletic participation. All teachers in the three countries believed that skill was the most important selection criterion for athletic teams.

The types and perceived consequences of financing for athletic programs differed across countries. In England, six types of financing were identified, with school and departmental funds mentioned most often. In Greece, the state was the only source cited. In the United States, athletic programs were funded by taxes and contributions by parents and students; commercial sponsorship, which has been drawing increasing interest in the United States in recent years, was not mentioned. All male teachers in the English sample, all male and female Greek teachers, and a majority of male and female teachers in the American sample said that limited funds restricted the scope of interschool competition. Female teachers in England expressed the least agreement with this idea, with less than half expressing agreement.

Caution must be used in generalizing from these results because they were based on limited sampling. The results, nevertheless, are interesting because they suggest some tentative generalizations about high school sports in different countries. In particular, the results indicate that even though Americans tend to place more emphasis on interscholastic athletics, American athletes, nonathletes, and teachers share some similar attitudes and beliefs about athletic participation with their counterparts in other countries. For example, even though American students tended to have a narrower conception of the value of athletic participation than students in England and Greece, students in all three countries generally agreed that successful athletic teams brought more community awareness of the school. In addition, teachers in all three countries favored broad rather than specialized sports participation, and most students in all three countries believed that parents approved of their child being selected to play on a school athletic team. Thus, these results suggest that even a relatively weak emphasis on interschool sports programs, as in the Greek case, does not necessarily mean that sports are seen as unimportant or as having no effects on participants or the school.

THE EFFECTS OF HIGH SCHOOL ATHLETICS ON MINORITY-GROUP STATUS AND INTEGRATION

Sports participation varies across cultural groups within nations as well as cross-culturally between nations. An important question raised by these differences in sports participation within nations is whether they contribute to, or merely reflect, differences in the status of cultural groups. Related questions are whether secondary school sports participation contributes to the integration or segregation, and status enhancement or status maintenance, of ethnic and racial minority groups and women. That is, does the social organization of high school sport in a community or a society help create or sustain cultural differences and social inequalities or does it minimize them? The case of Canada illustrates some of these issues.

Language-Group Differences in Sports Participation in Canada

In Canada, sports involvement is much greater for English-speaking (Anglophone) than French-speaking (Francophone) adults (White & Curtis, 1990a, 1990b). White and Curtis (1990a) tested the hypothesis that these differences in sports participation among English and French Canadians are at least partially caused by differences in the organization of competitive sports programs in the schools in English and French Canada. Their research relied on data from a 1976 sample of over 35,000 native-born Canadians who were twenty years old or older. White and Curtis observed that French Canada had been slower to develop interscholastic athletic programs and suggested that this might have been the result of historical differences in achievement values between Francophone and Anglophone cultures.

White and Curtis found higher participation levels for Anglophones than for Francophones in competitive sports and higher participation levels for Francophones than for Anglophones in noncompetitive sports. Although they expected some homogenization of cultural differences in recent years, they instead found greater Anglophone/Francophone differences in sports participation for younger than older age groups. The language group differences in competitive sports involvement were larger for the young than for the overall sample.

In comparing high school sports involvement within the younger age group in the sample, there were greater differences across language groups in intramural than interscholastic participation. Furthermore, intramural sports involvement had a stronger effect than interscholastic sports involvement on general and competitive sports participation and participation in sport with nonfamily members during adulthood. The stronger effects of intramural sports participation on adult participation suggested to the researchers that the ethos of intramural sport, emphasizing the processes of participation and competition more than the outcome, creates more positive sports experiences and leads to a more enduring interest in various types of sports involvement in adulthood. Former participants in high school athletics were more likely than nonathletes to participate in adult sports, but the effects were weaker than the effects of intramural participation. High school physical education classes had a minimal effect on adult sports participation.

White and Curtis acknowledged that they could not distinguish between the effects of socialization through high school sport and the effects of a selection process that led people with a prior interest in sports to high school athletic programs. They assumed that both processes operated, and that whatever way these effects combined, the Anglophones had more opportunities than Francophones for sports participation in high school. They recognized the need to consider possible effects on the two language groups of school or community sports experiences before high school. They also left open the question of whether historical differences in the organization of sport in English-Canadian and French-Canadian high schools persisted and, if so, whether they resulted from current value

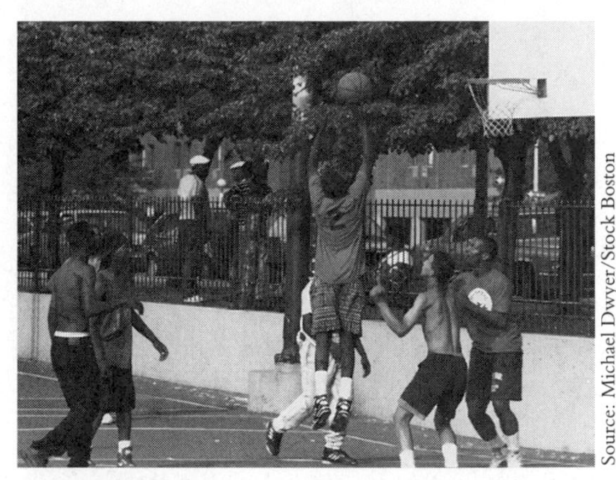

Source: Michael Dwyer/Stock Boston

FIGURE 7.2 Ticket to opportunity: basketball in black adolescent subculture.

differences or from the residual effects of traditional cultural differences in views of schooling.

Laberge and Girardin (1992) have questioned White and Curtis's suggestion that the difference in sports participation between French-speaking and English-speaking Canadians meant that Francophones did not value achievement as much as Anglophones. They have also questioned the equating of competitive with achievement values and proposed that the domination of organized sport in Canada by Anglophones along with possible discrimination, rather than differences in achievement values, may have limited sports opportunities or interests among Francophones. Laberge and Girardin also proposed that language-group differences in sports participation may have resulted from differences in cultural traditions of sports involvement—with Anglophones influenced by the British sports tradition—and not from differences in the desire to achieve.

In a similar critique of White and Curtis, McAll (1992) argues that more attention should have been paid to the role of sport in structuring and reinforcing inequalities. McAll proposed that the value-differences interpretation suggested by White and Curtis holds up only insofar as we ignore the question of discrimination and exclusion. Like Laberge and Girardin, he suggests that White and Curtis are "blaming the victim"—that is, Francophones—for their lesser involvement in sports. Also like Laberge and Girardin, he draws an analogy between the status of women and the status of Francophones in Canadian sport since women were less involved than men, and he suggests that we could draw the (erroneous) conclusion from White and Curtis that both groups had fewer opportunities because they had inappropriate values—that is, they were not sufficiently achievement-oriented.

McAll argues, on the contrary, that the school may have been instrumental in establishing traditions of sports participation for the dominant Anglophone group that reinforced their class dominance as they excluded the minority Francophones. He recalls the frequent argument in studies of British class structure that upper-class cohesion is constructed on the playing fields of elite English boarding schools. Close and enduring ties are thought to be formed in these intensely competitive sports settings. According to McAll (1992: 311), a network of private English Protestant schools that competed against each other in sports was established in Montreal by 1880, whereas Francophone Catholics did not establish such a network until 1966. He argues that these different histories of school sports do not reflect Francophone "cultural backwardness"—which, he says, White and Curtis are implying—but instead the class dominance of English Protestants in Quebec society. Thus, school sports can be seen as an important vehicle for maintaining class structure and solidarity.

Elite Secondary School Sports Participation and the Achievement or Maintenance of Elite Status

Armstrong (1984) developed the **class and status maintenance argument** of organized sports in boarding schools in Great Britain and the United States, but this function of sports was obscured, according to McAll, by White and Curtis's value-differences interpretation of sports differences between Anglophones and Francophones. According to Armstrong, participation in team sports at elite boarding schools in England and New England traditionally taught young men class and gender roles that would enable them to assume dominant positions in their school and, later on, at universities and in society. In Bourdieu's (1984) terms, such elite sports participation created valuable **cultural capital**, which included knowledge of ways of thinking, talking, and acting as well as the cultural and social awareness necessary to assume a position in the dominant stratum of society. Sports success brought acceptance and esteem at school, provided an avenue to school leadership positions, developed strong and lasting ties among the schoolboy athletes, created a strong loyalty to the school, and prepared athletes to attend and pursue sports at elite institutions of higher education.

Other studies (e.g., Eggleston, 1965; Berryman & Loy, 1976; Chandler, 1988) have also shown that boarding schools played an important role in preparing athletes to become students and athletes at elite universities and colleges. This role may be diminishing, but as Berryman and Loy (1976: 68) noted, the changing social backgrounds of Ivy League athletes mirror social change but do not necessarily imply that graduates of elite boarding schools have entirely lost their advantage in competing for the highest-level educational, social, and sports opportunities. At the time of their study, the early 1970s, an elite background still provided an advantage to athletes competing for positions on the major sports teams at Ivy League universities. A decade later, an extension of Eggleston's original study in England by Chandler (1988) revealed that private secondary school athletes still had an advantage over athletes from "state-maintained" schools in earning a letter—or "Blue"—in cricket and rugby, but the situation was reversed for the lower-status sport of soccer. The significance of possible class differences in lettering in elite sports at elite universities is, as Eggleston pointed out, the "likely association with social opportunities in university life and social and occupational prospects in later life" (1965: 241).

The idea that athletes at elite secondary schools have a special advantage over others with equal talent and motivation is at odds with the concept of sport as a democratizing force in society. It suggests that rather than serving as a vehicle for anyone with talent and motivation to achieve the American Dream, secondary school sport reinforces cultural and social advantages in society. Critiques of White and Curtis argue that they did not give enough consideration to this argument in their study.

SPECIAL FOCUS

Ethnic Resistance to the Dominant Cultural and Social Rituals in Texas High School Football

In his study of a high school football season in a South Texas town, Foley (1990) observed various ways that Mexicano, female, and working-class members of the community displayed resistance to traditional sports practices and rituals that demeaned them or contributed to the status of dominant groups. For example, a number of Mexicano males involved in antischool-oriented peer groups, often referred to as *vatos* (meaning "cool dudes"), participated in a drug-oriented lifestyle and rarely participated in sports, viewing them as "kids' stuff." Sports events, however, were used by the *vatos* to display their lifestyle and establish their reputations as fighters or tough guys. They also attended games to "hit on chicks from other towns," and they tried to show their courage through their efforts to"steal" women from another town while at road games. The *vatos* believed that they established their manhood more by their exploits off the field than they could by playing on the school football team.

The *vatos* were not the only example of ethnic resistance to the dominating influence and rituals of local high school football. Many Mexicanos, including a radical Chicano newspaper, reacted strongly, leveling charges of discrimination and *gringo* conspiracy, against school administrators and teachers when the homecoming queen and her court, all Mexicanas in the year observed by Foley, had to walk, rather than take the traditional open convertible ride, to the crowning ceremony. This departure from tradition was seen as a way of diminishing the significance of an event that had typically celebrated future Anglo leaders.

The case of the coach demonstrates how difficult dealing with local structures of privilege and power can be for aspiring ethnic minority-group members. Coach Roberto Trujillo was the town's first Mexicano head coach. In contrast to the stern disciplinarian Anglo candidate for coach, Trujillo was viewed as too friendly and soft on the players. He faced the usual criticisms of high school football coaches who are not successful or tough enough for the more rabid fans, but he also was caught between two cultures at a time of change. Trujillo sought ethnic and racial harmony but decided that powerful members of the dominant class in his community would never change their biased attitudes. Despite a record of 7–4 and second place in the conference, he could not earn widespread community acceptance or a comfortable niche for himself in his ethnically stratified community. Feeling constant pressure, having lost friends, and having developed an ulcer, Trujillo resigned as coach and left his hometown.

The coach could not use high school football to change ethnic relations in his community because football was part of the cultural and social rituals that maintained traditional inequalities. In the end, the entrenched social and cultural patterns of Anglo dominance in his community made it impossible for him to survive as a coach and try to create a meaningful change in the ethnic relations of his community.

High School Athletics and Intergroup Relations

It might not surprise us that sports at *elite* secondary schools contribute to the maintenance of the class structure and dominant class advantage, but the functioning of sport in a similar manner at *public* schools might be less expected. In fact, Bissinger's (1990) journalistic account of high school football in a Texas town and two other recent ethnographic studies by social scientists (Foley, 1990; Grey, 1992) have revealed that athletic programs in American public high schools can play a conservative role in reinforcing existing class, ethnic, racial, and gender inequalities in their communities.

The idea that sport reinforces existing social inequalities is at odds with the popular notion that social

SPECIAL FOCUS

Gender Relations and Texas High School Football

Foley's (1990) study of Texas high school football showed that the female cheerleaders and pep squad members were part of the traditional gender order where males perform and females cheer them on. The annual "powder puff" or women's football game probably best illustrates how sport reinforced gender inequalities in the school and the community. This game, which pitted senior females against junior females in touch football, took place each year on a Friday afternoon before the seniors' final game. A number of the seniors on the football team dressed up as females and performed as cheerleaders for the powder-puff game. Foley estimated that about one-quarter of the students at the high school, including mostly the more popular and successful Anglo and Hispanic students, attended at least a part of the game. More males than females watched, and the spectators generally did not view the game with much seriousness.

Foley was most impressed by the difference between the behavior of the male and female participants in the game. The males acted in a silly or burlesquelike manner, exaggerating or mocking female appearance and behavior. The female players, on the other hand, were serious about the game and tried to show the males, including their boyfriends on the football team, that they, too, could compete. Teachers saw the game as innocuous and a means to build school pride. For Foley, though, the game's cultural significance was that it served as a **ritual of inversion** by reversing usual cultural roles and practices. The silly manner in which the males played their roles made it clear that this event was a departure from the norm. They set the tone of the event, despite the female players' serious attempt to show their prowess, and in doing so demonstrated their physical and social dominance as males. The females may have been trying to challenge traditional stereotypes, but in the end, according to Foley (1990: 119), they were more likely playing an unintentional role in a ritual of male dominance and privilege.

contacts across racial and ethnic lines in sport breaks down social and cultural barriers and contributes to intergroup harmony. Foley's (1990) case study of Texas high school football (see the Special Focus discussion) showed that a Hispanic coach, the first of his ethnic group to be head coach at his high school, was able to make some progress in earning the respect of his Anglo players despite feeling ethnic prejudice from the larger community. Other research, with much more extensive sampling, offered little support for the argument that high school athletics improves ethnic and race relations.

For example, a study by Chu and Griffey (1985) of over 1,000 urban public high school students in upstate New York found that the racial attitudes and behavior of athletes and nonathletes were generally not significantly different. Yet their findings showed some interesting correlations, including statistically significant correlations between the percentage of blacks on a team and the frequency of interracial contacts in school. The percentage of blacks on a team also was correlated with the number of phone calls to students of a different race. More years of participation on a sports team increased the frequency of interracial phone calls as well as the desire to have more friends of a different race.

A surprising twist in Chu and Griffey's findings was that athletes in individual sports such as golf and cross-country running reported better racial attitudes and somewhat better social relations between races than athletes in team sports reported. Since social class differences did not explain these findings, the researchers suggested that the more limited interaction among teammates in individual sports might have made it easier to avoid interpersonal tensions that could have fueled racial prejudice. This kind of explanation may apply only to males, though. Female athletes in both individual and team sports held more positive racial attitudes than their male counterparts, and this relationship between gender and attitudes was stronger for team sport athletes.

Chu and Griffey's results are mixed. When they are considered in the context of other research of this type (see Chu & Griffey, 1985: 323–324), however, it seems clear that we lack consistent and compelling evidence that interracial or interethnic contact in high school sports significantly improves integration or reduces prejudice in meaningful and enduring ways. It has been proposed that interracial and interethnic social contact is most likely to contribute to prejudice reduction and more positive intergroup attitudes and behavior under conditions when the status of different groups is equal, when there are shared group goals, when there is cooperative interaction, and when the environment is supportive of positive social contact (Braddock, 1980: 179). In sport, a winning tradition could also create positive feelings that contribute to a support for intergroup harmony within teams. It may be, though, as Rees and Miracle (1984) have suggested, that high school sports teams, or at least team sports, often lack the conditions needed to nurture the kind of social contact between racial and ethnic groups that will cause real, deep, and enduring improvements in racial and ethnic attitudes and behaviors. The status rivalries, the structured inequalities between starters and bench warmers and between stars and the rest of the team, the heat of competition, the difficulty of sustaining consistent team competitive success, and community environments where the rituals and traditions of inequality are deeply entrenched all seem to be significant obstacles to better racial and ethnic relations.

HIGH SCHOOL ATHLETICS AND THE CULTURAL AND SOCIAL ASSIMILATION OF IMMIGRANT MINORITY GROUPS

Although watching or playing in ethnically or racially integrated sports may not significantly alter ethnic or racial attitudes and behavior, it can lead fans and teammates to accord more status to minority athletes. In fact, success in highly visible and popular sports can be one of the few avenues of status mobility in the school and community available to minority students. In his study of Garden City (Kansas) High School, Grey (1992) observed that sports participation was the major vehicle of upward social mobility for lower-status established-resident Hispanic students. This also was true for immigrant minority students, but the immigrants had their pursuit of status complicated by the use of popular American sports as the basis for status judgments. Indeed, their acceptance of mainstream American sports became a standard for assessing their willingness to assimilate into American life. When they favored soccer, which was not an established sport at Garden City High School, over the established American version of football and chose not to attend football games or try out for the team, they cast into doubt their "Americanism" and reinforced their already-low status as immigrants.

Nearly one-third of the 1,111 students at Garden City High School in 1988–89 were established-resident or immigrant Hispanic (21.4%), newly arrived Southeast Asian (7.8%), or African American (1.4%). About one-third of the students were transients, enrolling for less than the full school year. The average stay of these students was 4½ months. Athletics was the primary link between the high school and the community of Garden City, and since football season occurred at the beginning of the school year it had special importance in the school and community.

Athletics in this high school was a mechanism of school status attainment and identification for middle- and upper-middle-class Anglo males and for males from relatively more established and affluent Hispanic families. The number of immigrant Hispanic and Southeast Asian students involved in mainstream school sports programs was very limited, and participation of these groups in "established" extracurricular activities in general was minimal. The failure of members of the minority groups to participate in these mainstream school activities, and especially established athletic programs, symbolized their unwillingness to affirm the basic values of American life. Grey found that interaction with immigrants was uncomfortable for many local residents. Sport was a means of finding common ground to ease this discomfort, and when immigrants did not participate as athletes or spectators they opened themselves to criticism from established community members. They were viewed more negatively because they seemed to reject the important cultural symbols of dominant social groups in the school and community.

The one extracurricular activity that appealed to many Southeast Asian and Hispanic students was a school soccer club. In 1988–89, the teacher who helped organize the club and sponsored it was away for the year, and none of the remaining seventy full-time teachers was interested in sponsoring the club. The school principal, who had a football background and a son on the team, made little effort to find a sponsor. As a result, soccer disappeared from the school's campus. Even when it existed, its status was as a "nonestablished" sport, which meant that soccer had much less legitimation and support than the established Anglo sports.

Thus, at Garden City High School, the social organization of high school athletics was biased toward those from privileged backgrounds, provided little accommodation to those from immigrant or minority cultural backgrounds, and isolated those who preferred activities, such as soccer, that were outside the mainstream. In a sense, the established sports served a **cultural and social gatekeeping** function, letting in the accepted—that is, dominant class members—and keeping out the marginal or unaccepted—that is, minority-group members and especially immigrant minorities. When the immigrant minority-group members did not participate, they lost their chance for **cultural and social assimilation** and to become integrated into the mainstream and gain acceptance as "true Americans." Since soccer was not viewed by the school board or school authorities as a serious sport and was allowed to die, the immigrant students were deprived of a chance to become more attached to the school through sport, which is a function usually associated with participation in mainstream high school sports. The narrowing of sports alternatives to those favored by the dominant Anglo majority was a reflection of their cultural and social dominance in the school and community. Through these mainstream sports, the dominant class was able to dictate the terms of becoming "American."

CONCLUSION

High school sports clearly have implications that stretch far beyond the gymnasium or athletic field. In many communities, high school athletic programs are an arena for seeking out aspirations and dreams, for establishing or destroying reputations, for playing out social tensions, and for reaffirming the values and beliefs of the dominant classes. Grey's study, and several others we have considered, cast doubt on claims of democratizing or integrating functions of interscholastic athletics. At least the major male athletic programs are a component of the established cultural and social order of the school and the community and as such are unlikely to be at the forefront of major cultural or social change.

We also have pointed to a number of serious problems facing high school athletics and a few serious social problems involving high school athletes. Existing evidence leaves us with as many questions about the claimed character-building function of interscholastic athletics as we had about broader arguments discussed in an earlier chapter concerning socializing functions of sports. Evidence, however, indicates that sports at elite boarding schools in England and the United States may have a socializing function that prepares young men for elite universities and for participation in major or elite sports at those institutions. In serving this function, elite secondary sports participation may provide some of the cultural capital that these young men can use to establish their place in the upper strata of their society.

None of this means, though, that no benefits are to be gained from participation in public high school athletics by less advantaged students. Recent evidence has indicated that a variety of mostly academic benefits are to be gained for members of a variety of social groups when they engage in school sports. In particular, athletes generally appear to be more likely to do better than their nonathlete counterparts in their academic performance and to have higher educational aspirations. High school athletes seem to derive several kinds of benefits from interscholastic athletic participation with few educational costs.

We are directly or indirectly reminded several times in this chapter of the unusual or special emphasis on athletics in U.S. high schools. Students in other nations also participate in school sports, but their participation seldom receives the amount of public attention that interscholastic athletic participation receives in the United States. Despite the great symbolic and commercial importance of major athletic programs in many American high schools, the future of interscholastic athletics is on shaky financial ground in many communities. As school boards, school authorities, and other influential community members deliberate and battle over school and community priorities, serious debates have developed over the educational value of athletics. It is not clear how important these debates have been to parents, fans, and students. The debates continue, though, and we will see in the next chapter, on college athletics, that they are not restricted to high schools.

SUMMARY

The media have reported a number of significant social issues and problems related to interscholastic athletics in the United States that have echoed the concerns of athletic directors and school officials across the country. Among these issues and problems are budget shortfalls, recruitment and retention of qualified coaches and referees, violence, specialization and professionalization, legal liability, the intensity of competition, the limited number of female role models in coaching for female athletes, and off-field deviance by athletes. In general, high school athletic programs embody the cultural and social patterns and practices characterizing the community and society in which they operate.

Although there are questions about the educational value of interscholastic athletics, research has shown that high school students gain status and, in some cases, a boost in grades and educational aspirations from athletic participation. The most recent systematic evidence indicates that interscholastic athletic participation increases the ties that link students to their school and school values and that these ties make student-athletes more academically committed in a variety of ways. When females began to participate in school athletics in much greater numbers in the 1970s, researchers began to find some positive educational consequences of athletic participation for them. It appears, though, that the primary beneficiaries of high school athletic participation are white males. Although interscholastic athletics is emphasized a great deal more in the United States than in other countries, benefits of such athletic participation have been perceived and found in other countries as well.

The high school teacher-coach can have influence over the socialization of athletes, especially when athletes lack other influential adult influences in their lives. High school coaches often are very committed to coaching but can experience substantial role conflict and burnout. When a coach tries to teach citizenship through athletics, skepticism from athletes and backlash from the community can result.

High school sport is sometimes seen as a vehicle for breaking down cultural and social barriers in a community, but research generally has raised questions about this idea. It may instead reinforce existing cultural and social differences and inequalities. Minority-group members and especially immigrant minorities can find that established sports serve a cultural and social gatekeeping function, excluding members of minority groups.

CHAPTER 8

College Athletics

BEYOND INNOCENCE ABOUT COLLEGE ATHLETICS

Big-time college athletic events in the United States are major media and cultural attractions. They have been characterized as cultural spectacles representing all that is thought to be good and glorious in America: competition, fellowship, community, fun, and achievement (Jenkins, 1970). The Saturday afternoon college football game has traditionally been a celebration of American life, even a form of public art (Cady, 1978). This description may contain hyperbole, but few would disagree that college sports remain very popular in the United States despite recent controversies and criticisms.

Supporters of college athletics have claimed that it benefits the campus, community, and student body. We read about the way a successful college team can unify a campus and local community, generate revenue for the college and local businesses, sustain the loyalty of alumni and inspire their donations, and provide an exciting form of entertainment. At the same time, we read about assorted forms of recruiting violations, deviant behavior by athletes and coaches, and graduation problems for athletes. In reality, college athletics is a complex enterprise with a variety of possible positive and negative outcomes, which are summarized in Table 8.1.

Many questions have swirled around college sports in recent years. Some of the questions are relatively new; others have been asked repeatedly over the history of college sports. Gender equity and the role of television are products of the past few decades of development of college athletics, whereas concerns about institutional control over athletics; the financial viability of athletic programs; and the ethical, moral, educational standards for athletes and athletic personnel can be traced back many decades.

THE ORIGIN AND DEVELOPMENT OF U.S. COLLEGE ATHLETICS

No other country places as much emphasis on athletics in higher educational institutions as the United States does. College athletics was born in the United States during the Gilded Age, 1865–1900 (Noverr & Ziewacz, 1983), also dubbed the Era of Reconstruction, when major social institutions reorganized following the devastating impact of the Civil War. In this period sports became an integral part of everyday life after divesting itself of religious or puritanical restrictions (Betts, 1974).

College sports evolved from limited, student-based activities at a few institutions.The first intercollegiate athletic event is generally acknowledged to be a crew race on Lake Winnipesaukee, New Hampshire, between Harvard and Yale in 1852. A version of baseball was played by Amherst and Williams colleges in 1859. Track and field began in the 1870s when Columbia University students formed the Columbia College Athletic Association. On November 6, 1869, Rutgers defeated Princeton in the first football game, which actually resembled rugby more than the modern version of the game. In 1876, the Intercollegiate Football Association was founded by Harvard, Yale, and Princeton, making it the first predecessor of the National Collegiate Athletic Association (NCAA).

Early college sports were distinguished by student control, and they were not officially sanctioned by their universities. When sports such as football became

TABLE 8.1. Possible Outcomes of Intercollegiate Athletics

Individual or System Affected	Positive Outcomes		Negative Outcomes	
Education system	• Integration of student body • Reduction of conflict across class lines • Social control • Aggression displacement • Norm affirmation • Prestige • Community visibility • Academic community's visibility	• Generalized political and economic influence/support • Informal membership retention and support beyond formal break (i.e., graduation) • Attraction of membership • Promotion of "town-and-gown" affiliation	• Detraction from education mission	• Promotion of hostility of members
Athletic	• Character development (e.g., courage, humility) • Acquisition of social skills • Tension release • Education attainment • Occupational stress	• Educational opportunity • Physical fitness • Prestige • Tension/excitement • Identity formation • Affective association	• Character detraction • Negative aggression • Educational detraction • Exploitation of larger subsystems	• Role conflict and stress • Dehumanization and creation of delusions • Value distortion
Community	• Unity of members across social categories • Democratization of interpersonal relations ("talking sports") • Social control • Aggression displacement • Norm affirmation • Safety valve/tension management	• Entertainment—tension/excitement • Economic—flow of new capital • Opportunity for association with political and economic elite • Identity and visibility via boosterism • Community legitimacy—"big league"	• Creation of false consciousness • Diversion of attention from social problems	• Reinforcement of class distinctions
Society	• Socialization of accepted values, norms and conformity • Safety valve/tension release at class level	• Ritualistic expression of American life—norm and value affirmation • Reinforcement of gender, age, and racial role relations	• Socialization to norms and behavior that are demeaning or potentially disruptive (e.g., circumventing rules) • Intergroup/class hostility	• Diversion of attention from explitive social structure • Reinforcement of sexism, racism, and unequal distribution of resources

Source: UPI/Bettmann Newsphotos

FIGURE 8.1 Princeton versus Rutgers, 1914.

profitable enterprises, institutional administrators recognized these activities as potential sources of external political and economic support. As a result, alumni and administrators took over the operation of college sports, and full-time coaches and administrators were hired to control the program on behalf of the institutions.

Athletics had become a well-entrenched activity on many campuses by the turn of the century. The income and public notoriety that athletics brought to campuses established commercialism as the primary characteristic of college sport (Snyder & Spreitzer, 1989). The press made household names of Walter Camp, Amos Alonzo Stagg, Jim Thorpe, Pudge Heffelfinger, and Fielding Yost. Intersectional competition was stimulated by the first bowl game, the Tournament of Roses, in 1902, when Michigan trounced Stanford 49–0. By the 1890s, rivalries had evolved into the Big Game, where a win or loss against a traditional rival would reflect on the character of the institutions and the quality of their campus life (Cady, 1978).

College sport experienced continued growth in popularity and participation over much of the twentieth century. Except for a few incidents that sparked calls for reform, college sport, under the leadership of the NCAA, developed in a relatively unrestricted fashion. Early in the 1980s, growth and prosperity slowed because college athletics was experiencing financial chaos, despite lucrative television contracts and full arenas for the major college athletic powers. More significantly, college sport experienced a crisis in ethics and integrity. The place of these programs on campus and the relation of athletic goals to the academic mission of the institution were subjects drawing considerable scrutiny. In addition, scandals involving illegal payments from alumni to athletes, inadequate academic progress by athletes, criminal acts by athletes on and off the field, and favorable admission and grading practices for athletes led people to question the quality of the academic experience of athletes, issues of academic integrity, exploitation of poor and minority athletes, and financial accountability in college sport. Thus, various forms of social deviance and subsequent social criticism have been enduring elements of the history of college sport in the United States. Allegations of professionalization of players, lowering of academic expectations and standards for athletes, excessive violence on the field, illegal inducements to play, and the compromise of academic goals in favor of the needs and goals of athletic programs continue to be part of the college sports environment in the United States. They are similar to accusations made a century ago and persist despite the formation of the NCAA as a regulatory body designed to control the quality and character of college athletics.

THE ROLE OF THE NCAA IN COLLEGE ATHLETICS TODAY

The formation of the Intercollegiate Athletic Association in 1905 came as a result of the concerns expressed by faculties, administrators, and high-level public officials (including President Theodore Roosevelt) about the academic integrity, professionalization, institutional control, and players' physical safety in college sport. The name of this organization was changed in 1910 to the National Collegiate Athletic Association, but its purposes of preserving faculty control, operating in an ethical manner, and promoting educational principles were retained (Hardy & Berryman, 1982). The modern version of the NCAA still claims these purposes as the guiding philosophy for a amateur sports organization, but decades of controversy and challenge suggest that the NCAA has had a difficult time meeting these goals.

The initial NCAA organization of sixty-two institutions existed largely as a discussion group on football rules (Tow, 1982). The NCAA continued to develop championships and to generate specific rules for sports under its jurisdiction. Regulatory issues became a major concern, particularly after World War II, because the organization's **Sanity Code**, a set of guidelines for recruiting and compensating athletes and for exercising institutional control, was proving to be ineffective. In addition, television was making demands for athletic programming that could impact football attendance, and bowl games were increasing with no control or supervision (Tow, 1982). In other words, college athletics was becoming very complex, programs were moving in increasing numbers from a regional or local base to national visibility via television, demand for more championships was growing, and these conditions pressured institutions to engage in illegal recruiting and compensation practices. It was obvious that voluntary compliance to the Sanity Code was not working. Compliance could only be obtained with a strong enforcement program led by full-time professionals hired to follow the dictums of the association's membership.

In 1951, Walter Byers became the full-time executive director of the NCAA, and administrative offices were established in Kansas City. In 1953, the NCAA Council created an enforcement apparatus (the Committee on Infractions) and agreed to administer its first set of sanctions to three members of the association. Today, the NCAA is the largest and most prominent organization in intercollegiate athletics. A twelve-member executive committee elected from a larger executive council of forty-four members oversees the daily operations of the NCAA and supervises the Association's executive director. The most powerful NCAA members are the 130 institutions that operate large-scale football and basketball programs. They seem to be the prime beneficiaries of NCAA policies and practices, and these same programs cause most of the regulatory and enforcement difficulties. However, by virtue of its control of the distribution of millions in television dollars, the NCAA maintains significant control over member institutions.

The revenue potential from the NCAA's television contracts is substantial and contributes significantly to the power the NCAA has over member institutions. In December of 1994 the NCAA reached a multiyear agreement with CBS giving that network exclusive rights to broadcast Division I men's and women's basketball tournaments, for $1.75 billion, the largest contract of its kind ever signed. The commercial dependency of these institutions on the NCAA makes challenges to the structure of college sport and to the regulatory practices of the NCAA difficult to mount with any success. The corporate athleticism of college sport is directly related to the commercial demands of an entertainment business that is appealing to large national television audiences. Despite the revenues available from television, most athletic departments, except those with the most successful football programs, lose money. Many athletic departments must draw on funds from other university sources in order to balance the budget. Others must make even larger demands on alumni and booster constituencies, but this funding does not come without costs in control and governance.

Even though the NCAA has formalized an extensive body of rules and an elaborate enforcement procedure, it has been ineffective in controlling the abuses perpetrated by boosters, athletic administrators, coaches, and players. The NCAA currently suffers from lack of legitimacy because it has compromised its

SPECIAL FOCUS

The NCAA as a Cartel

A **cartel** is "an organization of independent firms which has as its aim some form of restrictive or monopolistic influence on the production and/or sale of a commodity as well as control of wages of the labor force" (Sage, 1982: 133). Cartels are prohibited by antitrust legislation. As a cartel, the NCAA limits output in the form of the number of games available for viewing either as a spectator or by television; it also restricts access to athletic inputs by controlling the competition for student-athletes (Koch, 1973; Becker, 1987; McCormick & Meiners, 1987; Fleisher, Goff, and Tollison, 1992). As a combine of sellers, NCAA members have agreed on the restrictive distribution of their product. The NCAA protects itself from prosecution under restraint-of-trade legislation by asserting that college athletics is not a business but an amateur voluntary organization whose procedures are determined by the consensus of its members. Koch (1971) argued that the NCAA is like a cartel because it (1) sets input prices for student-athlete services (e.g, financial aid limits); (2) regulates the extent to which these outputs or services can be used and limits the mobility options during an athlete's career; (3) regulates the number and type of game outputs that are available through scheduling restrictions and television contracts; (4) pools and redistributes a portion of the cartel's profits such as television monies; (5) makes information available to members about the business and political environment of college sports; (6) monitors or polices the behavior of cartel members; and (7) sanctions members of the cartel for rule violations.

Since the NCAA is the dominant force in the regulation of college athletics, its status as a cartel has evolved to that of a monopoly, particularly since it took over women's sports in the early 1980s. The NCAA has exclusive jurisdiction over financial aid to athletes, eligibility requirements, scheduling, and media appearances. If an institution does not like the way the NCAA operates, it can discontinue membership. If it takes this extreme measure, it effectively loses the chance for major athletic competition with big-time commercial rewards. The alternatives are membership in a small-college association such as the NAIA or the end of its intercollegiate athletic programs.

educational and amateur purposes in an effort to enhance the revenues of member institutions and of its administrative operations (Hart-Nibbrig and Cottingham, 1986). In the business climate of big-time college athletics, market values replace educational values as the driving force behind NCAA policies and practices. As a result, the NCAA can be described as a restrictive economic cartel rather than an educational association.

The NCAA has consistently denied allegations of restrictive practices, but a recent economic analysis of NCAA enforcement practices by Fleisher, Goff, and Tollison (1992: 5) demonstrated that the NCAA acts like a cartel, protecting the interests of its members through restrictions on product distribution practices and competition for players. The NCAA justifies its policies, rules, and operating practices in terms of the need to standardize rules, maintain amateurism, and uphold educational requirements among its members. Yet its general unwillingness to regulate more than labor inputs suggests that the NCAA is not fully committed to these standards. Other evidence also casts doubt on the NCAA's professed intentions. For example, the NCAA has had a monopolistic television agreement with one network, which recently was voided by the Supreme Court in the 1984 case of *The National Collegiate Athletic Association v. Board of Regents of the University of Oklahoma*; it has failed to regulate capital assets of athletic programs, such as stadia and training facilities; and it has permitted revenues for schools and income for coaches to increase while at the same time limiting the benefits to athletes to virtually the same level as 1950 benefits. The NCAA also has the ability to sustain a relatively low price for labor through limitations on financial aid and living-expense stipends even though the return on the productivity of this labor is approximately four times its cost.

The courts generally have been unwilling to enforce antitrust measures on personnel matters because of the voluntary nature of the NCAA. In the case of the *NCAA v. Jerry Tarkanian,* the former basketball coach at the University of Nevada, Las Vegas, the U.S. Supreme Court "found that the NCAA was a voluntary organization not subject to due process provisions" (Fleisher, Goff & Tollison, 1992: 15). Thus, the courts have monitored output practices of the NCAA but have ignored input or labor practices, which has enabled the NCAA to continue the practices of restricting competition for players and providing minimal compensation for athletes.

It is evident that NCAA enforcement practices do not treat everyone fairly. Fleisher, Goff, and Tollison (1992) asserted that the largest programs, with the most recognition, the largest investment in capital assets, and the most consistent winning records, stand to benefit the most from NCAA enforcement practices. Those with limited assets, whose records fluctuate dramatically, will more likely be the targets of investigation. Thus, any school whose football or basketball fortunes improve in a dramatic fashion over a few short years is most likely to be investigated by the NCAA. The NCAA punishes upwardly mobile programs that recruit athletes away from the perennial powers. Thus, according to Fleisher, Goff, and Tollison (1992: 106), the NCAA functions as an agent for the traditional big-time programs. Southeastern Louisiana, Oral Roberts University, the University of Nevada, Las Vegas, Cleveland State University, the University of Houston, Southern Methodist, and the University of Kansas represent examples of less established athletic powers that have been sanctioned by the NCAA. Western Kentucky, a Final Four participant in 1971, received two years probation and no television appearances for violations similar to those committed by UCLA at the same time. UCLA received a reprimand and a minimal one-year probation. Large-scale programs that have been consistent winners over the years are less likely to be sanctioned, particularly if these programs would be deprived of television revenues or postseason Bowl appearances. The consistent winners have an additional competitive advantage because their tradition of success and substantial capital assets make these program more attractive to the better athletes. Thus, programs that struggle to win are provided an incentive to cheat, an incentive that is the result of cartellike practices and biases of the NCAA.

THE NCAA AND WOMEN'S INTERCOLLEGIATE ATHLETICS

In 1978, Secretary of Health, Education, and Welfare Joseph Califano announced changes in Title IX of the Higher Education Act of 1972 that would extend affirmative action coverage to the opportunity for women and girls to participate in sport. Thus, college athletics, once virtually exclusively male, was to be open for all (Holland & Oglesby, 1983).

The Association for Intercollegiate Athletics for Women (AIAW), the first major governing body for collegiate athletics for women, was established in 1971 as a subunit of the American Alliance for Health, Physical Education, and Recreation (AAPHER). The AIAW eventually became an independent organization sponsoring national championships in thirteen different sports (Holland & Oglesby, 1983). The underlying principles of the AIAW included giving women the opportunity to strive for excellence in sport, to empower women in their sports experience by placing women in coaching and administrative positions, to encourage growth in the number of opportunities for women to participate in a variety of sports, to provide competitive opportunities at all levels including national championships, to use sport programs as educational and developmental activities that were not simply athletic and outcome oriented, and to consider the rights of athletes as the highest priority (Morrison, 1993).

AIAW founders sought to develop a model of athletics that would not reflect a compromise of the ideals of amateurism that had been experienced by the commercial, professional model dominating men's sports. Thus, AIAW programs offered no scholarships, no transfer limitations, no limits on number of sports offered at an institution, and no restrictions on the level of participation of an institution in athletics. The AIAW did not establish an enforcement division but

Source: AP/Wide World Photos

FIGURE 8.2 Women athletes in action.

instead encouraged self-policing and institutional control.

The programs of the AIAW struggled because of a lack of capable and experienced leadership, which can be largely attributed to the fact that few women had been able to reach leadership positions in athletics. In addition, financing women's programs and generating fan interest in these programs were problematic. Interpretations of Title IX were confusing. Some thought that women's programs should have a separate and independent structure to develop a program that did not reflect the philosophy or structure of men's programs. Others felt that women's programs should be under the same institutional and administrative policy and procedural guidelines as the men's programs, which meant that compliance could come only with a single merged men's and women's athletic program and a single national governing body. The NCAA was poised to take over.

In 1981, at its convention in Miami, the NCAA voted to sponsor championships for women. The NCAA offered to incorporate women's programs and the governance of these programs without any additional cost to the member institutions. The NCAA even "allowed" members to choose to participate in either the AIAW or the NCAA championship. In addition, the NCAA offered a television package that included both men's and women's championships and guaranteed a national audience for the women's game. Finally, to make it more difficult for the schools seeking to retain their AIAW affiliation, the NCAA scheduled its own national championships for women at the same time as those of the AIAW (Sperber, 1990).

The results were disastrous for the AIAW (Morrison, 1993). Institutions withdrew from the AIAW or drastically cut back their participation in AIAW-sponsored events, including championships. In 1982, the AIAW executive board voted to disband the organization. The NCAA had won its power struggle with the AIAW. Thus, the organization that had filed suit against Title IX in federal court in 1978, claiming that it did not apply to athletic opportunities for women, had successfully gained control over women's programs just four years later. The NCAA could create a model of women's sport that would reflect the characteristics of the men's version. It now controlled the potential financial benefits that could come from an enhanced television package and it could keep an independent organization of women's sport from being a threat to the NCAA's control of college sport.

Women's athletics entered the world of big-time commercial and corporate athleticism by adopting the professional model that many of the AIAW founders opposed. Men's officials complained about the financial drain their programs experienced when the women's version was added. Lobbying the Supreme Court to exclude athletics from coverage of Title IX, college athletic directors and the NCAA were successful in obtaining a favorable verdict in the *Grove City v. Bell* decision in 1984. This court decision stalled progress in women's rights in college athletics for at least four years. In 1988, Congress passed the Civil Rights Restoration Act (over the veto of President Reagan), which again required that the equal opportunity mandate apply to *all* programs in institutions receiving federal funding. This congressional action was strengthened by a 1992 Supreme Court decision stipulating that schools intentionally violating Title IX could face suits from women athletes and coaches for financial damages (Coakley, 1994: 210). Despite these actions to reverse the Grove City decision, women's college sports suffered substantial setbacks from which they still are trying to recover. Gender inequity in benefits, scholarships, and facilities in college athletics today are clearly revealed in Table 8.2.

The inequalities shown in this table demonstrate the basis for the evaluation of **gender equity** in the NCAA's programs. The battle over gender equity in college athletics looms as a major struggle in the coming years as women try to move toward more equitable allocations of resources and opportunities while men seek to protect what they have. Pressure is mounting to rectify the financial, facility, and participation differences between men's and women's sports. Men fear that if an equity arrangement is fully implemented, all sports, especially football, will be adversely impacted by cuts in budget and scholarships. Women are concerned that arguments by the men about adverse financial impacts and challenges to the viability of football programs are simply attempts to disguise a more fundamental desire among the men to retain a status quo that privileges men's programs over women's programs.

Efforts by the NCAA to resist gender equity may prove counterproductive if the federal government becomes impatient with progress on this issue and decides to intervene and impose its own equity solution. Radical feminist observers are advocating the creation of independently administered athletic programs for men and women and also support state or federal intervention to see that this happens. More moderate equity proponents call for the decreased power of those who currently run athletics—that is, male athletic directors—and increased power and representation of women and "those sympathetic to the feminist agenda in all governance structures" (Lopiano, 1993: 108).

TABLE 8.2 Comparisons of Average Resources, Events, and Participants in Division I Men's and Women's Athletic Programs

	Men	Women
Scholarships	$1,291,118	$505,246
Head coach salaries	$396,791	$206,106
Recruiting expenses	$268,996	$ 49,406
Scholarships awarded	143	59
Number of events	202	164
Number of student-athletes	323	130

Source: NCAA and *Women in Intercollegiate Sports 1992.*

A 1993 NCAA Gender Equity Task Force defined gender equity in the following terms:

> At the institutional level, gender equity in intercollegiate athletics describes an environment in which fair and equitable distribution of overall athletics opportunities, benefits, and resources are available to women and men and in which student-athletes, coaches, and athletic administrators are not subject to gender-based discrimination." (*NCAA News*, 1993: 14)

Any program would be considered "gender fair" only if the participants in both men's and women's programs would accept as fair and equitable the program of the other gender.

CORPORATE ATHLETICISM AND ATHLETIC DEPARTMENT DEVIANCE

Violations of NCAA regulations occur regularly. During the 1980s, an average of 16 of the 106 Division I-A schools with the most prominent football and basketball programs were under NCAA penalty annually. This represents an average increase of nine programs over the previous decade (Bailey & Littleton, 1991: 25). The NCAA heard over 1,300 cases of violations between 1950 and 1980. It put 150 institutions on probation between 1952 and 1985 for illegal recruiting practices such as excess visits or for offering illegal incentives such as "money handshakes" or incentive payments for accomplishments on the field, low-interest loans, jobs for relatives, free or low-cost automobile use, profits from the sale of game tickets, and travel reimbursement (Sack, 1991). During the last two decades, 50 percent of the Division I-A programs with the largest football programs and 30 percent of all of the 293 Division I programs were penalized by the NCAA. At any one time, between fifteen and twenty institutions were under some form of sanction. The extensive official and unofficial record of illegal activities led to the assertion that an underground economy in college sport exists that includes unreported payments and benefits to athletes (Hart-Nibbrig & Cottingham, 1986; Sack, 1991).

Sack (1991) studied the underground economy in college football and found that illegal payments to athletes were very common in the major football programs. He surveyed over 1,000 current and past professional football players by mail and found that up to 83 percent of the respondents knew other athletes who received improper benefits. Of these athletes, 31 percent admitted receiving benefits over and above what was allowed by NCAA regulations. Athletes from Division I programs were more likely to report knowing of or receiving illegal benefits than participants in smaller programs. In addition, a longitudinal analysis indicated that the incidence of payments increased over the years and most went to living expenses and meeting emergencies (Sack, 1991: 8). Blacks and highly recruited athletes were more likely to receive payments than whites and those less attractive to recruiters. Thus, the low level of financial aid allowed by the NCAA in addition to the limitation of not being able to work while competing forced a financial hardship on many athletes and helped make extralegal payments attractive to them. Table 8.3 indicates the extent of under-the-table payments to athletes in different athletic conferences as well as the general pattern of apparent increases in such payments in recent decades.

The minimal payments to athletes allowed by the NCAA and the competition for talented athletes create structural conditions conducive to the offering of improper payments to athletes and to the acceptance of these payments by athletes. Of course, these payments in excess of NCAA limits are seen as improper or illegal by the NCAA, but not necessarily by student-athletes who see them as scant payment for the services they provide to their institutions and athletic programs. Approximately 53 percent of the athletes interviewed by Sack (1991) did not see an ethical or moral problem in accepting illegal payments.

The problems associated with athletic departments cannot be explained in terms of character flaws in administrators or athletes. Social deviance in college athletics goes well beyond the misdeeds of a few mavericks, renegades, or outlaws. It serves the public image of the NCAA to have the public believe that only a few prominent wrongdoers, such as Jerry Tarkanian of UNLV or Jackie Sherrill of Texas A&M, are the primary cause of the problems in college athletics today (Schultz, 1989). In fact, it can be argued that

TABLE 8.3 Percentage of NFL Players Who Knew Athletes at Their Colleges Who Took Under-the-Table Payments, By Graduating Class and Conference

	Graduating Class		
Conference	After 1970	Before 1970	Total
Southeast	83	58	67
Big Eight	66	53	60
Southwest	65	47	58
PAC Ten	60	57	59
Big Ten	50	58	48
Major independents	49	40	45
Western Athletic	49	33	43
Atlantic Coast	35	29	32
Other[1]	27	22	25
TOTAL	50	45	48

[1]This category includes NCAA divisions for smaller schools and NAIA (from Sack, 1991:6).
Source: Sack (1991:6).

POINT-COUNTERPOINT

SHOULD COLLEGE ATHLETES BE PAID?

YES

College athletes should be fairly compensated for the wealth they produce for their college or university. Athletes work long hours under very demanding conditions, including the risk of serious injury, so that athletic programs can generate revenue and institutions can receive notoriety and visibility. The demands of college athletics make college athletes professionals. Paying these athletes a fair wage for their labor, which is in proportion to the revenue and other benefits they produce for their institution, would end the hypocrisy about college athletics as "amateur" sport. Athletes deserve to be paid.

NO

College athletes already receive very valuable benefits, such as a scholarship to attend college and substantial public exposure, that give them an advantage over their classmates. These advantages will give these athletes a big boost in their subsequent careers. Paying athletes would violate the concept of amateurism, which is a guiding philosophy of college athletics. It would almost impossible to decide how much to pay athletes, and richer colleges would have an advantage in being able to pay their athletes higher wages. Colleges cannot afford to pay their athletes.

How should this debate be resolved?

deviance is widespread and endemic in the structure of college athletics (Frey, 1994). Structural factors account for the pressures to break legal, regulatory, and ethical norms governing athletics. College athletic programs engage in a regular pattern of **organizational deviance** where organizations or their subunits, such as athletic programs, engage in deviant policies and practices that violate regulations or the law. These practices, such as recruiting incentives, are deemed to be necessary in order to be competitive or gain an "edge" on opponents.

When athletic departments engage in deviant behavior, such as violations of NCAA rules, they violate at least the professed expectations of their university and risk tarnishing its reputation. Yet such behavior, especially when it is undetected or unpunished, may be very functional for the athletic program, helping it become or remain successful. Organizations or organizational subunits such as athletic departments can be considered deviant if they adopt goals that are inconsistent with prominent values in society (Frey, 1994). Organizational subunits can also be considered deviant if they use means that they find acceptable in their own environment, but are not approved by the larger organization or society, to achieve societally approved goals (Sherman, 1982, Vaughan, 1983). For example, illegal practices, such as offering improper recruiting incentives, grade manipulation, and "handshake payments," are not approved by the university, the governing body for athletics, nor general society, but athletic departments may deem them acceptable or normal business in the athletic subculture.

Deviance can become the norm when cultural conduciveness is combined with an emphasis on performance and achievement (Finney & Lesieur, 1982; Gross, 1978). Deviance also is more likely to take place if the demands placed on the organization or on a particular subunit exceed that unit's ability or capacity. There is considerable pressure on athletic departments to be successful, but the level of resources supplied from legitimate sources such as student fees and legislative appropriations are often inadequate to meet these demands. To make up the difference, deviant measures or strategies to gain more resources, such as talented athletes and dollars, are employed.

Over the years, college athletics has evolved from a student-controlled activity that existed for the purposes of fitness and diversion to a activity best described as a major commercial and corporate entertainment enterprise. A distinctive aspect of the corporate world of college athletics is that big-time athletic departments often operate autonomously.

The scandals and corruption resulting from their actions have reflected on the reputation of the university in general. University officials, viewed as helpless in the face of a strong athletic coalition of boosters, alumni, and coaches, have had difficulty monitoring and controlling these activities. The Knight Commission, the latest effort to evaluate college athletics, maintained in its final report that "college athletic programs have lost their bearings.... they threaten to overwhelm the universities in whose names they were established and to undermine the integrity of one of our fundamental national institutions (Knight Foundation, 1991: vii).

Athletic departments can operate relatively free of university control for several reasons. First, the university has forced the athletic department to seek the resources it needs to operate from external sources. Universities do not provide a major share of athletic department budgets. Legislative appropriations for athletics have peaked or are on the decline. Gate receipts and student fees have reached their upper limits. This has led to the formation of the coalition of "alumni, coaches, athletic administrators, and community representatives who exchange resources in the form of money, materials and influence for the right to associate with coaches and athletes, for the status or prestige the association brings, and for the access to other persons like themselves who may possess political and economic resources they need or want" (Frey, 1991: 8). This "booster coalition" produced an alliance of persons who may or may not share the larger interests of the university but want to exercise some influence over the athletic programs they have provided resources to support.

A second major reason for the independent status of university athletic departments is inherent in the organizational nature of the university. Universities are generally portrayed as "loosely coupled" organizations characterized by diffuse goals, decentralized and fragmented decision making, and weak linkages among departmental units (Aldrich, 1979; Kerr, 1963). Control seems to be the antithesis of university life (Frey, 1994). Academic departments have strong ties to the administration; they typically depend on it for critical personnel, capital, and operating resources. Athletic departments, which do not depend on the university for critical resources, are weakly connected administratively to the larger institution and therefore are able to operate with less scrutiny.

In addition, athletic departments operate relatively free of university control because of the orphan status assigned to athletics. Very often the lines of communication and the administrative chain of command between university officials and the athletic department are ill defined and confusing (Gilley & Hickey, 1986). Presidents have acted with indifference or have simply ignored the athletic departments, despite continual warnings that they need to be more involved in the governance of athletic departments (Frey, 1987–88; Bok, 1985). Athletic departments are viewed as noneducational units that exist for entertainment and public relations purposes and operate in accordance with values and ideals that run counter to educational and intellectual principles. Thus, college faculty do not involve themselves in the governance of athletics because they are uncomfortable with the profit-entertainment and nonacademic nature of athletics; athletic personnel feel faculty do not have anything of value to contribute to athletic governance; rewards for diligent oversight of athletics are few and of little consequence; and they fear being coopted because faculty known to be friendly to athletics will lose stature among their colleagues (Weistart, 1987).

The commercialism of athletics and its dependence on external sources of funding are actually very consistent with the practices of the modern university. Universities are large-scale educational corporations that respond more to the dictates of the student, legislative, research-granting, and benevolence markets than to the internal educational needs of students and faculties. Presidents become fundraisers and public relations figureheads, and they leave academic matters in the hands of a senior vice president or provost. Thus, athletic departments essentially are doing what is expected of them—responding to crucial constituent markets in order to garner resources that are inadequately supplied by legislative appropriations or student fees. The corporatization of the athletic department is an extension of the corporatization of the university. In this corporate context, market and business principles are likely to prevail over ethical considerations and a life of intellect. As we will see in the next section, nowhere are the pressures of college sport more prevalent than in the occupational life of the coach.

College Coaches

The structure of the college sports system, with its emphases on winning, money, and entertainment, puts tremendous pressure on coaches that, in turn, produces a high turnover rate and an increased risk of burnout. Coaches are seldom hired or retained for their ability to produce good students with admirable character. They are typically hired because they are proven winners or are seen as having the potential to become big winners; they are fired because their won-lost record does not satisfy fans, alumni, or university officials. These parameters are true today for men's and women's programs.

Coaching is more complex and demanding than most observers understand. Coaches must do more than run practices and develop game strategy. They also raise money, develop community support, and manage staffs and budgets, and they must spend considerable time and energy recruiting potential athletes with little assurance that they will be successful in obtaining a commitment.

The environment of college coaches produces considerable stress and anxiety as they face internal and external pressures to win and maintain the financial viability of the program. These pressures are likely to come from internal constituencies such as university officials and from external booster and alumni groups who see the college program as a measure of the success of their community and a source of pride and excitement in their lives. Coaches are told to "run a clean program," "graduate your players," and "stay within the budget." Of course, they are also told to win games. These demands produce role strain and role conflict. As we have seen, role strain occurs when competing demands are associated with a single role, such as when boosters to expect a coach to win, academic administrators and the athletic director expect the coach to stay within budget, and faculty expect the coach to reduce the time athletes spend in their sport and increase their time studying. A coach can experience role conflict when he or she has to balance competing expectations of the coaching role with spouse, business, and parental roles.

Most of the role conflict research concerning coaches has centered on the perception and resolution of conflict between the coaching role and the teacher role. Decker's (1986) analysis of research on role conflict found that (1) males experience less role conflict than females, (2) coaches of two or more sports feel greater conflict than coaches of one sport, (3) coaches of individual sports experience less role conflict than coaches of team sports, (4) coaches who teach physical education experience more conflict than teachers of other subjects, and (5) those who prefer either to teach or to coach experience more role conflict than those who demonstrate a preference for both roles. Her findings for small-college coaches were consistent with the results of studies by Massengale (1974) and Locke and Massengale (1978) showing that small-college coaches experienced less role conflict than their counterparts in larger institutions.

Since few coaches in Division I teach in the classroom, the role conflict they experience might represent disagreement between the expectations of the coach's role and those of family, friendship, and administrator roles. Hastings's (1987) case study of the social world of the college swim coach documents the role strain and conflict a coach can experience. Adaptations to role stress include dropping out of the coaching profession, disaffiliating from other roles such as parent or spouse, prioritizing roles, or becoming superorganized to balance the demands of all occupied roles. Hastings suggests strategies of managing cross-pressures that were effective in his multiple roles of coach, professor, and parent.

Conflicts can generally be resolved by voluntarily or involuntarily leaving the position, by accepting another position either as a promotion or demotion, or by trying to gain greater control over the facets of a program that can produce the greatest impacts on role performance (Massengale, 1974). Turnover data for college coaches are not readily available, but even the most casual observation suggests that the coaching profession does not have much job security.

Shuster's (1991) study tracked every coaching change between 1975 and 1990 for 292 Division I men's basketball programs. Only twenty-four programs em-ployed one coach for the fifteen-year time period, whereas forty-five employed two coaches, 102 em-ployed three, sixty-eight employed four, and over fifty employed five or more during this time span. An average of 3.26 coaches were employed by each of the schools evaluated, making the average tenure of a coach a little more than four years. The average number of coaches employed by the fifty-six teams to reach the Final Four the last fifteen years was 2.25. Sperber (1990) reported that the American Football Coaches Association found that the average Division I head football coach was in his job 2.8 years. Tenure expands as the level of play decreases to an average of 6.8 years.

Despite reforms, the pressure to win and generate revenue is great as ever. Coaches also are held responsible for the off-the-court behavior as well as the academic performance of their athletes. These extra responsibilities increase role demands and are a major source of stress. If a coach does win more games than are lost or if the athletes are exemplary persons and students, there is still no guarantee of job security. Thus, the threat of termination coupled with the pressure to win and perceptions of inadequate resources and insufficient control will create a temptation for coaches to use any means, legal or illegal, to be successful. Whether coaches leave a program voluntarily or involuntarily, they are not without social supports from the coaching subculture, which tries to make sure that a fired or prematurely retiring coach has a job. This subculture is also described as an "old boys' network" designed to promote loyalty to peers in the face of disloyalty from institutions and programs.

College coaches tend to move under the auspices of a sponsor or someone who recommends or endorses a candidate for a position (called *sponsored mobility*) rather than because they have successfully competed for a position strictly on the basis of their skill and won-lost record (called *contest mobility*). Educational credentials, athletic success as a player, coaching experience and success, and the human capital factors often have little effect on the first coaching job. These factors would be important if contest mobility prevailed in the coaching occupation (Loy & Sage, 1978). Head coaches tend to hire their former athletes or promote the hiring of loyal long-term assistants who served their apprenticeship under the head coach. In addition, players who received notoriety through their exploits on the field stand a greater chance of being hired because of the favorable publicity and recruiting advantages a program receives with a famous player as coach. Finally, institutional prestige is a factor when little is known about a coach's ability but character traits can be generalized from what is known about the program. Since most coaches are white and male, they tend to hire people like themselves when they have the opportunity. In this environment, women and black males have some difficulty finding satisfactory opportunities in college athletics, particularly as head coaches.

Longitudinal analyses of opportunities for women in college coaching and administration by Acosta and Carpenter (1990) showed that women rarely coach men's teams, that over half of women's teams are coached by men, and that men are in charge of 85 percent of women's programs. There are fewer coaching and administrative opportunities for women today than was the case twenty years ago. Annelies Knoppers and her colleagues found that the decline in the number of women coaches was not because of a scarcity of qualified coaches (the human capital argument), but could be attributed to hiring practices and the characteristics of those doing the hiring (1989: 357). The human capital tenets were conveniently used to justify not hiring a woman or assigning a lower salary than might have been commanded by a male coach. The authors of this study found that athletic directors, mostly male, serve as gatekeepers of the profession and will hire persons like themselves. That is, the patriarchal dominance of a male subculture of coaching is the most significant factor explaining why there are fewer women coaches and why women who coach receive lower salaries than their male counterparts. Males control the coaching profession in order to maintain its prestige and to match that status with higher rewards. Women remain marginalized in the coaching profession by only having opportunities to coach the less prestigious men's sports or to be relegated to an assistant's position in the high profile women's sports (Kane & Stangl, 1991).

Brooks and Althouse (1993) analyzed coaching opportunities for African Americans in the college ranks. While there are a few prominent African-American

POINT-COUNTERPOINT

SHOULD COLLEGE COACHES BE TENURED?

YES

Tenuring college coaches in the university would give these individuals job security, making it impossible for them to be fired simply on the basis of their won-lost record. Coaches would be less likely to encourage or engage in illegal and deviant practices if they knew that their position was secure. Coaches would also devote more time to educational purposes of sport and be less worried that the financial life and death of their program rested on their shoulders.

NO

Coaches should not be tenured because they generally lack the academic and professional credentials required for tenure. They have no classroom experience and no understanding of the norms of the professoriate in most cases. Unsuccessful coaches, whom institutions want to replace, will be a financial burden to the institution. Most important, tenure is not granted for job security. Its purpose is to assure academic freedom to faculty members.

How should this debate be resolved?

head coaches, such as Nolan Richardson, Ron Dickerson, and George Raveling, the percentage of black head coaches of Division I programs still hovers at 5 percent. The percentage of black assistant coaches is even less. African Americans aspiring to be coaches suffer from not being a part of the coaching subculture that serves as a referral network for coaching positions. Many African Americans did not occupy central positions as players, and it is from these positions that coaches and mangers tend to be recruited. That is, blacks were not generally found in positions such as quarterback in football or catcher in baseball that were defined as requiring leadership and intellect and as more crucial to the outcome of the game.

Brooks and Althouse recommend that African-American coaches expand their personal networks and affiliations in order to enhance the possibility for consideration for future jobs. That is, African Americans need to develop their own subculture to refer potential coaches for positions and to serve as emotional support for coaches who are separated from a coaching job. They also need to link their own subculture to the powerful white male coaching subculture that tends to exert substantial control over occupational opportunities.

In 1994, the Black Coaches Association (BCA) threatened to organize a walkout of the NCAA Final Four basketball tournament unless their demands that academic requirements for athletes be reevaluated and that blacks be given more access to coaching and administrative opportunities. After an intercession by the National Department of Justice, the NCAA agreed to establish negotiations with the BCA in order to accommodate their request.

Recent developments suggest that the role of the coach is changing. Players at several institutions have challenged the authority of the coach or demanded changes in coaching style. A dictatorial or authoritarian style and motivation by fear and verbal and physical intimidation seem less prevalent in college coaching today. Many athletes are no longer afraid to speak out if they feel their rights and personal dignity are being challenged.

A highly publicized recent incident illustrating the rights and power of players versus coaches is the firing of University of California basketball coach Lou Campanelli near the end of the 1993 season. He was observed by the athletic director using excessively abusive language in a postgame locker room tirade. Several members of the team had previously approached the athletic director asking that Mr. Campanelli be directed to change his behavior or these athletes would transfer. Tom Miller of Army also was fired for alleged mistreatment of athletes, as were Central Michigan's Keith Dambrot, Lamar's Mike Newell, and Colorado State's Earle Bruce. Players at

Memphis State, Oklahoma, Drake, South Carolina, Morgan State, and the University of Nevada, Las Vegas have challenged the coaching styles and interactional skills of the head coaches of these institutions. Players are forcing a change in the role of coaches and the one-time common autocratic relationship between coaches and players. Of course, members of the coaching subculture have protested these changes and expressed resentment about firings coming as a result of player demands. It seems clear, though, that in the current era intimidation and denigration are no longer popular or accepted by players and many administrators.

In addition to player revolts, college coaches face potentially greater threats to their authority if the $10 million law suit of Bryan Fortay is upheld. Fortay accused University of Miami coach Dennis Erickson of promising to make him a starting quarterback on the Miami team in 1991. The disgruntled quarterback, who later transferred to Rutgers after failing to earn the starting job, asserted that his chances for a professional career were immeasurably damaged by not being able to start at quarterback for Miami, a school well known for the quarterbacks it has placed in the NFL. The legal justification for this argument is likely to hinge on whether the athlete is considered an employee and the school his employer. If recruitment results in an implied legal contract between an athlete-employee and institutional employer, coaches could be legally bound to provide the opportunities implicitly or explicitly promised when a blue-chip athlete is "hired." Such legal requirements could create major constraints and obligation for coaches, which could substantially limit their options in player assignments. Also being monitored is the case of Kent Waldrep, who filed for workman's compensation after being paralyzed in an on-field accident in 1974 while a running back for Texas Christian University. Waldrep won in an administrative court, but the ruling is being appealed.

ATHLETICS AND ACADEMICS IN HIGHER EDUCATION

A significant amount of NCAA legislation in recent years has been directed at the assertion that athletes can no longer be legitimate college students because of the demands and pressures placed on them by their participation in college athletics. Intersectional schedules that require transcontinental and even intercontinental travel, out-of-season practice sessions, specialized position meetings in addition to regularly scheduled practices, public relations and community involvement expectations, and the expectation of carrying a full-time class load make it very difficult for college athletes to do well in the classroom. Graduation within a reasonable length of time, such as five years, is even more difficult.

Prospective athletes frequently do not receive a realistic idea of the academic demands of college when they are being recruited (Adler & Adler, 1991). The pressure of contending with student-athlete role conflict has intensified as college athletics has evolved into a commercial entertainment enterprise. Role conflict is greater for those athletes in the high profile, big-time NCAA Division I programs; it is less of a factor for those in lower divisions. Role conflict is also more of a problem for male athletes on scholarship than for female athletes and those not receiving financial aid (Sack & Thiel, 1985). In most cases, the role conflict of the student-athlete is resolved by devaluing or deemphasizing the academic role (Purdy, Eitzen & Hufnagel, 1982; Adler & Adler, 1991).

Athletes participating in revenue-producing sports (football and basketball) have academic performances well below athletes in nonrevenue-producing sports and well below the academic achievement of the general student body. Scholarship athletes have also shown lower academic achievement than their nonscholarship counterparts (Adler & Adler, 1985; Lance, 1987). Women athletes, who at one time performed academically at the same level as the general student body, more recently have tended to do worse academically when they have been involved in revenue sports such as basketball. Reports showing graduation rates and academic achievement rates of athletes to be higher than the rates of nonathletes may not give a totally accurate picture of the academic performance of athletes because they may fail to control for differences in sports programs. Basketball and football players in big-time programs have tended to experience the most academic difficulties.

In some cases, athletic departments have engaged in questionable practices, such as registering athletes in courses where the instructors are known to be "friendly" to athletes, placing athletes in community colleges where course curricula are less demanding, pressuring professors to change grades, placing athletes in courses that fulfill eligibility requirements but do not contribute to the student's academic progress toward graduation, and clustering athletes into certain majors at a rate disproportionate to the enrollment of the rest of the student body. The latter was found to be especially true in Division I basketball programs and more so for black athletes than white athletes, male athletes than female athletes, and nonscience than science majors (Case, Greer & Brown, 1987). In a sophisticated statistical study, Leonard (1986) found that Division I basketball players, when compared to basketball players in smaller programs, were more likely to: (1) take a less demanding major; (2) take perceived easy courses; (3) miss important exams; (4) hustle professors for grades; (5) take fewer courses per semester; (6) miss taking courses they really wanted to or needed for their major; (7) cut themselves off or socially isolate themselves from the rest of the student body; and (8) rank their athletic participation higher than student activities.

Concern in the early 1980s about dubious and illegal academic practices led to a call for academic reform in college athletics. A major move in this direction was the formation of the NCAA Presidents Commission in 1981. This group had limited power in the NCAA Convention, but it exerted considerable influence on reform matters. Prior to the efforts of this commission, athletes could remain eligible by earning 96 credits over a four-year period with a grade point average of 1.96. Athletes could be eligible by NCAA standards even if their academic performance was below what was required by the institution for regular standing as a student. In addition, the credits earned did not have to be in a major or department or represent progress toward a degree. It was not unheard of for an athlete to earn well over 100 credits after four years of athletic participation but only be a second semester sophomore academically. Reform was needed.

Proposition 48, Proposition 42, and Proposition 56

The Presidents Commission took the lead in the reform of academic standards by developing NCAA Bylaw 5.1(j), known as Proposition (or Prop) 48 (Lapchick, 1989). Prop 48, effective fall 1986, focused on the academic preparation of athletes for college. It stipulated that to be eligible to practice or compete in a sport in any NCAA Division I or I-A program, a college freshman had to have a 2.0 grade point average in eleven core high school courses in English, mathematics, social sciences, and physical sciences, *and* a minimum score of 700 (out of a possible 1600) on the SAT or 15 (out of a possible 36) on the ACT. Athletes meeting only one of the requirements and earning a 2.0 GPA in all their courses were designated "partial qualifiers." They still could receive athletically related financial aid, even though they were not allowed to participate in athletics. If they met Prop 48 standards after the first year of college, they were left with three years of athletic eligibility.

The requirement of minimum SAT and ACT scores has been controversial, particularly for black athletes. Among blacks taking core courses, the mean ACT score is 18.2. The mean ACT scores for Native Americans and Mexicans is 19.9, for Puerto Ricans 20.9, and for Asian Americans 22.8. The average SAT score for whites is 925–930, and for blacks is 710–715 (Leonard, 1993: 314). Many black educators argued that the use of these scores were discriminatory because of a racial bias in standardized tests that contributed to lower scores among black students. Thus, a large proportion of the potential black student-athlete population would be ineligible under Prop 48 (Uehling, 1983). Edwards (1986) noted, for example, that 51 percent of black students scored less than 700 on the SAT and 72 percent scored less than 15 on the ACT. Furthermore, an NCAA study of the class of 1981 indicated that if these rules were implemented, six of seven black male basketball players and three of four black football players in the largest programs would have been ineligible as freshmen. The study also found that one of three white male basketball players and one of two white football players would not have met the Prop 48 standards (Lapchick, 1989: 20–21).

A fear was expressed that the implementation of these standards could effectively prevent many black—and white—athletes from pursuing a career in professional athletics. College sport was the only avenue available as a track to the pros for aspiring young athletes in many sports. Proponents of Prop 48 argued that these minimal standards were necessary to send the message that a serious academic commitment was required to play college sports, to discourage athletic recruiters from pursuing athletes with little chance of being academically successful in college, and to provide partial qualifiers a chance in their freshman year to establish themselves academically (Coakley, 1994: 403). An important spokesperson for the higher standards was the widely known sport sociologist and activist Harry Edwards, who stood apart from many other black educators in supporting these higher standards. He agreed with opponents of Prop 48 that the standards were arbitrary but also contended that they were too low. Edwards believed that it was important for black youths to have athletic role models and socialization opportunities that emphasized academic achievement as well as athletic success (Edwards, 1983, 1986).

When Proposition 48 was implemented in the fall of 1986, less than 10 percent of the football players and only 13 percent of the basketball players were ineligible. The NCAA study showed that over 69 percent of all black male athletes the class of 1981 would not have met the Prop 48 standard, but in 1986–87 less than 20 percent actually were ineligible. Furthermore, a survey of Division I institutions in 1987–88 (with 202 respondents) revealed that only 4.5 percent of all enrolled freshmen were partial qualifiers and 1.4 percent were nonqualifiers. Football had 34 percent of all partial qualifiers, and basketball had 13 percent of the partial qualifiers. Only 15.5 percent of blacks in the revenue sports of football and basketball were partial qualifiers or nonqualifiers. Although blacks were disproportionately represented among athletes who were partial qualifiers or nonqualifiers (i.e., 58 of 60 in basketball and 141 of 152 in football), the total number of blacks in these categories were less than one-quarter of the number predicted by the NCAA study (Lapchick, 1989: 21).

The NCAA has not limited its reform efforts to Proposition 48. At its 1989 convention, it approved Proposition 42, which imposed a stricter penalty for failing to meet the Prop 48 standards. Under Prop 42, athletes who did not meet the GPA *and* standardized test score standards of Prop 48 could not receive an athletic scholarship in the first year. This action created even more controversy in the black community. Temple basketball coach John Chaney labeled Prop 42 "racist," and Georgetown University basketball coach John Thompson walked off the court in protest in the first game after this convention. His reaction led to a debate that resulted in a decision to table Prop 42 until issues raised by this proposed measured could be studied more extensively (Lapchick, 1989: 21).

The ultimate impact of Proposition 48 and related academic preparation measures can be evaluated in terms of the graduation rates of athletes. Since 1986, approximately 75 percent to 80 percent of all partial qualifiers have done well enough academically to remain in college. In addition, the graduation rates of athletes improved from 51 percent in 1983/85 to 57 percent for those who entered in 1986, the year Proposition 48 was implemented. The graduation rate of the general student body of Division I schools was 55 percent. Even schools with the lowest graduation rates for athletes, such as Texas Southern University, the University of Houston, and Long Beach State, improved their rates. Graduation rates for black male athletes increased to 41 percent from a 33 percent rate prior to Proposition 48 (Blum, 1993a).

The success of these reforms can be attributed to increased attention to academic qualifications by high school recruits and increased academic support services for athletes provided by universities (Coakley, 1994: 404). Critics of Proposition 48 and similar measures remain concerned about the negative impact on black athletes. They point out that fewer black athletes are enrolling at Division I institutions. In addition, 54 percent of the black athletes in the 1981 NCAA study who would have been ineligible under Proposition 48 graduated, and they graduated at a higher rate than athletes in their cohort who met the Prop 48 standards (Lapchick, 1989: 22). A recent analysis of the latest graduation figures demonstrated that black athletes graduate at a higher rate than black males in the same cohort (64% to 61%) and that the lower graduation rates for athletes is the result of white athletes' "under-

POINT-COUNTERPOINT

SHOULD ACADEMIC STANDARDS BE REDUCED FOR ATHLETES?

YES

The requirements set forth in Propositions 48 and 42 reflect a racial bias and are an additional obstacle to academically motivated athletes who often are black and come from disadvantaged family and educational backgrounds. The SAT and ACT tests are culturally biased in favor of middle-class whites. Many prospective black college athletes will be denied an opportunity to get into college and better themselves unless NCAA admission requirements take into account the cultural biases in tests used to judge admissibility. Athletes can be academically successful in college even with low STAT and ACT test scores if they receive sufficient academic support.

NO

NCAA academic standards need to be maintained or even raised to send a clear message that athletes must meet reasonable educational expectations to be able to play. Too many athletes enter college without adequate academic preparation and with no intention of getting a degree and a desire only to prepare for a professional sports career. The current or higher admission standards make it more likely that they will be serious students who will be able to take advantage of their education in college. In addition, current policies permit partial qualifiers to be admitted and establish themselves academically before competing.

How should this debate be resolved?

performing their white cohorts" (Siegel, 1994: 215). These results could be interpreted to mean that test scores and GPAs can be flawed predictors of academic success for black college athletes and that other or additional measures should be used to determine the academic qualifications of black high school students.

In addition to new admission requirements, the NCAA implemented Proposition 56, which required that an athlete make academic progress toward a degree by taking courses that meet general curriculum requirements for all students at the university, college, and departmental level. The faculty representative to the NCAA and/or a college representative would certify academic progress each year.

Despite this series of academic reforms, big-time college athletic programs continue to impose heavy demands on athletes. These demands can distract academically motivated athletes from pursuing their classroom goals. The difficulties college athletes have in developing a serious academic commitment were shown in a systematic and perceptive study of a Division I basketball program by sociologists Patricia and Peter Adler (1985; 1991). The Adlers discovered that most athletes arrive on campus with high academic expectations; the more they become involved in athletics, however, the greater the role conflict they experience and the more intense the pressure to resolve this role conflict in favor of athletic demands. The basketball players switched their primary role identity from academics to athletics because the athletic role gave them power and notoriety they could not derive from academics. This switch was reinforced by the encouragement and approval of the coach, the most significant person in the athletes' role set. In addition, athletes became isolated from the rest of campus life because the demands of the athletic role made it impossible to fulfill the academic and social roles normally associated with being a student. These athletes eventually switched to easy courses and less demanding majors.

Meyer (1990) found that while the academic idealism of male athletes deteriorated over time, the idealism of women athletes continued and was enhanced during their college experience. The athletic subculture of women, while still very important to the participants, and the influence of the coach did not detract from their academic goals in the same manner as it did for men. The difference is attributed to the lower priority, revenue, and career pressure of athletics for female than male athletics. The women could also draw on a more supportive family and educational background than could the men. In addition, the women may not have received the same encouragement in performing the sports role as men since women's participation is typically not seen as important or as a career commitment. According to Adler and Adler (1991), male athletes are "engulfed" by the athlete role. It becomes a master status overriding all other roles and statuses. There is no place for the athlete to disengage from the role without serious personal or social consequences.

CYCLES OF REFORM EFFORTS IN COLLEGE ATHLETICS

We have seen that reform has frequently been a subject in U.S. college athletics since shortly after its inception in the nineteenth century. Most of the early criticisms and calls for reform came from college faculty and journalists with leading newspapers and magazines who were concerned about the demeaning impact of athletics on the character of the student body, the distraction from academic work, the professionalization of the athlete, the lowering of academic standards for athletes, and the excessive violence and occasional deaths, especially in football. These early expressions of concern largely fell on deaf ears, as administrators and alumni continued to promote athletics as beneficial to the financial and social climate of the school.

In the early 1900s—"football's ugly decades"—the faculties of many institutions, including Harvard, Stanford, and Columbia, led an abolition movement seeking a ban of football because of the game's brutality (Moore, 1967). In addition, a White House conference of prominent coaches, faculty, and alumni from Harvard, Yale, and Princeton was called by President Theodore Roosevelt to discuss ways to reduce the violence of the sport. The conference produced a weak resolution calling for violence to be curtailed, but it had no impact, as the 1905 season was as brutal as any other. Eighteen players died and 159 were severely injured (Rader, 1990: 180). In 1909, thirty players were killed and over 200 were injured in football (Moore, 1967: 61). The regulation of mass play by the newly formed Intercollegiate Athletic Association and the increasing popularity of the new and more open forward passing game eventually reduced the deaths in football.

The Intercollegiate Athletic Association was formed in December 1905 as a regulatory body to set standards for competition; it later became the National Collegiate Athletic Association. Since the founding of this organization, various investigative bodies, commissions, task forces, and review panels have proposed a variety of reform recommendations. Many of these recommendations have kindled hope of genuine reform. Like the NCAA itself, however, they typically have had little effect on the basic structure and practice in big-time college sports.

According to Hanford (1979), the Roosevelt inquiry was the first of four major "benchmarks of external scrutiny" of college sport between 1905 and the 1970s. The second was sponsored by the Carnegie Foundation for the Advancement of Teaching and supervised by Howard Savage. This investigation, begun in 1926, surveyed 148 college and secondary schools in the United States and Canada, and it resulted in a series of reports issued in 1929 and 1931 (see Savage, 1929). The Carnegie Commission was concerned with the growing professionalism of college football linked to the increasing use of nonfaculty coaches and trainers and the special treatment accorded athletes. The commission concluded that the NCAA had been ineffective in curbing professionalism. Presidents were called upon to turn this situation around and to return college football to its original amateur status. Hanford observed that the report "had little effect on the direction in which men and events were moving intercollegiate athletics" (Hanford, 1979: 69). The NCAA responded with formal regulations governing questionable behavior, such as recruiting practices, but it did not make major changes in the structure or operation of major college sports themselves. The organization expanded its administrative structure, however, by creating special departments to enforce regulations and to oversee the conduct of athletics.

Administrative control was not returned to the students as Savage had recommended, but the pattern of reform was established: cosmetic rules would be adopted, public statements would praise the results of the inquiry, change would be promised. In the end, little reform took place. The commercial, corporate, and professional nature of college athletics remained intact.

In 1951–52, a third inquiry into the nature of college athletics was launched by a committee of college presidents and sponsored by the American Council on Education (ACE). Still concerned about the professionalism of football but now alarmed over the scandals associated with gambling on college basketball, the ACE decided that the presidents needed to get more involved in the governance of intercollegiate athletics (Hanford, 1979). Among other items, this group recommended that spring football be dropped and postseason bowl games be discontinued (ACE, 1952). The report's suggestions were totally disregarded.

In 1973, the ACE instigated another probe into college athletics by suggesting a thorough investigation of the nature of college athletics (Hanford, 1974). The ACE appointed a Commission on Collegiate Athletics in 1977, but the work of this committee gained little public attention and had little impact on college athletics. Some years later, the ACE again expressed concerns about the governance of college athletics in a recommendation to the NCAA to create a board of presidents with final authority and veto power over NCAA legislation. Walter Byers, the executive director of the NCAA at the time, led the resistance to this suggestion. It was soundly defeated by the 1984 convention. In its place, a presidential commission with advisory power was created in the hope that this at least would give the appearance that presidents were in control.

The 1970s saw some limited probes into college athletics by the federal government. In 1974, Congress created the President's Commission on Olympic Sport to mediate a longstanding dispute between the NCAA and the Amateur Athletic Union (now The Athletic Congress). The NCAA also came under scrutiny by the House Subcommittee on Oversight and Investigation amid reports that the NCAA enforcement practices were arbitrary and violated constitutional provisions of due process. Neither inquiry changed the nature of the NCAA or the structure of college sport. More recent federal inquiries may have different results, as many states and their representatives in Congress are very unhappy with what the NCAA has done to their universities and to athletes.

Congress has been concerned that several of the investigations into sport have not produced anything but superficial reforms (McMillan, 1992). In 1990, for example, Congress passed the Student Right to Know Act that forced colleges and universities to reveal their graduation rates to prospective students. In addition, legislation has been introduced to guarantee due process to coaches who are arbitrarily dismissed, to force universities to publish the income and expense associated with athletics, and even to establish a National Commission on Intercollegiate Athletics. Then Congressman Tom McMillan introduced a comprehensive sport reform bill in 1991 known as the Collegiate Athletic Reform Act. This act was designed to restore education as the main priority of college campuses. The bill also gives the NCAA a five-year antitrust exemption for the negotiation of television contracts in exchange for its promise to institute certain reforms such as gender equity, due process for athletes, downsized programs, and five-year, rather than one-year, renewable scholarships. The NCAA is likely to resist such efforts at external control by Congress or any other regulatory body for fear that its political and economic standing will be compromised. It is not clear where McMillan's legislation will lead because he did not win reelection in the 1992 national elections, but his effort is part of a pattern of increasing external scrutiny into the nature and practices of big-time college sports.

The most recent large-scale inquiry into college athletics was sponsored by the Knight Foundation, a Washington, D.C. philanthropic organization. This investigation could be called the fifth major benchmark of external scrutiny of college athletics in this century. Composed of well-known academic administrators and athletic officials, the Knight Commission was charged with the task of developing a reform agenda for intercollegiate athletics. The commission was particularly concerned that the activities of athletic departments would reflect badly on the perceived integrity of the larger institutions and call into question the proper relationship of athletics to the academic mission of those institutions. In addition, the commission was concerned about the quality of academic life for student-athletes facing increasingly demanding athletic roles. These concerns were no different than those of previous reform efforts.

As in the past, the most recent reform recommendations began with the admonition that strong and decisive action by the presidents was needed as the key element in successful reform. "They must be in charge—and must be *understood* to be in charge—on campuses, in conferences and in the decision-making councils of the NCAA" (Knight Foundation, 1991: vii). Presidential control became the "one" in the **one-plus-three model of reform**. The "three" were academic integrity, financial integrity, and independent certification. Presidents would have the same degree of control over athletic programs that they had over other areas of the campus, and the role of the president would be expanded in the governing structure of the NCAA. Athletes would have to be making progress toward a degree without cutting any corners on admission or eligibility. "No pass, no play" would be the guiding principle of athletic participation. Athletic departments would not be independent subsidiaries of the university nor would their financial operations be separate from normal university control. Finally, each NCAA-affiliated university would have to undergo an annual certification to verify that the athletic department was acting in a manner consistent with the goals of the larger institution, that there were acceptable and effective fiscal controls, and that student-athletes were representative of the larger student body in terms of admissions, academic progress, and graduation rates.

Despite the far-reaching nature of these recommended changes, the Knight Commission contained no recommendations for changes in the basic structure of college athletics as a corporate and commercial enterprise. Winning, profit, and entertainment values were not challenged as fundamental elements of the world of big-time college sports. National championships, national and international scheduling, a dependence on television, and the power and practices of the NCAA were not directly addressed by these reform proposals. Furthermore, there is no indication that they will be any more influential than recommendations from the major inquiries of earlier eras. Congressman McMillan, who served on the Knight Commission, called the commission's report "timid." As a result, he initiated federal legislation to implement measures of reform that would have a more significant impact on athletics (McMillan, 1992). Since NCAA control and presidential influence have failed to lead to significant changes in the past, governmental intervention or its threat may ultimately have a stronger impact on the nature of college athletics.

New Reform Recommendations

Reform messages have generally focused on four areas of concern: (1) the economics of college sport, or the maintenance of fiscal integrity; (2) unethical recruiting practices and the special on-campus treatment of athletes; (3) the opportunities for women in college sport; and (4) the relation of athletic goals to academic goals at the institutional and individual level (Hanford, 1979). The issue that has received the most attention has been how to reconcile the potentially conflicting demands of academic and athletic roles so that athletes may have a satisfactory academic experience.

The NCAA is not likely to approve changes that might weaken the stature of big-time college athletics or its own power in the governance of intercollegiate athletics. At the same time, universities that want to retain or seek the status of a major athletic power without further draining their budgets are not likely to accept a larger internal financial responsibility for the funding of their athletic programs. They know that being competitive in big-time college sports can cost a lot of money, and the current budget crunch in higher education does not permit most universities to spend substantially more on athletics. As a result, they must rely on external support, which can foster a loss of university control over athletics to people and groups outside the university.

The most radical reform suggestion is to give commercialized intercollegiate athletic programs autonomy. This amputation of athletics from the formal structure of the university would mean that universities would not be expected to exercise administrative control over athletic programs or their participants as athletes (Bailey & Littleton, 1991). In this case, the

SPECIAL FOCUS

A Radical Reform Proposal

Frey and Massengale (1988) argued for the restructuring of college and high school athletics in terms of a *club*, rather than a *professional*, model, in which the quality of the participation experience is paramount and winning and revenue production are secondary concerns. The professional model is most accepted at the higher levels of college (and high school) athletics. As observed earlier, big-time athletic programs operate relatively autonomously, and participants place a lower priority on the student role.

Frey and Massengale proposed that a college sports system based more on the club or amateur model would require the following reforms of the current system:

1. Mainstreaming of athletes and the reduction of the star system by encouraging full participation of student-athletes in the social and academic life of the university.
2. The elimination of postseason tournaments at all levels.
3. Sports in distinct seasons, with only regional competition.
4. No athletic events, team meetings, or practices be held in prime class time during the day or during exam periods.
5. Part-time coaches with faculty appointments and salaries commensurate with their faculty rank.
6. No full-time athletic administrators.
7. No connection between booster clubs and university athletic programs.
8. The opportunity for student-athletes to transfer to another college or university without penalty such as a loss of a year's eligibility.
9. Value-training exercises in each sport.
10. No athletic scholarships at any level.

The implementation of these reforms would mean a basic restructuring of athletics and the demise of the NCAA as the governing body. It would mean deemphasizing the commercial aspects of college sport and restructuring its financial basis. For example, if all Division I institutions were to get an equal share of the income, we would expect a reduction in the instances of unethical behavior because such behavior would not result in an income advantage.

athletic team would rent the school affiliation, athletes would not be required to attend class or meet any academic standards, athletes would be paid a salary and be covered by worker's compensation, coaches would have no connection with the institution as either faculty or staff, and amateurism would no longer apply as the philosophical basis of athletic participation.

Sperber (1990) suggested that it is time to stop pretending that the athletic department has anything to do with the academic mission of the university. If the fifty largest "franchises" in what Sperber calls "College Sport, Inc." existed as independent entities, they would not have to deal with issues of academic integrity and could operate more openly as businesses, much like professional sports franchises. The financial and organizational separation of athletics from the academic institution would mean an immediate end to most illegal and unethical activities related to the tension between the academic and athletic roles of student-athletes and the effort to treat these student-athletes like ordinary students. It also would mean that the oversight and enforcement functions of the NCAA would be largely unnecessary. Institutions still playing college sports would offer substantially scaled-down programs in which commercialism and professionalism would have little meaning.

CONCLUSION

College sport was born in a controversy of commercialism and professionalism when the first intercollegiate event was sponsored by a businessman who wanted to use a regatta between Harvard and Yale to stimulate interest in his lakeside resort and railroad line. Despite calls for reform and many commissions to recommend change, the modern version of big-time college athletics reflects similar characteristics. It is a commercial enterprise driven by market values rather than educational purposes. College presidents have difficulty gaining control over the operation and direction of their major athletic programs. Yet few would deny that college sports have captured the imagination of the American public. Although many decry the current state of affairs, few are willing to support drastic change.

A study conducted nearly a decade ago found that over 90 percent of presidents felt there were problems in college athletics, but only one-third felt their own institutions had problems in this area. These same presidents saw themselves as "in control" and believed that alumni and boosters were not a problem in the governance of their athletic programs (Gilley & Hickey, 1986). These are surprising results since almost every commentary, report, and commission study has cited the need for *more* involvement of the presidents in the governance of athletics. In fact, though, it is often very difficult for presidents or anyone else to alter the well-entrenched structural patterns of the world of corporate commercial college athletics (Frey, 1982, 1987–88). If the obstacles to internal change within the college sports system cannot be overcome, however, college athletics may be tempting much-dreaded external intervention by the federal government.

Future of College Sport Simulation Game

Background (Nixon, 1985): As a result of financial, legal (especially Title IX), social, and political (from NCAA reformists) pressures, administrators at Big-Time University (BTU) in Campusville are faced with the prospect of having to make major changes in the structure of their athletic program. The policy changes decided by the BTU Athletic Board could produce sweeping changes in the athletic life of their campus and could have a significant impact on other intercollegiate and high-level amateur programs. These changes could even affect professional sports, since BTU supplies a large number of athletes to pro sports leagues each year. In a sense, BTU can be viewed as a test case that could set a precedent for other universities with big-time athletic programs or at least big-time aspirations. Thus, the deliberations of the BTU Athletic Board are being followed closely by a variety of interested observers.

The Proposal: A coalition of students and faculty calling themselves SPORTS (Society of People Organized for a Radical Transformation of Sport) have come before the BTU Athletic Board with a proposal recommending five basic changes in the structure of the BTU athletic program. The changes recommended by SPORTS are:

1. To eliminate athletic grants-in-aid (or scholarships) immediately and to base all future financial aid decisions for athletes on the same academic and need criteria applied to nonathlete candidates for financial assistance.
2. To provide the same number of varsity sports for women and men and to provide equivalent staffing, operating budgets, and access to athletic facilities for female and male teams in the same sport.
3. To charge admission to home athletic contests only to spectators not affiliated with the university as students or employees.
4. To fund the athletic program from gate receipts, an optional annual student fee of $200, and the general fund of the university as a whole and to eliminate television coverage of BTU athletic events.
5. To require coaches to devote one-half of their work schedule each year to classroom teaching.

Roles (to be randomly assigned): The main groups and individuals involved or interested in the BTU Athletic Board deliberations are:

1. The BTU Athletic Board Members
   a. President of BTU
   b. Athletic Director for Men's Sports

c. Athletic Director for Women's Sports
d. Male BTU faculty member
e. Female BTU faculty member
f. Male BTU graduate and former star BTU athlete
g. Female BTU graduate and former star BTU athlete
h. Male undergraduate athlete at BTU
i. Female undergraduate athlete at BTU
j. President of BTU Board of Trustees
k. Wealthy local businessman and BTU booster

2. Interested Parties
   a. Male BTU coaches
   b. Female BTU coaches
   c. Local chapter of the BTU Alumni Association
   d. BTU Boosters Club
   e. SPORTS representatives
   f. Presidents of ABC, CBS, NBC, Fox Sports, ESPN
   g. President of USOC
   h. Executive directors of NCAA and CFA
   i. Commissioners of NFL, NBA, NHL, and Major League Baseball
   j. Sports editors of Campusville News and BTU News
   k. Sports commentator on local TV news program
   l. President of BTU Student Association
   m. President of BTU Faculty Senate
   n. President of BTU Black Student Union
   o. President of student chapter of NOW
   p. President of Campusville Little League
   q. President of Campusville Merchants Association
   r. Male BTU athletes
   s. Female BTU athletes
   t. Organized crime figure (gambler/fixer)
   u. Local residents of Campusville
   v. Influential state legislator
   w. Parents of BTU students
   x. President of College Coaches Association
   y. President of Association of University Presidents
   z. Executive Director of FANS (Fight to Advance Nation's Sports, a consumer advocacy group founded with support of Ralph Nader).

Action: The BTU Athletic Board must vote on each of the five items brought before it by SPORTS. Until the vote, interested parties (specified above) may use whatever resources likely to be at their disposal *in reality* to try to influence board members. These forms of influence could range from direct appeals, material inducements (legitimate or otherwise), and friendly persuasion to threats or even blackmail. That is, you should try to simulate the kind of decision-making process you would expect to see if this proposal were actually before the athletic board of a real college. The vote will take place 15 minutes before the end of the period, and it will be by secret ballot. After secret balloting has been completed, board members will announce and explain, in succession, their votes. Class discussion follows.

SUMMARY

No other nation in the world has placed as much emphasis on athletics in higher educational institutions as the United States has. Intercollegiate athletics emerged in the United States in the middle of the nineteenth century. Despite its origins as a limited, student-based activity with little institutional recognition or support, college athletics quickly became immersed in commercialism and professionalism. When sports such as regatta and football demonstrated their capacity to attract public interest and generate revenue, college athletics came under the control of university and college administrators.

Concerns about violence in football and the relationship of athletics to the academic mission of higher education, along with a series of scandals, prompted faculty and administrators on many campuses to challenge the role of athletics at their institution. These concerns and issues have also periodically fueled major reviews of college athletics. The National Collegiate Athletic Association (NCAA)—initially called the Intercollegiate Athletic Association—was organized at the beginning of this century to regulate and preserve college athletics in the face of serious challenges and problems. The NCAA has overseen the corporate and commercial development of college sports, but it has not been fully effective in limiting problems and abuses.

Major challenges in contemporary college athletics include cost containment, ethical problems in recruitment and special treatment of athletes, the mandate of gender equity, and clarifying the relation of the goals of big-time athletic programs to the academic mission of higher education. Pressures to win and make money in big-time college sports regularly made the role of coaches highly stressful, spawned various forms of cheating, and created role conflict for student-athletes. College athletics is facing public and congressional scrutiny, and college presidents are again being encouraged to assume strong control over their athletic programs. Among the most radical reform recommendations, though, are those that propose the separation of major college athletic programs from their institutions. Despite calls for reform, college athletics remains a highly popular part of American popular culture.

CHAPTER 9

The Professional Sports Business

THE PARADOX OF PROFESSIONAL SPORTS

Phil Elliot is a flankerback for a fictitious Dallas professional football team in the book and film *North Dallas Forty* (Gent, 1973). A talented player and a free spirit, Phil has gotten himself into trouble with team management on many occasions for his freewheeling conduct, and he has been benched a few times each year during his career for what his coach called a lack of maturity and discipline. In a conversation with a teammate, Phil talks about his perception of the nature of his sport.

> We're not the team, man, they [coaches and management] are.... We're just the ... equipment to be listed along with the shoulder pads and headgear and jockstraps.... People don't talk about football teams anymore, they talk about football systems, and the control long ago moved off the field. (Gent, 1973: 279–280)

What Phil Elliot loves about football is the chance to exhibit his special athletic talents in the heat of competition, and the thrill he feels when he does something well transcend the cheers of the crowd and dollars in his paycheck. For Phil, the motivation is the game itself and the thrill or joy it provides. In recent years the fun has diminished because of the demands of the business and the rationalization of the game. He continues to sacrifice his body every week, and he keeps catching passes and scoring touchdowns. He also continues to clash with his coaches and the management of his team. In the end, Phil is released because he cannot and will not follow their rules and because they do not want to honor his three-year contract in the waning years of his career.

The **paradox of professional football and of all professional and commercialized sports** is evident in this tale. Professional sports are games and businesses at the same time, and the thrills and joys of one can be at odds with the demands and constraints of the other. In seeking their escape in sport, fans prefer to see the game and eschew references to the business. In pursuing careers, professional athletes dedicate themselves to the game, reap the sometimes very lucrative rewards of the business, but may resent being treated as commodities or subordinates. In managing, manufacturing, promoting, or owning sports, sports officials, managers, entrepreneurs, promoters, and investors talk effusively about the drama and heroics of the game while taking a cold and calculating approach to the generation of victories and profits. Chapter 8 showed that these apparent contradictions can be found even in realms of sport, such as intercollegiate athletics, that call themselves amateur.

Although we will concentrate on the business of professional sport in this chapter, we should not lose sight of the mythic qualities of "the Game." These qualities, and related dreams of fame and fortune, have inspired fantasies and aspirations of professional sports careers for many boys and young men in American society. These fantasies and aspirations survive despite the fact that only a tiny fraction of the population ever achieves professional athlete status (Eitzen & Sage, 1993: 241–242). A 1993 Lou Harris Survey of High School Athletes revealed that 51 percent of black high school male athletes and 18 percent of whites believed that they would become professional athletes (*Charlotte Observer,* 1993). The reality is that white males fifteen to thirty-nine years of age have no better than a 1/25,000 chance of becoming a professional athlete

FIGURE 9.1 Mass commercial sport in modern America.

and black males in this age group have no better than a 1/50,000 chance of becoming a professional athlete (Leonard & Reyman, 1988). The odds against females becoming professional athletes are even more staggering: 1/250,000 (Leonard & Reyman, 1988). Unlike their male counterparts, however, girls seem to understand the odds against them. Girls and young women recognize that except for a few sports, such as golf and tennis, the professional sports world has been for men only.

The fact that professional sports careers are elusive, fleeting, and disillusioning is largely irrelevant to the popular appeal of professional sport as long as spectators and fans follow the game. Most do not like to think about sport as a spectacle and believe that the spirit of the game has been hurt by placing too much attention on entertainment and not enough on athletics (Miller Lite, 1983: 82–83). Ironically, it is the commercial dimension of professional sport as a business, which many spectators and fans prefer to ignore, that has made it possible for them to witness or watch a wide assortment of professional sports.

The proliferation of professional sports leagues and teams in recent decades and the accompanying growth in the range of sports spectator viewing opportunities are striking facts. In the four major North American sports leagues alone—the National Basketball Association (NBA), the National Football League (NFL), Major League Baseball (MLB), and the National Hockey League (NHL)—the number of franchises has expanded from 61 in 1967 to 102 in the early 1990s (Baldo, 1993). The expansion of these and many other sports has happened in large part because of the financial investment and coverage by television (see Wenner, 1989), with cable networks and superstations such as TBS and WGN joining the major networks as prominent actors.

In North America, people can watch over 7,500 hours of sports programs in a year if they have access to the variety of cable stations (Coakley, 1994: 338). Television networks and cable companies use sports to attract viewers and respond to viewers' substantial appetite for sports by broadcasting many hours and types of sports. While television executives and producers keep an eye on their ledger and account books, they remain mindful that fans are attracted by the mythic, heroic, and exciting qualities of "the Game." The balance between the paradoxical elements of business and game is an important element of the dynamic tension in the highly commercialized and rationalized realms of professional sport today.

THE DEVELOPMENT OF MODERN PROFESSIONAL SPORTS: THE CASE OF BASKETBALL

Each professional sport has its own distinctive history. In trying to understand the history of American professional sports in the modern era, however, it seems especially instructive to consider the history of professional basketball. In economist Roger Noll's words, basketball is "the sole popular and financially significant sport that is truly modern in origin" (1991: 19). Noll's analysis of the development of the professional basketball business draws attention to economic, financial, and management dimensions of this sport that can be found in virtually all professional team sports.

Basketball was invented by Dr. James Naismith at the International YMCA College in Springfield, Massachusetts in 1892 (see the timeline in Table 9.1). The

TABLE 9.1 Timeline: Professional Basketball in the United States

1892	Invention of basketball by Dr. James Naismith at International YMCA College in Springfield, Massachusetts
1914	New York Celtics (later the "Original Celtics") formed as first professional basketball team
1927	Harlem Globetrotters organized by Abe Saperstein
1930	Rens (black professional) basketball team formed
1937	Establishment of National Basketball League (NBL) as first stable professional basketball league
1946	Formation of Basketball Association of America (BAA)
1949	Merger of BAA and two other professional basketball leagues to form the National Basketball Association (NBA)
1954	Seventeen original NBA teams reduced to eight
1961	Formation of American Basketball League (ABL) as competitor to NBA
1963	Death of ABL; nine teams in NBA
1967	Formation of American Basketball Association (ABA) as competitor to NBA
1976	ABA merged into NBA, adding four new teams to NBA in New York, Denver, Indiana, and San Antonio
1980	Right of first refusal replaced option clause
1983	Adoption of salary cap
1990	Twenty-seven teams competed in four regional divisions

middle-class origins of this sport distinguish it from other sports with their origins in the upper, working, or lower classes (e.g., see Gruneau, 1983; Dunning, Maguire & Pearton, 1993). Basketball became an intercollegiate sport before the beginning of the twentieth century and was professionalized twenty-two years after its invention. The first successful professional basketball team was the New York Celtics, which formed in 1914 and later became the "Original Celtics." Like other professional teams of this period, the Celtics were mainly a barnstorming team, one of the many during the early period of professional basketball that played intermittent exhibition games against each other and against amateur teams from the same areas as their professional opponents. The origins of professional basketball are similar to the early history of baseball between its creation in the 1850s and the establishment of the National Association of Professional Baseball Players in 1871, the first modern professional sports league.

The establishment of the first stable professional basketball league, the National Basketball League (NBL), in 1937 created the foundation for the emergence of professional basketball as a big business. Noll (1991) has noted the importance of the change from barnstorming to stable league organization in management and marketing strategies. Stable leagues had parity in schedules, a clearer basis for comparing teams and players, standard rules, and a single champion. The possibility of a league champion and the pursuit of the championship offered important marketing opportunities to the teams and league. With an established league, more emphasis was placed on the competitive balance among teams than was the case among barnstorming teams.

The differences today between the competition in the National Basketball Association (NBA) and the bungling opponents, such as the Washington Generals, typically faced by the touring Harlem Globetrotters illustrate the greater seriousness given to real competition, rather than entertainment, by established professional sports leagues. Of course, we should not underestimate the importance or pervasiveness of the various forms of entertainment used to promote the

Source: Lecture materials of Amos Alonzo Stagg

FIGURE 9.2 The origins of basketball.

NBA today. According to Noll (1991: 20) professional sports consumers are mainly interested in, and will pay more money to see, organized championships involving serious, highly intense competition. The historical importance of the NBL is that it was the first professional league in the sport to offer stable organization and relative competitive balance, and these facts help explain why the American sports public began to view professional basketball more seriously in the late 1930s.

Three professional basketball leagues formed after the lull of the World War II, with two of these leagues basing their teams in small towns and cities where amateur teams were popular and most of the players could be found. The sport was overdeveloped, though, and most of the small-town teams failed, causing the three leagues to consolidate into one, the National Basketball Association (NBA), in 1949. Most of the teams in this new league were part of the Basketball Association of America (BAA), which formed in 1946 and was located mainly in large eastern and midwestern cities. For this reason, the founding of the NBA is traced from the founding of the BAA in 1946 rather than from the merger in 1949. In five years, the seventeen original teams in the NBA were reduced to eight; in the 1963–64 season, nine teams played. Teams also relocated. The major reasons for the changing composition of the NBA and subsequent expansions were its increasing popularity in the largest cities and the growing involvement of television, which wanted to position teams in the most lucrative media markets.

The NBA's success attracted competitors. The first was the American Basketball League, which formed in 1961 but died in the middle of its second season because of organizational problems and a lack of credibility among fans (Noll, 1991: 20–21). A more successful competitor, the American Basketball Association (ABA), formed in 1967. It had teams in some major cities, but most were in smaller cities and many were located in the South. The ABA was an innovative league, using a red, white, and blue basketball and three-point shots and pioneering the idea of a regional franchise such as the Virginia Cavaliers, or Carolina Cougars. The lack of a clearly defined home did not work, however, and when the league merged with the NBA in 1976, its surviving six teams had distinct homes in New York, Denver, Indianapolis, San Antonio, Louisville, and Salt Lake City. Four of these teams entered the NBA as the New York Nets, the Denver Nuggets, the Indiana Pacers, and the San Antonio Spurs.

The ABA pursued a merger strategy early in its existence, but its plans were held up several years by litigation. The players' union fought the merger in an antitrust suit because the competition between leagues for their services inflated their salaries and gave them more leverage in contract bargaining. The antitrust laws reflect a general economic policy that there should be open and free competition in the marketplace so that consumers will be able to buy the best possible product at the lowest possible price and so that the resources of society will be allocated as efficiently as possible (Ross, 1991: 152). This open and free marketplace is also one where suppliers can try to sell their services to the highest possible bidder. Although the merger reduced the competitiveness in the market for player services, the players' association ultimately agreed to it and it took place in 1976.

The merger agreement between the NBA players and their league was followed by a series of changes. The rookie draft was retained, but players were allowed to become free agents and could sign with any team if they did not sign with their drafting team after two years. The requirement that veterans play an option year before trying to switch teams was dropped for experienced players, but the compensation clause, which applied to players who signed with another team, remained until 1980. At that time, it was replaced by a *right-of-first-refusal* provision, which allowed a player's old team to retain him by matching the offer made to the player by a new team. In 1983, the salary cap was adopted. This cap stipulated a minimum and maximum each team could pay in total player salaries, but these stipulations had numerous qualifications. By the end of the 1980s, an agreement was reached to phase out the right of first refusal and to reduce the number of rounds of the rookie player draft to three and then two (Noll, 1991: 37–38).

The delay of the merger cost the new NBA entries six years of financial losses as members of the ABA. It also cost both leagues much higher player salaries because of the interleague competition for players. Despite such costs, the owners of these teams ultimately realized substantial financial gains as the value of their franchise increased from their initial investment of $25,000 to join the ABA to approximately $50 million ten years after joining the NBA (Noll, 1991: 24–25).The financial growth of the NBA as a business during the 1980s was generated by increased attendance and increased television broadcast revenues. Fan and television interest in this period was stimulated by the presence of great stars, such as Larry Bird of the Boston Celtics, Magic Johnson of the Los Angeles Lakers, and Michael Jordan of the Chicago Bulls, who not only displayed great skills on the court but also pushed their teams to the top of the NBA during this era. By the 1990s, professional basketball players were the best-paid professional team-sport athletes, and many of the owners and coaches were enriched by their sport as well.

Although professional basketball players earn very high salaries today, we still might ask why the NBA players' association has been willing to accept a number of restrictions of its members' economic freedom that may have suppressed their earning power. For example, as noted earlier, the players settled their antitrust suit in the 1970s by agreeing to the NBA-ABA merger that permitted the NBA to have monopsony power over them; the NBA was the only buyer for their services. In addition, they have agreed to the rookie draft, the right of first refusal, salary caps, restrictions on competition for top players, and a long-term contract with the owners (Noll, 1991).

According to Noll (1991: 39), a popular explanation of the apparent acquiescence of the players is that they recognized the need to make such concessions so that competitive balance among teams and the prosperity of the league could be preserved. In this argument, the stars in a league would move to the best teams in the biggest markets or to teams with the most extravagant owners if the marketplace were completely free and open. Under such conditions, it is assumed, a number of teams would be consistently weak and a number probably would fail. Having some restrictions on the players' economic freedom presumably helps the players by maintaining fan interest in the sport, which keeps the league financially healthy and allows the players to earn high salaries.

Noll (1991) contends that this argument is wrong as an economic theory and is not supported by basketball history. He points out that the leaders of the NBA players' association were aware that no connection existed between monopsony in the player market and competitive balance in the league. They had made this argument before Congress and in court. Thus, the competitive balance argument is an explanation used by league officials to defend monopsony, but it does not explain why players have accepted limited economic freedom (Leonard, 1993: 355).

Noll believes that a more compelling explanation is that the players and their union thought that certain benefits offset the possible suppression of their salaries in a restricted marketplace. The players' association has been dominated by veterans who were not stars and thus has been willing to accept restrictive policies that might have penalized the rookies and established stars but benefited themselves. For example, they have been more interested in protecting a minimum salary, negotiating fringe benefits for all players, and maximizing the share of playoff revenues enjoyed by all members of playoff teams than in eliminating the rookie draft and restrictions on the salary potential of top players. Of course, the players' association did not want to tempt entrepreneurs to form another league and lure disgruntled college players or established stars away from the NBA. Because it has been relatively easy to form a competitive league, the union and league have negotiated agreements reducing the restrictions on the economic freedom of players and have permitted stars and rookies to sign increasingly lucrative contracts. For these reasons, over the past two decades players and management in professional basketball have generally been able to avoid the serious conflicts that have characterized management-labor relations in other professional sports, such as baseball and football.

THE FINANCIAL STATUS AND FUTURE OF PROFESSIONAL SPORTS

The future of professional basketball and other professional team sports depends on whether leagues will continue to expand or will be forced to contract by declining fan interest in less successful franchises and by reduced television investment. Successful professional sports leagues may face future competition from new leagues, especially in the case of basketball, where startup fixed costs are relatively limited and arenas are readily available. A major uncertainty in future development is player costs.

The current high salaries of professional athletes make it difficult for new leagues to compete openly with the established leagues for stars or even journeymen. Indeed, many observers question how long the established leagues will be able to afford escalating salaries (Serwer, 1993). Some argue that even player contracts of $84 million signed by Charlotte Hornet star Larry Johnson and $74 million signed by Chris Webber as a Golden State Warrior rookie are justified by NBA revenues. They are affordable in a market that has seen revenues increase from $200 million in 1984 to approximately $1 billion in 1993. The loosening of the salary cap for teams in a new collective bargaining agreement between players and owners is likely to fuel further salary escalation as long as the financial foundation of the sport remains intact. NBA executives reportedly are becoming nervous, though, about the length of new contracts and the amount of money paid to untested rookies (Serwer, 1993).

The high salaries of professional team athletes and the equally high tournament and commercial endorsement earnings of the stars of individual sports such as heavyweight boxing, auto racing, golf, and tennis raise questions whether the salaries are deserved. For example, income from salaries, tournaments or competitions, and commercial endorsements in 1992 was $34 million for heavyweight boxer Evander Holyfield, $29 million for Chicago Bulls star Michael Jordan, $21.9 million for Formula One racer Ayrton Senna, $14.5 for fellow Formula One racer Nigel Mansell, $10.1 for golfer Arnold Palmer, $9.7 for San Antonio Spurs center David Robinson, $9.2 million for golfer Jack Nicklaus, and $9 million for tennis player Monica Seles (Kiersh & Buchholz, 1993: 70). Both Palmer and Nicklaus are legendary figures in golf, and both are members of the senior tour and long past their prime playing years. Nicklaus has been called the wealthiest sports figure in the United States, with his company Golden Bear International grossing approximately $50 million and his licensing business earning $300 million in worldwide sales in 1992 (Kiersh & Buchholz, 1993: 70). The general pattern of escalation in sports-related earnings of athletes in twelve professional sports is shown in Table 9.2.

When the athletic marketplace is relatively free of trade restraints, athletes earn salaries and purses reflecting the dynamics of the law of supply and demand and the persuasiveness of their agents. In this marketplace, athletes with more modest talent and accomplishments may be overpaid by desperate or unwise owners, but for established professional sports teams even such financial mistakes do not necessarily prevent owners from reaping large rewards. *Financial World* estimated in 1991 that the NBA, NFL, NHL, and MLB had a combined operating profit margin of 17 percent on $3.7 billion in revenues (Baldo, 1993). Furthermore, the average franchise value of an NFL team in 1991 was $132 million; of a MLB team, $121 million; of an NBA team, $70 million; and of an NHL team, $44 million (Baldo, 1993).

These profit margin, revenue, and franchise value figures help explain the high salaries earned by the athletes in these sports. Some financial analysts (e.g., Baldo, 1993) have warned that established sports leagues and teams may be entering their middle-age

TABLE 9.2 Top-Paid Professional Athletes, 1983–1991[1]

Top Money Earner in	1983	1987	1991
Baseball (MLB)	Mike Schmidt, Phillies $1.65 m	Jim Rice, Red Sox $2.41 m	Darryl Strawberry, Dodgers, $3.8 m
Basketball (NBA)	Moses Malone, 76ers $1.55 m	Moses Malone, Bullets, $2.15 m	John Williams, Cavaliers, $5.00 m
Football (NFL)	Tom Cousineau, Browns, $.67 m	Jim Kelley, Bills, $1.40 m	Joe Montana, 49ers $4.00 m
Hockey (NHL)	Marcel Dionne, Kings $.45 m	Wayne Gretsky, Oilers, $.95 m	Wayne Gretsky, Kings, $3.00 m
Boxing	Larry Holmes $9.4 m	Michael Spinks $4.00 m	Buster Douglas $19.56 m
Horse Racing	Angel Cordero $.97 m	Jose Santos $1.13 m	Gary Stevens $1.39 m
Men's Tennis	Ivan Lendl $1.63 m	Ivan Lendl $1.99 m	Stefan Edberg $1.995 m
Women's Tennis	Martina Navratilova $.81 m	Martina Navratilova $1.91 m	Steffi Graf $1.92 m
Men's Golf (PGA)	Craig Stadler $.74 m	Geg Norman $1.15 m	Jose Maria Olazabal $1.63 m
Women's Golf (LPGA)	JoAnne Carner $.31 m	Pat Bradley $.50 m	Beth Daniel $.86 m
Rodeo	Chris Lybert $.12 m	Lewis Feild $.17 m	Ty Murray $.21 m
Bowling (PBA)	Earl Anthony $.13 m	Walter Ray Williams, Jr. $.15 m	Amleto Monacelli $.20 m
First Place, Sport Top 100 Salary Survey	Larry Holmes, Boxing $9.4 m	Michael Spinks $4.00 m	Buster Douglas $19.56 m
Last Place, Sport Top 100 Salary Survey	11-Way Tie #98 $.60 m	Hector Camacho, Boxing $1.00 m	2-Way Tie #99 $2.33 m

[1]Compiled from *Sport* Magazine annual salary surveys for 1983, 1987, 1991, based on contract salaries and prize money; figures based on specified year in baseball, basketball, and hockey and on immediately prior year in all other sports.

years of economic maturity, in which further expansion and revenue growth and opportunities are curtailed. We can expect that with huge differences in the worth of franchises within leagues, there will be battles among owners over revenue sharing. In the extreme case of Major League Baseball, there was a difference in 1991 of over $190 million between the value of the New York Yankees ($225 million) and the Seattle Mariners ($34 million) (Baldo, 1993).

The future financial success of the professional team sports industry depends on the capacity of owners and management to limit the growth in player salaries and/or to increase revenues. League officials also must be able to deal with government attempts to remove the special legal advantages that professional sports have enjoyed, which have allowed them to avoid regulation of their monopolistic (single-seller) and monopsonistic (single-buyer) practices. In recent

POINT-COUNTERPOINT

ARE PROFESSIONAL ATHLETES OVERPAID?

YES

Professional athletes are paid much more than they deserve. They are merely entertainers, and even the lowest paid professional athletes earn far more than teachers, public officials, and others who have spent their lives working at more demanding and important jobs. Professional athletes have become greedy, spoiled, and more concerned about making money from their sport than pursuing excellence and victories. Huge guarantees of prize money and long-term salary contracts diminish the motivation of athletes to perform at their peak every day, which robs fans of the quality of play they pay to see. Owners also complain that increasing salary escalation will make poorer teams less able to pay big stars and remain competitive with the richer teams. They fear that big salaries will bankrupt their sports.

NO

Professional athletes are a tiny elite in the very risky *business* of sport. Fans are willing to pay to see them, and television is willing to invest large sums of money for the right to broadcast the sports they play. The growing revenue generated by professional sports deserves to go mainly to the athletes because they are the reason for the appeal of sport. Legitimate stars who are paid or earn millions each year are a smart investment for sports owners, promoters, and investors because they draw huge audiences and generate much more revenue than they are paid. The mistake owners and promoters make is to pay large salaries and guarantees to non-stars. Athletes should not have to apologize for all the money in their sports. Where will the money in pro sports go if it does not go to the athletes?

How should this debate be resolved?

years, Congress has become more interested in considering the repeal of the exemption from antitrust legislation that Major League Baseball was granted by the Supreme Court in 1922 (Anderson, 1992; Smith, 1992; Associated Press, 1993b). Senators have expressed concern over a number of baseball practices, including the increased number of games on pay television at the expense of regular television broadcasts into homes; the reserve clause; the refusal of Major League owners to approve the sale of the San Francisco Giants to a Florida group; allegedly racist hiring policies and treatment of players; repeated reprieves from ostensible lifetime bans for players disciplined for drug use; investment of local tax dollars to pay for the construction of stadia; the forced resignation of the baseball commissioner and the failure to replace him for more than a year. Ohio Senator Howard Metzenbaum said:

> Baseball has become high and mighty and the question is, is this really a sport any longer or is this just a plain business? . . . Who are these people who own baseball that they can do all of these things and not have any sense of the community or public responsibility in the nation's favorite pastime? (quoted in Smith, 1992)

We know, of course, that professional sports and commercialized sports in general *are* businesses while also being sports. The future success of sports as businesses will depend on the ability of teams, leagues, and sports to respond effectively to a variety of financial, political, and legal pressures as they also maintain the interest of fans. Senate Judiciary Committee Chair Joseph Biden remarked that "unless baseball gets its act together in a way that is monumentally different from where they are now, this committee will be back with the votes that will change the status of baseball" (quoted in Associated Press, 1993a). In fact, on January 6, 1995 Senator Daniel Patrick Moynihan of New York introduced legislation to repeal the antitrust exemption enjoyed by baseball since 1922.

For both fans and U.S. senators, the mystique of professional sport is jeopardized by the arrogance of owners and players, their inattention to local fans and communities that support them, and public exposure of too many of the social and economic realities of its business aspects. An irony of the commercial success of professional sports is that this very success has made it more difficult to deflect attention from the business aspects and keep fans—and politicians—focused on the game. Closer scrutiny of the business of professional sport by politicians points to the close tie between the political and economic aspects of sports.

THE POLITICAL ECONOMY OF PROFESSIONAL SPORT AND THE LAW

Major League Baseball is unique among professional sports in the United States in enjoying an exemption from the **antitrust laws**, which prohibit businesses from engaging in forms of contact with each other that restrain trade or commerce. It derived this special monopolistic privilege from the 1922 decision of the U.S. Supreme Court in *Federal Baseball Club of Baltimore Inc. v. National League of Professional Baseball Clubs* (Freedman, 1987: ch. 3). Justice Oliver Wendell Holmes wrote in his opinion:

> The business is giving exhibitions of baseball which are purely state affairs.... the fact that in order to give the exhibitions the Leagues must induce free persons to cross state lines and must arrange and pay for their doing so is not enough to change the character of the business.... the transport is mere incident, not the essential thing. That to which it is incident, the exhibition, although made for money, would not be called trade or commerce in the commonly accepted use of those words (quoted in Freedman, 1987: 32).

In effect, the Supreme Court deemed that Major League Baseball was not engaged in interstate commerce or trade and was not a commercial activity. Later regarded by courts and legal experts as a dubious decision by one of the most esteemed justices of that era (Freedman, 1987: 32), Holmes's ruling nevertheless has had far-reaching effects on baseball and other professional sports in the United States. One very significant effect was legal protection of the player reservation system or **reserve clause**, which bound players to a team in perpetuity or until traded, sold, or released by the team owner.

Subsequent court challenges to the reserve clause occurred, but the courts were reluctant to overturn the original decision and urged Congress to consider legislation to remedy possibly illogical or inconsistent aspects of Justice Holmes's opinion. The lack of judicial and legislative action has not prevented professional athletes from using other means to weaken reserve, option, and other restrictive clauses in their contracts. Interleague competition, collective bargaining, and strikes opened up the marketplace in a number of professional team sports. In February 1976, a U.S. district court decision upheld a ruling by arbitrator Peter Seitz in 1975 that freed major league pitchers Andy Messersmith and Dave McNally from their contracts. The Seitz ruling was a landmark in creating the legal basis for **free agency** (which allows professional athletes to sell themselves to the highest bidder). Free agency has spread from baseball to other professional sports. The NBA dropped its requirement of compensation for lost free agents in 1980 and removed the requirement of an option year before moving to another team (Nixon, 1984: 171). The NFL, after years of restrictive rules and court battles, entered free agency before the 1993 season (King, 1993). Players periodically have had to fight, through their union or agents or in the courts, to withstand efforts by owners to restrict free agency and salary increases.

Major League Baseball enjoys an exemption from the antitrust laws, but a variety of other professional sports have had their restrictive economic practices curbed or modified by the courts or legislation. For example, the courts decided that the NFL's reliance on radio and television meant that teams in the league were engaged in interstate commerce and therefore were subject to the antitrust laws. In 1961, Congress passed the Sports Broadcasting Act, which granted a limited exemption from antitrust liability so that professional sports leagues could enter into television pool arrangements with the major networks. In 1973, Congress prohibited agreements that prevented the broadcast of games involving a professional football, baseball, basketball, or hockey team in the local area, if they were sold out 72 hours or more before the start of the game (Freedman, 1987: 82). Professional football established a blackout policy to prevent overexposure of its product, but Congress recognized the interest of the public in televised professional football and enacted legislation to assure the public of access to televised games that they could not or did not want to attend in person. In effect, the NFL had become a victim of its own popularity, which was generated in large part by television (Nixon, 1984: 191).

Court decisions challenged other NFL practices as unreasonable restraints of trade or commerce that violated the antitrust laws, including the Rozelle Rule, which allowed the commissioner (then Pete Rozelle) to decide what was "fair compensation" for a lost free agent; the player draft system, which restricted the right of college players to play for a team of their choice; the prohibition on tampering with players, which kept one team from negotiating with or making an offer to a player whose rights were owned by another team; a cross-ownership ban, which prevented NFL owners from investing in any other professional sports league team; and control over the franchise market, which included the right of the league to prohibit the relocation of existing franchises, to deny transfers of franchise ownership, to refuse to add new teams, to assess "compensation" when one team entered the territory of another, and to require teams to share gate receipts and television revenue and provide financial support for less successful franchises (Freedman, 1987: chs. 3, 4).

Professional basketball, hockey, boxing, wrestling, tennis, golf, soccer, bowling, automobile racing, and harness racing are some of the other sports that were involved in court cases over alleged antitrust behavior (see Freedman, 1987: ch. 3). As a result of judicial pressure and legal bargaining, professional sports leagues were forced to make accommodations to the antitrust laws, discontinuing policies prohibiting athletes from signing contracts until four years after graduation from high school and revising or rescinding their reserve clauses. In individual sports, such as professional boxing and wrestling, bookers and promoters were prohibited from conspiring to limit competition.

In general, two types of antitrust cases exist against professional leagues or their member teams (Roberts, 1991). The first type involves conflicts between two different leagues or between members of different leagues. Such cases often involve attempts by less established leagues to challenge more established ones. To win, plaintiffs bringing such suits have to prove that the defendant had monopoly power that restricted their trade or commerce and that the defendant acted improperly in achieving or maintaining their monopoly power (Roberts, 1991: 136). In 1982, for example, the North American Soccer League (NASL) challenged the NFL ban on persons owning or controlling teams in their league from investing in teams in other professional sports leagues. A federal appellate court reversed a district court decision and ruled that the cross-ownership ban was a violation of the Sherman Antitrust Act, especially since it was a ban by a stronger league in relation to a weaker league (Freedman, 1987: 83; Roberts, 1991: 137–138).

In another case, the U.S. Football League (USFL) sued the NFL in 1986 for $1.69 billion in damages for conspiring to restrict their operation in the major league professional football market in the United States. Here the jury agreed that the NFL was a monopoly but that it was not an unreasonable restraint of trade, and it awarded the USFL $1 (Freedman, 1987: 39; Roberts, 1991: 136). While the NASL cross-ownership case generated some legal controversy, the interleague cases have had little impact on the law or the structure of professional sports (Roberts, 1991: 138).

The second type of antitrust cases involves intraleague conspiracies, and a number of these cases have had an impact on professional sports. In the intraleague cases, parties have disputes with a league and claim that a league rule, decision, or action is an unlawful conspiracy among clubs in the league to limit competition among themselves. Cases in this category involve a wide range of types of league conduct: reserve rules, player drafts, rules barring entry into the league by players, control of the franchise market and team relocation, and broadcast contracts and practices. Equipment manufacturers and players have even disputed some playing-field rules (Roberts, 1991: 140).

The defendant leagues have usually won these intraleague cases, but when they have not, some significant changes in the application of the law to professional sports and the operation of specific sports have resulted. For example, the John Mackey and Yazoo Smith cases in the mid-1970s invalidated, respectively, the NFL reserve system in place at the time and the college player draft as it was then structured. Some have argued that these two decisions fundamentally

POINT-COUNTERPOINT

ARE PROFESSIONAL SPORTS LEAGUES SUFFICIENTLY DIFFERENT FROM OTHER BUSINESSES TO WARRANT THEIR EXEMPTION FROM ANTITRUST LAWS?

YES

The Supreme Court decided in 1922 that baseball was not like other trade or commerce in its essence, even though it had business aspects. That decision has not been overturned. Professional sports require a competitive balance on the field to attract fans and remain financially viable. Competitive balance can only be maintained by allowing teams in professional sports leagues to cooperate in mutually beneficial ways. Leagues need to assure that talent is distributed as evenly as possible among teams, that franchises are geographically dispersed and located in markets that can sustain them, and salaries are not so high that they destroy the financial viability of the teams and league.

NO

Sports leagues operate primarily as businesses, being mainly concerned about making a profit. Decisions about player contracts and salaries, franchise location, media contracts, and the entry of new franchises into the league are made entirely on the basis of business calculations and without any concern over loyalties to players or local fans. The corporate commercial nature of modern sport justifies its treatment as a business by the courts and justifies the application of the antitrust laws. Owners have no right to enjoy privileges unavailable to other businesses, which allow them to restrict the economic freedom of players. Sports leagues have profited in recent years despite losing special legal privileges.

How should this debate be resolved?

changed the nature of labor relations in professional sports (Roberts, 1991: 140). In the Los Angeles Coliseum case brought by the Oakland Raiders, which challenged the right of the NFL to require the Raiders to play its home games in Oakland instead of Los Angeles as the Raiders wished, the NFL was found to be engaging in an unlawful conspiracy. Along with having a significant impact on professional football and other professional sports, these kinds of decisions pointed to the confusing and inconsistent ways in which antitrust laws were interpreted. According to Roberts (1991: 141), "the legacy of these cases is that today there is virtually no conduct of any sports league (other than baseball) involving any matter that cannot conceivably be challenged successfully in the right court." Professional sports leagues, teams, and owners have sought protection from the antitrust laws for a number of reasons. The competitive advantage enjoyed in labor relations, in negotiations with the mass media, and over potential rival leagues and owners is fairly obvious. Owners and league officials have also argued that their business is unique and that some restrictions on the normal operation of the market have been necessary to enable leagues to maintain competitive balance among teams on the field and to remain profitable. Along with economists such as Noll (1991) and the players, the courts have questioned these arguments, but the public has remained unaware of or unconcerned about such matters.

Social Consequences of Professional Sports Monopolies

According to Ross (1991: 167), public enjoyment of professional sports and their major place in our popular culture have caused us to overlook the social costs of the monopolistic practices of prominent professional sports such as baseball and football. He argued that these social costs justify breaking up the professional sports league monopolies. Among these costs is the harm to players, fans, and taxpayers. Sports monopolies harm the economic interests of players and limit fan enjoyment by artificially limiting the salaries and free movement of players. A free market operating within a league and according to the law of supply and demand allows players to seek employment offering the best possible compensation. It also provides an incentive to owners to assemble and field the best possible team their money can buy. Highly competitive teams vying for playoff positions and league championships are most likely to excite fan interest. Sports leagues operating in the public interest in an open market are less likely to be able to replace free television broadcasts with pay television, which limits the access of less affluent sports fans to professional sports contests.

Furthermore, according to Ross (1991), a free and competitive market environment for professional sports leagues would correct some of the inefficiencies in management that occur in monopoly situations. He cites inefficiencies such as owners who call plays from the owner's box despite a lack of technical expertise in the sport; active management of sports franchises by family members of owners who lack sport knowledge and experience; owners who sell star players merely to cut their payroll; and owners who squander team financial resources to meet gambling obligations or to pay personal debts incurred in other enterprises.

Sports Monopolies, Public Subsidies, and the Public Interest

Monopoly sports leagues obtain large public subsidies from local taxpayers by taking advantage of their monopoly position and rabid civic interest in having a major sports franchise in their city or area. The most common subsidies are rent reduction and tax breaks on stadium property. With many cities typically interested in franchise opportunities in established professional sports leagues, league owners can extract substantial costs, for instance, in entry fees and foregone revenue sharing and television revenue, from prospective new franchise owners. The owners, in turn, can pass on a portion of their costs to local taxpayers by demanding public subsidies. When local officials balk, the owners can threaten to locate elsewhere.

According to a study by Okner (1971) approximately 70 percent of professional sports facilities (including stadia, arenas, and rinks) were publicly owned and financed (with revenue bonds issued by city, county, or state governments). The average team at that time paid less than 10 percent of its home gate receipts for its lease (Leonard, 1993: 350). Once built, publicly subsidized sports facilities are usually expensive to maintain and burden the taxpayers with a large debt (Sage, 1990b: 142–151).

Sports Franchises and Community Development

A study of fourteen stadia across the United States conducted by economist Dean Baim (cited in Holland, 1993) revealed that one—privately owned Dodger Stadium in Los Angeles—repaid its original investment and that a few others had that potential in the future. Overall, though, the fourteen stadia collectively were \$136 million in debt, with the publicly owned New Orleans Superdome at \$70 million in debt, in the worst financial condition (Holland, 1993). When franchises are relocated, the public incurs an extra burden because it not only has to bear the continuing indebtedness from stadium construction and maintenance, it has also lost the tenant that may have generated the most or only income for the facility. Thus, public subsidies benefit private franchise owners at public expense, and despite these subsidies franchise owners may choose to relocate when they are dissatisfied with their profits, leaving the public with a large debt burden.

The large costs many municipalities incur in supporting a professional sports franchise warrant examining the justifications given by public officials and owners for such public investment. One powerful argument asserted by civic boosters in many cities was expressed by Indianapolis Mayor William Hudnut: "The image of a city is certainly affected by the presence of a professional franchise.... If you ask people what the great cities of America are, I'll bet 99 out of 100 cite an NFL city" (quoted in Whitson & Macintosh, 1993: 221). The notion that a world-class city sponsors world-class sports events (such as the Super Bowl and the Olympics) and has major professional

SPECIAL FOCUS

Local Subsidies of Professional Sports Franchises

When George Shinn, owner of the NBA Charlotte Hornets, made his bid for an NBA franchise, his lease for the Charlotte Coliseum was a major selling point. He impressed NBA owners with his $1-a-game rent deal. When a long series of sellouts produced higher-than-expected profits for the coliseum, Shinn complained that he was not getting a fair share of the profits from the deal he proposed. He wanted a share of the parking and concession revenues from Hornets games, and he threatened to move the team to South Carolina if he did not get a more favorable deal. Shinn's original lease allowed him to keep all proceeds from tickets, novelty sales, and television rights and to share advertising and skybox revenues with the coliseum authority. The renewal option of his lease would have increased his rent from $6 per game in the sixth year to $10,000 per game in the tenth year and would not have given him any revenue from parking or concessions. The Hornets made a profit of $4 million in their first season and were expected to make a profit of $8 to $12 million in their second season, when owner Shinn started his public posturing about the lease (Rhee, 1990).

The public uproar caused by George Shinn's comments (Leland, 1990; Sorensen, 1990) died down after Shinn seemed to back away from his original threats, and the improvement of the team and the popularity of its young stars in subsequent seasons tended to deflect attention from this controversy. It nevertheless points to the special bargaining advantage of an owner who is the only one permitted to have a big-league franchise in town. It indicates how owners of professional sports franchises are able to profit individually with the assistance of public subsidies from public and taxpayer support and tax breaks (Baade & Dye, 1988).

George Shinn has not been alone in gaining such benefits at public expense. For example, the Indiana Pacers of the NBA had a flexible twenty-year lease that did not require them to pay rent until they generated a profit from both the Market Square Arena and the team (*Charlotte Observer,* 1990). Professional sports teams also received nearly $1 *billion* from cities to build municipal stadiums between the 1950s and late 1980s.

sports teams is often part of the economic development strategy of ambitious government officials, business leaders, and other civic boosters in many cities. Whitson and Macintosh (1993) note that many North American cities are trying to enhance their images and restore the prosperity of their downtown areas by luring major league sports teams and events. Public-private partnerships are forged to pursue these ends in what David Harvey (1989) called **entrepreneurial cities**.

It has been claimed, along with the world-class city argument, that professional sports franchises generate employment, consumer sales, and taxes from sporting events; provide recreation for residents of a community; and improve civic pride (Leonard, 1993: 350; Coakley, 1994: 315). There is evidence that the presence of professional sports teams and major league sports events in cities can add millions of dollars to the economy from consumer expenditures and revenue from items such as ticket taxes, concessions, employee taxes, and stadium rental fees. For example, it has been estimated that the Yankees add more than $50 million annually to the New York City economy and that the Pirates add over $20 million to the Pittsburgh economy. The Super Bowl is thought to add over $100 million to the economy of its host city (Leonard, 1993: 350–351).

On the other hand, economist Robert Baade did a national study of the economic impact of new stadia on their local communities and found no significant changes in retail sales, personal income, or other economic indicators after the construction of a new stadium (cited in Holland, 1993: 181). According to Baade, sporting events do not generate new spending. They merely shift spending from other leisure activities and events to sport.

In general, definitive proof of the claims made by civic boosters about the economic, prestige, and morale benefits of sports franchises and events is difficult to find. It is clear, however, that being a world-class city and having professional sports franchises have very different meanings and implications in the lives of the powerful and wealthy as opposed to less powerful and less affluent community members. Poorer local citizens cannot afford the cost of attending professional sports events, and families with average incomes see their budgets severely stretched by such costs. Team owners may argue that public subsidies enable them to keep down the price of tickets, but there is little evidence of such benefits for fans. Ticket prices seem to be at the highest levels that the local market will bear (Eitzen & Sage, 1993: 253–254). Baim observed that government subsidies from an urban tax base may be seen, in effect, as transfer payments from poorer taxpayers to rich team owners (Holland, 1993). In addition, poorer citizens lack the power to resist urban renewal projects in or around their neighborhoods for stadium or arena construction.

Lipsitz (1984) concluded from case studies of the construction of sports stadia in St. Louis, Los Angeles, and Houston in the 1960s that such construction was not associated with significant net gains or losses for the economy or quality of life in these cities. St. Louis got Busch Stadium and downtown office buildings, but it also lost industry, population, and federal funding. Los Angeles enjoyed the successes of the Dodgers and was happy to earn new tax revenue from the Dodger owner, but the city experienced serious racial conflicts, the departure of major industries and the tax base and employment opportunities they provided, deteriorating schools, violent gangs and rising crime rates, insufficient housing, a shortage of mass transit, and social cleavages between the rich and poor. Residents of Houston touted their Astrodome as the "Eighth Wonder of the World" and saw tremendous growth of their city during the years of migration in the 1970s and 1980s to the Sun Belt. Yet the rapid development brought with it great pressure on Houston's infrastructure of human services, utilities, and transportation systems.

None of these cities experienced even a slight trickling down of benefits to the masses from the construction of new stadia or the presence of major league sports franchises. Even the financially successful Dodger Stadium has generated less revenue than other types of commercial enterprises on the same site. In St. Louis, the departure of the Cardinals' NFL franchise reflects the problems the city has had with its sports investment and with its downtown economy in general. In Houston, the Astrodome has been a drain on taxpayers. Furthermore, in 1988, Houston Oilers owner Bud Adams used a threat to move his franchise to Jacksonville, Florida to obtain an additional 10,000 seats for the Astrodome, seventy-two new skyboxes, and more favorable lease terms (Eitzen & Sage, 1993: 250–251).

Relatively few cities or municipalities have the opportunity to be the home of a major league professional sports franchise. Small and medium-sized cities become interested in minor league or less popular sports for many of the same reasons that motivate larger cities to seek a franchise in the NFL, NBA, NHL, or MLB or try to become host of major sports event. Johnson (1993) examined case studies of fifteen locales, from Fresno, California to Buffalo, New York, to clarify the political process by which communities decided to invest in stadiums for minor league baseball teams.

Johnson suggests that local citizens play a more active role in the politics of professional sports investment in smaller cities boosting a minor league franchise than in large cities with major league aspirations or commitments. He found in all his cases a tension between a demand for popular control and an effort to develop the local economy. Local officials tried to shield their negotiations with team owners from the public until they had finished and received support from elected officials for their stadium agreement. Public hearings usually had little significance. Even when they attracted substantial public attention, they still had little impact on the final decision. Local officials tried to find ways to pay for stadia without relying on general-fund revenues or new broad-based taxes. Financing strategies such as hotel room taxes, lease-purchase agreements, and guaranteed loans were "off budget" and did not require voter approval.

As with major league franchises in large cities, public investment in minor league teams in smaller cities

SPECIAL FOCUS

The Political Economy of Major League Baseball in Colorado

Sage (1993) used the example of the financing of a major league baseball stadium in Denver to show how various forms of power can be used to influence taxpayers to subsidize such construction projects. He drew from the work of Rothstein (1986), who identified three faces of power: public and institutional (exercised by public decision makers and institutions in public decisions); the erection of barriers or obstacles to the political participation of those who would raise issues or exert pressure challenging the status quo or dominant groups; and the creation of a sense of powerlessness and illegitimacy among subordinate groups or classes. The combination of all three faces of power results in a political environment in which only the issues and concerns of the elites and major institutions get raised, and the issues and concerns of subordinate groups are suppressed or are never raised.

The power of MLB owners to require the construction of a new stadium, despite the existence of 76,000-seat Mile High Stadium, illustrates how local political agendas can be set by groups outside the local community and refined and implemented by local political and business elites, leaving out the input of most local taxpayers. In this case, local taxpayers were being asked to subsidize the costs of private enterprise. The sales tax initiative was jointly supported and pushed by Major League Baseball officials and very powerful political, economic, and local groups in Colorado. Their cause was aided by favorable publicity and explicit endorsements from the mass media. Faced with such well-organized and powerful pressure, the city council, legislature, and voters were not inclined to resist. Many of the typical justifications were offered in defense of this public financing of the stadium, including economic benefits, neighborhood redevelopment opportunities, and civic pride.

In the campaign to pass the stadium tax initiative, attention in public discussion tended to focus on issues and actions encouraging passage. The focus of public discourse was shaped by politicians, business leaders, the mass media, and social elites. The terms of the debate were set by the strong desire of members of the Colorado elite to have a new major league franchise in Denver. Little or no attention was paid to potentially adverse consequences of the new tax for taxpayers or of the construction of the new stadium on Denver residents. In addition, a very favorable lease for the Colorado Rockies was negotiated out of the public eye. Public disclosure of the terms of the lease initially prompted some questions and criticisms, but the public and institutional face of power was quickly mobilized in defense of the lease and the team.

As in the campaign for the sales tax initiative, the fusion of power and economics worked against the interests of ordinary citizens and effectively suppressed their voices. Lacking powerful advocates on their side, those opposing the public financing of the stadium were virtually powerless. We see in this case how the three faces of power identified by Rothstein and applied by Sage may conspire against a fully open discussion of the public interest in public decisions. This case also demonstrates that much of the political process that shapes such decisions occurs outside the public eye and beyond the reach of ordinary citizens.

---

has a relative small economic impact on the local community. Johnson found, with a few exceptions, no significant corporate funding of stadium construction costs. In those cases, corporate investors had a financial stake in the development of the stadium or site. Local businesses seldom invested directly in the stadia, preferring instead to buy advertising space on the outfield fence, rent skyboxes, buy season tickets, or purchase a large number of tickets for a promotion during specific games. State funding was usually based more on the capacity of local politicians and developers to exert political influence on state legislators and officials

POINT-COUNTERPOINT

DO PROFESSIONAL SPORTS FRANCHISES BENEFIT LOCAL COMMUNITIES?

YES

Having a professional sports franchise enhances a community's reputation and enables it to attract more business and local investment. Sports arenas can restore the prosperity of downtown areas by stimulating business and general economic development, generating more consumer spending, expanding the local tax base, and increasing local employment opportunities. Professional sports teams are a source of local pride and can unite a city behind local development efforts. Professional sports franchises provide opportunities to forge fruitful public-private partnerships for development.

NO

Professional sports franchises drain public resources, diverting them from essential public spending on local social problems. Public investment in professional sports franchises benefits private franchise investors and more affluent members of the community who can afford to attend sports events. Investment in local sports facilities cannot stop general economic downturns that cause a loss of industry and jobs, that puts added pressure on the local infrastructure of utilities and transportation, and that pushes poorer residents from their homes to make room for stadium construction.

How should this debate be resolved?

than on the pure economic benefits to be gained from the stadium construction.

Johnson (1993: 246–253) proposed that if a minor league team and stadium are to stimulate local economic development, local officials must use a development logic that clearly identifies measurable economic benefits for the local community. There must be a plan for the stadium to become self-supporting; communities must obtain long-term agreements that tie the team to the facility and community for a long time; and the stadium should be located in a geographical area where it serves as an anchor to stimulate other kinds of investment and business development. Johnson recognized that it would be expensive to build and maintain a stadium large enough and possessing enough amenities both to attract large crowds to an area and to serve as a stimulus for economic development.

Large crowds are an essential ingredient in the financial success of a sports franchise. People are more likely to attend sports events when their local team does not have to compete for attendance with teams in other sports, has star players and a consistent winning record, and competes for league or divisional championships (Noll, 1974; Baade & Tiehen, 1990). Baseball attendance has also been affected by the size of the city where a team is located but, perhaps surprisingly, not by stadium size. Fans may be more likely to attend games in stadia that are small enough to give spectators a sense of proximity to the action on the field and that have a charm associated with their design. For example, Baltimore Oriole fans have found great appeal in the new Camden Yards stadium, which captures some of the ambiance of stadia of the past.

THE GLOBAL EXPANSION OF NORTH AMERICAN PROFESSIONAL SPORT

In recent years, professional sports investors in North America have expanded their vision to explore the possibilities of sports investment beyond the North American continent (Coakley, 1994: 305–307). This expansion is much like the pattern of development of multinational corporations, which have extended their operations and markets around the world in pursuit of greater profits. The global expansion of North American professional sports has included the sale of broadcast rights to television companies in a number of different countries and the sale of licensed products

SPECIAL FOCUS

American Football in England and Europe

The emergence of American football in England in the early 1980s occurred in a context with little popular awareness of this sport, beyond the Super Bowl, and with very limited press coverage. Although there was criticism of the growing influence of American culture on England and Europe, this criticism did not center on the global expansion of American sports culture (Maguire, 1988a, 1990). The development of American football in England was strongly affected by a "network of interdependency" linking the newly created Channel 4 in England, the marketing strategy of the NFL, and the involvement of Annheuser-Busch, the makers of Budweiser beer (Maguire, 1990).

Channel 4, with limited funds and limited air time, turned to American football and English basketball to help deal with its problems. Their regular coverage of American football was a departure from the normal, more straightforward and journalistic British television sports coverage. They used a British production company with video and commentary from the major American television networks, CBS, ABC, and NBC, to produce the games. Their programming included edited highlights, popular announcers, and rock 'n' roll title music and graphics to create a family entertainment experience. The viewing audience, which climbed steeply between 1982 and 1986 and then declined somewhat by the end of the decade, appeared to be largely affluent, male, and younger.

The composition of the audience was important because it affected marketing strategies of Channel 4, the NFL, and Annheuser-Busch. In the first year, the production costs for American football broadcasts were underwritten by Annheuser-Busch. They also had commercial ads for Budweiser during breaks in the football telecast. The ads sometimes featured American football players in an effort to sell the American beer to British viewers. The beer company was especially interested in establishing a connection between their beer and the stereotypically masculine football players because they needed to create a more masculine image for their beer in the British market. Beer and masculinity are closely linked in British society, but British men generally have found drinking the dark, heavy bitters more masculine than drinking the lighter, carbonated, foamy lagers such as Budweiser. Annheuser-Busch wanted to use the American football telecasts as a means to overcome the image of their beer among British men as weak and unmanly.

Annheuser-Busch extended its marketing strategy in 1986 to the establishment of the Budweiser League. In doing so, it directly challenged leagues established earlier in the three years since the inaugural Channel 4 telecasts. Rival leagues consolidated in 1986 as the British American Football League (BAFL). BAFL included thirty-eight teams by the end of the 1980s, and the Budweiser League had seventy-two teams. The number grew to 105 in 1987, with 60 percent in southern England. The BAFL teams were mainly located in the midlands and north of England. The Budweiser League involved a ten-game season and playoffs leading to the Budweiser Bowl, with league play from late April to late August (thus, not directly competing with the NFL). Teams had a minimum of forty-five players, and the number of American nationals allowed on each team was reduced from four to three in 1989. Sponsorship from multinational corporations, such as Johnson & Johnson, Paul Masson Wines, American Express, Hitachi, Minolta, and Trusthouse Forte, helped teams pay the expense of recruiting American players and coaches

In August 1986, the first American Bowl game, which pitted NFL teams against each other in the preseason, was played between the Dallas Cowboys and Chicago Bears. The first game was played at London's Wembley Stadium and was jointly sponsored by TWA, American Express, and Budweiser. The game was a sellout of 80,000 within seven days, despite ticket prices as high as $33. The success of this game inspired Annheuser-

continued

Special Focus *(continued)*

Busch to invest more money in American football in England. Related to the success of football was a significant increase in the sales of Budweiser beer in England. Annheuser-Busch initially owned 51 percent of the shares in its leagues and in all its teams. By late 1988, ownership was turned over to the teams, but the beer maker still was involved in sponsorship, which focused mainly on an elite group of twenty teams, with ten in southern England.

The NFL has benefited in a number of ways from the growth of interest in American football in England and Europe. NFL Films and NFL Properties have been able to enter into potentially lucrative overseas distribution and licensing agreements. The basis for their marketing strategy in England has been the audience profile of Channel 4 football viewers. Having achieved this foothold in England, the NFL sought further international expansion opportunities, including assorted merchandising ventures and a spring league in Europe, called the World League of American Football, involving NFL reserve team players and European teams (King, 1991). The World League began in 1991 with ten teams and funding from twenty-six of the twenty-eight NFL owners. There were ten teams, with franchises in London, Barcelona, Frankfurt, and Montreal, as well as six American cities. The league was suspended after its second season with losses of $20 million. The problem was not generating interest in Europe, where the league was popular, but in U.S. cities. Thus, any possible revival of the league was to have an all-European cast. To foster an environment conducive to the revival of this league, the NFL played exhibition games in Barcelona, Tokyo, Berlin, and London during an eight-day period in the summer of 1993, drawing 205,377 spectators (Rushin, 1993).

The NFL's vice president of operations, Roger Goodell, said,"We do not see the World League as an instrument of profit. We see it as a tool to make American football more popular" (quoted in Rushin, 1993: 20). He added that the NFL wanted to see the World League expand to East Asia and Latin America. There were also plans for an American Bowl game in Moscow. Furthermore, in 1992, NFL Commissioner Paul Tagliabue formed the NFL World Partnership to promote American football as a participation sport in Europe.

Despite Roger Goodell's disclaimer, the marketing and financial motives behind the global expansion of NFL competition cannot be overlooked. Maguire (1990) observed that the development of American football in England is not strictly an economic phenomenon, but he also pointed out that capital investment and entrepreneurial activity have been interwoven in a network of commercial and sports interdependencies that has driven the globalization of American football. The business of the network is profit, even though the intentions of the sports promoters may be broader, as Mr. Goodell was suggesting.

such as apparel with team logos to people around the world. It also has included games by NFL, NBA, NHL, and MLB teams in England, France, Germany, and Japan; the formation of the World Football League by the NFL; and the subsidization of an American football league in England. The chance for worldwide television coverage motivated the National Basketball Association to approve the Olympic participation of the Dream Team in the 1992 Summer Games, despite the risks to the health of their top players. The NBA finals now are seen by people in over eighty nations every year. Of course, professional tennis players, golfers, auto racers, boxers, and athletes in a number of other individual sports have been competing in countries around the world for a long time.

The global expansion of professional sports serves the capitalist growth imperative of the businesses of sport, television, and their commercial partners. Maguire's (1990) analysis of the "making of American football in England" between 1982 and 1990 reveals the symbiotic relationship between large corporations and professional sports organizations in the global expansion of sport. This sports development occurred in the broader cultural context of English society and incorporated more general elements of Americanization and cultural change in Europe and England.

Source: AP/Wide World Photos

FIGURE 9.3 Baseball: The global sport.

Although American influence is strong in the globalization of sports today, Americanization is not the only influence in the diffusion of sport. The expansion of American football into England shows how **cultural diffusion** can be reversed: American football itself was the result of a cultural diffusion process that drew from rugby and soccer games imported from England (Riesman & Denney, 1951). Today, both of these "foreign" games are becoming increasingly popular as participant sports in the United States, and soccer interest in the United States developed enough to enable American soccer promoters to host the World Cup in 1994. In addition, European sport has been exported to Latin America (Arbena, 1989), Americans have been involved in the development of its sports in Latin America (Klein, 1991a; 1991b), and Japanese investors have become involved in American professional sports teams (Steinbreder, 1993).

The globalization of sport often is driven by economic interests, but new sports from other countries are accepted in a society because there is a basic resonance between the culture of the sport and the culture of groups in the society accepting it. Commercial enterprises attuned to this cultural resonance have been able to exploit the connections among sport, culture, and consumerism for profit in the global marketplace. One example is Annheuser-Busch's use of American football to sell their beer in England. Another example is Nike's use of a sports-based marketing strategy to sell a wide assortment of athletic shoes and apparel throughout the world (Katz, 1993). In 1993, Nike sold an estimated 100 million pairs of shoes; one in five of its shoe sales was outside the United States, mostly in Europe. Company officials expected foreign revenues to exceed U.S. revenue within a few years. When Nike opened a small outlet store in Shanghai, hundreds of prospective customers waited in the dark for the chance to be among the first to own Nike shoes in their country.

Nike chairman Phil Knight has long had a keen sense of the cultural linkages among sport, the sports fantasies of consumers, and the success of his company. This sense was founded on his belief that sport was "the culture of the United States," which ultimately would define the culture of the world (Katz, 1993: 56). The association of this cultural diffusion with American culture and corporate interests is part of the reason that foreign critics are especially wary of the linking of the global expansion of American sports with the expansion of American multinational corporate interests (Maguire, 1990).

CONCLUSION

We have seen in this chapter how sport has operated as a business, and how it has enjoyed a peculiarly privileged status in the business world. Professional sports team owners have taken advantage of legal exemptions, tax loopholes, and public subsidies to make money from sport for themselves. For many years, owners disproportionately benefited from sport in comparison to athletes. Over the past three decades, however, athletes have become more militant in asserting their legal and economic rights, and as a result they have achieved more rights and much higher incomes. In professional team sports, players associations have negotiated collective bargaining agreements that have formalized these rights and opportunities. In individual sports with professional associations, such as golf, tennis, and rodeo, official policies have been supportive of athletes' rights because the athletes have had some organizational control (Coakley, 1994: 320).

Although many fans believe that athletes today are overpaid and question their dedication to the game and although fans may be disillusioned or angry about the frequent movement of players and the occasional relocation of franchises, no evidence suggests that support for professional sports is dying. Spectators continue to attend professional sports events in large numbers, and fans continue to watch a variety of sports on television. Competition among cities for professional sports franchises continues to be intense. Many communities still view having a major league franchise as a mark of a world-class city. As long as people buy the tickets and sit in front of their televisions to watch professional sports, the business of professional sport will be financially stable; that is, it will continue to generate the revenue from gate receipts, television and radio, licensing agreements for apparel sales, and other ancillary sales that form the financial base of modern professional sports.

We have seen the corporate world, sport, and the mass media come together to develop professional sports as global phenomena. This partnership has been mutually beneficial for business, sport, and the media, as we saw in the case of American football in England. Foreign critics, however, have argued that this global expansion threatens to replace the cultural distinctiveness of their nations with cultural values and practices that are shaped by the economic values and interests of American or multinational business corporations. In fact, the global expansion of North American professional sports has accompanied a broader global commercialization of sports.

This commercialization has affected sports at the amateur and intercollegiate levels as well. An important question is how well the public, investors, and the mass media will continue to support the increasing number of sports and sports teams at these various commercialized levels. For example, it has been suggested that the one of the new expansion teams in the NFL, the Charlotte Panthers, could pose an economic threat to big-time—and big-money—intercollegiate teams in the Carolinas. These teams formerly had a monopoly on fans, press coverage, radio and television air time, corporate sponsors, and advertising money in their areas. As Clemson University athletic director Bobby Robinson observed:

> We don't yet know the effect the Panthers will have on the collegiate market, but it will be significant. Any way you look at it, it is competition. We'll just have to develop different marketing strategies and concepts, and put a successful product on the field to continue a viable program (quoted in Blum, 1993b: A39).

We know that sport is competition, but this statement clearly reflects the kind of competition it has become as a business. For the owners, managers, and officials of the professional, intercollegiate, and amateur sports businesses, the competition is for the fans' dollars. We will have to see how this competition shapes the competition on the field, and how it affects the interest of fans. Fans have been willing to watch and pay despite the many changes in sport that have accommodated the interests of the business over the interests of the game and their own as fans. If the interest of fans wavers or declines, though, we can expect to see the same kind of consolidation or redirection of professional sport, and commercialized sport in general, that has characterized many big corporations in recent years.

SUMMARY

Professional sport is paradoxical because it is both a game and a business, and elements of these two forms of professional sport can clash. This chapter focused on the business of professional sport. Over the past few decades, we have seen a proliferation of professional sports, sports leagues, and sports teams, including a global expansion of professional sport. The case of professional basketball in the United States illustrates how economics and the law can be interwoven in the development of a professional sport in this century.

Many professional sports today have become very lucrative, with investors, administrators, athletes, and the mass media reaping large financial rewards from them. Although Major League Baseball still enjoys an exemption from the antitrust laws, the monopolistic practices of all North American professional sports have been undercut in recent decades by a variety of challenges, including competition from new leagues, the unionization of players, and court decisions. As a result, owners and promoters have lost power to athletes, and the athletic marketplace has become more open and free.

Sports franchises have often received public subsidies to attract them to cities and keep them there. Questions have been raised about the benefits cities derive from this kind of public investment in sport, but many entrepreneurial cities continue to seek new professional major and minor league sports franchises to enhance their image and to stimulate economic development. The growth of professional sports has become a global phenomenon, and the influence of American sports and multinational corporations in this diffusion of sport around the world has been powerful and controversial. The case of American football in England and Europe shows how sports and business interests merge with the mass media in this global diffusion process

CHAPTER 

Owners, Athletes, and Fans in Professional Sport

PROFESSIONAL SPORTS ROLES AND MODERN REALITIES

Teenage fantasies of being a professional athlete or even of owning a professional sports franchise must be tempered by realities that seem to have little to do with sport—realities of contract negotiations, strikes, agents, lawyers, and constant pressure under the ever-present eyes and ears of the mass media. These kinds of realities have been the major focus of this book. We have learned that the world of professional sport cannot be fully understood without understanding that it both transcends reality in fans' dreams and is deeply embedded in the real life of society for those more directly involved in its production as athletes, owners, and management. While fans try to preserve their symbolic refuge by fending off the intrusion of too much real life, athletes, owners, and management officials cannot escape the powerful reality of professional sport as a highly commercialized corporate world strongly influenced by money and the mass media. The roles, relationships, and experiences of athletes, owners, and sports managers—as well as fans—bear the deep imprint of the *business* of professional sport. The evolution of professional sports roles over the twentieth century has been increasingly marked by this fact.

OWNERSHIP PATTERNS AND PROFITS IN PROFESSIONAL SPORTS

Ownership patterns in professional sport have varied over time, and even within the same league today we can identify a number of different possible ownership arrangements (see Nixon, 1984: 157–160). In nineteenth-century baseball, players unhappy with owner dominance of their sport formed their own leagues several times—the International Association in 1876, the American Association in 1880, the Union Association in 1884, and the Players' League in 1890. All of these efforts failed, however, because the players lacked capital to build their own stadia, political influence to establish mass transit to their playing fields, business and administrative experience, and entrepreneurial skills needed to make their leagues profitable (Nixon, 1984: Figler & Whitaker, 1991: 191). Fledgling leagues in many professional sports have failed over the years because the organizers have lacked the requisite capital, influence, experience, and skills to run successful businesses.

The legendary owners in the early history of professional sports in the United States were drawn to their sports by a love of the game. They learned the business of sport well enough to put their teams and sports on a relatively solid financial and organizational footing. These owners grew with their sports. They included men such as George Halas in football, Eddie Gottlieb in basketball, and Clark Griffith, Connie Mack, and John McGraw in baseball. Totally dedicated to their sports, they immersed themselves in all aspects of the operation of their franchises, including promotion, financial management, and even coaching. Their only business was sport.

In a number of cases, once professional sports leagues became established, the pioneer owners were joined by a new group who had become wealthy in businesses outside sport. For this next generation of owners, who in baseball included brewery owners August Busch of the St. Louis Cardinals and Jacob Ruppert of the New York Yankees and chewing-gum maker William Wrigley of the Chicago Cubs, sport offered a means to promote their other businesses. At the same time, though, these owners generally manifested a strong commitment to their sport and team.

It is difficult today to become an owner of a team in an established professional sports league, to invest in a new league, or to become a professional sports tour investor or sponsor without substantial financial capital. Although the return from an investment in professional sports may be lucrative, a variety of stresses, risks, and other costs are associated with professional sports involvement. The tax advantages associated with the opportunity to operate a self-regulating monopoly or cartel have been eroded or eliminated by the courts, Congress, or the involvement of player agents and players' associations. The financial risks of investing in a new league or less established sport are obvious. In addition, owners of new franchises in established leagues are asked to assume very large financial burdens to gain entry into the league. Once they gain entry, they face formidable obstacles competing against established teams.

Of the major professional sports leagues established in basketball, hockey, soccer, tennis, football, track, and lacrosse between 1966 and 1980, only the North American Soccer League and the Women's Professional Basketball League were still in existence in 1980. Of course, both of these leagues died in the 1980s. Investing in a failed league can be expensive, too. Owners of the original franchises in the World Football League reportedly lost $34 million (Nixon, 1984: 159).

Investors in more established professional sports have not had to face the collapse of their league, but they have had to deal with uncertainties about future television revenue, steeply escalating salaries or prize money, unhappy athletes and coaches, dissatisfied fans, and a critical press. These assorted costs may offset the prospects for financial gain and the joy of sport itself. In the current environment of stresses, risks, and problems in professional sports, the once dominant sportsman-owner has become less prevalent, and absentee owners motivated mainly or exclusively by financial considerations have become increasingly common. In many cases, the owners have been corporations, syndicates, or municipalities rather than individuals or families (Nixon, 1984: 158).

Financial factors in sport today create unstable ownership because owners often realize their greatest financial gain from selling their franchise rather than from operating it (Ozanian, 1994). From 1993 to 1994, revenues increased 14.6 percent to $5.1 billion, but player costs increased 19.4 percent to $2.85 billion, resulting in a decline of 19.1 percent, or $374.6 million, in operating income (revenues minus operating expenses). Despite falling profits, owners who sell their franchises can still reap huge gains. For example, Norman Braman paid $65 million for the NFL Philadelphia Eagles in 1985, saw operating earnings drop over 40 percent between 1990 and 1993, and yet was offered $185 million for the franchise in 1994.

Of course, franchise values can fluctuate a great deal from year to year. Between 1993 and 1994, the estimated market value of the NBA Seattle Supersonics increased 90 percent and the estimated value of nine other NBA teams increased by 46 percent or more. In the same time period, the estimated value of eight MLB and two NHL franchises *decreased* between 9 percent and 19 percent (Ozanian, 1994: 58). The top, bottom, and average franchise values for the four major team sports leagues in North America are shown in Table 10.1.

Leonard (1993: 335–338) identified five general patterns of ownership that can be found in professional sports today, which are listed in Table 10.2. The *first* type is illustrated by the New York Knicks (NBA) and Rangers (NHL) and by the Chicago Cubs and St. Louis Cardinals (MLB). They are subsidiaries of larger corporations, such as Paramount Communications (NBA Knicks and NHL Rangers), the Tribune Company (MLB Cubs), Annheuser-Busch (MLB Cardinals), the Walt Disney Company (NHL The Mighty Ducks of Anaheim), the Ewing Kauffman Trust (MLB Royals), Comsat Denver (NBA Nuggets), the Japanese Kokusai Green (NHL Tampa Bay Lightning), the Canadian Molson Companies (NHL Canadiens), Labatt (MLB Blue Jays), and Maple Leaf Gardens (NHL Maple Leafs). The NHL St. Louis Blues are owned by twenty local corporations, including Annheuser-Busch, Emerson Electric, May Department Stores, and Southwestern Bell (Steinbreder, 1993).

The *second* type of ownership pattern includes individuals or families whose major or exclusive business

TABLE 10.1 Professional Sports Team Franchise Values, 1993[1]

Franchise Values	MLB	NBA	NFL	NHL
Top Team	NY Yankees $166 m	LA Lakers $168 m	Dallas Cowboys $190 m	Detroit Red Wings $104 m
Bottom Team	Montreal Expos $75 m	Indiana Pacers $67 m	Detroit Lions $138 m	Winnipeg Jets $35 m
League Average	$107 m	$99 m	$153 m	$61 m

[1]Franchise values are based on the value of gate receipts, media revenues, stadium revenues, and other revenues such as licensing and merchandise (Ozanian, 1994)

is sport. Contemporary owners such as Al Davis of the Los Angeles Raiders (NFL), Calvin Griffith of the Minnesota Twins (MLB), the O'Malley family of the Los Angeles Dodgers (MLB), and the Rooney family of the Pittsburgh Steelers (NFL) remind us of owners in earlier eras of professional sport.

The *third* pattern of public ownership is illustrated by the Boston Celtics (NBA), who are a publicly traded franchise listed on the New York Stock Exchange and the Green Bay Packers (NFL), who are a community-owned, nonprofit corporation. Sixty percent of the Celtics stock is owned by three men, Alan Cohen, Don Gaston, and Paul Dupee. They bought the team in 1983 for $19 million, and in 1986 they sold 40 percent to the public, raising $48 million and maintaining control of their franchise (Steinbreder, 1993).

The *fourth* pattern of entrepreneurial ownership, the most prevalent type, includes owners who have invested money from other successful businesses outside sport. As we previously noted, owners of this type were present in the earlier stages of a number of professional sports. One of the most visible today is George Steinbrenner, who became wealthy in the shipping business and invested part of his fortune in the ownership of the New York Yankees (MLB).

The *fifth* type combines elements of public and private ownership. The best example is the Pittsburgh Pirates when they were jointly owned by private investors, corporations, and the city of Pittsburgh. Their

TABLE 10.2 Contemporary Professional Sports Ownership Patterns[1]

Type of Ownership	Examples
Corporate subsidiary	Gulf and Western Industries (NBA Knicks)(NHL Rangers); *Chicago Tribune* (MLB Cubs); Annheuser-Busch (MLB Cardinals)
Primary financial interest	Al Davis (NFL Raiders); Rooney family (NFL Steelers); O'Malley family (MLB Dodgers); Calvin Griffith (MLB Twins)
Public ownership	Publicly traded franchise on New York stock exchange (NBA Celtics); community-owned nonprofit corporation (NFL Packers)
Entrepreneurial ownership (most prevalent type)	Ship builder George Steinbrenner (MLB Yankees); McDonalds Joan Kroc (MLB Padres); Publisher Nelson Doubleday (MLB Mets); Oil investors Lamar Hunt (NFL Chiefs), John Mecom (NFL Saints), Bud Adams (NFL Oilers)
Public-private combination	Private investment group of corporations and individuals and city of Pittsburgh (MLB Pirates)

[1]This table was adapted from Leonard (1993: 335–339).

subsequent ownership syndicate included seven Pittsburgh-area businesses (including Alcoa and Carnegie Mellon University) and three businessmen who owned equal shares in the team (Steinbreder, 1993).

In a study of 141 owners, co-owners, and majority partners of U.S. professional teams in football, basketball, baseball, hockey, and soccer between 1982 and 1984, Flint and Eitzen (1987) found that 48 percent derived their wealth from communications, transportation, real estate and land development, or oil production industries. These are industries emphasizing entrepreneurial activity and risk taking, in the mold of **classic entrepreneurial capitalism**. Approximately 20 percent of the owners in this study had their primary business interests in what Flint and Eitzen called **monopoly capital enterprises**, including the automotive industry, banking, finance and insurance, brewing and liquor, industrial and manufacturing firms, and sales and merchandising. About 10 percent were sports executives primarily involved in franchise operations and management, and nearly 6 percent were involved in the law or medicine.

Flint and Eitzen's research indicated that the professional sports team owners they studied were more likely to own local rather than national businesses and to have social and business ties that were more local than national. Although they generally lacked connections to a national economic or social elite, many owners nevertheless were very wealthy. In 1993, *Sports Illustrated* ranked all the owners in the four major professional sports leagues in North America according to their estimated net worth (Steinbreder, 1993). The net worth of individual and family owners ranged from $3 million (George W. Bush of the Texas Rangers baseball team) to $3.2 billion (Paul Allen of the Portland Trailblazers basketball team). The top seven were all worth more than $1 billion and also included Richard DeVos of the NBA Orlando Magic ($3 billion), Ted Arison of the NBA Miami Heat ($2.5 billion), Ted Turner ($1.9 billion) of the NBA Atlanta Hawks and MLB Atlanta Braves, George and Gordon Gund of the NBA Cleveland Cavaliers and NHL San Jose Sharks ($1.5 billion), Hiroshi Yamauchi of the MLB Seattle Mariners ($1.4 billion), and Bob Tisch of the NFL New York Giants ($1.1 billion). Allen was a cofounder of the software giant Microsoft Corporation, and his fellow billionaire owners were associated with businesses ranging from Amway (DeVos); Carnival Cruise Lines (Arison); broadcasting (Turner); real estate, cattle, and financial investment (Gunds); Nintendo (Yamauchi); and theaters, insurance, tobacco, watchmaking, and CBS (Tisch). Another ten individual and family owners were worth $500 million to $900 million, and a total of fifty-nine were worth $100 million or more.

Fun, excitement, ego gratification, public visibility, the chance to be close to elite athletes, vicarious identification with athletes, and community service have been cited by sports observers and owners themselves as motives for sports team ownership (Nixon, 1984: 159; Leonard, 1993: 339). In recent decades, however, economic motives have obviously been very important or most important to many owners and prospective owners. Owners today are typically investors who expect a profitable return on their investment along with psychological and social rewards.

Put in perspective, the major professional team sports are not one of the largest industries in North America. The total value of the NFL ($3.64 billion), MLB ($3.60 billion), NBA ($2.16 billion), and NHL ($1.42 billion) was $10.82 billion in 1993. The most highly valued franchise in professional sports, the Dallas Cowboys, was worth an estimated $190 million in 1994, and the least valued, the NHL Winnipeg Jets, was worth an estimated $35 million (Ozanian, 1994). In 1991, the average operating income per team in MLB was $7.1 million; in the NFL, $8.8 million; in the NBA, $4.8 million; and in the NHL, $3.1 million (Leonard, 1993: 344).

In comparison, the largest individual industrial corporation, General Motors, had sales of nearly $126 billion and assets worth over $173 billion in 1989. The thirty-ninth largest industrial corporation, Goodyear Tire and Rubber, had sales of over $11 billion and assets worth nearly $8.5 billion in 1989. The market value of the leading stock in 1989, Exxon Corporation, was almost $91 billion, and the market value of the fiftieth stock in 1989, Pfizer Inc., was nearly $11.5 billion in 1989 (*Information Please Almanac,* 1991: 48–50).

Even though the professional team sports industry is not equivalent to General Motors and individual franchises are small businesses, professional sport has nevertheless become too expensive and potentially lucrative to be only a hobby for sportsmen. A number of examples of franchise appreciation indicate how profitable the purchase *and sale* of a professional sports franchise can be. The Denver Nuggets (NBA) were bought in 1985 for $20 million and sold in 1989 for $65 million. The Dallas Cowboys (NFL) were purchased in 1984 for $60 million and sold in 1989 for $140 million. In 1979 the Baltimore Orioles (MLB) were bought for $12 million and in 1988 they were sold for $70 million. The Portland Trailblazers were purchased as an expansion franchise in 1970 for $3.5 million, and in 1988 they were sold for $70 million (Leonard, 1993: 344). In addition, the Cubs (MLB) were last sold in 1981 for $20.5 million and were estimated to be worth $135 million in 1990. The Mets (MLB) were sold in 1980 for $21.1 million and were estimated to be worth $175 million in 1990 (Leonard, 1993: 345). Furthermore, franchises in established leagues rarely become bankrupt, and most are highly profitable (Leonard, 1993: 345).

In this context, it is easy to understand why economic motives might be preeminent among contemporary professional sports team owners (Beamish, 1991). Economic considerations also have been a major feature of relations between owners and players, but this fact has been true since players formed their own leagues in the nineteenth century to try to gain more control over their salaries and employment. In the next section, we will examine more closely major aspects of labor-management relations in professional sports.

LABOR-MANAGEMENT RELATIONS IN PROFESSIONAL SPORTS

Beamish's (1991) study of the varied (individual, small-group, family, and conglomerate) forms of ownership of the seven Canadian teams in the NHL revealed that each owner was a large corporate entity with a diversified portfolio of economic holdings that included a hockey club, that these owners individually and collectively controlled a great deal of economic capital, and that they were personally tied to the corporate elite of their country. Although many of the American owners in Flint and Eitzen's (1987) study were entrepreneurial capitalists with local or regional rather than national ties, it is likely that they shared the dominant-class ideology of the Canadian owners, which was antagonistic toward unionization and labor negotiations. According to Beamish, Canadian NHL team owners tended to view collective action by workers as "abnormal, unnatural, and, indeed, pathological" (1991: 217). The irony of professional sports ownership is that many ostensible free-enterprise capitalists have found professional sports investment appealing because it offers the chance to participate in a cartellike self-regulating monopoly that neutralizes the risks of economic competition and offers special financial benefits (Flint & Eitzen, 1987: 22). In this monopolistic setting, owners have had numerous political, legal, and economic mechanisms, including draft systems, reserve and option clauses, salary caps, and rights of first refusal, to stifle, suppress, and otherwise control players.

Sports historian David Q. Voigt (1991) has characterized the history of baseball as a century of labor strife pitting "serfs versus magnates." This characterization implies the lengthy class struggle between major league players and owners in which players have sought economic freedom and owners have tried to suppress or circumscribe it, relying on their special exemption from antitrust laws. In this struggle, players formed protective associations, fraternities, guilds, and even leagues of their own to gain more economic freedom. The balance of power between players and management did not begin to shift toward players until the players' associations developed a more assertive philosophy of unionism in the 1960s.

Arguably the most influential union leader in the recent history of North American professional sports was Marvin Miller, an expert on labor law and negotiations with a background working for the steelworkers. Miller was appointed in 1966 as the first full-time executive director of the Major League Baseball Players Association (MLBPA), and his initial challenge was to convince MLBPA members that there were genuine differences between them and owners, that there were ways to deal with labor-management issues, and that the MLBPA was only effective means of improving the status of players (Korr, 1991: 131). Within a decade of his assuming leadership of MLBPA, the century of serf-magnate relations in baseball was radically overhauled by the creation of free agency. As indicated earlier, free agency allowed players to seek employment in a competitive marketplace and to sign with the team of their choice, within the constraints of the collective bargaining agreement negotiated by their union and the owners. This radical change was aided by an arbitration ruling on pitchers Andy Messersmith and Dave McNally that was favorable to the players, and it occurred despite the persisting reluctance of the Supreme Court and Congress to take away baseball's unique exemption from antitrust laws.

FIGURE 10.1 On strike.

Source: AP/Wide World Photos

Although the relationship between players and owners was fundamentally transformed with the advent of free agency and the assertiveness of the players' associations in baseball and a number of other professional sports (Sage, 1990: 163–166), owners were not left without power. They still had substantial resources with which to contend with player demands. Furthermore, in baseball, after a number of years of fierce bidding against each other for free agents, the owners engaged in collusion in 1985 and 1986 to suppress the further escalation of salaries (Sage, 1990: 162–163). A reflection of the increased power of the players, though, is that after their union filed grievances against the owners, arbitrators ruled in favor of the players in both 1985 and 1986 because they found that the owners had acted improperly in conspiring to restrict salaries (Figler & Whitaker, 1991: 195).

Despite their substantial resources, owners are no longer able to dictate salaries and employment conditions to players. Players are represented by the potentially formidable collective power of their unions, and they also can rely on agents to represent them as individuals in contract negotiations with owners and management. If owners, a league, or tour officials refuse to bargain in good faith with the players' union, they face a possible strike. Professional sport has seen a number of strikes and threatened strikes in baseball (by umpires as well as players), football, hockey, and soccer since 1972 (Leonard, 1993: 462). The NBA is the only major professional sports league that has seen relatively little labor strife during this period, and we have discussed the reasons for the lack of significant labor-management conflict in this league. If owners fail to deal seriously with the contract demands of player agents, players may engage in holdouts and sit on the sidelines, sometimes for an entire season. In professional basketball, players unhappy with the team that drafted them or with their proposed contract have sometimes signed to play in Europe for one or more seasons. Unhappy NFL prospects have gone to the Canadian Football League, and unhappy MLB players have gone to Japan. Of course, players have had increased bargaining leverage during the times when rival professional leagues existed.

One of the first cases where players were represented by agent-lawyers involved the two Los Angeles Dodger pitching stars of the 1960s, Sandy Koufax and Don Drysdale. Dissatisfied with their contracts and learning that Dodger management had been manipulating them in individual contract negotiations, the two pitchers decided to have agents represent them as

POINT-COUNTERPOINT

ARE UNIONS, STRIKES, AND AGENTS GOOD FOR PROFESSIONAL SPORTS?

YES

Prior to the unionization and legal representation of players and player militance, owners exploited players. Players were treated as *chattel*, the legal term for a movable piece of property: they were drafted, bought, paid, traded, sold, and released almost entirely at the discretion of owners. Chattel also means slave, and even though some players were paid well for their services, they were still virtual slaves of owners before free agency. The players *are* sport; without them, the enterprise would collapse. Thus, they should be in control and receive the majority of rewards. Agents, unions, and even strikes have been necessary at times to assure players of the legal rights, economic freedom, financial rewards, and longterm economic security to which they are entitled. Without countervailing power, owners would continue to exploit players and take most of the rewards for themselves. In the words of former pitcher Jim Bouton, "While the players don't deserve all that money, the owners don't deserve it even more" (Will, 1994).

NO

Unions, strikes, and agents have caused a corruption of the essence of sport. The focus has shifted from the field to the bargaining table, law office, courts, the wallet, and the streets during strikes. Player demands are insatiable. Having earned free agency and huge salaries, players continue to want more. They seem more concerned about their contract and the size of their salary than about how well they play the game. In this climate of negligible loyalty to the sport and local fans, fans have little reason to feel enthusiastic about the players or their sport. In 1994, baseball fans faced the eighth strike in their sport in twenty-two years. The long-term financial health of a sport and the livelihood of players ultimately depend on the continued interest of fans. Players and owners cannot continue to overlook the welfare of fans. They need to redirect their attention from economics, collective bargaining, and the law to the playing field if they want to assure the future of their sport and their livelihood from it.

How should this debate be resolved?

a single entity. Their novel approach to contract negotiations led other professional athletes to demand the kind of legal representation and multiyear contracts that Koufax and Drysdale had (Figler & Whitaker, 1991: 192–193).

Agents, the Law, and Economics

Sports agents have assumed an important and pervasive presence in professional sports over the past three decades. The number of sports agents has grown from a few dozen in the mid-1970s to an estimated 20,000 in the early 1990s (Steinberg, 1991). This growth may be attributed to factors including the growth of the four major professional team sports and a variety of other major and minor league professional and commercialized amateur sports; the perceived need for agents among players; legal changes, such as the elimination of the reserve clause and the establishment of free agency, which have increased opportunities for representation; increased economic incentives for both agents and players fueled by the infusion of huge sums of media money into commercialized sports; and the perceived glamor of being an agent (Steinberg, 1991; Miller, Fielding & Pitts, 1992). Although many agents are competent and responsible in their representation of their client's interests and the interests of the sport he or she plays, a number of agents have been incompetent and irresponsible. They have misrepresented their clients' value to them, pressed college athletes to risk their college eligibility by signing early with them, given bad or inadequate advice about business investments or legal rights, and offered poor preparation for life after or outside sport.

Efforts at state and self-regulation have had relatively little influence so far. The Association of Representatives of Professional Athletes (ARPA), a voluntary self-regulating group, has developed a code of ethics for their members, but only a tiny fraction (150/20,000) of the agent population are members. One of the leading sports and entertainment agents, Leigh Steinberg (1991), has proposed the need for a nationwide group to oversee the qualifications of agents for rookies as well as for union certification of agents for both rookies and veterans. Steinberg has set high ethical standards for himself and his clients and is well regarded for insisting that his athletes' contracts include provisions to return a portion of their earnings to their community or educational institution (Staudohar & Mangan, 1991: 16).

Miller, Fielding, and Pitts (1992) are professors of sports management who have recommended a uniform code to regulate athlete agents. Their recommendations identify six key areas for this code: (1) a general definition of both the professional player and the agent; (2) a "quasi-judicial legislative system"; (3) specific agent qualifications and certifications; (4) posting of a bond by agents prior to registration approval; (5) criminal penalties for code violators; and (6) a list of prohibited activities for all relevant parties, including players, team managers, team owners, coaches, and agents.

The attempt to regulate player agents is part of the still unfolding story of the increasing complexity of player-management relations in professional sports. In this story, players and owners continue to battle over the legal rights and economic freedom of players and the need for restrictive covenants in players' contracts and restrictive practices by owners in the athletic marketplace (Wilson, 1991; Meggyesy, 1992; Staudohar, 1992; Steinberg, 1992). Players seek to maximize their financial rewards, long-term economic security, and freedom of choice within the constraints of their perception of the welfare of their sport. Owners seek maximum, sustained performance from players who often press for long-term, no-cut contracts while they also try to operate their businesses as efficiently as possible and at the lowest possible cost. Although both players and owners may acknowledge and accept the need for concessions for their mutual benefit or the welfare of their team or sport, a basic structural conflict remains between the economic interests of players and owners that is embedded in the traditional player contract (Wilson, 1991). What has changed since the 1960s is the increased awareness among players of their legal and economic rights and their increased desire to be assertive or even militant in their pursuit of those rights.

PROFESSIONAL SPORTS AND THE FANS

While players and owners have battled with one another, the sports public has questioned the motivations and actions of both. The Miller Lite survey of the American public (Miller Lite, 1983), which was conducted in 1982, showed that 50 percent of the over 1,300 respondents aged fourteen and older disagreed that professional athletes were more dedicated to the game than they were to their own gain. In addition, 50 percent believed that the cost of tickets for sports events was too high, 45 percent believed that athletes should not strike because strikes threatened the spirit of the game, and 40 percent said that strikes by players decreased their support for a particular sport. These people were not necessarily sympathetic toward owners, though. Thirty-seven percent said they agreed that in disputes between owners and professional athletes, owners tended to be more reasonable than the athletes, but 48 percent said they disagreed.

Fans could question the motives of owners as well as players. Table 10.3 shows the escalation in cost of a day at a Major League Baseball stadium for fans in relation to the increases in the top salaries of players and the U.S. minimum wage over the past four decades. Owners and promoters have made ticket prices quite expensive in many sports and have moved or threatened to move their franchises and events without regard to local fan loyalty. On many occasions players have spurned autograph seekers, shown disdain for local fans and the sports media, and left behind adoring fans by jumping to a new team with a better salary offer. In the late 1970s and early 1980s, fans began to see the changing state of the economics and politics of sport with the large increases in player salaries and ticket prices and the rise of militant player relations with owners. Yet at that time their love of sport still transcended their dissatisfaction with owners and players and their cynicism about the commercial influences creeping into sport. In 1982, 51 percent of the Miller Lite respondents said that their interest in watching a sporting event was very high (31%) or somewhat high (20%), and 75 percent said that athletes were good role models for children.

Fans have not shown interest in mobilizing to assert their rights. A consumer movement organized by Ralph Nader in 1977 to defend their rights, called Fight to Advance the Nation's Sports (FANS), died a year later from lack of interest. FANS sought to rectify various kinds of disregard and abuse of fans, including lack of fans' participation in the governance of sports, limited public information about the operation of the sports they watched and supported, high ticket prices, expensive and poorly prepared food at stadium concession stands, the special legal and economic advantages of professional sports teams owners, and the limited concern about spectator interests displayed by those who televise sports (Nixon, 1984: 226–227; Figler & Whitaker, 1991: 197–199). Sports fans have not yet taken seriously their role and rights as consumers, apparently preferring instead to see sport essentially as a place of fantasy and escape.

When fans begin to see sport more as a business than as a game, their interests, rights, and loyalty will be ignored or abused at the peril of players, owners, and the mass media. The three primary sources of revenue for professional sports teams are gate receipts; radio and television rights; and income from ancillary sales of various items such as concessions, parking, the right to use team logos, game programs, and souvenirs (McPherson, Curtis & Loy, 1989: 120–122). All three types of revenue depend on the interest and dollars of fans.

With the large investment in sports facilities made by owners and local residents, professional sports teams must try to maximize attendance at live events. Fans expect owners to try to assemble winning teams, and a team's league standing is the variable that researchers find most often related to attendance (e.g., Baade & Tiehen, 1990). Winners attract more fans. Economists suggest that a limit exists on how much a team can increase its attendance by enhancing the quality of its players. At some point, the addition of one more expensive superstar will not pay off enough in the purchase of additional tickets—and perhaps also the improvement of the team's record—to justify the investment (Leonard, 1993: 340).

Another important factor affecting attendance is the amount of competition from other professional sports teams in the local market. For example, Major League Baseball teams were found to have lower attendance in local markets in which more sports teams

TABLE 10.3 Cost of a Day at the Ballpark, 1952–1992[1]

	1952	1962	1972	1982	1992	Total Percent increase, 1952–92
Costs						
Cost of Box Seats	$3.00	$3.50	$4.00	$8.50	$14.50	383 $11.50
Cost of Yearbook	$ .50	$ .50	$1.00	$3.00	$ 7.00	1300 $6.50
Earnings						
Top Salary	$45,000 Rizzuto, Raschi	$85,000 Mantle	$78,000 Stottlemyre	$1.3 m Winfield	$5.3 m Tartabul	11,678 $5.255 m
U.S. Minimum Wage	$ .75	$1.15	$1.60	$3.35	$4.25	467 $3.50

[1]Figures from *Sport* (1993): 37.

were competing for spectators (Baade & Tiehen, 1990). Other research on baseball showed that three factors accounted for most of the variations in attendance for two teams, the Texas Rangers and the St. Louis Cardinals, in 1982 (Marcum & Greenstein, 1985). The three factors are the quality of the opponent, the day of the week that games are played, and promotional activities.

The researchers found that promotions and the opposing team had substantially more effect on Rangers' attendance than on Cardinals' attendance, which could be explained by the very poor season the Rangers had in 1982. Promotions had the strongest effect on attendance in the Ranger data, and day of the week was most important factor in the Cardinal data. When a team was doing well, it did not seem to matter who the opponent was or what kinds of promotions were being offered. On the other hand, when a team was doing poorly, fans seemed to need the extra incentives of giveaways and first-place opponents to induce them to show up at the ball park.

The current financial structure of professional sports depends on television coverage and television revenue (e.g., see Bellamy, 1989). Television coverage is especially important because it draws new fans to a sport, encourages their interest in attending live sports events, and rekindles fan interest. Television is also the primary vehicle for expanding interest in professional sports around the globe.

SOCIAL ASPECTS OF THE CAREERS OF PROFESSIONAL ATHLETES

Fans continue to be drawn to professional sports by their fascination with elite athletes. Sports merchandisers such as Nike have had great success when they have tapped the popular fascination with elite athletes to sell their products. Much of this fascination is based on fantasies and myths. For many fans, the fascination, fantasies, and myths seem to have survived questions about the amount professional athletes are paid, their dedication to the game, their willingness to strike, and the amount of emphasis placed on entertainment in their sports today (Miller Lite, 1983). The social realities of professional athletes' lives are much less glamorous—and more problematic and complex—than the fantasized images fans want to believe.

Life on the Run

Senator Bill Bradley wrote about his experiences as a professional basketball player in a book called *Life on the Run* (Bradley, 1976). The life that he described was not the one fans romanticize in terms of glamor, constant excitement, wealth, and fame. In his portrayal, instead, were such elements as the weariness from constant travel; the sameness of a seemingly endless series of hotels and airport lounges; the loneliness and sense of impermanence associated with being on the road so often and hopping from one city to the next with little time to rest anywhere; the invasions of his privacy by curious or rude fans; the sometimes hostile behavior of opposing fans in the stands and on the street; the dangers and presumed obligations of being famous and constantly in the public eye; the exaggerated or false sense of self-importance tied to being a celebrity; the constant risks of pain and injuries; and the related uncertainty about retirement, which could be hastened by injuries (see Nixon, 1984: 175). Bradley also wrote about the rewards of being part of a team, confronting stimulating competitive challenges, making friends among teammates, and winning a championship.

Origins and Occupational Experiences of Professional Athletes: The Case of Professional Football

The origins and occupational experiences of professional athletes have changed a great deal over this century. For example, we can consider the case of professional football (Riess, 1991). Riess noted that the first professional football player was Pudge Heffelfinger, a former All-American at Yale University, who received $500 in 1893 to compete for the Allegheny Athletic Association against the Pittsburgh Athletic Club.

Before the establishment of the American Professional Football Association in 1920 (which became the

NFL in 1922), there were other men, including active college players who played under pseudonyms to protect their eligibility, who were paid to play football. A number of teams consisted entirely or largely of blue-collar workers, such as steel workers and railroad mechanics. In these earliest years of professional football, players were usually college stars and local factory workers, and their job as professional football players conferred low status and paid them low wages. This fact did not change with the emergence of the new professional league in 1920. There were exceptions, such as Red Grange, a major college star who was signed by George Halas of the Chicago Bears. His contract linked his pay to gate receipts, and he was able to make approximately $250,000 in one year. In general, though, professional football in the 1920s was a part-time weekend job that supplemented a primary job. At a time when Major League Baseball players were making $5,000 a season, the average professional football player of the 1920s made $75 to $100 per game.

Professional football became a full-time occupation in the 1930s, and players were largely middle-class college graduates who came from across the country but were heavily recruited from the "football belt" of Pennsylvania, Ohio, and then Texas. By the end of the 1940s, 70 percent of the NFL players were from the Midwest, Mid-Atlantic states, and the Southwest. Professional football players have generally been less urban in their background than boxers or basketball players and about as urban as Major League Baseball players. In the 1950s, NFL players were about half as likely as their counterparts in earlier decades to have come from rural areas and very small towns. The areas most likely to produce professional football players were small working-class towns and cities in the football belt where the Friday night high school football game was the main attraction in local social life. Professional football players from large cities were proportionately overrepresented in the 1930s, but since then they have been underrepresented. There have been relatively fewer chances to play football in urban areas because playing space has been inadequate, community support has dwindled because of competition from other recreational and sports activities, and the cost of football has been too high.

Middle-class college graduates were attracted to the sport in the 1930s because it offered the chance for relatively good pay and other opportunities during the Depression. Their presence helped elevate the status of the sport. In the 1940s, the NFL attracted players who were well-educated but were from less affluent families and were second-generation Americans from eastern and southern Europe. In 1946, blacks reentered the NFL after a twelve-year informal hiring ban. Between the 1950s and early 1980s, most professional football players have had working-class backgrounds. Although they generally earned better than average salaries compared to their counterparts outside sport, it was not until the recent past that salaries in the NFL could make players wealthy. This has been true across all the major team sports in North America. In 1967–68, for example, the average salary in the NFL was $21,000; in the NBA, $20,000; in the NHL, $20,000; and in MLB, $19,000. In 1992–93, the averages were $496,000 in the NFL; $1,250,000 in the NBA; $379,000 in the NHL; and $1,090,000 in MLB (Coakley, 1994: 321). A further steep rise in NFL salaries could be anticipated with the initiation of free agency before the 1993 season.

Riess (1991) observed that players who were active in the NFL in the 1930s and 1950s did well in their careers after professional football, largely as a result of their high level of education rather than their fame from football. Football was more likely to be an indirect than direct vehicle of upward social mobility. That is, mobility probably resulted more from the chance to attend college with an athletic scholarship than from skills, recognition, or opportunities provided by playing professional football. With larger salaries, commercial endorsements, and financial in-vestment opportunities available to NFL players—and other professional athletes—in the past decade, professional sport probably is more directly responsible for the long-term upward economic mobility of these athletes. This is true to the extent that athletes receive good financial advice and invest their money wisely for the future. Long-term occupational mobility is still likely to depend on educational attainment and career preparation, though (Nixon, 1984: 182).

Uncertainty and Insecurity in Professional Athletic Careers

Team sports generally offer more financial security than individual sports. Team sport athletes sign contracts, but except for the stars, individual sport athletes must often vie for the chance to compete in tournaments or races and cannot depend on lucrative appearance or exhibition fees or commercial endorsement contracts to support them when they do not perform well in competition. In boxing, much of the money earned by the athletes traditionally has gone to the promoters, trainers, managers, and sponsors, and in other individual sports, athletes may share their earnings with investors who sponsored them in the early years of their careers (Coakley, 1994: 323). In general, professional athletes in individual sports must deal with what sport sociologist Nancy Theberge (1978) called **structured uncertainty** or built-in uncertainty in their sports careers (cited in Nixon, 1984: 165).

Individual-sport athletes face variability and indeterminacy in tournament settings, uncertainty about their own performance, challenges trying to fraternize in the locker room with opponents, and itinerant lifestyles, as well as uncertainty about income (see Leonard, 1993: 93). For young athletes in their teens, such as tennis player Jennifer Capriati, professional sports tours can cause difficult adjustment problems (Leerhsen, 1990). They experience many of the stresses and problems associated with life on the run. In addition, they may confront the problems of identity formation and control that Coakley (1992) proposed as sources of adolescent burnout. They live out their adolescent development in the public eye, which can magnify the normal difficulties they experience and create new ones. They may be expected to display the maturity of adults before they have a chance to finish their childhood.

A major fact about most professional athletic careers is that they are relatively brief. Much attention has been paid to the longevity of pitcher Nolan Ryan's career—along with its greatness. Ryan remained highly productive as a professional athlete for an unusually long time, but it is noteworthy that his retirement happened before his fiftieth birthday. There are few occupations where retirement occurs at this relatively young age. In most cases, of course, professional athletes retire at much younger ages, usually in their twenties. The average length of careers in the four major team sports is between three-and-a-half and seven years (Coakley, 1994: 321).

PROFESSIONAL SPORTS CAREERS AND MARRIAGE

A study of twenty-eight former NFL players by William Lide, a former NFL player himself and a sport sociologist (reported in Leonard, 1993: 96–99), indicated that 80 percent of their marriages remained stable after a period of initial readjustment to retirement. Those who had difficulties with post-NFL social adjustment often had their adjustment complicated by financial stress. These were players who were in the NFL from one to four years.

It would not be difficult to imagine the stress of professional sport affecting an athlete's marriage. Wives of professional athletes have occasionally written about the problems that caused their marriages to break up (e.g., Bouton & Marshall, 1983; Torrez & Lizotte, 1983), and there has been at least one noteworthy effort by an athlete's wife to publish a quarterly journal, called *The Waiting Room*, for "women in professional baseball" (Wolff, 1987). *The Waiting Room* was edited and published by Maryanne Simmons, wife of former major leaguer Ted Simmons, and it survived for nine issues in the early 1980s before it went out of business. Simmons said about her experience:

> When we started *The Waiting Room*, we were looked at with great skepticism and a little bit of fear. Front offices were not cooperative (in supplying addresses and distributing copies). We were talking frankly about things. Most wives still believe that if they conform, dress properly and do a lot of charity work, it will help their husband's career. It will not. (quoted in Wolff, 1987)

Though there have been a few publications or stories in the popular literature about wives or families of professional athletes, this topic has been largely ignored by sociologists. The work of Brandmeyer and Alexander (1983) is an exception. They used Papanek's idea of the **vicarious two-person single ca-**

Source: AP/Wide World Photos

FIGURE 10.2 Family life in the fishbowl: The Joe Montana family

reer to show what it is like to be the wife of a major league baseball player. This concept describes a situation where both the employer of the husband and the husband expect the wife to conform happily to the stereotypical roles of supporter, comforter, helpmate, backstage manager, homemaker, and mother. In these roles, the baseball wife is supposed to build her life around being a "husband-oriented wife" devoted to helping her husband in his career (see Lopata, 1971). The justification is the benefits both would derive from his success as he pursues "their" American Dream through sport. Of course, this pursuit of success involves sacrifices, especially when there are years in the minor leagues where the pay is poor, the job security and prospects worse, and the lifestyle nomadic (see Berler, 1993, on life in minor league baseball).

For wives of professional athletes, there is the hope of stardom for their husband, reflected glory, and financial security. For themselves, though, there is the ambivalence associated with dependent roles that are both required and devalued. They have often had to live their private lives in the public domain as "the wife of," with no identity of their own. They have had to adjust to the transition between the season when their husband is rarely home, and the off-season, when he may be home and expect to take charge of the family life he ignored for many months as he played his sport. In many respects, the wives of professional athletes who have children live a substantial portion of the year as single mothers. When their husbands play at home during the season, wives are expected to attend games, showing support and concern.

A great frustration for many wives is that their husbands' success, on which they are so dependent, rests on performance on the field that they are unable to influence. Meanwhile, they must sit in the stands and listen to the abuse heaped upon their husbands by other spectators, read criticism of their husbands in the press, and, occasionally, even read or hear about their husbands' infidelities, indiscretions, or misbehavior off the field. Even when success comes, it is not without problems. Reflecting on her estrangement from her husband Joe, who was a star quarterback for the Washington Redskins, Shari Theismann said, "Success ruined our marriage, or perhaps Joe's inability to handle that success. He lost his values" (Maxa & Elliott, 1984: 142).

Once her husband's career is over, a husband-oriented sports wife who has lived her life vicariously through her spouse may have as much difficulty adjusting as he does because she has not been able to pursue a career of her own to achieve some financial or personal independence. In general, the lives of these women are structured at the intersection between two "greedy" institutions, the family and her husband's work (Coser, 1973). When he retires, the hold of one of these powerful institutions diminishes for both spouses. Thus, the adjustment for both husband and wife can require a great deal of negotiation at the time of retirement from professional sport.

PROFESSIONAL SPORTS RETIREMENT

The effects of professional sports retirement are difficult to generalize because the experiences of former professional athletes have varied (Nixon, 1984: 179–183; Leonard, 1993: 95–100). Careers can end in a

SPECIAL FOCUS

Athletes and Domestic Abuse

Despite frequent publicity about domestic violence incidents involving prominent athletes, this issue has failed to touch the reverence of fans for most of these athletes (Bruscas, 1994). Indeed, when O. J. Simpson was being pursued as the suspected murderer of his former wife, Nicole, and her friend, a great deal of public sympathy was expressed for him, which caused the prosecutor to remind the public who the victim in this crime was. The fans who sympathized with him seemed to forget or not to care that Simpson had pleaded guilty to 1989 charges of domestic violence against Nicole and that she had called police on two subsequent occasions to report abusive verbal and physical intimidation by her former husband.

Simpson's fame and highly sanitized public image may explain public unwillingness to condemn him for his violent behavior and, in many cases, to believe he was capable of such violence. The violence or extreme aggressiveness that is part of many sports and sport socialization could explain why athletes engage in violence off the field and use it in response to problems in their most intimate relationships. According to Jama Clark, a Seattle therapist and expert on relationships, attractive prominent athletes such as O. J. Simpson, who are accustomed to public adoration, may suffer from a syndrome she called the **star effect**, which leads to expectations that a spouse will be dutiful, submissive, and totally devoted (quoted in Bruscas, 1994: 3B). The failure to meet such expectations could provoke violent outbursts.

A University of Arizona study of college athletes revealed that male athletes were more likely than their non-athlete counterparts to act violently toward women (cited in Bruscas, 1994: 3B). Playing a violent sport, in particular, may teach men that being violent is being masculine and that violence is an acceptable means to resolve problems. Men who learn these lessons in sport are prime candidates for domestic abuse and other acts of violence outside the sports arena.

number of different ways, but in the most general terms, they end because athletes decide to retire voluntarily or are forced to retire because they are released. In team sports, the decision to retire frequently is involuntary and forced by management. Athletes often delay retirement as long as they can make a good income and are able to perform at a competitive level. When retirement decisions are voluntary, they may involve a great deal of ambivalence because athletes can often cite a long list of attractions in their professional sports career along with the costs (Leonard, 1993: 95, 99-100). In individual sports, athletes may retire because they cannot perform well enough to make a good living at their sport (Nixon, 1984: 179–180). Even stubborn or persistent athletes ultimately understand the implications of their consistent failure to make it through the qualifiers or early rounds of a tournament or to finish among the leaders in a race.

Retirement, Injuries, and Disability

Many players are driven from sports they love by chronic pain and injuries that diminish their skills and destroy their bodies. All athletic careers wear down the body, but the more combative and violent contacts are especially damaging. The titles of three of the feature articles in a four-part *Chicago Sun-Times* series on professional football careers reinforce this point: "A Game of Pain" (Hewitt, 1993a); "Playing at Any Cost" (Hewitt, 1993b); and "Through the Tears" (Hewitt, 1993c). The fourth article, "Opposing Views" (Hewitt, 1993d), conveys the willingness of some former players and the unwillingness of others to criticize the game that disabled them.

This series reported the results of a study commissioned by the National Football League Players Association (NFLPA) in 1990 to find out whether former

Source: UPI/Bettmann Newsphotos

FIGURE 10.3 OJ on trial: the fall of the hero.

players believed that their careers were worth the pain and injuries. The results were based on the responses of 645 players whose careers spanned the early 1940s to 1986. More than one-third said they retired because of a disabling injury, and nearly two-thirds said they lived with a permanent injury from football. Although most players interviewed for the newspaper articles publicly said they would play again despite the injuries and disabilities, the survey results revealed a different story. According to the survey, players in recent years have increasingly questioned the value of their physical sacrifice in football (Hewitt, 1993a).

The survey indicated that professional football was becoming more violent and disabling. Whereas 38 percent of the players who retired before 1959 said they had a permanent injury from football, the percentage increased to 60 percent for those retiring in the 1960s and to 66 percent for those retiring in the 1970s. The percentage seemed to stabilize at the 1970s level, with the figure 65 percent at the end of the 1980s. There were striking cases of players, such as former Detroit Lions guard Mike Utley, who have been paralyzed by football but insisted they would play again if given the chance. On the other hand, a preliminary report of an update of the NLFPA survey indicated that the percentage of players who listed disabling injury as the main reason for retiring rose from 37 percent in 1990 to 41.4 percent in 1993. The percentage saying they retired with a permanent injury dropped somewhat, from 65 percent to 61.1 percent. In comparison, Occupational Safety and Health Administration statistics on the high-risk construction industry indicate that injuries happen to 12.8 workers out of every 100. Another indicator of the physical risks of professional football is the average life span of NFL players, which, at sixty-two years, is ten years less than the life span of the average American male. The NFL disputed these figures, but what was more compelling than possible increases or decreases in injury *rates* was a pattern of increasing severity and cost of injuries (Hewitt, 1993a).

More statistics could be cited about football injuries and disabilities, and more data could be cited about disabled former players who continue to extol the virtues of a professional football career. The reasons for such reactions can be understood in terms of the patterns of positive deviance and the normalization of pain and injury that we discussed in an earlier chapter. Note here that former players are becoming increasingly more likely to reconsider their acceptance of pain and injury as part of the game. As they reflect on their football injuries and resulting disabilities, more may initiate lawsuits against the NFL for malpractice or negligence or to gain worker's compensation. In recent years, multimillion dollar awards have been granted to players in such suits (Hewitt, 1993a; 1993c).

One of the ways that NFL players have met the unusual physical demands of their sport has been drug use. According to the NFLPA survey, former players reported using the following drugs to deal with injuries or enhance performance: Novocaine, cortisone, anti-inflammatories, amphetamines, caffeine tablets, alcohol, steroids, marijuana, and cocaine (Hewitt, 1993b). The use of street drugs, such as alcohol, marijuana, and cocaine, was reportedly fairly small. It is interesting, and troubling, that 9 percent of the former players said, "I don't know what I took." The team doctor and the team trainer were the source of drugs for more than half of the respondents.

Being forced to leave professional football because of injuries has emotional effects. The NFLPA study revealed that 70.6 percent of the players who retired because of injuries cited emotional problems some time during their first six months after retirement. Of those who did not leave because injuries, 52.6 percent said they had similar emotional problems. The study's authors suggested that the existence of emotional difficulties raised questions about the relationship of retirement to matters such as divorce and financial stain.

It is easy to understand the emotional anxiety and financial worries of professional football players, in particular, about their careers. Unlike the NBA, NHL, and MLB, where 75 percent of player contracts are guaranteed, less than 1 percent of NFL contracts are guaranteed. The premiums for contract insurance are extremely high, which means that NFL players are vulnerable. Their players' association has negotiated financial compensation for career-ending injuries into their latest collective bargaining agreement, but the physically disabling effects of playing professional football and the emotional problems of having to retire with permanent injuries and disabilities must still be addressed.

An interesting irony is that Major League Baseball players, whose bodies tend to be much less ravaged by their sport, receive much more generous pension benefits than NFL players. In one example, a twelve-year NFL veteran will receive $28,056 in annual pension benefits at age fifty-five, whereas a twelve-year veteran of Major League Baseball who played at the same time will receive $64,464 in annual pension benefits along with money from a variable investment fund. Before 1959, there was no pension plan for former NFL players, and with the average NFL career estimated to be between 3.2 (NFLPA figure) and 4.5 years (NFL Office figure) according to Riess (1991: 239), it is difficult for most players to accumulate much in their pension account. The most recent NFL collective bargaining agreement significantly increased the amount of pensions and permanent-disability benefits. The old timers who were formerly without pensions are now scheduled to receive improvements in pensions that players' associations in other sports will try to duplicate. The problem for many former players, though, is how to survive financially for the thirty-five or more years between retirement and their pension (Hewitt, 1993e). A college degree, job training, and off-season job experience, along with wise investment of their sports income, make life after sports retirement easier to handle.

Retirement and Failure

Part of the emotional difficulty of retirement, especially when it is forced by injuries or management, is the sense of failure or loss that may accompany it. Ball (1976) wrote that the ever-present possibility of failure is one of the problems of a professional athlete's occupational life that never goes away. The possibility of being viewed as a failure by others inside and outside their sport when they are sent down to the minor leagues, are released outright, or are no longer able to qualify for tournaments or races adds to the personal sense of failure felt by athletes in these circumstances.

Ball's research revealed that reactions to an athlete's downward mobility—or "failure"—varied according to the structure of the sport. In double-tiered sports with minor league systems, such as hockey and baseball and tennis and golf with their satellite tours, those who fail at the higher level are demoted to a lower level in the sport. At this lower level, they remain visible to the public and other players, which contributes to their sense of marginality and failure. Ball proposed that athletes in these situations are likely to face degradation and embarrassment as a "deadman" or "nonperson." They carry a stigma of failure, which makes other players who are anxious or uncertain about their own status reluctant to associate with them. Former major league pitcher Jim Bouton (1971) wrote about feeling like a leper when he was sent down to the minor leagues.

Gallmeier (1989) found in his study of minor league hockey players that the meaning and experience of failure differed according to whether they were "gassed" or released, traded, or placed on waivers for other teams to bid on their services. Failure was experienced most acutely when players were gassed.

POINT-COUNTERPOINT

IS A PROFESSIONAL SPORTS CAREER THE BEST POSSIBLE CAREER CHOICE FOR YOUNG ATHLETES?

YES

Very few careers provide as much opportunity for fame, influence, and money as a professional sports career does. For those who love sports, it is a dream fulfilled to be able to play the game you love and be paid well for it. In even the less publicized and lucrative sports, athletes today can make at least as much money as they could in nonsports employment, and they are able to travel much more extensively than they could in most other jobs. In the more commercially successful sports with powerful players' associations, athletes earn a lot of money during their careers and benefit from generous pension arrangements later in their lives. All professional athletes have a chance to pursue a second career at a relatively young age, and they begin their second career with the headstart provided by their fame from sport.

NO

Although lucrative and glamorous for some, professional sports careers are filled with tensions and pressures associated with performing before the public and mass media. These tensions and pressures can lead to marital and family adjustment problems and substance abuse, and they are exacerbated by constant mass media attention. Professional athletes also know that their careers rest precariously on the state of their bodies and health. Even the relatively few whose careers last more than a few years often face lifelong disabling conditions from their sports injuries. Professional athletes accustomed to big paychecks and cheering often find that life after athletic retirement never measures up to their athletic careers, especially if they did not prepare with a good education and alternative job experience.

How should this debate be resolved?

Being traded was disappointing or shocking to some players, but it offered some consolation because former teammates were often sympathetic. According to Gallmeier, being placed on waivers was rated in between being gassed and being traded on a continuum of failure.

In other professional leagues without a minor league or lower level in their structure, players are released outright. Thus, those forced to leave the sport do not suffer the public embarrassment or degradation of the "skidders" who fall to the minor leagues in more hierarchical professional sports structures. Players in these sports tend to receive more sympathy from fellow athletes because they know that being released means leaving the professional sport entirely in most cases. For these athletes, there also is likely to be less temptation or opportunity to hang onto their major league career aspirations than in cases where they are able to continue playing at a minor league level. In situations of less opportunity, failure is more frequent and tends to be more expected and less traumatic (Nixon, 1984: 180–181).

Kristin Seaburg, an undergraduate at the University of Vermont and a student in Howard Nixon's sport sociology class in 1983, wrote a poem that captures the possible lament of many retired athletes:

Alone.
No more attention.

Only memories are left,
Loneliness surrounds him. Once a
Dedicated, perfectly toned,

All-American
Trying to be the best.
How he remembers
Learning the game,
Ecstatic cheers from the crowd.
Terrible how it has all
Ended in the deafening sound of silence.

CONCLUSION

The emotional fall in retirement is very difficult for many professional athletes because the climb has been so steep—to wealth, fame, and success with the whole world watching, in some cases. The business of professional sport offers athletes the chance to climb this mountain by playing a game. It also offers owners and fans the chance to bask in the reflected glory of the accomplishments of great athletes and athletic teams. Owners can also reap substantial financial gain from their association with sport. For athletes, though, life on the run can have significant stresses and problems. Owners and management officials also experience significant stress in running their sports and dealing with athletes, fans, and the mass media. Fans who become aware of these realities of sport often are robbed of their symbolic refuge. The spirit of the game is typically secondary to the business of sport. Despite these realities, including the rising cost of a day at the ballpark, fans still persist in their loyalty.

For many years, owners disproportionately benefited from sport in comparison with athletes. Over the past three decades, however, athletes have become more militant in asserting their legal and economic rights, and as a result, they have achieved more rights and higher incomes. In professional team sports, players' associations have negotiated collective bargaining agreements that have formalized these rights and opportunities. In individual sports with professional associations, such as golf, tennis, and rodeo, official policies have been supportive of athletes' rights because the athletes have had some organizational control (Coakley, 1994: 320). We are still waiting for fans, whose interest is the foundation for the survival of all professional sports, to seek rights and influence consistent with their importance in these sports.

SUMMARY

Professional ownership patterns have varied over the past century, but a major aspect of the evolution of professional sports team ownership is the increasing emphasis on investment and profit. Until the past few decades, team owners were able to operate virtually as cartel-like self-regulating monopolies, and they were able to dictate salaries and employment conditions for athletes. Labor-management relations in professional sports have changed dramatically with the increasing militance and power of player unions and associations, the entry of agents into individual contract negotiations, collective bargaining, free agency, and competition from other leagues. Athletes in many professional sports are now very well paid, but they continue to battle with owners, management, and promoters over their relative share of the revenue in their sport, playing conditions, and long-term economic security.

The price of a day at the ball park has steeply escalated over the past two decades, and the realities of sport, such as contract disputes, strikes, and drug rehabilitation programs for addicted athletes, have increasingly distracted attention from the action on the field. Yet fans still watch, and they still pay to attend sports events. Efforts to organize fans as a consumer movement so far have failed.

Professional sports careers have evolved into full-time occupations over this century, and a professional sports career has become a life on the run. It combines opportunities for excitement, fame, and money with tensions, stresses, pain, and injuries, and it is all played out in full view of the mass media and the public. Spouses and families of professional athletes share in the joys and the pain. In many cases, spouses feel the frustration of their husband-athletes in the form of domestic abuse. Unless it is voluntary and is followed by a fulfilling job or career, retirement can produce a sense of failure in athletes.

CHAPTER 11

Social Class, Status, and Leisure Sports Participation

LEISURE PHYSICAL ACTIVITY IN AMERICA

Adults in the United States spend more time watching than playing sports, but many actively participate in sports for leisure, fitness, and competition. For example, it was estimated that in 1990 this country had over 70 million swimmers, 66.5 million people who walked for exercise, 57 million bicycle riders, 41 million bowlers, 31.5 million people who used exercise equipment, 26 million basketball players, 25 million runners and joggers, 25 million volleyball players, 23 million golfers, 22 million softball players, 19 million tennis players, 16 million skiers, and 8 million racquetball players (U.S. Bureau of the Census, 1991: 238). It is estimated that between 50 percent and 70 percent of American adults participate in some form of physical exercise or sport each week (Eitzen & Sage, 1993: 33; Leonard, 1993: 5).

These statistics suggest a vigorous population of athletically oriented people in the United States. In the late 1970s, in fact, there was even talk of an American "fitness boom." Between 1961 and 1978, the percentage of Americans reporting daily participation in physical exercise had doubled to 47 percent (Nixon, 1984: 227). These statistics and images are misleading, however. No more than 20 percent of the current adult population in United States participates frequently or intensely enough in physical activity to gain health and fitness benefits (Coakley, 1990: 379). It also appears that future generations of adults will be less physically active than their parents. Only 2 percent of 18 million American youths in the mid-1980s passed the President's Physical Fitness test of strength, endurance, and flexibility (Leonard, 1993: 196). Furthermore, although more Americans are involved now than in the past in physical activities, a disproportionate number are from the privileged strata of American society.

Adult participation in physical exercise and sports in the United States is greater for rich than poor, executives than blue-collar workers, and college graduates than high school graduates (Leonard, 1993: 196). The uneven picture of participation in physical activities reflects an important reality of sport and society in the United States and other countries: participation is stratified. By this, we mean that there are basic inequalities, which reflect the major institutionalized inequalities among social groups—age, gender, race/ethnicity, and social class. These inequalities become readily apparent in analyzing the nature of participation in high-level, competitive sports. They exist in access to programs, facilities, and equipment; in the quality of programs, facilities, and equipment; and in participation rates. Furthermore, the types of sports pursued by people from different social backgrounds vary, in part for reasons of historical segregation and discrimination and in part because of cultural differences in sports preferences. The relationship of social inequalities and cultural differences to leisure sports participation will be explored in more depth in this chapter.

SOCIAL STRATIFICATION, SPORT, AND LEISURE

Social stratification is a structural feature of all societies and groups that is based on an unequal distribution of valued resources among members. Inequality is distributed by social class, occupation, race, gender, religion, and age categories forming distinct patterns of ranking that persist over generations. In general, persons with higher class standing are more likely than people in lower classes to have more of the things that society values, such as material goods, attractive and important jobs, an affluent and exciting lifestyle, interesting and enjoyable recreational opportunities, and positions of leadership or authority. The institutionalized nature of stratification makes it very difficult to change the system. It also makes **vertical social mobility**, or movement up or down the stratification ladder, very difficult. Rigidly stratified societies, with more enduring structures of inequality, have less social mobility up or down the social class hierarchy than societies where class boundaries are open and fluid.

The facts of social stratification and limited social mobility are contradictions of an American Dream in which many Americans express great faith (Nixon, 1984). The American Dream encourages people from modest social origins to believe that they can move up the social and economic ladder of success with ambition and hard work.Thus, the American Dream implicitly tells women, disadvantaged racial and ethnic minority-group members, and poor, elderly, and disabled persons that their lower status is a product of their inability to exercise the ambition, effort, or talent to do better. This view legitimizes the persistence of inequalities and conveys the message that people in the lower classes have only themselves to blame for not doing better.

An emphasis on motivation characterizes the structural-functionalist view of stratification. This view assumes that the jobs that are important for a society but are difficult to do because they require special talent or skill or are very demanding are paid more so that people will be motivated to seek them. Thus, according to structural functionalists, inequality is necessary for the good of society. In contrast, Marxists and conflict theorists assert that inequality is not necessary, but an outcome of the practice of dominant classes wanting to maintain their position of prominence in society. In the Marxist perspective, economic status, specifically the possession of property and wealth, determines who has power and who does not.

Weber offered a multidimensional perspective of social stratification, incorporating social and political dimensions along with an economic or class dimension. The social dimension involves prestige or *status,* which refers to the amount of social honor, respect, or deference accorded to positions, groups, or activities in society. For example, different positions on a sports team, such as starter and substitute, are accorded different levels of social honor, just as different sports, such as golf and roller derby, are associated with different levels of prestige. Prestige levels are a reflection of the values in a group or society. The political dimension of stratification concerns positions of authority and the exercise of power. Economic, status, and political advantages are associated with the dominant strata of society. The most disadvantaged people in society are low in economic class, social status, political standing, and power. People in the middle and even upper-middle strata of society can vary in their relative positions of class, status, and power. For example, professional athletes may earn great sums of money and have relatively high status if they play a major sport, but they may have limited authority or power.

A myth of American sports has been they are open to all. In fact, American sports of various kinds have erected barriers to keep out certain categories of people, such as blacks, women, and members of the lower classes. This reality is at odds with the American Dream and the democratic notion of American society and sport. The stratification of sport reflects broader patterns of stratification in society.

Class, Status, and Leisure Sports

Leisure and sports participation patterns reflect the impact of social stratification on lifestyle choices. For example, people in higher classes and strata generally having higher participation rates. Also, the types of leisure sports that people pursue or enjoy most vary by social class and status as well. Sports favored by the upper class include polo, yachting, and equestrian competition, while golf and tennis tend to be played mainly by members of the upper and upper-middle classes. Blue-collar workers are more involved in team sports, such as basketball, volleyball, softball, and baseball (Curry & Jiobu, 1984: 68; Nixon, 1984: 232–236; Eitzen & Sage, 1993: 302–304; Leonard, 1993: 191). Upper classes prefer active participation over

SPECIAL FOCUS

Class and Status in Leisure Sports Across Nations

Class and status differences in active leisure sports participation among adults are not confined to the United States (McPherson, Curtis & Loy, 1989: 181–182). For example, Renson (1976) found that people in the higher strata of Belgium were overrepresented among skiers, golfers, tennis players, and fencers. Participants in gymnastics, track and field, judo, boxing, soccer, and team handball were more likely to come from the lower classes. Looking at gymnastics participation in different nations shows that the status of a sport can vary across nations. Gymnastics is a low-status sport in Belgium, a middle-status sport in West Germany (Lüschen, 1969), and an upper middle-status sport in the United States (Loy, 1972). In addition, the social class and status backgrounds of the participants in a given sport may vary over time within a nation or community as well as across nations and communities (McPherson, Curtis & Loy, 1989: 182). For instance, Spanish women have shown a pattern of increasing involvement in physical recreation and sport over the past two decades, which parallels what has happened in other Western societies (Bunuel, 1991). In general, the percentage of Spaniards participating in physical recreational activities increased from 12.5 percent in 1968 to 34 percent in 1985 (Martinez del Castillo, 1968), and during the same period, the participation of Spanish women increased from 6.8 percent to 23 percent (Garcia Ferrando, 1986). By approximately 1990, the male participation rate was up to 48 percent and the female participation rate was up to 29 percent. Bunuel believed that these participation rates reflect changing definitions of the appropriateness of physical activity for Spanish women and increased pressure on male-dominated sports associations to provide more opportunities for sports participation, in addition to the appearance-enhancing exercise activities that these organizations tended to support.

spectatorship; middle-class and lower-class people generally favor watching sports over direct participation (Leonard, 1993: 191). In addition, upper-class and upper middle-class people tend to play individual sports in exclusive clubs or with expensive equipment, while people in the lower-middle and working classes tend to play team sports sponsored by community recreation departments, churches, employers, unions, and local businesses (Eitzen & Sage, 1993: 302).

Zelman (1976) argued that in general, sports preferences differ across social classes in terms of the nature of the social experiences and opportunities the sports provide, the personal needs they meet, and the environments in which they occur. For example, upper-class sports are more costly, tend not to involve physical contact, are often pursued for social rather than competitive reasons, and include as participants both men and women of all ages. In contrast, lower-middle and working-class sports are not only less expensive but are also likely to include physical contact, be highly competitive, and involve younger male participants (Leonard, 1993: 193).

Segregation or Democratization in Leisure Sports?

The idea that leisure sport, like sport at other levels, helps reproduce patterns of social stratification raises questions about whether class and status distinctions in leisure sports are diminishing and whether those that still exist result from voluntary choice or discrimination. The answers to these questions are complex. For example, when elite private secondary schools channel students from different social backgrounds into different types of sports on the basis of differential prior exposure to these sports, they maintain class and status distinctions in these sports (Eggleston, 1965; Berryman & Loy, 1976). Similarly, traditional class and status distinctions are maintained and class systems are reproduced when public schools, parents, friends, and community recreation programs encourage children and adolescents to pursue sports typically played by people of their social class, race, ethnic group, or gender. When they are socialized in this way, people are likely to make sports choices as adults that follow

Source: AP/World Wide Photos

FIGURE 11.1 Mass leisure sport: the New York Marathon

traditional class-related patterns. Whether these adult choices are voluntary or constrained is an interesting question in this context. In either case, the outcome is a perpetuation of existing stratification patterns in sport and society.

Even though socialization into sport often reproduces past patterns of stratification in sport, a number of sports have become democratized in Western societies over this century. **Democratization** is the opening up of sports to people from more varied social class and status backgrounds, which means that more middle and working class people are involved in formerly elite sports. For example, the construction of public tennis courts, golf courses, and swimming pools has made these sports, which were once confined to country clubs and estates of the wealthy, accessible to people of the middle and lower strata. Access to sport and other institutions has been based relatively less on ascription than in the past. When **social ascription** operates, access to positions is based on qualities such as race, ethnicity, class background, and gender that are defined by parental characteristics, attributes possessed at birth, or social class origins.

Two distinct democratic alternatives to an ascribed aristocratic opportunity structure are based on egalitarian and meritocratic philosophies that assume participation should not be restricted on the basis of social background (Gruneau, 1975: 130). These democratic philosophies differ, however, in the extent to which sports participation opportunities are based on ability or performance. **Egalitarianism in leisure sports** implies that everyone is welcome, and it is the guiding philosophy of many public or community recreational programs. Egalitarianism is the ethos of mass sports in which maximizing the number of participants is more important than developing champions. The idea of **meritocracy** usually applies to more serious amateur sports programs and clubs in which there is an emphasis on competing for positions and rewards and moving to increasingly higher levels of competition. Meritocratic programs embody the middle-class ethos of the American Dream, which emphasizes achievement, success, and continual competitive striving for upward mobility. Anyone can get into the race but only a few can be winners.

Democratic changes based on egalitarian and meritocratic criteria have increased opportunities for a wide range of members of society to participate in sports that once were restricted to participants from the privileged classes of a society. As we have seen, however, there remain a number of class and status distinctions in sport. In addition, formerly elite or upper-class sports that have become more democratized still remain segregated (McPherson, Curtis & Loy,

SPECIAL FOCUS

Golf, Democratization, and Social Inequality

In considering recent statistics about the popularity of golf worldwide, it is easy to accept the claims of the game's democratization. Some have referred to a "golf boom" in recent years and have argued that it is the sport of the 1990s. Stoddart (1990) noted that there are over 25,000 golf courses around the world, with over 25 million registered players in the United States alone. Despite the burgeoning of interest around the world, golf remains a sport marked by social class and status distinctions and segregation. There has been an overdevelopment of expensive private courses in nations such as the United States and Canada, and in other nations, such as France, Spain, Italy, Great Britain, and Australia, the sport continues to be oriented primarily toward the upper socioeconomic levels (Stoddart, 1990).

Davidson (1982) examined the relationship of golf to social inequality in the United States and argued that even with the democratization of golf that has been occurring for several decades, there remain social class and status differences between those who play regularly and those who do not. It still tends to be a sport more regularly played by white men in higher-status occupations and in mainline Protestant denominations. There are further and deeper social distinctions based on where people play. Wealthy people of high status can choose to play at private or public golf courses, but golfers who have less wealth or lower status are restricted to public courses.

Membership in private golf clubs is often quite expensive, but money alone cannot earn a person membership. The elite private clubs in the United States have a history of excluding nonwhites, non-Christians, and other minorities. Memberships in private golf clubs are generally family based, but women experience discrimination and segregation within these clubs by being allowed to play at certain times less popular with men and to play only with other women. Men typically have enjoyed the most convenient tee times, have had more chances to play, have played on longer and more demanding courses within clubs, and have expected their wives to supervise children while at the club. Thus, status distinctions exist within exclusive private clubs as well as between private clubs and public golf courses.

Davidson (1982: 217) argued that golf reflects and is influenced by significant social divisions in society. He also argued that in some ways, the game legitimizes and directly contributes to social differences and inequalities. Exclusive private golf clubs help maintain the privileged status of their members by means such as admission policies that restrict membership to people in dominant racial, ethnic, religious, and socioeconomic strata. They also offer opportunities for interaction in a social network in which people mutually reinforce their prejudices toward members of outgroups and their sense of their own social superiority. In the social networks of these clubs, children form friendships with others of similar class backgrounds, and acquire the "cultural capital" of ways of thinking, talking, and acting (Bourdieu, 1984) that will enable them to maintain their class standing in the future.

1989: 183). Golf is a good example of how a "democratized" sport continues to contribute to social inequality. Few people with minority or disadvantaged backgrounds are found participating at the highest levels of golf competition.

The social organization of golf shows how people from prominent families who control the dominant institutions of their community are able to extend their influence to leisure sports. They use their resources to organize, finance, and maintain elite sports clubs (Davis, 1973; McPherson, Curtis & Loy, 1989: 183). In addition, they are the most likely people to attain leadership positions in major voluntary or amateur sports associations (Beamish, 1985). When elite clubs and amateur sports associations use exclusive social and economic membership criteria, they reinforce historical patterns of segregation in their sport. Indeed, the appeal of certain sports or athletic clubs to dominant class members and the upwardly mobile may be that they provide a basis for social distinction (Nixon, 1984: 234–236).

The Leisure Class and the Limits of Democratization

Veblen (1899) observed that the late nineteenth-century upper-class industrialists of the United States were a *leisure class* who distinguished themselves from the masses by conspicuous consumption and conspicuous leisure. **Conspicuous consumption** is a public display of material goods, lifestyles, and behavior in a way that ostentatiously conveys privileged status to others for the purpose of gaining their approval or envy. **Conspicuous leisure** is public involvement in expensive or extravagant leisure pursuits to convey the appearance of privilege for approval or envy. According to Veblen, when people are freed from the ordinary concerns of survival, they are strongly motivated by a desire for esteem or honor. This desire can be fulfilled in a capitalistic society by spending money on expensive sports endeavors and sports club memberships. Furthermore, by engaging in nonproductive pursuits such as leisure sports, people can show a detachment from the productive work associated with manual labor and the masses. Thus, in Veblen's terms, conspicuously participating in such leisure sports and exclusive athletic clubs is a form of status display that creates an invidious distinction between the privileged and the masses (Gruneau, 1975; Figler & Whitaker, 1991: 7).

Veblen's analysis was insightful but incorrect in many ways (Gruneau, 1975: 126–127). For example, Veblen underestimated interest in conspicuous display among the middle and working classes and overestimated such interest among the aristocratic classes. In addition, he failed to recognize the amount of work that could be involved in serious leisure pastimes and the attraction of the privileged classes to serious leisure sports such as America's Cup racing. Despite these shortcomings, a basic insight from Veblen's work remains compelling: that is, participation in expensive, time-consuming, and exclusive leisure sports separates participants from the masses.

Gruneau (1976, 1983) challenged Stone's (1955) assertion that the leisure class had become a democratized "leisure mass" in the United States. He also challenged Betts's (1974) argument that sport had made a significant contribution to American social democracy. Gruneau (1983: 81) cited research showing a continuing pattern of underrepresentation of the lowest strata among Canadian leisure participants. He interpreted democratization as a process of meritocratic *bourgeoisification*—or extension of middle-class privilege—rather than *massification* (Gruneau, 1976: 110). It was a process that involved the middle class and some skilled manual workers who used the middle class as a reference group and used middle-class striving for achievement as a guiding philosophy. In Gruneau's view, members of the privileged classes will continue to have more and better opportunities in leisure sports as long as prominent families are able to confer socioeconomic advantages and the widest range of life chances on their children.

Gruneau (1983: 82–83) believed that limiting the analysis of social inequality in leisure sports to questions about the breadth of opportunities for participation, or democratization, misses the more important questions of power and domination. He pointed out that Veblen's analysis of the leisure class failed to examine the complex nature of the relationships of sports participation and control of sports associations to class inequality and the political order (Gruneau, 1975: 127). Thus, while the opportunity to participate in sport is ideologically and practically made available to everyone, the conditions of participation are set by an elite who control the nature or structure of the sport. This is not unlike the manner in which all of society's associations and institutions operate. The political order is reinforced by these organizations.

LEISURE SPORTS SUBCULTURES

When leisure sports players interact frequently with each other, they form a social network, and as members of this sports network they are likely to develop a distinctive culture. The type of group they form, called a **leisure sports subculture**, is a network of people who are directly or indirectly tied to one an-

POINT-COUNTERPOINT

HAVE WE BECOME A LEISURE-ORIENTED SOCIETY?

YES

The work week now is 40 hours or less, making it possible to have many hours each week for leisure. Labor-saving technology, such as microwave and self-cleaning ovens, dishwashers, and clothes washers and dryers, make household chores less demanding and time consuming. In addition, the increased use of home and domestic services, such as commercial gardening and lawn care, grocery purchase companies, home shopping networks, fast-food restaurants, child-care agencies, home cleaning and maid services, create more leisure time. Advertising, marketing, and the popular culture all glorify leisure activities and related merchandise.

NO

The work week is expanding to more than 40 hours for average working people, who must have two jobs to meet their daily needs or maintain their lifestyle, and people in professional and managerial occupations routinely work more than 40 hours per week. Labor-saving technology is a reflection of the increased amount of time that individuals, couples, and parents must work. While children may be watching more videos and television and playing more video games, their parents are spending less time at home and involved in purely leisure pursuits. The leisure industry is a response to the needs of an overworked society, unable to find enough time for leisure.

How should this debate be resolved?

other by the distinctive culture of a leisure sport in which they actively participate (Donnelly, 1981, 1985; Donnelly & Young, 1988; McPherson, Curtis & Loy, 1989: 250–251; Ferrante, 1992: 109–110).

Subcultures are social networks in which participants interact with each other and exchange resources, messages, and influence. Subcultural dimensions include special meanings, symbols, clothing, equipment, values, beliefs, norms, attitudes, identities, language, rituals, and types of behavior that link subculture members and distinguish them from other people in society (Nixon, 1992: 128). When subcultures resist or challenge basic norms and values of the dominant culture of a society, they are referred to as *countercultures,* and when they are seen as threatening or violating societal norms and values, they become *deviant subcultures* (Ferrante, 1992: 109, 255). Subcultures once defined as deviant can evolve to become part of the larger or mainstream culture. This has been the case with sports such as hot-dog skiing and surfing that have moved from fringe sports to Olympic and national status.

Since leisure sports participation often reflects the stratification of society, many leisure sports subcultures may be seen as expressions of the values, attitudes, tastes, interests, style, or experiences of a social class or status grouping (McPherson, Curtis & Loy, 1989: 250–251). Leisure sports subcultures develop when participants establish a communication network focused on their shared sports interest. Communication may be face to face or more indirect and formal (e.g., a newsletter). This communication sustains the social network and produces a distinctive culture as network members create or rely on a language, set of beliefs, type of dress, equipment, and cultural practices that are uniquely associated with their sport. The subculture becomes more influential as participants become more avid about their sport and interact more frequently with other enthusiasts. As they become insiders, the subcultural network and culture become an increasingly important frame of reference for members' identity, thinking, and behavior (McPherson, Curtis & Loy, 1989: 251–254). The subculture becomes a community, bonding individuals in a sense of

SPECIAL FOCUS

The Development of the Surfing Subculture as a Social Scene

Surfing has been dubbed a *social scene* because it represents a set of well-known patterns of behavior for its participants, who believe that surfing is "where the action is" (Irwin, 1973). Important features include the voluntary nature of participation, shared meanings of the surfing lifestyle, a noninstrumental orientation to involvement in the surfing scene, variable commitment to the scene, ranging from a fad to a permanent way of life, and a sense of personal identity associated with being a surfer.

According to Irwin (1973), there were four distinct stages in the history of surfing. In the initial *articulation* stage, people with a common interest and free time were drawn together and developed a new lifestyle focused around some specific focal elements, such as wave riding and beach life. During this stage participation was intense, with substantial experimentation, innovation, and spontaneity in surfing style and equipment features. Surfing entered its second stage of *expansion* after distinctive subcultural practices and shared meanings evolved along with a cohesive lifestyle. The invention of a new, more maneuverable surfboard and the prospect of a distinctive beach-oriented lifestyle contributed to acceleration of the growth of the subculture in the 1950s. The influence of surfing-oriented movies, magazines, and music in addition to an appeal to lower-and middle-class southern California adolescents stimulated surfing's growth in the late 1950s and 1960s.

In its third stage of development the surfing, subculture was *corrupted* by its increased popularity because it led to more competitiveness, more concern about status display, and less spontaneity and experimentation. Surfing became more dangerous as the number of surfers per wave increased drastically and inexperienced newcomers created safety problems for their more experienced counterparts. Newcomers became objects of scorn as they invaded the surfing scene in greater and greater numbers. Their increasing numbers made surfing more competitive and more status conscious about differences in skill. The growth in numbers made it more difficult to maintain a recognized position in surfing, and as a result it became necessary for surfers to prove themselves with skillful but conventional maneuvers and with conspicuous displays of the insider language and the appearance of being "real surfers." Most experienced members of the subculture did not want to risk being labeled a "kooky" newcomer by experimenting with new surfing techniques or styles of behavior. Only the best-known and most skillful surfers were secure enough to try new things. Thus, the burst in popularity of surfing made the subculture less cohesive by emphasizing invidious status distinctions, and it made the surfing community less dynamic by stifling creativity. In these ways, the surfing subculture was severely corrupted.

Corruption of the surfing subculture was followed by *decline*. Decline does not mean the disappearance of surfing. It refers instead to the virtual disappearance of the cohesive network of enthusiasts and their distinctive culture. According to Irwin (1973), by 1960 surfing was becoming so popular that it could no longer maintain its integrity as a distinctive subculture. Ironically, the forces that made it popular also corrupted it and caused its decline. Many of the newcomers were seen as "pseudosurfers" by veteran members of the subculture because they lacked the special knowledge, skills, and commitment to the subcultural practices that would have made them authentic surfers. Thus, shortly after 1960, the original surfing subculture became diluted and divided and effectively died. Irwin speculated that all subcultures—or social scenes—probably lose their distinctive identity while they are still growing.

SPECIAL FOCUS

Running as Serious Leisure

An apparent boom in running occurred in the United States and worldwide in the decades of the 1970s and 1980s. Those who became serious competitive runners often went through a series of stages of socialization that transformed their identities while deepening their commitment to running (Nash, 1976, 1979). This socialization process began with casual, part-time jogging motivated by a desire for weight control, health, longevity, or relaxation benefits or a wish to escape from everyday routine. As joggers become more committed to running, they may become regular runners and rearrange their schedules to assure that they run regularly, invest in expensive running shoes and special clothing for runners, believe in the intrinsic value of running, find it a pleasurable activity, and begin to associate part of their identity with running. When regular runners join running clubs and routinely participate in weekend races, their involvement in the running subculture becomes more intense and their associations with other runners more frequent. They are at the stage of being *serious runners,* transformed from casual joggers to competitive distance runners (Lutz, 1991).

Serious runners rearrange their entire lives to accommodate their training and competition. They are immersed in the far-reaching subculture of running, which includes an international network of races, racers, running experts, and running clubs; an assortment of running-related equipment and clothing; special running knowledge, language, and publications; and a lifestyle and identity focused on running. Running can take a higher priority than work, family, or other roles for many at this level. For the most serious of these leisure runners, running becomes—in the words of Dr. George Sheehan (1978), the erstwhile self-acknowledged guru of running—"a self-renewing compulsion" (Nixon, 1984: 227–231). This compulsion is a spiritually fulfilling experience mixing play, personal rediscovery, and joy. For others, it is an addiction that has positive or negative consequences (Nixon, 1984: 230).

Leisure sports obsessions become *negative addictions* when participants, who tend to be ambitious and successful males, continue to train or compete despite injuries or illnesses that could become more serious, despite interference with job obligations, and despite interference with marital and family relationships (Brown & Curtis, 1984; Yates, Leehey & Shisslak, 1983). In fact, serious running can be very time consuming. The reordering of priorities to facilitate running does not necessarily diminish work or career goals, but it does tend to cause marital disagreements (Robbins & Joseph, 1980). Potential marital and family disruption may be why married people have been underrepresented among serious runners (Brown & Curtis, 1984). It is not clear what percentage of male runners— or other leisure sports participants—actually become negatively addicted. It may be, though, that the most vulnerable are those most caught up in competitive striving.

Initially, running serves as an ephemeral role that offers a chance for affirming male identity as well as for escape from the rat race of competitive striving. If runners develop close personal relationships in the running subculture, they find the social support and sense of fun and fulfillment they lack in their pursuit of occupational success. That is, running becomes an eventful experience, contributes to a sense of well-being, and offers relief from the pressures of everyday life (Nash, 1979). The subculture becomes the basis of participants' most important communal bond, an understated role of leisure subcultures (Frey & Dickens, 1991).

---

commonality that may command fierce loyalty (Frey & Dickens, 1991).

If mass media such as television, movies, music, or magazines create new interest in a leisure sports subculture, they may fuel a substantial growth in participants. In some cases, interest may grow to the point where the subcultural network reaches around the world. This happened with surfing (Irwin, 1973; McPherson, Curtis & Loy, 1989: 253).

Identity and Obsession in Leisure Sports Subcultures

For some, the surfing subculture provided an **ephemeral role**, which Steele and Zurcher (1973) defined as a temporary role chosen to meet social or psychological needs incompletely satisfied by the more central and enduring roles of everyday life. Members of the surfing subculture experienced a temporary departure from dominant role obligations of family, work, and school that did not provide the source of gratification they found in riding the waves. Steele and Zurcher (1973) found a similar pattern with bowling in which this activity provided a source of "catharsis" and "separation" for white-collar participants and status and self-esteem enhancement for blue-collar workers. Both groups enjoyed bowling because it provided an experience not duplicated in other roles.

Irwin (1973) found that surfing offered its most devoted adherents an identity, a source of status recognition, and a lifestyle that made participation in the surfing subculture more than an ephemeral matter. Surfing was a **serious leisure activity**. The lives of these devotees were immersed in the surfing subculture, and surfing defined who they were. The process through which initially casual, then ephemeral, leisure sports roles become central has not been confined to surfing. The conversion from a casual, ephemeral role to a more serious pursuit occurs when participants become increasingly involved in the subcultures of their leisure sports and the satisfactions and rewards from their leisure sports roles become more important than the consequences of other roles. Running has often demonstrated this transformation from ephemeral to serious leisure participation (see the Special Focus discussion).

Identity, Masculinity, and "Prole" Sports Subcultures

Although the demonstration of masculinity through sport is very important to frustrated white-collar professionals and managers, it may be even more important to male blue-collar workers, whose subordinate roles in the workplace can lead to a sense of emasculation. Working-class men may be attracted to *prole*—for proletarian or working-class—sports subcultures because they symbolize toughness, asceticism, and hard manual labor (Renson, 1976). Sports such as motocross racing and bodybuilding appeal to working-class men because these activities symbolize power.

Martin and Berry (1987) studied a predominantly working class motocross racing club and concluded that participation in this sport served as a substitute for alienating work for these working-class males. The Sunday races were the highlight of their lives because the events and the subculture represented an opportunity to express traditional masculine values of rugged individualism, aggressive activism, competition, achievement, and success that was not possible in jobs that were routinized, boring, and lacking decision-making discretion (Martin & Berry, 1987: 273).

Competitive bodybuilding is another example of a leisure sports subculture that offers lower-class and working-class men a chance to assert power and traditional masculinity (Klein, 1986; Randall, Hall & Rogers, 1992). Muscles, the traditional emblem of masculinity, are the heart of the subculture of bodybuilding (Wilkinson, 1984; Glassner, 1989: 311). Randall, Hall & Rogers (1992) argued that bodybuilding provides men with a chance to express traditional masculinity in terms of control, rugged individualism, independence, competition, and domination, as well as the sensitivity and creativity of more contemporary masculinity. One can see posing in competition as a more stereotypically feminine aesthetic activity, just as the grueling weight training needed to build muscles can be seen as men's work. In either case, the bodybuilding subculture offers men a chance to enhance their overall sense of masculinity as it feeds their self-image. However, the bodybuilding subculture has been labeled by outsiders as an expression of narcissism or vanity, and these images have raised questions about the sexual orientation of male bodybuilders. In the same way, outsiders question the femininity of female bodybuilders who develop large "masculine" muscles.

In the past, homophobic suspicions about bodybuilders, coupled with suspicions of widespread use of steroids in the sport, cast doubt on the respectability of this subculture. Klein (1986) suggested that these suspicions may have been true, but in an ironic twist the suspected vanity, narcissism, and sexual hedonism of one sort or another became fashionable in the popular culture of the 1980s (Lasch, 1979). As a result, bodybuilding became more popular. This development is interesting because powerful commercial interests in bodybuilding have sought to bring the sport into the cultural mainstream by emphasizing wholesomeness and fitness rather than these more trendy aspects of popular culture. It has become fashionable for middle-class and professional men and women to work out, lift weights, and build muscles in the gym (Stevenson, 1986; Brubach, 1993). The bodybuilding subculture, once defined as deviant, is now included in the larger culture and is a good example of cultural diffusion, or the dispersion of elements of a unique subculture to the larger, mass culture.

The popularization of formerly prole leisure sports can corrupt the subculture. This is what has happened to both surfing and running (Irwin, 1973). Commercial interests in these subcultures are responsible for popularization and corruption, and these subcultures lose their distinctiveness as a result. They lose their special meaning and appeal for their original members when they become part of trendy popular culture or the cultural mainstream. A tension exists within growing leisure sports subcultures between the retention and dilution of their original values and cultural practices (Klein, 1986).

Identity and Risk in Sport Subcultures

All sports pose potential risks (Frey, 1991). For example, there is an obvious risk of physical injury in motocross racing; in other sports, the physical risk is present but less evident. Sports also pose potential risks to personal reputation and self-worth. In certain sports, such as mountaineering, white-water kayaking, deep-powder skiing, big-wave surfing, skydiving, and hang gliding, physical risk is a central element of the activity. Such high-risk sports generally require a high degree of control over environment, action, and self. Failure to exercise such control leads to serious injury or death (Vanreusel & Renson, 1982).

Many high-risk leisure activities do not seem to have the usual characteristics of sport. Some observers (e.g., Hilliard, 1986) have referred to them as *frontier challenge activities* because participants must rely on great physical skill and courage to deal with the challenges of extreme environmental conditions. We will treat these activities as sports, however, because they often require substantial amounts of physical skill; they are organized; and there is at least competition with the physical environment, if not also with other persons or groups within the subculture.

Donnelly's (1982) study of mountain climbing in Britain found that the first participants had elite social origins. Young English university graduates, mainly from Oxford and Cambridge, began to climb the Alps for recreational purposes in the middle of the nineteenth century. The early climbers may have been motivated by growing interest in athleticism and muscular Christianity in the elite boarding schools at that time. These elite climbers relied on professional guides and were very concerned about safety. By the twentieth century, the sport had become democratized as middle- and working-class climbers formed climbing groups in Britain.

Following World War II, the increase in the number of working-class climbers continued. This trend was the result of increased affluence of the working class and the inspiration of the ascent of Mount Everest provided by Edmund Hilary in 1953. These working-class climbers formed their own clubs because they were unable to join more established clubs, and they pushed climbing to higher levels of difficulty, risk, and competition.

Working-class climbers did not try to emulate climbers from the middle and upper classes. Instead they established their own clubs and their own standards, and they eventually came to dominate the sport for a period of nearly two decades in the middle of this century. It is especially interesting that while their more privileged counterparts relied on guides and emphasized safety, the working-class climbers struck out on their own and attempted more and more difficult climbs. The greater appeal of risk to working-class climbers may be consistent with the idea, suggested by Renson (1976) and by Martin and Berry (1987), that different social classes construct leisure sports and their subcultures in different ways to reflect different class-related social and cultural experiences.

Vanreusel and Renson (1982) studied the high-risk sports of rock climbing, spelunking, and scuba diving.

They found that among rock climbers and spelunkers, and to a limited degree among scuba divers, the perception of social stigma contributed to the formation of a high-risk subculture and the reinforcement of the value of risk within the subculture. The stigmatizing and discrediting of these activities derived from a negative stereotyping of risk-sport participants as irresponsible "thrill seekers," "crazies," or "fanatics." The stigma was reinforced by an apparent enjoyment of such labels by the participants themselves.

The fact that outsiders discredited participants in these high-risk activities tended to have a positively reinforcing effect on the commitment of rock climbers and spelunkers. Each of these subcultures exhibited an elaborate set of formal and informal norms designed to ensure safety and survival. These norms were frequently violated by rock climbers and spelunkers, but as long as this deviance was an individual act and did not jeopardize the safety of others, it was generally overlooked. In fact, it appeared that daring behavior, either as a group or by individuals, was valued over safe behavior in the rock climbing and spelunking subcultures.

Although risky behavior was valued among rock climbers and spelunkers, new members of these subcultures did not earn any special recognition for risky behavior. New members could be dismissed for failing to adhere to the formal norms of their sport. In the beginning, they were supposed to focus on learning the technical and physical skills needed to be competent in the sport. At the same time, however, they quickly learned that the reputations of experienced subculture members were built on risky acts. Thus, they came to understand the ambivalence toward risk taking that characterized the subcultures of rock climbing and spelunking.

Vanreusel and Renson (1982) found that scuba divers were significantly less likely than rock climbers or spelunkers to violate their subcultural norms of safety and thus were less likely to accept the stigma of risk taking as a positive source of commitment or cohesion in the subculture. In general, the status of scuba divers and rock climbers was determined by how well they measured up to international performance standards in their respective subcultures.

Even though participants in high-risk sports subcultures seem to tempt death or serious injury or engage in suicidal behavior, these risk takers are likely to hold quite different views of their sport than the general public holds (Hilliard, 1986). Participants in risky sports do not feel they are taking unnecessary chances with their lives or physical well-being. They believe they are in control, particularly of the physical preparation (e.g., tying a parachute, checking the rope) for the activity. MacAloon and Czikzentmihalyi (1983) found that over two-thirds of the thirty rock climbers they interviewed said that their sport was not dangerous, and only one said that thrill seeking was a major motivation for climbing. Rather than being threatened by physical danger, climbers tended to be stimulated by it and to experience "flow" as a result of it. **Flow** is an intense experience that fuses self with activity and enables the participant to make decisions and act almost unconsciously. The body seems to take over the action and give the participant a sense of "pleasant total involvement" as he or she makes the right decisions and moves without thinking (MacAloon & Czikentmihalyi, 1983: 372).

Frey (1988b) found similar results in a study of skydivers. Parachutists perceived a minimal amount of risk existed when they jumped from airplanes and engaged in individual or group freefalls. They had confidence in their training and in their ability to physically prepare themselves and their equipment for ensuing jumps. The risk that existed was worth the feelings of exhilaration and self-fulfillment they experienced.

Arnold (1976) found that skydivers had a realistic idea of the slight chance of an accident during their jumps. He contended that rather than dwelling on death or danger, the skydiving subculture focused mainly on skydiving as an escape and interaction with other enthusiasts as an opportunity for community. Arnold reported that skydivers tended to be highly committed to their sport, making an average of sixty or more jumps per year and spending much of their free time with other jumpers. They were very interested in the technology and equipment of their sport, carefully logged their jumps, and competed both individually and as members of team formations. While he found no consistent patterns in the class backgrounds of skydivers, Arnold noted that there was a clear bias against women. Women participated in the sport, but a number of male jumpers and a few instructors said

POINT-COUNTERPOINT

ARE PEOPLE IN HIGH-RISK SPORTS CRAZY?

YES

People who participate in risky sports, such as skydiving or rock climbing, have a secret death wish that motivates their involvement. They derive a thrill putting their life on the line and taking significant chances that they will be hurt or killed. They are not the type of people who fit in the mainstream of society or accept the norms accepted by normal people.

NO

People in high-risk sports do not focus on the risk. They are in control of the activity and themselves because they are mentally and physically prepared. They participate because they enjoying challenging their personal skills and courage, not because they want to risk death. They are also attracted by the opportunity to interact with other people sharing the same passion for their sport.

How should this debate be resolved?

that they did not trust women or did not believe that women were capable of becoming as skillful as men in the sport (Frey, 1988b).

It is easy to understand why family members of participants in high-risk sports might oppose such participation. If they do not share a passion for such sports and are outside the subculture, they are likely to see risks and dangers that are minimized within the subculture. For example, Brannigan and McDougall (1987) studied hang gliders and found that almost without exception, the flyers' families opposed their participation in this sport. These family members had great fears about the safety of their loved ones in this sport. Like the skydivers, though, the flyers themselves were aware of cases of deaths in their sport but widely believed that "good flyers" had the expertise needed to avoid danger. They developed their sense of competence from experience in the air and from the positive reinforcement they received from members of their hang-gliding subculture.

The mix of occupational backgrounds found in other high-risk sports also characterizes hang gliders. An important distinguishing characteristic of this and other high-risk sports subcultures is the preponderance of young (in their early to mid-twenties) unmarried males. Drawn to the sport by friends or siblings and motivated by a passion for flying, these young men have not yet developed career, family, or lifestyle commitments that could discourage them from risky activity. The status, identity, and sense of competence developed within their sport's subculture sustains and deepens their commitment to hang gliding.

A key to understanding involvement, noninvolvement, and withdrawal is the subculture. As novices develop skill in a high-risk sport and spend an increasing amount of time with others involved in the sport, their commitment to the subcultural interaction and culture of their sport grows and deepens. As subcultural involvement increases, contacts with outsiders decrease, and participants derive more and more of their identity, status, and associations from the subculture. Subcultural beliefs that minimize danger combine with an increased sense of competence to enable flyers, skydivers, rock climbers, spelunkers and other high-risk sports enthusiasts to participate in their sport without being gripped by fears that outsiders, such as family members, feel.

Status, Competition, and Cooperation in Leisure Sports Subcultures

It should be evident by now that commitment to leisure sports is sustained through the interaction of participants who make up the subculture of that sport. Members of these subcultures share a strong passion for their sport and depend on each other for support. Leisure sports subcultures also produce conspicuous and invidious status displays and intense competition among members. For example, the development of divisive status relations and competitive behavior contributed to the corruption and decline of the original surfing subculture (Irwin, 1973). Divisive or not, status and competition are normal aspects of subcultures as organized social networks and social groupings. The nature and survival of leisure sports subcultures are affected by how members organize status relations and how they balance competition and cooperation within the subculture.

Status in the Dojo In certain leisure sports subcultures, the outline of the status hierarchy is clearly visible to casual observers. In a martial arts dojo, or school, status distinctions are readily seen in the colors of the belts and the condition of uniforms worn by students. Jacobs (1987) noted that the belts enable martial arts students to see who is a status superior, equal, or inferior. The amount of wear seen in a uniform further reveals the seniority of a dojo member within a rank. Weekly competition is very intense as dojo members attempt to move up the status hierarchy. As karate students progress up through the status ranks of their dojo, they earn certain privileges, such as special access to the dojo master-proprietor and the chance to learn esoteric karate knowledge and new weaponry techniques. Upper-level belts also have an obligation to teach, which is both a responsibility and a form of authority over lower-ranking members of the dojo.

The stratified structure and social relations of the dojo, with its multilevel hierarchy of statuses, tangible evidence of status ranks, clear-cut authority relations, cliques, and competitive striving, appear to parallel the social structure and dynamics of the larger American society. Jacobs suggested, however, that participation in the dojo has a different meaning than everyday life for its members. The status hierarchy of the dojo lies at the core of its social organization and social solidarity. With its roots in the somewhat mysterious lore and myth of karate, this hierarchy is an alternative to the status systems of everyday life and a source of special meaning for participants in this subculture. Furthermore, unlike the sometimes ambiguous statuses of everyday life, status in the dojo is clearly marked and status relations are clearly defined. Thus, members of this subculture know precisely who they are, where they stand, and what they have accomplished.

Status Hierarchies and Social Order in the Water Status hierarchies organize social relations in other leisure sports subcultures, but the nature of the hierarchy and its effects may be much less evident than in the dojo. For example, Nixon (1986) observed the informal subculture of faculty and staff recreational swimmers in a university pool and found that the swimmers had established a subtle but influential system of social control to maintain orderly and sociable patterns of interaction in the pool (McPherson, Curtis & Loy, 1989: 8–9). The social organization of this pool subculture was especially interesting because the swimmers had little personal knowledge about each other—they were "familiar strangers"—and because their behavior was governed largely by an unwritten and largely unspoken structure of social norms, statuses, and roles.

An important part of the maintenance of social order in the water was the establishment of a status hierarchy among swimmers. There were four levels of status in the hierarchy of the observed pool, based on frequency and intensity of involvement in this pool setting. They were serious-regulars, casual-regulars, serious-nonregulars, and casual-nonregulars. The serious-regulars were positively or negatively addicted to swimming and were committed to maintaining peak performance and to pushing their bodies to high stress levels. The stars in this category swam exclusively in the fast lane. They were envied by the strivers, who were not as fast as the stars but were dedicated to improving their strokes, speed, and/or endurance.

Casual-nonregulars were at the bottom of the hierarchy. They lacked the long-term commitment to swimming characterizing the serious swimmers. Among casual-nonregulars, novices had limited swimming experience and skills. Faddists swam because swimming was recommended by a friend or was thought to be the latest trend in fitness or recreation. The usually much faster serious-regulars rarely interacted in the pool with the casual-nonregulars because they swam in different lanes. The casual-nonregulars appeared infrequently at the pool because higher-priority activities filled their schedules, they did not like making a regular trip to the pool, or they were intimidated or frustrated by the crowded pool and its patterns of laps and circles.

Between the serious-regulars and casual-nonregulars were swimmers in the other two status categories. Casual-regulars swam regularly for health, fitness, or relaxation reasons, but lacked the competitiveness of the more serious swimmers. Serious-nonregulars were like serious-regulars, but usually lacked their endurance because they swam less frequently. This category, however, included crossover athletes who were very good swimmers, were in excellent shape, and were in the pool to prepare for a competition, such as a triathalon. Regulars generally had higher status than nonregulars in the pool subculture because they were more committed to the maintenance of the social structure of pool interaction and were more familiar with its norms, statuses, and roles.

The informal code of behavior in the pool included expectations about which lane to swim in; the path to follow in a lane; how to pass other swimmers; and the kind of attention to pay to other swimmers. In the last category, swimmers were expected not to stare at others, especially members of the opposite sex. The general rationale for these norms was incorporated in the mini-max principle of interaction, which is that pool interaction ought to minimize personal involvement and maximize order.

By conforming to the informal code of the pool, swimmers could expect to swim without interference and with the assurance that their personal privacy would be respected. In this social order, other swimmers usually deferred to the highest-status serious-regulars, who were given space in the chain of swimmers in the fast lane or in another relatively uncrowded lane. Conflicts and disruptions of the normal sociability and order of the pool subculture usually occurred because nonregulars did not understand or conform to the informal pool norms. Regulars used bumps, taps, or occasional words of instruction to get deviants into line. They tended to be more verbal, blatant, and forceful in dealing with nonregulars than with other regulars. In fact, even when regular members of the subculture were reminded by subtle cues of their violations of the pool code, they tended to be embarrassed or apologetic. Such reminders seldom involved swimmers who were regulars in the same lane. Lifeguards, who were students and younger than most of the faculty and staff members who swam, rarely became involved in social control. The maintenance of the informal social organization of the pool was implicitly understood to be the responsibility of the swimmers themselves, and they usually maintained it with very few words and without developing close personal relationships with others in their leisure subculture.

This case of a pool subculture demonstrates how subtle nonverbal cues can establish cooperative interaction patterns among familiar strangers. This cooperative environment also incorporated instances of sociable informal competition among members of the subculture. Both cooperation and competition coexist in interesting ways in many leisure sports subcultures. In some based on high-risk activities, such as mountain climbing and team skydiving, successful competition with the environment or another team, and even survival, depend on substantial cooperation *within* the team. Partners or teammates must also trust each other to be competent and composed so that their collective effort can be successful (Donnelly & Young, 1988). In leisure sports subcultures in which competition is more formal and sportlike, there may be a need to extend cooperation and trust *between opponents* to facilitate the process of competition and avoid serious danger. Thus, perhaps paradoxically, cooperation is needed between opponents to make competition better or to moderate its demands on all competitors. This idea is illustrated by the case of bicycle racing.

Competition, Cooperation, and Bicycle Racing

Albert's (1984, 1990, 1991) analysis of the subculture of bicycle road racing provides a good example of how competition and cooperation can coexist in subtle and dynamic ways to moderate the stresses of this very demanding and risky sport. Although fans are excited by attempts to break away from the pack, long road races—in contrast to shorter *criteriums*—often involve extended periods in which all or most of the riders are massed together in one or more tight packs. In the *peleton,* or main group, riders leave the start line together and may pedal within inches of each other, sometimes at high speeds, for a large part of a race or for the entire race. In the peleton and trailing packs, there is much conversation, and team members break wind, or draft, for teammates who need to conserve their energy. Thus, there will be tensions within teams about whom the team will support in a race, and some team members will be expected to sacrifice their chance of winning for the good of the team and its best rider or riders. In addition, competitors call to opponents who are seeking to push the pace or move out in front to "take it easy" or "slow down." Riders on different teams typically have a shared commitment to try to maintain a pace that is more comfortable for the entire peleton and closely trailing packs.

A mistake of novice racers is to try to move out in front prematurely when they lack the strength or endurance to maintain the pace and lead. When a group of riders believes that too much of a gap has opened between themselves and an individual rider or pack of riders, they will work together, taking turns drafting and pushing at the front of a paceline to try to close the gap. In effect, riders work together to try to punish the rider or riders who have broken from the pack. Within groups, though, there always is the tacit understanding and trust that riders will have sufficient bike handling skills, strength, and composure to hold their place in line or pull their share of the load when called on to do so. These tacit agreements break down as teams get closer to the finish line. Previously cooperative behavior is transformed into intense competition.

Albert (1991) points out a number of sociologically important features of the bicycle racing subculture that involve competition and cooperation. First, socialization into this subculture requires an understanding of insider knowledge about informal rules of competition and cooperation. Second, novices often face difficulties because they fail to learn the delicate balance between competition and cooperation. In fact, according to Albert (1991), North Americans often have competitive and social interaction problems in this European-based subculture because they fail to understand this delicate balance. Third, a dominant part of the racing subculture is an informal pattern of mutual assistance that usually works against those who try to succeed on their own. Riders construct these cooperative patterns, such as pacelines, as situational constraints on aggression and competition, and new riders who are effectively socialized into the subculture learn these patterns as part of their socialization. In general, then, the usual North American view of sport as a relentless battle for individual competitive dominance fails to capture the nuances of the bicycle racing subculture, with its mix of informal cooperation with competition.

Status and Power in Leisure Sports Subcultures

In leisure sports subcultures, as in society in general, status differences often imply power differences. For example, we saw that the most powerful members of the karate dojo subculture tended to be the master-proprietor and highest level black belts; the serious-regulars tended to have the most power in the pool; and veteran racers tended to exercise collective power over novices in bicycle racing. Stevenson (1986) also found that status and control were highly correlated in the subculture of a university weight room.

Klein (1986) found a small caste of powerful entrepreneurs ruling bodybuilding. Male and female professional bodybuilders were in the next caste, and they competed with each other both for prize money and closer political connections to the powerful entrepreneurs and mentors who dominated the sport. Amateurs tried to gain some access to these same figures as mentors, but had less status, influence, and success than their professional rivals. Filling in the base of this status and power hierarchy were the noncompetitive bodybuilders, who were primary consumers of the products and performances manufactured by the other strata.

A small Canadian community-based track and field club studied by Pitter (1990) showed two levels of authority, consisting of the general membership and the board of directors. There was one board member whom everyone in the club saw as the central figure in the club. He was the technical director and also a coach, and formal responsibilities for budgeting and travel were concentrated in his hands. In effect, this board member occupied one level of formal authority and everyone else occupied a second level. He tended to make decisions that favored the best runners in the club or the runners he coached, and the general membership could do little about this practice. Those who could not make a substantial commitment to the club or who were of lesser ability assumed a second class status with little or no power in the club.

Stratification of Tennis Subcultures Other settings in leisure sports subcultures display these forms of stratification. For example, Muir (1991) found that members of a tennis club used a kind of *performance principle* (Pitter, 1990) to include or exclude other players as doubles partners. This club showed clear patterns of segregation and status boundary maintenance based on age and gender, but Muir was especially interested in the structuring of relations on the basis of perceived playing ability. Even potential partners rejected by stronger players tended to accept this "implicit-tournament" structure of relations. That is, they recognized their ability dictated their standing among other players and their chances of being selected as a doubles partner. Rejection was often polite, indirect, or subtle, but the result of interaction among tennis players in this club was that they generally knew their place in the hierarchy of the club's competitive subculture. Muir attributed these patterns of interaction to general acceptance among both stronger and weaker players of a fiercely competitive ethic.

Muir found evidence of these patterns of status segregation at other country clubs in his community in the Deep South of the United States. As he suggested, these patterns are not likely to be confined to his community, owing to the pervasive commitments to seriousness and winning in tennis clubs and the subcultures of other ostensible recreational sports. In the Canadian amateur track and field club studied by Pitter (1990), the performance principle was the primary basis for coaches' willingness to share their time and expertise with runners and to listen to input from them about their training programs. It also determined who got the most and best resources.

SPORT IN SPECIAL POPULATIONS: DISABLED AND ELDERLY ATHLETES

In sport, being disabled generally means that an athlete is at least temporarily out of action. In fact, high-level sports often cause disabling injuries, and lingering disabilities from injuries are a major reason for retirement from sports careers (Lerch, 1984; Nixon, 1993b). Moderate to severe permanent disabilities usually exclude people from participation in serious sports in the mainstream of society. There are, of course, some notable exceptions. Major league pitcher Jim Abbott has one arm. He followed outfielder Pete Gray, who played one year with the St. Louis Browns in the mid-1940s, as the only other one-armed player in modern major league baseball history. At least seven Major League Baseball players have been deaf. In addition, there have been amputee triathletes as well as blind and visually impaired athletes who have successfully competed with and against outstanding sighted athletes in triathalons, wrestling, judo, karate, swimming, crew, track and field, marathons, powerlifting, gymnastics, tandem cycling, sailing, and football (Buell, 1986).

Disability and Sport

Wheelchair racing illustrates both the possibilities of athletic achievement for people with permanent physical disabilities and the problems that may arise when they compete in integrated settings. Because of the prejudice, discrimination, and stigmatization of **handicapism**, people with disabilities historically have confronted obstacles trying to participate in the mainstream of society and sport (Nixon, 1991b). Opportunities in segregated disabled-sports settings have substantially increased since Sir Ludwig Guttmann organized the first sports event for physically disabled athletes in Stoke Mandeville, England, in 1948 (Sherrill, 1986; Steadward & Walsh, 1986). Disabled athletes have made substantial progress since then. However, even though in 1978 George Murray became the first wheelchair racer to defeat the able-bodied winner of the Boston Marathon, the path to opportunity, acceptance, and respect for disabled athletes has remained cluttered with obstacles. This is especially true when disabled athletes try to cross the line into integrated sports settings.

Despite much admiration for athletes like George Murray, Craig Blanchette, and Doug Heir, there still is resistance to the participation of wheelchair racers in

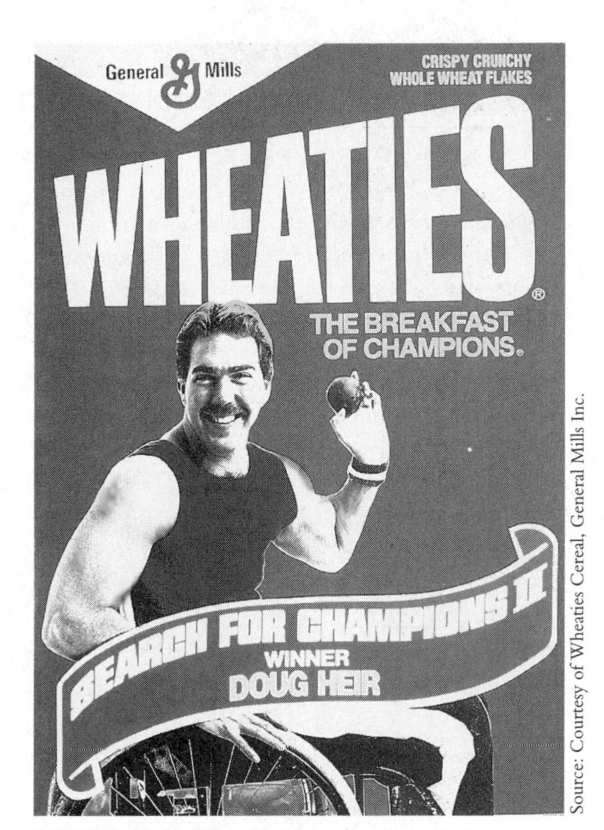

Source: Courtesy of Wheaties Cereal, General Mills Inc.

FIGURE 11.2 Wheaties Champion Doug Heir

road races with able-bodied runners. According to Brandmeyer and McBee (1986), resistance has tended to be based on questions about safety, spectacle, and authenticity. Race directors have claimed that wheelchairs are too dangerous and that accidents on wet or uneven pavement at high speeds could lead to serious accidents. Some race officials have also failed to see wheelchair racers as serious athletes and have worried that their participation might be intended as a spectacle to draw attention to some political message about disabled people. Others, such as Fred Lebow, the director of the New York City Marathon, have questioned the appropriateness of wheelchair racers competing against runners on foot.

It appears that outstanding athletes with permanent disabilities have been denied full respect, as well as opportunities in integrated competition, because many people continue to see the Special Olympics as the appropriate model of sport for disabled people. There is no question that this event's controlled competition and everyone-is-a-winner philosophy have resulted in rewarding experiences for many Special Olympics participants and their families (Orelove, Wehman & Wood, 1982; Orelove & Moon, 1984; Coakley, 1990: 68, 70–71). At the same time, the application of this image of to all sports for disabled persons limits the opportunities for disabled persons with outstanding talent and serious sports intentions (Brandmeyer & McBee, 1986). Furthermore, this segregated model of sports would seem to do little to enhance the integration of people with disabilities into the mainstream of their community (Orelove, Wehman & Wood, 1982; Orelove & Moon, 1984).

Some athletes with permanent disabilities wish to participate in high-level competitions for disabled persons, but other athletes, such as wheelchair racers, have a desire for the competitive push and recognition that come from competing against the best able-bodied athletes in their sport. The issue raised by disabled athletes is integration at appropriate levels of competition (Nixon, 1991b: 186–188). Boxing, wrestling, and weightlifting try to equalize competition by having different weight divisions, and other sports have gender and age divisions. Similarly, wheelchair road racing has developed classifications based on the degree of physical impairment and disability (Brandmeyer & McBee, 1986: 184). In this context, athletes with disabilities, such as Jim Abbott, George Murray, and Doug Heir, compete against able-bodied opponents because such competitive settings match their ability.

Thus, one could argue that if persons with disabilities are encouraged to develop their athletic as well as other abilities, some are likely to be capable of participation at the highest levels of certain mainstream sports. We have seen that specific physical impairments do not preclude outstanding performance in some high-level sports. In other cases, people with disabilities may be able to participate at lower levels of integrated sports. There may also be cases where people with disabilities find that participation at some level of disabled sports is most appropriate and appealing to them. The principle that seems relevant in this context is that all persons, disabled or not, ought to have a chance to participate in sports at levels that are appropriate for their ability, motivation, and interest.

POINT-COUNTERPOINT

DO WHEELCHAIR RACERS BELONG IN THE SAME RACE AS ABLE-BODIED ROAD RACERS?

YES

Wheelchair racers are serious athletes who train hard to become strong, fit, and highly competitive. Even though their bodies function differently than those of able-bodied runners, the athleticism of top wheelchair racers rivals that of their able-bodied counterparts. Just as able-bodied runners compete in different age divisions in the same race, wheelchair racers compete in a distinctive division against each other. Preventing wheelchair racers from sharing the road with able-bodied runners denies them the respect and visibility they deserve as high-level athletes.

NO

Although wheelchair racers have overcome many obstacles to be able to compete, they should not be in races with able-bodied runners. If they lose control of their chairs, serious accidents and injuries can result. They also undermine the credibility of races as major sports events, making them appear more as spectacle than sport. They cannot be compared to able-bodied runners as serious athletes because they rely on the technology of their chairs more than their bodies themselves to compete. It is similar to racers on foot competing in the same race with auto racers, motorcyclists, or bicyclists.

How should this debate be resolved?

Older Athletes

Along with disabling injuries, the aging process contributes to decisions by athletes to retire from high-level sports and to drop out of leisure sports. Aging is not just a physiological process, though. It is also a social process that may involve the prejudice and discrimination of **ageism**. Just as handicapism has restricted sports opportunities for disabled people, ageism has limited the opportunities of older people, both in high-level sports and in recreational or leisure sports. As a result, athletes in professional and high-level amateur sports typically have been considered old at relatively young ages. In addition, there have been relatively few organized recreational sports programs and activities for people in the later years of life. The absence of organized programs is important because formal organization tends to facilitate and encourage leisure participation (Fine, 1989).

In a recent survey of over 400 adults in Illinois, Fishwick and Hayes (1989) found no significant difference in participation in individual physical activities by age. They also found, however, that older people engaged in much less strenuous activities and were much less likely to participate in competitive activities or team sports. Older people biked, fished, walked, and gardened, while younger people were involved in aerobics, jogging, racquetball, swimming, tennis, and a range of team sports. Although women generally participated less than men in team sports, those women who were involved in team sports when they were younger dropped out more gradually than men as they got older.

Although older and elderly people in the United States may be more physically active than in the past, the proportion of elderly people engaged in strenuous or competitive physical activities and sports remains relatively low (Rudman, 1989). Among those studied by Fishwick and Hayes (1989), there was virtually no participation in team sports among women by the age of sixty or among men by the age of seventy, and the decline in such participation was steady and steep from the age of twenty for men. There are various possible explanations of declining frequency and intensity of physical activity and sports involvement as people age (Harootyan, 1982; Figler & Whitaker, 1991: 208–210). In general, though, wherever ageist ideas and practices about the physical capabilities of older people are dominant, we can expect to see a substantial underrepresentation of these people in organized physical activities and sport. Furthermore, a self-fulfilling prophecy results when people believe they cannot do as much physically as they get older, proceed to do less because they hold this belief, and ultimately become less physically capable because they have chosen to do less and less (Harootyan, 1982: 143; Figler & Whitaker, 1991: 218).

Since health does not appear to be a reason for limited participation in vigorous physical activity or sport for many older Americans (Harootyan, 1982), the idea that these activities are only for the young help explain reduced involvement in them as people age (Figler & Whitaker, 1991: 211). When older people do not believe that physical recreation and sports programs are appropriate or necessary for them, there is unlikely to be pressure to create such programs, which will reinforce patterns of nonparticipation.

Rudman (1989) found that people of different ages tend to participate in sports and fitness programs for different reasons. Younger people have distinct competitive and social reasons, but as people get older the distinction between these reasons tends to blur. Thus, older people seem to enjoy friendly competition. Rudman argued that in order to attract middle-aged and older adults into sports and fitness programs, it is necessary to diversify the types of programs and methods of promoting them. As in the case of sports and disabled people, programs for elderly people need to match their levels of ability, motivation, and interest (Snyder & Spreitzer, 1989: 117–119). Therefore, along with Masters' and Senior Olympics programs, there needs to be a variety of more informally oriented programs that gives people a chance to meet and casually interact with a variety of other people as they get their exercise or compete.

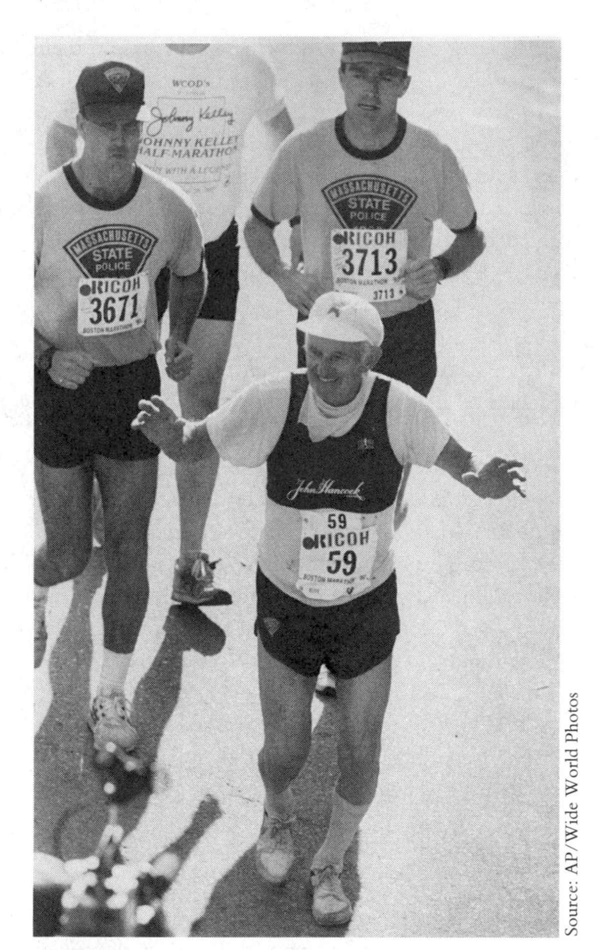

Source: AP/Wide World Photos

FIGURE 11.3 Never too old: Johnny Kelley completes his 56th Boston Marathon

CONCLUSION

Sport is stratified in a number of different ways, and evidence of stratification can be found in a variety of sports settings and subcultures. These patterns of social class and status differences have persisted despite processes of democratization that have opened up a number of formerly elite or aristocratic sports to participants from other classes and strata. In many cases, though, democratization of opportunities for participation has been limited to the middle class, as many sports continue to be out of the financial or practical reach of average working-class and poor people.

Class and status distinctions can also be found between and within leisure sports subcultures. Many of these subcultures do not involve sport in the precise sense that we have defined it, but those we have considered in this chapter contain some degree of physical challenge or competition involving either the environment or other people. They are also similar in linking people with a strong interest in a particular kind of physical activity. Some of these subcultures carry a stigma because their members engage in high-risk activities of dubious value to outsiders, but we have seen that leisure sports subcultures often perform functions and meet needs that are not readily understood or appreciated by outsiders.

For those most involved in these subcultures, their leisure often has a special meaning that is greater than anything else in their lives. The greatest threat to the special meaning and appeal that subcultures have for their devotees may be their growing popularity, which corrupts the values and cultural practices that originally defined the subculture. The cohesion and survival of leisure sports subcultures often depend on how status relations are organized and how competition and cooperation are balanced within them.

Stratification is clearly evident in the underrepresentation of disabled and elderly people in sport in the United States. As disabled people have begun to move into the mainstream of society and the population has become older, we have seen increased involvement of disabled and elderly people in various kinds of sports. Inequalities persist, though, along with the residue of well-entrenched patterns of handicapism and ageism. The discussion of stratification and the experiences of disabled and older people in sport foreshadows the discussion of racial, ethnic, and gender relations in the next two chapters. Sport offers a useful vantage point for understanding how racism and sexism have been constructed and changed in society.

SUMMARY

Although many Americans participate in leisure sports, the patterns of participation reflect the social stratification of American society—with rich participating more than poor, executives more than blue-collar workers, and college graduates more than high school graduates. Many sports formerly associated with the elite strata of society, such as golf and tennis, have become more democratized, with more middle-class participants. Class differences in active leisure sports participation remain, however, with certain sports such as polo and yachting still confined to the upper classes and other sports such as softball and bowling associated more with the working class. Class differences are also correlated with sports viewing and can be found in countries around the world. Conspicuous sports consumption and participation become forms of status display that create and reflect a desire for clear distinctions between the privileged classes and the masses but that invite envy or resentment from the less advantaged.

The imprint of status, as well as power, can be seen in various leisure sports subcultures. The case of surfing illustrates how physical activities evolve into leisure sports subcultures, which in turn become corrupted and decline. The cases of running and a number of other sports subcultures show how identity becomes tied to subcultural commitment. Masculine identity is especially emphasized in prole subcultures, such as motocross racing and bodybuilding. Popularization of these subcultures in the middle class can dilute their appeal to their working-class participants. In some leisure sports subcultures, such as rock climbing and skydiving, the element of risk is a distinguishing characteristic. Risk is viewed differently by participants than by outside observers, though. The stimulation of these sports can lead to an experience of flow for participants. In subcultures such as the martial arts and lap swimming, strict formal or informal status hierarchies organize social relations. In bicycle racing, we see how competition and cooperation coexist in subtle and dynamic ways to limit the stress of a demanding and risky sport.

Handicapism and ageism have restricted the participation of disabled and older people in sport. There have been exceptions in high-level sports in the past, and in recent years an increasing number of disabled and older people have become actively involved in sports at various levels. Controversy has surrounded the participation of disabled people in certain high-level sports with able-bodied participants, however, as in the case of wheelchair racers. Disabled and elderly athletes counter stereotypical myths about the limitations of disability and age.

CHAPTER 

Race Relations and Sport

RACIAL STRATIFICATION IN THE UNITED STATES

Perceived differences in skin color have had profound effects in American society. Being a person of color—that is, other than white—has generally meant less economic privilege, prestige, and power. This stratification of white and nonwhite races and ethnic groups has been a persistent fact in American history, and it has reflected deep divisions and caused violent conflicts in American society (Giddens, 1991: ch. 8; Hess, Markson & Stein, 1991: ch. 11). What is especially interesting about **racial stratification** and conflict is that apparently obvious biological differences of race are not as obvious or biologically pure as people typically think they are.

Though we would not want to deny the imprint of biological factors on human behavior, biology does not determine race relations. From a sociological perspective, race relations are social and cultural constructions, which are based on *perceived* differences in physical appearance and *beliefs* about the origins and implications of those differences. People believe and act as if there are distinct racial groups based on skin color, and in this way racial difference becomes a social fact and race is assumed to be an ascribed status characteristic (Giddens, 1991: 300–301; Hess, Markson & Stein, 1991: 251–252).

Racism occurs when people attribute social significance to perceived racial differences and treat members of other racial groups with prejudice and discrimination. **Racial prejudice** involves an unfavorable feeling or attitude toward members of another racial group. **Racial discrimination** involves unfair or arbitrary treatment of members of another racial group. More specifically, discrimination occurs when people are systematically underrewarded or denied opportunities solely because they are perceived as belonging to a particular racial group. Thus, prejudice is attitudinal, and discrimination is behavioral. Racist attitudes and behavior tend to be shaped by negative preconceptions or stereotypes of racial groups. Stereotypical thinking involves a fixed or rigid cast of mind, and racial stereotyping leads people to see members of other racial groups in terms of preconceived negative characteristics and behaviors (Giddens, 1991: 301–305).

In the United States, the clearest racial distinctions tend to be white, black, and Asian. We also tend to think of other categories of people, such as Native Americans, Hispanics, Arabs, and Polynesians, as racial groups, but their members may vary a great deal in physical appearance, which makes it very difficult to fit them into distinct racial categories (Phillips, 1993: 150). The term **ethnic group** is more helpful than race in capturing the distinct cultural identity, background, and life of many types of people in a society. It refers to categories of people who share a common cultural identity and cultural heritage.

Racial and ethnic groups outside the mainstream of a society often face prejudice and discrimination, and they have difficulty becoming accepted and integrated in the mainstream. The sociological notion of a minority group implies a lack of full integration. A **racial minority group** is a category of people who (1) are perceived as genetically distinct and as possessing distinctive physical characteristics; (2) are targets of prejudice and discrimination; (3) have an awareness of a common inferior status and identity that they share with other members of their race; and (4) think, feel, and act in terms of this shared sense of inferior status

SPECIAL FOCUS

Restaurants, Racism, and The NFL

In March 1993, a class action suit claiming racial discrimination was filed by thirty-two black customers in California against the corporation owning Denny's restaurants. Two months later, a suit was filed in Baltimore federal court by six black U.S. Secret Service agents, alleging that they waited an hour for service in an Annapolis, Maryland Denny's restaurant while they watched other customers being quickly served (*Charlotte Observer,* 1993a; Clarke and Greiff, 1993). These suits were seen in the Charlotte, North Carolina area as a possible threat to a bid for a National Football League franchise, thus revealing the sometimes complex webs linking sport to society.

This web of connections is untangled by understanding that the president and chief executive of Flagstar Cos. Inc., the corporation that owned over 1,400 Denny's restaurants at the time of the suits, was Jerry Richardson, and that his Richardson Sports was spearheading the drive to bring NFL football to the Carolinas. In an effort to rectify possible damage to the Charlotte NFL bid, Flagstar negotiated two highly publicized "Fair Share" agreements with civil rights leaders of the NAACP to provide minority jobs, contracts, and franchises. Lawyers representing Denny's employees and customers suing the company contended that the agreement provided much less than claimed and was virtually unenforceable (Menn, 1993). Regardless, this case demonstrates how sport can be deeply enmeshed in the political economy of race relations in the United States. It shows how important the *image* of good race relations has become for the NFL and corporate America.

---

and identity. Minority implies perceived inferior status or being powerless, but minority groups are not necessarily numerically smaller than other groups in society. In South Africa, for example, whites are a numerical minority but are the dominant group, whereas the larger subordinate nonwhite population includes a variety of racial minority groups.

The black or African-American minority in the United States represented over 12 percent of the American population or nearly 30 million people in 1990 (Macionis, 1993: 324). The disadvantaged status of black Americans is well documented (Macionis, 1993: 335–337). Whether the disadvantaged status of black Americans stems from racial prejudice or a background of poverty is debatable. That racism has been a significant obstacle to black integration, achievement, and mobility in American history is unquestionable.

HISTORICAL PATTERNS OF BLACKS IN SPORT IN THE UNITED STATES

Race and sport has been one of the most researched topics in sport sociology, and most of this research has focused on African-American males, which will also be the focus in this chapter. The prominence of black athletes in professional basketball, football, and baseball in recent decades obscures the more enduring historical facts of racial prejudice and discrimination against blacks in American sport. Despite their success, blacks still experience racism in sport.

Jackie Robinson's entry into Major League Baseball in 1947 broke a significant color barrier, but it was not, however, the first time that a black athlete competed in a racially integrated sport in the United States. It was not even the first time that a black athlete played Major League Baseball. An appreciation of the significance of Jackie Robinson's breakthrough requires that we place it in historical context (see Ashe, 1988).

The First Three Historical Stages: Exclusion-Inclusion-Exclusion

The history of black males in American sport has been marked by several stages of exclusion and inclusion (Guttmann, 1988: ch. 9; Eitzen & Sage, 1993: 323–

327). Significant stages, events, figures, and accomplishments in African-American sports history are summarized in Table 12.1. Blacks were almost totally excluded from sport in the first stage, which extended to the Civil War. Notable exceptions were slaves who became boxers in the early years of the nineteenth century; Tom Molyneux and Bill Richmond are two examples. In addition, plantation owners in the Deep South promoted boat races between crews of their slaves in the early 1800s. They placed bets on these races and secured their wagers with future crops (Betts, 1974: 334).

In the second stage of African-American sports history, from the Emancipation Proclamation until approximately 1920, there were some major breakthroughs and accomplishments by black athletes in integrated sports, including boxing, horse racing, baseball, bicycle racing, and football. The general pattern in this second stage, however, was toward diminishing opportunity for black athletes in integrated

TABLE 12.1 Time Line of Black Sports History in the United States[1]

Historical Stages and Years	Significant Black Sports Events, Figures, and Accomplishments
Pre–Civil War	Exclusion with exceptions in boxing (e.g., Tom Molyneux, Bill Richmond) and boat racing between crews of slaves
Emancipation Proclamation—1920	Inclusion and accomplishment followed by increasing segregation with separate-but-equal doctrine and Jim Crow
1872	Bud Fowler is first paid black baseball player
1875	Oliver Lewis wins first Kentucky Derby
1883–1884	Moses Fleetwood Walker and Welday Walker become first black Major League Baseball Players; Isaac Murphy rides first of three Kentucky Derby winners (third in 1891)
1887	Cuban Giants are first black professional baseball club
1900	Marshall Taylor is professional world and American bicycle racing champion
1904	George Poage becomes first black Olympian
1908	Jack Johnson wins world heavyweight boxing championship
1911	Henry McDonald becomes first black professional football player
1915	Jack Johnson loses heavyweight title; the title is not regained by a black boxer until 1937
1916	Fritz Pollard is first black player to compete in Rose Bowl and is first black All-America football player
1918	Paul Robeson elected to All-America football team
Post-World War I—Post World War II	Virtual racial segregation of most major sports
1922	Fritz Pollard is first black head coach in NFL
1936	Jesse Owens wins four gold medals at "Nazi" Olympics in Berlin
1937	Joe Louis wins world heavyweight boxing championship
Post-World War II—present	Black breakthroughs, accomplishments, and dominance in major professional and college sports coupled with limited opportunity in a number of sports (especially for women) and residual racism

(continued)

Table 12.1 (*continued*)

Historical Stages and Years	Significant Black Sports Events, Figures, and Accomplishments
1945	Kenny Washington and Woody Strode are first black NFL players
1946	Marion Motley and William Willis are the first black players in the All-American (football) Conference
1947	Jackie Robinson is first black to play modern Major League Baseball
1949	Jackie Robinson is first black winner of the MVP award in Major League Baseball
1950	Three black players break NBA color bar
1952	PGA permits black golfers to compete under "approved entry" classification
1957	Althea Gibson is first black to win Wimbledon tennis championship
1959	Althea Gibson is first black female professional tennis player
1960	Charlie Sifford is first black on PGA tour
1966	Bill Russell is first black NBA head coach
1968	Emmit Ashford is first black MLB umpire
1971	Wayne Embry is first black NBA general manager; Satchel Paige is first Negro League player inducted into Baseball Hall of Fame
1974	Frank Robinson is first black MLB manager
1975	James Harris is first black regular starting NFL quarterback; Lee Elder is first black golfer to compete in Master's tournament
1985	Eddie Robinson (Grambling) becomes all-time winningest college football coach
1986	Proposition 48 instituted
1987	Al Campanis creates furor with racial slur on *Nightline*

[1]Compiled from Betts (1974) and Leonard (1993).

sports after a period of visibility and even dominance, in certain sports. Horse racing illustrates this pattern (MacDonald, 1981: 94). Jockey Isaac Murphy rode three winners in the Kentucky Derby between 1884 and 1891, which established a record that was not broken until the 1960s. The best jockey in an era dominated by black jockeys, Murphy became a horse owner and earned a considerable amount of wealth and respect during his life. The era of black dominance in horse racing ended because entrenched segregationist practices prevailed, and by the 1920s few black jockeys were left.

The third stage of exclusion stretched from the early part of the nineteenth century until after World War II in most American sports. It was fueled by segregationist laws and practices that emerged at the end of the nineteenth century, such as the separate-but-equal doctrine endorsed by the Supreme Court and the Jim Crow laws. By the time the World War I was over, newly erected racial barriers excluded blacks from most major professional sports in the United States and greatly restricted black participation with whites in amateur sports and intercollegiate athletics. Even in boxing, traditionally the sport of the less privileged classes (Wacquant, 1992), a virtual color bar existed after Jack Johnson lost his title in 1915. The Golden Age of Sports in the 1920s was marked by the reign of two white boxing champions, Jack Dempsey and Gene Tunney, in the 1920s. It was not until Joe Louis defeated James Braddock in 1937 that the heavyweight championship was won again by a black fighter.

At the college level, a few black athletes successfully competed in football, basketball, baseball, and track at white institutions before World War II, but they tended to be token representatives of their race (Eitzen & Sage, 1993: 325). At universities and colleges in the South and Southwest, there were no black athletes before World War II because campuses in these regions were racially segregated. Thus, until the middle of the twentieth century, with relatively few exceptions, black colleges provided the only intercollegiate athletic opportunities for blacks. However, for even the stars, playing for a black college meant relative obscurity outside the black community (Grundman, 1986).

Public Reactions to Black Athletes in a Changing Society

We gain valuable insights into the complex and changing racial attitudes of the American people during the twentieth century by considering public reactions to five prominent black athletes whose careers span the century: Paul Robeson, Jack Johnson, Jesse Owens, Joe Louis, and Muhammad Ali. These athletes offer striking contrasts in personality and demeanor. More important, though, they responded differently in different eras to being black athletes in America, and their responses led to different kinds of public reactions.

Paul Robeson has been called one of the most talented figures in American history (Tygiel, 1983: 334). Along with being elected to Walter Camp's All-American football team in 1918, he earned Phi Beta Kappa honors at Rutgers and later became a highly acclaimed dramatic actor and singer. White America turned against Robeson, though, when he expressed sympathy for communism and told an audience in Paris shortly after World War II that American blacks would not participate in a war against the Soviet Union (Tygiel, 1983: 334). Caught up in a growing wave of anticommunism, his challenge to American racism was branded "anti-Americanism." As a result of his politics, Robeson's name was withdrawn from the listing of All-American football teams; he became an object of contempt; and he lost his fame, wealth, and career.

Jack Johnson was bold, flamboyant, and assertive during a period when racial segregation was increasing in American society. His open defiance of social and cultural conventions for black men of his time provoked racist taunts from boxing crowds on both American and foreign soil, inspiring a search for a **great white hope** to replace him (Guttmann, 1988: 122–125). Johnson's most brazen defiance of racial conventions was his public courting of white women, three of whom he married. When he married a former prostitute, an outraged member of the U.S. House of Representatives asserted to his colleagues on the floor of the House that intermarriage "between whites and blacks is repulsive and averse to every sentiment of pure American spirit" (quoted in Guttmann, 1988: 123). Facing racist-inspired legal and social pressure, Johnson fled the United States, only to find racial prejudice and discrimination in Europe and England as well. He ended his seven-year exile five years after losing his championship to a white man, Jess Willard, in Havana in 1915.

Like Jack Johnson, Jesse Owens was a great athlete. Unlike Jack Johnson, he became an American hero because he avoided personal controversy and achieved athletic prominence against the backdrop of the rise of Nazism. Owens was the star of Hitler's "Nazi Olympics" in Berlin in 1936, winning four gold medals. In doing so, he posed an embarassment and challenge to the Nazi ideology of Aryan supremacy and fueled nationalistic feelings at home (Mandell, 1971; Guttmann, 1992: 67–69).

Joe Louis, called the "Brown Bomber," was also a quiet man whose athletic prominence was magnified by the social and political tensions of the pre–World War II period. His accomplishments in the ring had great symbolic significance to black Americans, who needed a new black hero. Louis also was cheered by white Americans, who saw in his defeat of the German Max Schmeling a patriotic victory over the rising Nazi threat. Joe Louis secured his place as an American hero by presenting the appearance of a well-mannered "good Negro" who was a "credit to his race" (Guttmann, 1988: 124). He was tutored to avoid the public outspokenness and excesses that brought disrepute to Jack Johnson and the kind of political rhetoric that destroyed Paul Robeson's image and career. Although Louis, like Johnson, had affairs with white women, including Norwegian Olympic figure skater and movie star Sonja Henie, he conducted them out of the public eye.

The unpopularity of Paul Robeson and Jack Johnson and the popularity of Jesse Owens and Joe Louis among white Americans suggest that black athletes risked public disapproval, or worse, if they challenged conventional standards or appeared to criticize their country. On the other hand, even with strong currents of racism pervading society, it seems that black athletes can become heroes among whites as well as blacks as long as they remain personally uncontroversial and express or symbolize pro-American sentiments. Guttmann (1988: 120) cautions us, however, to recognize that the relative influence of the two conflicting impulses of nationalism and racism will depend on the combination of social forces that are at work in American society at the particular time black athletes are achieving prominence. The case of Muhammad Ali, viewed alongside the other four we have examined, demonstrates how complicated generalizing about public reactions to prominent black athletes can be.

Like Jack Johnson, Muhammad Ali was outspoken, outrageous, and defiant. Like Paul Robeson, he expressed strong and controversial political sentiments. He did not consort with white women, but he became a Black Muslim, recited verses to extol his own prowess and mock his opponents, and refused induction into the army during the Vietnam War for religious reasons. Ali's stance against the war touched the political sensitivities of white America, much as Paul Robeson did in an earlier era.

Ali fostered racist sentiments among many white Americans by his words and actions. Yet unlike Robeson and Johnson, Muhammad Ali gained the admiration of a significant segment of the white population in the United States and of many in the mass media. He also became a hero to blacks in the United States and abroad. Part of the reason for the more mixed reactions to him than to Johnson and Robeson among whites was the very different social and political era during which Ali was in his athletic prime. The 1960s and early 1970s were inflamed by the Vietnam War and by passions for civil rights, human rights, and demands for significant social change. Prior to this era, change was not popular; the country was very conservative on social issues. Thus, the political and social conditions of their respective eras shaped public reactions to these five great black athletes, Paul Robeson, Jack Johnson, Jesse Owens, Joe Louis, and Muhammad Ali.

Stage 4: The Post–World War II Era of Black Opportunity

Jackie Robinson ushered in a new era of black opportunity in sport when he put on a Brooklyn Dodgers uniform in 1947 (Tygiel, 1983). Kenny Washington and Woody Strode became the first blacks to play in the NFL in 1945, and Marion Motley and William Willis were the first blacks to play in the All-American Conference when they joined the Cleveland Browns in 1946 (Leonard, 1993). These breakthroughs did not have the impact of Jackie Robinson's entry into Major League Baseball, though, because professional football had not yet achieved the prominence it enjoys today. In the late 1940s, baseball was touted as the U.S. "national pastime," one that had been white for nearly the first half of the twentieth century.

Blacks played Major League Baseball in the nineteenth century despite a vote in 1867 by the National Association of Base Ball Players to exclude clubs with "colored" players (Guttmann, 1988: 121). Recurring hostility from white players and racist badgering from fans ultimately proved too much for black players, and by the end of the 1890s professional baseball was segregated once again (Voigt, 1976; MacDonald, 1981: 95; Tygiel, 1983). From that time until after World War II, blacks played in the Negro Leagues. The great stars spawned by these leagues did not enjoy the fame and incomes of less talented whites in "Organized [white] Baseball" (Peterson, 1970). In addition, the

SPECIAL FOCUS

Branch Rickey's Great Experiment

Unlike many of his counterparts in Major League Baseball, Branch Rickey was a modernizer who believed in ongoing rational research, experimentation, and refinement to improve his sport. Before joining the Dodgers, he had developed an innovative farm system for player development for the St. Louis Cardinals. Rickey was also motivated by financial incentives. He believed that after integration, new black fans, with their increasing purchasing power following the war, would outnumber lost white fans at the box office (Guttmann, 1988: 129–130).

Sociologists Leonard Broom and Philip Selznick (1963: 529–530) viewed Rickey's experiment as a social action program with a number of explicit and carefully formulated long-term goals. These goals included generating support for his signing of a black player from Dodger directors and stockholders; selecting an individual who could handle the pressure both on and off the field; creating favorable media and public reactions; gaining support and understanding of his intentions from the black community to avoid misinterpretation, misunderstanding, and other pitfalls in his program; and producing acceptance among Dodger players (Leonard, 1993: 208). Branch Rickey found in Jackie Robinson the player and person he sought for his "great experiment" (Tygiel, 1983). With his superb play and exemplary behavior, Robinson was able to overcome initial protests from Dodger teammates and much racist sentiment from other players in the league as well as from fans to achieve the goals of Rickey's program. Note, however, that twenty years after his historic breakthrough, Jackie Robinson wrote that he knew that he "never had it made" (the title of his autobiography) and that he still felt that he was "a black man in a white world" (quoted in Leonard, 1993: 208).

franchises in their league were financially unstable, equipment and facilities were often poor, and schedules were uncertain.

A number of factors combined after World War II to open the door to blacks in Major League Baseball. A formidable obstacle to racial integration was Commissioner Kenesaw Mountain Landis. When Landis died in 1944, his successor, Albert "Happy" Chandler, was much more sympathetic to racial integration, especially after African Americans had fought and died for their country during World War II. In addition, the war had created a scarcity of white major leaguers, which encouraged a few club owners to approach stars of the Negro Leagues. When Brooklyn Dodger President Branch Rickey brought Jackie Robinson to the major leagues, however, it followed a 15-1 secret vote by owners in 1946 to maintain segregation. That year, they defended their segregationist policy before a U.S. Senate subcommittee on the grounds that there were not enough qualified black ball players; that the Negro Leagues would be hurt by integration; and that opponents of Major League Baseball were trying to damage its reputation by calling it racist (Voigt, 1976: 116–117). Branch Rickey was a determined dissenter. He signed Jackie Robinson to play for a Dodger farm team in Montreal in 1946, before promoting him to the Dodgers the following April (Guttmann, 1988: 125–130).

An ironic consequence of the integration of modern Major League Baseball by Jackie Robinson, predicted by the white major league owners, was the destruction of the Negro Leagues. Black fans stopped following Negro teams and sometimes rode hundreds of miles to sit in segregated stands to watch blacks in Major League Baseball (Guttmann, 1988: 130). The crumbling of the Negro Leagues greatly overshadowed, though, by the transformation of white sports by integration, which Jackie Robinson's name symbolizes. His influence spread beyond sport. According to sportswriter Leonard Koppett, Jackie Robinson's highly publicized integration of Major League Baseball confronted ordinary Americans with the reality of racial segregation and prejudice, which they had found easy to ignore or accept in the past (quoted in Tygiel, 1983: 344).

After Jackie Robinson The list of breakthroughs and accomplishments of black athletes in the United States since World War II and the time of Jackie Robinson has become very long (e.g., see Ashe, 1988; Eitzen, 1989). A statistical analysis of contemporary black sports participation in the United States reveals at least four striking patterns. First, black males now are overrepresented in certain sports and underrepresented in others. Second, male black athletes are the stars of many of the sports they play. Third, black females do not match the levels of participation and success of black males in sport. Fourth, there is residual racism in the sports in which black males have excelled in recent decades. These patterns are the main focus of most of the remainder of this chapter.

BLACK SPORTS AND WHITE SPORTS

It may seem strange in this era of integrated sports to refer to black and white sports. These labels refer to the fact that today there are a number of sports in which blacks are overrepresented and many others in which very few black faces can be seen (Phillips, 1993: 174). The black sports for males are basketball, football, track and field, baseball, and boxing. Black females are disproportionately found in basketball and track and field and on the national women's volleyball team. In view of their success on the national team, black females surprisingly are *under*represented on major intercollegiate volleyball teams.

The list of sports in which blacks are underrepresented is long, and includes many of the white country club and resort area sports such as swimming, diving, tennis, golf, and skiing. Economics and access tend to account for high participation rates in some sports and low rates in others for blacks. According to Phillips (1993: 175–176), the sports opportunity structure, defined in terms of access to facilities, coaching, and competition, is a better explanation of black sports participation than alternatives such as the idea that blacks are especially attracted to sports that offer future financial payoffs or careers. For example, the payoff explanation does not account for the failure of blacks to shift to certain sports with new or existing occupational prospects, for the ostensibly lower interest of white athletes in the potentially lucrative sports that are very appealing to black athletes, or for the appeal of the sport of track and field, where lucrative financial opportunities tend to be restricted to a tiny elite.

The special appeal of top-level sports such as basketball, football, track and field, baseball, and boxing to black males is explained by the greater access they have to facilities, coaching, and competition in these sports than in other sports. The special attraction of boxing to lower-class Hispanic males in the United States may be explained in the same terms. Whites have many more sports choices, and a number of these sports choices appear to be largely restricted to affluent whites. For reasons of economics and discrimination, whites more often than blacks have the choice of participating in school sports or in sports that are offered outside school and may require special fees, personal investment in equipment and lessons, and private coaches. Black athletes typically have to rely on public recreational and school sports programs. Community and school athletic programs tend to be more limited or poorly funded in the economically depressed inner city and rural areas where many blacks live rather than in the predominantly white suburbs.

Although economics and the sports opportunity structure seem to explain much of the racial variation in sports participation in the United States, certain patterns remain largely unexplained. For example, it is not clear why there have been no top-level African-American runners in long-distance events or why there are rarely white faces among the top American sprinters. Biological arguments (see Kane, 1971; Phillips, 1993: ch. 7) have been advanced, but they seem dubious in view of the recent dominance by black Africans in distance events on the track and on the road and the appearance of white Europeans among the world's great sprinters every year. Compelling alternative explanations are lacking, though (Phillips, 1993: 178–179).

Snyder and Spreitzer (1989: 204–206) proposed that variations in the proportional representation of blacks and whites in different sports have important

TABLE 12.2 Interaction and the Racial Composition of Sports Teams[1]

Racial Composition: Proportional Black Representation	Sports	Type of Black-White Interaction and Minority Status
Uniform white teams/sports Zero blacks	Pre-integration white sports	None
Skewed white teams/sports Approx. 15 percent or less	Gymnastics, golf, tennis, swimming, ice hockey	Black tokens are socially isolated, placed under intense scrutiny, and held to high standards
Tilted white teams/sports Approx. 25 percent to 33 percent	College football, Major League Baseball	Intergroup majority-minority relations; blacks have allies and can form minority cliques and coalitions and have some influence over team subculture
Balanced teams/sports Approx. 40 percent to 60 percent	Professional football (pre-1992)	No longer a clear minority, blacks can become more comfortable, expressive, and assertive in relations with white teammates
Tilted black teams/sports Approx. 67 percent to 75 percent	Professional basketball	Intergroup majority-minority relations; whites have allies and can form minority cliques and coalitions and have some influence over team subculture
Skewed black teams/sports Approx. 85 percent or more	Boxing, sprints	White tokens may be socially isolated and touted as "Great White Hopes"
Uniform black teams/sports Zero whites	Negro Baseball Leagues; Harlem Globetrotters	None between players, but white owners may control team or sport (e.g., Globetrotters)

[1]Adapted from Snyder and Spreitzer (1989: 204–206), who derived their conceptual framework from Kanter (1977b).

consequences for black-white interaction and minority status in those sports. We can see from Table 12.2 that racially uniform, skewed, tilted, and balanced teams and sports could produce quite different social experiences for black and white athletes. The visibility of individual blacks, their social isolation, and their sense of being "trapped" by rigid conceptions of their role behavior are likely to diminish as the proportion of blacks increases on a team and in a sport. Blacks may also feel that they can be more expressive and assertive, both on and off the field, when they have more social support from members of their own racial group (Eitzen, 1989: 311). Snyder and Spreitzer (1989b) have speculated, though, that white sports fans in the United States may become disenchanted with or alienated from sports where whites become the minority or tokens. Thus, it is possible that as the proportion of blacks in a sport increases and black athletes become more comfortable, assertive, and outspoken in it, they may face either increasing racist reactions from white fans or a loss of white fan support.

There is not yet any systematic evidence of such cause-and-effect relationships (McPherson, Curtis & Loy, 1989: 200), but the perception or fear of possible lost fan support, coupled with their own racist attitudes, may have motivated sports team owners in the past to maintain racial quotas. We know that the black quota in Major League Baseball and a number of other sports was zero for nearly half of this century, and when the color bar fell, many teams were slow to integrate. In this context, a story told by former NBA star Bill Russell seems especially relevant. He said he

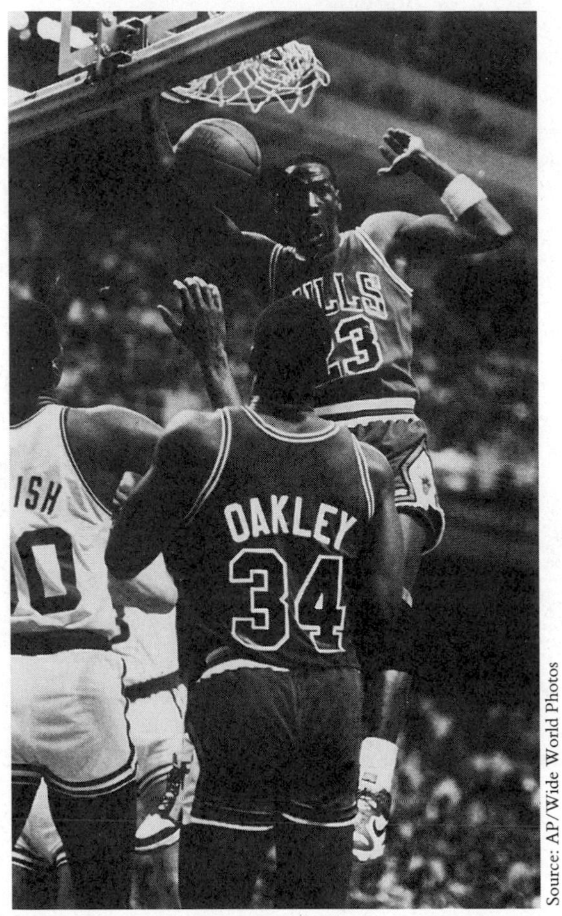

Source: AP/Wide World Photos

FIGURE 12.1 The dominance of the black male athlete: Michael Jordan.

knew a college basketball coach who started two blacks for home games, played three on the road, and put five in the lineup whenever victory was in doubt (Figler & Whitaker, 1991: 290).

BLACK SPORTS ACHIEVEMENT AND SUCCESS IN THE UNITED STATES

Many black male athletes have taken advantage of their opportunities to participate and excel in integrated sports. Indeed, the trend of increasing black athletic accomplishment in basketball, football, baseball, and boxing has led some observers to refer to the "black dominance" of these sports (e.g., Bledsoe, 1973; Edwards, 1973). In the mid-1950s, less than 10 percent of the players in the National Basketball Association, National Football League, and Major League Baseball were black. In fact, ten years after Jackie Robinson's breakthrough in baseball, there were only twelve black and six Latino major leaguers. By 1990, approximately 80 percent of NBA players, 70 percent of boxers, 55 percent of NFL players, 25 percent of the top track and field athletes, and 22 percent of Major League Baseball players were black (Leonard, 1993: 218). Blacks surpassed their percentage representation in the American population in baseball in 1958, in basketball in 1958, and in football in 1960 (Leonard, 1993: 210).

Black athletes have become stars in a number of sports, and these black stars are among the highest paid athletes in professional sports, especially in boxing, baseball, basketball, and football (Leonard, 1993: 232). In addition, black track and field stars, such as Carl Lewis, command large appearance fees. A few black superstars, such as Michael Jordan, Magic Johnson, David Robinson, and Shaquille O'Neal, have also earned millions of dollars from clothing and shoe contracts and other commercial endorsements (Plagens, 1992).

The exceptional achievement and success of black male athletes in a number of highly visible sports have been the subject of much debate in sport, sport sociology, and elsewhere in society. Some sport sociologists argue that a strong interest in sport and a special celebration of sports excellence may be distinctive aspects of a black subculture in America (Phillips, 1993: 87). Spreitzer and Snyder (1990) presented evidence to support this argument and to dispute the notion that the commitment to sport among African Americans is merely a result of their economically disadvantaged position (e.g., see Rudman, 1986). If there is a distinctive subcultural emphasis on sport (for males) among African Americans, it may provide a source of special encouragement, direction, and motivation that help explain why blacks have become so good at the sports they play (Phillips, 1993: 87). Socioeconomic factors also may play an important part in the equation of black sports success. Sports are highly valued in black subculture because black occupational and financial opportunities have traditionally been restricted by racism. Thus, young black males are encouraged to

SPECIAL FOCUS

Black Superstars and White Attitudes

The recent success of black athletes, especially in basketball, in gaining lucrative commercial endorsement contracts implies an improvement in racial attitudes in the United States. This is not necessarily true, though (Swift, 1991). Although the companies who hire them may deny it, success with endorsements may require that black athletes sacrifice their racial identity. Marketing people and product manufacturers seem to have a special interest now in black superstars who can be crossovers and have an appeal that crosses racial lines. If the white sports public in the 1990s is color blind about its sports heroes, it probably is because advertisers and sports agents have made them appear racially and politically neutral and unthreatening (Swift, 1991: 58). The white sports public seems to admire the athletic gifts of black superstars, but there is no evidence that this admiration spills over into racial attitudes toward blacks in general or that white fans would continue to admire and buy the products endorsed by black athletes who appear too militantly black (Swift, 1991a).

think of sports as a primary route to fame and fortune, and with such encouragement they put much of their talent and energy into becoming successful athletes.

Not surprisingly, sociologists favor explanations of black sports success focusing on subcultural and, especially, social structural factors over ones stressing race-linked physical, physiological, and psychological differences (Eitzen & Sage, 1993: 330–332; Phillips, 1993: chs. 7, 8). Subcultural and social structural explanations have also been more persuasive among sport sociologists than explanations focusing on the disproportionate number of female-headed households in black America (Leonard, 1993: 229–231). In the black family explanation, black males are assumed to develop strong attachments to coaches to compensate for the absence of their father. This attachment to the coach is supposed to motivate blacks to outperform whites. This argument seems dubious, though, because the absolute number of whites in female-headed households is greater than the number of blacks in such households and because it is difficult to believe that black males will become strongly attached to their coaches as father figures when most of their coaches are white (Leonard, 1993: 229). If the subcultural and social structural arguments are correct, black and white performance differences in sport should diminish as racist barriers fall, as blacks gain opportunities for success in a wider range of sports and in a wider range of occupations, and as sport becomes one of a large variety of perceived career paths among blacks.

THE ALMOST FORGOTTEN BLACK WOMAN IN SPORT

With all the attention focused on black male athletes, we must be reminded that there also have been some outstanding black *female* athletes. African-American women, like their male counterparts, have experienced racism. In addition, though, they have experienced prejudice and discrimination based on gender (Alston, 1980). The combination of racism and sexism has conspired to limit sports opportunities for black female athletes, to limit the visibility and rewards of those black women who have excelled in sport, and limit the amount of research on black women in sport (Winbush, 1987; Birrell, 1989). These athletes have been overrepresented in basketball, track and field, and international volleyball. Like all women's sports, however, these sports tend to receive relatively limited mass media attention except for every four years during the Olympics, which has generally made these black and white stars relatively invisible. In addition, these sports have not offered the kinds of occupational or commercial opportunities that top black male athletes have enjoyed.

Recent statistics give us some idea of the effects of double discrimination against black women in college

POINT-COUNTERPOINT

DO BLACKS HAVE A NATURAL ATHLETIC ADVANTAGE OVER WHITES?

YES

Blacks are born with more athletic ability than whites, especially in particular sports and sports events. Since blacks make up less than 15 percent of the American population and are often overrepresented at the highest levels of the sports where they have had opportunities, they must have more natural talent than their white counterparts. In sports such as basketball, which require a wide range of athletic abilities and skills, black athletes dominate, displaying speed, quickness, grace, and jumping ability that few white players can match. White players often are known more for their shooting, dribbling, and passing, which are skills that can be developed from constant practice and do not require as much purely natural ability.

NO

Black athletes often outperform white athletes because sport is more important to blacks than to whites, especially among males. Thus, they spend more time than whites practicing and playing sports. In addition, the black (male) population concentrates on fewer sports than the white (male) population because fewer sports are readily accessible to blacks. As a result of greater commitment to a narrower range of sports, blacks become better at those sports. The fact that white superstars can also be found in professional basketball, baseball, and football, where black athletes excel, indicates that whites are capable of athletic excellence, too, and are not naturally inferior in sports.

How should this debate be resolved?

sports. For example, a study by the NCAA gender-equity task force revealed that approximately 17 percent (685) of the 4,044 women who entered Division I institutions on athletic scholarships in 1984 were black, and that of those 685, nearly 84 percent (575) were on basketball or track teams. NCAA statistics from the 1990–91 academic year showed that 2,760 black women received athletic scholarships at Division I institutions and of this number 83 percent participated in basketball or track (Blum, 1993a). These kinds of statistics prompted James J. Whalen, president of Ithaca College and co-chair of the gender-equity task force, to remark that "black women bear the consequences of having the fewest opportunities and the least amount of support [in college athletics]" (Blum, 1993a: A39).

The opportunities for black women in athletic administration and coaching are even more restricted than they are for athletes. Anita DeFrantz, who is president of the Amateur Athletic Foundation of Los Angeles, a member of the International Olympic Committee, and a black former Olympian, has had a special vantage point from which to view such opportunities. She cited the following statistics to demonstrate the racism and sexism that prospective black women athletic leaders confront (DeFrantz, 1991). First, a survey of 106 NCAA Division I institutions with women's basketball teams showed that 11 were coached by black women. Second, there was one black female athletic director—and seven black male athletic directors (Blum, 1993a)—at those 106 institutions. Third, there were no black women serving among the executive directors of the fifty governing bodies for U.S. Olympic sports. Fourth, there has never been a black woman on the coaching staff of a U.S. Olympic basketball team.

Anita DeFrantz is a rare exception to the pattern of limited access for black women to administrative positions in sport. Another recent exception is Bernadette Locke-Mattox, a former basketball star, who was hired by the University of Kentucky as the first female assistant coach in men's intercollegiate basketball (Finn, 1992). With relatively few black women having sports experience, black women face an especially big hurdle in trying to secure a position of leadership or authority in a sports world where leadership opportunities for women have been shrinking (Coakley, 1990: 187; DeFrantz, 1991).

The limited amount of involvement of black females in sport in general can be explained by the relatively small number of visible athlete role models for females in general and black females in particular and by the limited amount of financial support for sports for women (Blum, 1993a). DeFrantz proposed that money is the greatest barrier to sports participation for minority women (*CQ Researcher,* 1992a). She contended that the financial obstacle was getting worse, with economic problems causing cutbacks in the public school and recreational programs that have been most accessible to minority women (DeFrantz, 1991).

An additional obstacle for black females has been the perceived attitudes of peers and parents. A study in 1988 showed that black girls were more likely than white girls to think that "boys make fun of girls who play sports" and to have parents who believed that sports were more important for their sons than their daughters (cited in *CQ Researcher,* 1992b). These results clearly indicate that if a special subcultural emphasis on sport exists among African Americans, it is a *male* emphasis.

A lack of family and community support may explain why black females were found by Sabo, Melnick, and Vanfossen (1989) to derive virtually no academic or mobility benefits from athletic participation. In fact, urban black female athletes seemed to be burdened by their athletic participation. They had problems securing employment and getting to college after high school. Compared with black males, Hispanic males and females, and white males and females, black females were the only category for which athletic participation provided no educational mobility benefits. More specifically, participation in high school athletics by black females was not related to higher grades, achievement test scores, school dropout rates, later college attendance, having a goal of earning a degree after high school.

Despite the obstacles, some black female high school athletes go on to college sports, especially in basketball and track. Some also play volleyball, although they seem to have relatively more restricted opportunities in intercollegiate volleyball than on the national team. Eitzen and Furst (1989) found that black women who play volleyball at the college level experience some discrimination. This discrimination occurred in the form of **stacking**, a term coined by Harry Edwards in 1967 to refer to racial segregation by position in sport in which whites are disproportionately concentrated in certain positions and blacks are disproportionately concentrated in others (Edwards, 1973: 205).

Source: AP/Wide World Photos

FIGURE 12.2 Breaking the racial barrier in elite women's sports: Althea Gibson.

In volleyball, the setter is a crucial and central position. Eitzen and Furst found in their study of Division I college volleyball teams that black women were *under*represented at the setter position. They also were *over*represented at the hitter position, which was a very "athletic" position requiring jumping, agility, and reaction. No statistically significant racial differences were apparent among blockers or defensive specialists. An implication of the stacking findings is that black players were seen as relatively better suited for very athletic positions, while white players were seen as relatively better suited for positions requiring intelligence, coolness, and leadership ability. Implicit racial stereotypes are embedded in such thinking. To the extent that coaches and athletes believe such stereotypes, black and white athletes may be sorted or sort themselves into different positions without being conscious of the racist implication of such positioning. Indeed, such racial segregation might even seem "natural."

According to Berghorn, Yetman, and Hanna (1988), the likelihood of racial stacking diminishes when the proportion of minority group members in a sport increases to the **tipping point**—which is the proportion needed to change the dynamics of interaction between races. The proportional representation of black players in women's intercollegiate volleyball is clearly below the tipping point because it is skewed, and this explains why they are so susceptible to prejudice and discrimination. In women's intercollegiate basketball, however, there seemed to be a reduction in the amount of stacking as proportional representation of black players increased from 24 percent in 1985 in all three NCAA divisions (Berghorn, Yetman & Hanna, 1988) to 35 percent in Division I in 1986 (Furst & Heaps, 1987).

Stacking in women's basketball has generally involved the center position, where height is a primary consideration. Since there are more very tall white women than black women, this stacking seems to be a result of population characteristics rather than racism. Women stack less than men in basketball, perhaps because women have been able to watch men integrate their teams and as a result, avoid some of the racist practices found in the men's sport (Berghorn, Yetman & Hanna, 1988). Stacking in men's sports will be discussed later in the chapter.

RESIDUAL RACISM IN MEN'S SPORTS IN THE UNITED STATES

The idea of lingering vestiges of racism in American sport today seems inconsistent with the picture of black achievement and success painted earlier. We have seen, though, that the rosy picture of blacks in sport applies to men and to relatively few black women, who have had to also contend with the effects of sexism as well as racism. We have had hints of continuing prejudice and discrimination against black men in sport, with mention of black sports and white sports, the restricted black sports opportunity structure, racial stereotyping, racial quotas, limited endorsement opportunities, and stacking. However, looking back forty years in 1987, it was easy to conclude that American sport had come a long way in eradicating racism since Jackie Robinson broke into Major League Baseball. Evidence that it still had a distance to travel toward racial equality ironically came during an event meant to celebrate Jackie Robinson and racial progress.

Al Campanis and Report Cards of Racial Progress

Al Campanis was Los Angeles Dodger vice-president for player personnel when he appeared on Ted Koppel's ABC-TV *Nightline* program in April 1987 to pay tribute to Jackie Robinson. Campanis, who had been in the Dodger organization since 1943 and contributed to the Dodger's reputation for recruiting minority ball players, stunned Koppel and his viewers when he said that baseball did not have black managers or general managers because blacks "may not have some of the necessities" for these jobs. He continued, "Why are black men, or black people, not good swimmers? Because they don't have the buoyancy" (Associated Press, 1987a; Gammons, 1987). Campanis apologized the day after making his remarks, and called the incident "the saddest of my career" (Associated Press, 1987b). He was nevertheless forced to resign by the Dodgers, whose president said that "comments given by Al Campanis are so far removed from what the organization believes that it is impossible for Al to continue his responsibilities" (Associated Press, 1987b). During Campanis's tenure with the Dodgers, they never had a black in the position of general manager, field manager, pitching coach, or third base coach, who is second to the manager in authority on the field (Coakley, 1990). Although Campanis's comments created a furor among African Americans, some welcomed them because they drew attention to the limited management opportunities for blacks in sport and the racial prejudice related to this fact. At the time he made his comments, less than 2 percent (17 of 879) of the top administrative jobs in baseball were held by blacks, while over 20 percent of major league players were black (Gammons, 1987).

The Center for the Study of Sport in Society (CSSS) of Northeastern University has published an annual *Racial Report Card* for several years, documenting the minority presence in professional basketball, football, and baseball in the United States. Highlights of the 1993 report (CSSS, 1993) are given in Table 12.3. The contrasting percentages of blacks in playing versus other key positions in the NBA, NFL, and MLB probably are the most striking elements of Table 12.3. In general, the NFL and MLB lag far behind the NBA in key nonplaying positions for blacks.

Baseball increased its percentage of minority employees in the front office (other than managers, trainers, scouts, coaches, and instructors) from 2 percent at the time of the Al Campanis incident in 1987 to 17 percent in 1992. The 17 percent figure was a 7 percent increase over the percentage in 1988, but only a 1 percent or 2 percent increase over the percentages between 1989 and 1991. In 1992, 9 percent of the

TABLE 12.3 Percentages of Blacks in Key Positions in the NBA, NFL, and MLB, 1992–93[1]

Position	NBA	NFL	MLB
Player	77	68	16
Owner	0	0	1
Chair of board	0	0	0
President/CEO	0	0	0
Vice president	15	6	4
General manager[2]	32	0	0
Head coach	26	7	14
All coaches	27	19	13
Head trainer	4	4	0
Radio and TV announcer	12	2	5

[1]CCCS, 1993.

[2]Note twelve teams in the NFL and eleven teams in the NBA did not list a general manager.

Major League Baseball front office employees were black, which is the same percentage as the previous three years (CSSS, 1993).

Richard Lapchick, director of the CSSS, attributed the delayed progress in minority hiring in baseball to the lack of a commissioner. He explained the trend of declining black major leaguers to a reduced number of inner-city baseball programs. A reason for the reduction in inner-city programs may be related to a decreased number of black fans at major-league ball parks. For example, the Atlanta Braves, located in a city that is 60 percent black, discovered in 1991 that blacks fans bought only 4 percent of their tickets. They hired a multicultural marketing manager, who succeeded in increasing black attendance by 4 percent. He noted that the effort to increase black attendance, which has been undertaken by other teams in the league, was motivated by a business incentive rather than a moral commitment to minorities (Starr, Barrett & Smith, 1993).

The commitment of Major League Baseball to black fans during the two decades following Jackie Robinson's breakthrough is called into question by Sager and Culbert (cited in Starr, Barrett & Smith, 1993), who conducted research to learn why ten of the original sixteen major league teams relocated to a new stadium or city between 1950 and 1970. They found that such moves were more influenced by the racial composition of the neighborhood where the team played than by stadium age, team record, or even attendance. Teams moved from areas with an average black population of 49 percent to locations that were 16 percent black. Teams that did not move were located in areas that averaged 24 percent black. They have a similar motivation to leave areas with a high percentage of low-income Hispanic residents, despite the fact that baseball had as many Latin players as black players in 1993 (Starr, Barrett & Smith, 1993). How much has baseball's "escape" from black communities affected the pool of black major-league prospects? Some clubs are beginning to realize that if they want to restore their black talent pool to prior levels and bring more black fans into the ball park, they will need to make a conscious effort to reach out to black young people, black fans, and the inner cities where many live.

Although the NBA leads the other major sports leagues in opportunities for blacks and has a reputation for progressiveness in race relations, it still does not have a management or ownership structure that approximates the percentage of its minority players. Coaching and administrative opportunities for blacks and other minorities are also much less than playing opportunities in college sports as well. The Black Coaches Association reported that in 1991, approximately 10 percent of NCAA Division I basketball teams had black head coaches, and that figure was down almost 3 percent from nearly 13 percent in 1989 (Leonard, 1993: 237). A *USA Today* survey of sixty-three major Division I athletic programs in 1991 revealed that 12.5 percent of 3,083 key positions in these programs were held by black, Hispanic, Native American, or Asian American minorities, and that 19 percent of the jobs held by these minority-group members were positions of authority (*USA Today*, 1991). These minority groups represented 20 percent of the U.S. population at that time. In the sixty-three Division I programs surveyed, 2 athletic directors, 28 assistant athletic directors, 43 head coaches, 236 assistant coaches, 21 athletic trainers, 44 academic advisers, and 10 sports information directors were minority-group members. Although these numbers are small, it should be remembered that they are still much greater than the number of minority females in key support or executive positions in athletics.

Stacking Patterns

The most researched form of racism in sport is stacking (see recent summaries of research and explanations in Eitzen & Sage, 1993: 333–338; Leonard, 1993: 211–224; Phillips, 1993: 179–190; Stebbins, 1993). Concentrating black players in a limited number of specific race-identified positions forces them to compete among themselves, rather than with members of other racial groups as well, for membership on the team and for playing time.

The seminal research on stacking was done by Loy and McElvogue (1970), who hypothesized that racial segregation in professional team sports in the United States was positively correlated with the centrality of positions. They found that blacks were overrepresented in baseball in the outfield positions and in football in the defensive (especially defensive back) and offensive backfield positions. There was a high concentration of whites in the catcher, pitcher, shortstop, second base, and third base positions in baseball and in the quarterback, center, offensive guard, and linebacker positions in football.

Other sports besides baseball, football, volleyball, and basketball have shown evidence of stacking. In British soccer, for example, blacks, especially West Indian blacks, have been stacked in the outside-support forward positions, and whites have been overrepresented in the central positions of goalkeeper, center-back, and center midfield (Maguire, 1988b; Melnick, 1988). Furthermore, Maguire (1988b) found that none of the ninety-two soccer teams he studied had a black team captain. Stacking in Australian rugby has placed whites disproportionately in central positions and Aborigines in wide positions (Hallinan, 1991). In Canadian professional football, the racial pattern in American stacking is reversed, with American blacks favored over white Canadian players (Stebbins, 1993). In the NHL, blacks are nearly invisible, and stacking relates to Canadian ethnic-linguistic differences. The French-speaking ethnic minority is both significantly underrepresented and stacked (Lavoie, 1989a).

Explanations and Implications of Stacking

Loy and McElvogue (1970) drew from the work of Grusky (1963) and Blalock (1962, 1967) in formulating a centrality explanation of stacking. Grusky assumed that more central locations in social networks involve more coordinative tasks and higher rates of interaction with others in the network. Blalock assumed that racial discrimination is less for people in positions that involve less purely social interaction, that depend less on skill in interpersonal relations, and that enable individuals to succeed mainly on their own. Loy and McElvogue concluded from the Grusky and Blaylock assumptions that black players would be stacked in non-central positions because they would engage in less leadership and coordination of potentially resentful white teammates and less dependent interaction and interaction in general with whites. Thus, stacking black or minority players in less central positions oc-

SPECIAL FOCUS

Stacking in Canadian Sports

In the Canadian Football League (CFL), teams have been required since the early 1980s to restrict the number of "imported"(foreign) players on the active roster of thirty-five to sixteen (Stebbins, 1993). (In the CFL, the game is played with twelve rather than eleven players.) Since 1970 imported players increasingly have been black Americans, and since the early 1980s relatively few whites have been imported. These imported players have tended to be stacked in the quarterback as well as defensive back, running back, and wide receiver positions. Canadians tend to be stacked in the positions of offensive lineman, kicker, and punter. Overall, most of the stacking of imported players has been on the offensive team, but there is some stacking on the defensive team as well. White Canadians mainly play the positions left after black American imports have filled the skill positions. Thus, in Canada, black American quarterbacks, for example, get an opportunity to play central positions that they find much less accessible in the National Football League; in the CFL, white Canadians tend to face the kind of prejudice black players face in the United States. The main dimension of stacking in the CFL may be national rather than racial, with coaches believing that Americans are better trained than blacks or whites who learn the game in Canada.

Stacking in the National Hockey League (NHL) reflects the powerful dynamics of ethnic relations between Francophone (French-speaking) and Anglophone (English-speaking) Canadians. Lavoie (1989a) found that the proportion of French Canadians in the NHL steadily decreased between the early 1970s and the 1980s, in part because of the influx of European and American college players. By the 1983-84 season, approximately 10.5 percent of Canadian players were Francophone, nearly 70 percent were Anglophone, and about 19.5 percent were European or American. At that time, approximately 25 percent of the Canadian population was Francophone. Thus, unlike some of the major American sports, in which racial minority-group members were significantly overrepresented, in the top sport in Canada, the ethnic minority of Francophones was significantly underrepresented.

In addition, Francophone players in the NHL were stacked. During the period between 1972 and 1984, Francophones were significantly overrepresented in the position of goalie. By 1983–84, they also were underrepresented among defensive players. Even in French-speaking Quebec, Francophone players experienced stacking. According to Lavoie, this was due the fact that all high-level hockey in Canada tended to be controlled by English Canadians.

Lavoie found an interesting pattern in comparing stacking with performance results. He discovered a negative correlation between the proportion of Francophones in a position and the performance advantage of Francophones over Anglophones at that position. That is, for the position of defenseman, where Francophones were underrepresented, their performance was substantially better than Anglophone performance levels. For the position of goalie, where Francophones were overrepresented, there was little difference in performance between them and Anglophones.

Lavoie interpreted these results as implying that Francophone defensemen faced barriers of discrimination, with only very talented Francophones making it to the NHL in this position, while Francophone goalies found it much easier to make it to the NHL. He proposed that Francophone hockey players in Canada were victims of prejudicial myths conveyed to club officials and coaches by Anglophone scouts, for example, that Francophone players were too small, not aggressive enough, too lazy, and too offensively oriented to play defense. As a result of such prejudiced beliefs, NHL teams were discouraged from drafting Francophones as defensemen and, to some extent, as forwards, unless they were outstanding, and these teams tended to be inclined to release marginal Francophone players.

curs because management and coaches fear that white or dominant-group players will not accept the authority, influence, or outcome control of minority teammates.

The Canadian football and ice hockey results urge caution in overgeneralizing the centrality explanation of stacking, since the Canadian cases suggest that stacking could result more from stereotyping players' abilities by coaches, club officials, and scouts than from the centrality of positions (Lavoie, 1989a; Stebbins, 1993). Nevertheless, racial segregation in a number of sports seems to be a social ecological or spatial distribution phenomenon, with minority group members overrepresented in peripheral positions and majority group members overrepresented in central positions. Of course, the stereotyping that appeared to operate in the CFL and NHL also helps explain why whites and nonwhites are assigned to central and noncentral positions, respectively, in a number of other sports.

The prohibitive cost hypothesis advanced by Medoff (1977, 1986, 1987) is an economic explanation of stacking in baseball. Medoff proposed that blacks and whites tend to play different positions because they differ in their ability to pay the costs of learning certain positions, such as pitcher, catcher, and shortstop. As the economic status of blacks has improved, more blacks have been able to afford the extra costs of training, instruction, equipment, and facilities needed to learn these positions. With less resources, blacks turn to "less costly" positions, such the outfield or first base. Thus, in Medoff's perspective, it is economics, rather than racial discrimination, that explains stacking.

Medoff's economic interpretation has stirred much criticism (e.g., Yetman, 1987; Lavoie, 1989b). Phillips (1993: 189–190) made three criticisms of this explanation: (1) Aspiring baseball players do not freely choose their positions, as Medoff assumed. Players move or are moved around to meet changing team needs at different positions. (2) Player development costs are typically subsidized by programs, leagues, or clubs, often beginning with Little League. Players only need a glove. (3) Hitting skills are more important than fielding skills in player development, and everyone has to learn to hit to play baseball. Thus, there are no development cost differences in the skill area that matters most. In this context, contended Phillips, trying to explain the differences in positions played by blacks and whites on the basis of different development costs seems irrelevant.

Another explanation of stacking focuses on racial differences in socialization or role modeling (McPherson, 1975). and assumes that aspiring young black athletes try to imitate their black sports heroes and learn the positions they play. In this way, initial stacking patterns are reproduced over generations by a kind of self-selection process. This explanation loses credibility when it is recognized that about half of the early major leaguers in baseball played central positions and that the proportion of blacks in these positions *declined* after these early years (Phillips, 1993: 188). Furthermore, many black athletes who play noncentral positions at the professional level played central positions earlier in their careers, as Eitzen and Sanford (1975) showed in football. The socialization explanation also does not take into account Phillips's point that athletes often do not actually choose their position. Team needs or coaches' biases may weigh much more heavily than the individual desires or even past experiences of athletes in position assignments.

Explanations of stacking based on race-linked physical or psychological differences have no more research support than similar arguments meant to explain black overrepresentation in certain sports or the black dominance of these sports (Coakley, 1990: 216–220; Leonard, 1993: 222–223; Phillips, 1993: ch. 7). On the other hand, stereotyping based on myths about these differences may play an important role in stacking and opportunities in general for blacks and other minorities in sport. Stebbins (1993) made a strong argument favoring stereotyping over centrality in stacking explanations, and the most general explanation may involve stereotyping.

The matching of players to positions on the basis of stereotypes of the players' abilities may not be intentionally racist, but its effects are racist. Coaches seem most likely to make position assignments on the basis of stereotypical racist myths when the racial composition of teams is skewed or tilted toward a particular race and when they believe there are clear differences in position requirements and in the capacity of members of different racial or ethnic groups to meet them. We have seen in the cases or the CFL and NHL that such stereotyping may operate in complex and curious ways. The five major types of explanations for stacking are summarized in Table 12.4.

The persistence of stacking patterns in professional sports has important implications for black athletes (Eitzen & Sage, 1993: 338; Leonard, 1993: 218–219). First, to the extent that black athletes are excluded from positions requiring thinking or leadership, stereotypes of their "natural inferiority" are reinforced. Second, black athletes are often stacked in positions that require abilities, such as speed, dexterity, and quickness, that decline faster than the abilities needed for other positions. As a result, stacking causes many black athletes to have shorter careers, to earn less money in their careers, and to qualify for less retirement benefits than white players in their sport (Best, 1987). Third, since players in central positions are more likely to become managers and coaches, stacking deprives blacks of a chance to get the perceived necessary background for such careers (Leonard, Ostrosky & Huchendorf, 1990; Eitzen & Sage, 1993: 340–341; Leonard, 1993: 218–219).

Other Forms of Residual Racism in American Sports

Although racial progress has occurred in U.S. sports over the past several decades, recent progress has been slow or stalled. Even though African-American athletes in a number of sports now make millions of dollars, we cannot be sure they are paid fairly. By fairness, we mean that black and white athletes earn equal pay for equal ability or performance. In the 1980s, black players were paid less than white players in professional baseball, basketball, and football (Leonard, 1993: 232–233). Since blacks have tended to outperform whites in these sports, lower black salaries could reflect salary discrimination. Increasingly sophisticated statistical analyses of racial and ethnic salary comparisons by position have indicated that there still may be some salary inequities for certain positions (Leonard, 1988a, 1989; Lavoie & Leonard, 1990), but no clear trends in salary discrimination have been uncovered for positions where inequities seemed to exist.

TABLE 12.4 Stacking Explanations

Type of Explanation	Major Assumptions
Centrality	Minority players are stacked in less central positions because they engage in less leadership and coordination of potentially resentful dominant-group players and because there is less intergroup interaction in such positions.
Stereotyping	Members of a particular racial or ethnic group are stacked in certain positions because coaches and others involved in personnel selection and training associate the perceived requirements of the positions with stereotypical characteristics of members of the racial or ethnic group.
Prohibitive cost	Blacks and whites play different positions because they differ in their ability to pay the costs of learning certain positions.
Socialization or role modeling	Aspiring young black athletes try to imitate their black sports heroes and learn the positions they play, and in this way initial racial differences in position preferences are reproduced over generations by self-selection.
Race-linked physical and psychological affinities	Blacks and whites inherit or have certain race-linked physical and psychological traits that make them more able to play or excel at some positions than others.

POINT-COUNTERPOINT

DOES THE RELATIVE ABSENCE OF BLACK QUARTERBACKS, PITCHERS, AND CATCHERS IN PROFESSIONAL SPORTS REFLECT RACISM?

YES

Since black athletes perform well at these positions at lower levels of sport, the relative absence of black quarterbacks, pitchers, and catchers in professional sports must reflect racial prejudice and discrimination. Coaches, team officials, and scouts are not intentionally racist when they stack black players in certain positions, but they are acting on the basis of racially biased stereotypes and myths when they deny black players access to more visible and central positions with more decision-making responsibility and control. If the stereotype of black athletes is that they are naturally gifted with various athletic abilities, it does not make rational sense that they are denied access to any position on the athletic field. If there is no difference between the abilities of black and white players and if there is no racism in position assignments, then blacks and whites should be randomly distributed across positions.

NO

Black and white athletes differ in their ability to play particular positions in a number of sports. This difference in ability may be due to race-linked physical or psychological traits or to racial differences in the desirability of different positions nurtured by culture and socialization. Coaches, team officials, and scouts are motivated to acquire and develop the best possible talent for every position in their sport. It would be disadvantageous for them to stack black and white players for racist reasons. There is too much scientific testing used today to allow prejudice to enter into decisions about the abilities, skills, and potential of athletes. Black athletes are more likely to play other positions than quarterback, pitcher, and catcher because they are better suited for these other positions, because they have voluntarily chosen to play these other positions, and because they want to play the positions of black heroes.

How should this debate be resolved?

In previous decades, black players faced a higher performance standard than their white counterparts to make the team in baseball, basketball, and football. Rosenblatt (1967) first demonstrated this phenomenon of **marginal discrimination** in Major League Baseball. He found that between 1953 and 1965, black major leaguers had a mean batting average 20 points higher than whites'. Yetman and Eitzen (1984) showed that black professional basketball players outscored whites between the 1961–62 and 1974–75 seasons. The difference steadily decreased during this period, however, and by the mid-1970s it was only 1.5 points. It appears that marginal discrimination more or less ended in Major League Baseball in 1988 (Phillips, 1993: 195). It also seems that by the mid-1980s only a minimal amount of marginal discrimination remained in college basketball (Berghorn, Yetman & Hanna, 1988). The tendency for blacks to outperform whites at certain positions in the NFL found over twenty years ago (Scully, 1973) probably would be difficult to identify today, with nearly 70 percent of NFL players being black.

NCAA Propositions 48 and 42, which were discussed in Chapter 8 on college sports, have been seen by many black civil rights leaders and higher-education officials as a form of racial discrimination. At the same time, other prominent figures in the black community have applauded these efforts for conveying the message that academic achievement must accompany the pursuit of athletic success. The joining of the issues of racism and academic standards for college athletes has created great controversy in black America as well as the NCAA.

NCAA research reported in 1993 showed that graduation rates rose 8 percent for black male athletes (41% versus 33%) and 10 percent for black

SPECIAL FOCUS

Apartheid Sport in South Africa

The character of the South African **apartheid** system is best described by the notion of a racial caste system. A comparable system existed in the southern United States during slavery. A racial caste system places people in a stratum in society on the basis of their race. It denies people in lower castes the opportunity to move up in society unless they are able to change others' perceptions of their race. People in lower castes generally are viewed as inherently inferior in intelligence, motivation, personality, morality, and a variety of other valued characteristics. Until recently in the South African caste system of apartheid, upper-caste whites saw themselves and were seen by others as having an immutable natural right to their superior status, and nonwhites were treated as though their inferior caste status was natural and unalterable (Ferrante, 1992).

Apartheid, which means "apartness" in Afrikaans, established separate and unequal public facilities for Europeans (whites) and the three nonwhite races of Coloreds (mixed-race), Indians, and Africans; mandated that nonwhites live in separate and inferior "homelands"; justified menial employment and wages for nonwhites; restricted the movement of nonwhites; limited their social contact with whites; banned nonwhite-led opposition groups; denied nonwhites the right to vote on national issues; and required or permitted other practices that all had one aim: to maintain separate white and nonwhite areas. Underlying this aim was the desire of whites to maintain their superior caste status, economic privilege, and political dominance (Ferrante, 1992: 272–275).

A curious fact about sport in South Africa is that interracial sports and teams were not officially banned. In fact, though, such legal prohibition was unnecessary, with so many other laws stipulating residential segregation and the segregation of public facilities and restricting interaction between whites and nonwhites. As a result of the multitude of apartheid laws governing nonwhite behavior in other realms of society, it was virtually impossible to have interracial teams or competition, nonwhite sports participation in white-only areas, racially mixed foreign teams or nonwhite individuals traveling and competing freely in South Africa, or the racial mixing of spectators at sports events (McPherson, Curtis & Loy, 1989: 107). The deepest division in South African sports has separated "establishment sport"—which has included the white-controlled network of clubs, national organizations, and broad federations associated with the national apartheid regime—from the antiestablishment, antiapartheid, nonracial sports movements led by nonwhites. Each sector has had its own facilities, events, champions, and history (Kidd, 1991a).

Under the strict segregation of the Nationalist Party's "petty apartheid," from 1948 until the late 1970s, this sports segregation meant greatly inferior opportunities, resources, and recognition for nonwhites in sport. "Grand apartheid," which allowed some liberalization of contact between races, gradually replaced petty apartheid in the late 1970s and the 1980s. Reflecting this theme of liberalization was an article in *Sports Illustrated* magazine that referred to "cracks in the racial barrier" in South African sport and offered examples of racial mixing of some sports teams (Hawthorne, 1976).

This ostensible liberalization did not really create integrated sports (Kidd, 1991a). In 1983, another *Sports Illustrated* article heralded further examples of integration of South African sports, but it also asserted that these examples created an "illusion of progress where little change exist[ed]" (Gammon, 1983: 78). Grand apartheid and the government's multiracial sports policy maintained a sports system in South Africa that was still very unequal.

The illusion of change quickly became reality in early 1990 when the newly elected South African president Frederik Wilhelm de Klerk announced in parliament that black opposition leader Nelson Mandela would be

(continued)

Special Focus *(continued)*

freed after being a political prisoner for twenty-seven years and that his militant African National Congress movement would be legalized. De Klerk also promised constitutional changes that were meant to create racial equality in the law and repeal of the webs of laws, regulations, and orders that maintained the structures of apartheid.

The changes de Klerk promised caused resistance, a jockeying for power, and disruptions from various sides, but the International Olympic Committee and other international sports bodies were sufficiently impressed by the progress and promise in South Africa to set in motion a process to lift the bans that had virtually isolated South Africa from the world sports community for many years (Johnson, 1991b). Buoyed by hope, nonracial sports leaders expressed cautious optimism but were aware that a society and government that had been so oppressive for so long were unlikely suddenly to be truly egalitarian and fair minded toward the nonwhite majority of the population.

It is far from reality today because the distribution of resources in South Africa has been so unequal for so long. For example, in 1982 it was reported that the 15 percent of South Africa that was white controlled 73 percent of all running tracks, 83 percent of the swimming pools, and 82 percent of the rugby fields (Human Sciences Research Council, 1982). Furthermore, Kidd (1991: 36) noted that in Natal, the 330,000 blacks in the townships of Umlazi and Lamontville had access to six soccer fields and two swimming pools, while the 212,000 whites who lived in nearby Durban had 146 soccer fields and fifteen swimming pools for their use. In addition, government spending on nonwhite sports was a fraction of the spending on white sports, and much less than what was needed was being spent on the upgrading of nonwhite sports facilities. In this context, equality in sport, like equality in the rest of South African society, appears to be a tiny specter on the horizon for a country just beginning to confront the dynamics of fundamental changes in race relations and stratification.

female athletes (54% versus 44%) after Proposition 48 was adopted in 1986 (Blum, 1993b). The report did not satisfy critics such as John Chaney, though, who said, "This report may be just another indication that opportunities are being taken away from youngsters, and many more black youngsters in particular. If you want to have a great graduation rate, just keep raising the standards" (quoted in Blum, 1993b: A42).

What disturbed Chaney and other opponents of the new NCAA academic eligibility standards was that approximately 600 fewer black athletes were enrolled in Division I institutions than in the three previous years (Observer Staff and News Services, 1993). These critics have complained that these measures are biased against students from disadvantaged backgrounds and poorer high schools and that they rely too heavily on the results of standardized SAT and ACT tests, which were thought to be culturally biased against blacks (Kroll, 1989). Even Harry Edwards, who initially supported Proposition 48 as a means to encourage more academic commitment from young black athletes (Edwards, 1986), opposed the more stringent standards, set to go into effect in 1995, as an "elitist, racist travesty" meant to keep poor blacks out of NCAA institutions (quoted in Kroll, 1989: 57).

When the status of the black athlete in the United States was re-examined in 1991 (e.g., Johnson, 1991a), twenty-three years after the groundbreaking series *The Black Athlete—A Shameful Story* (Olsen, 1968), the results produced a picture of racial progress in some areas and residual racism in others. The forms of residual racism that characterize American sports today are often subtle, difficult to substantiate, and complex, unlike the blatant racism that Jackie Robinson and his contemporaries faced nearly a half-century ago. The prejudice and discrimination faced by black female athletes, combining racism and sexism, is more like what black male athletes faced in the past than what they face today.

CONCLUSION: A FINAL ASSESSMENT OF RACE AND SPORT

We have concentrated in this chapter on race and sport in the United States, and especially, on the experiences of black males in American sport. We can see that sport has sometimes been at the edge of social change, but that it generally mirrors the entrenched patterns of racial inequalities, relations, and attitudes in the larger society. We have noted that blacks in some other countries, such as Canada, may not confront the same kinds of racism that blacks in the United States have faced and now face. Indeed, in Canadian professional football, there may a kind of racism that *favors* black imports over native white players. On the other hand, nonwhites in South Africa have faced an intensity and extent of racism that have had few parallels in other nations.

We could say much more about race, racism, and sport in the United States and in other nations. For example, if the research were available, we would have much more to say about the experiences of black female athletes and athletes who have minority status but are not black (Birrell, 1989). At this point, though, we know that black male athletes have become dominant in a few sports but are largely absent from most others. We know that few blacks can be found in positions of authority and almost none in positions of ownership in the sports in which they excel as athletes. We know that black female athletes in the United States have become somewhat more visible to the public in the past decade, but not much more visible. We know that America continues to honor the memory of Babe Ruth, but made Henry Aaron a "prisoner of memory," taunting and threatening him as he pursued the Babe's home run record (Capuzzo, 1992). We know that various forms of residual racism remain in sport and that race relations may be regressing in the larger society.

In this context, it may be most revealing to cite a comment by Arthur Ashe shortly before his death in 1993. Ashe said that having AIDS was difficult, but it was not nearly as difficult as being black (quoted in Deford, 1993b). Just as Jackie Robinson may have spoken for his generation of black athletes when he said that he "never had it made" (in the title of his autobiography), Arthur Ashe may have conveyed a sentiment felt by even the most honored and respected black athletes of the 1990s in the United States. That is, race still matters, and black superstars still can be made to feel like members of a minority group.

SUMMARY

Race relations and racial inequality in American sports generally reflect entrenched patterns of racial stratification in the larger society. Throughout sports history in the United States, black athletes have experienced racial prejudice and discrimination, and for long periods of this history blacks have been segregated from whites in nearly all sports. The biggest crack in the color bar in sports came with Jackie Robinson's entry into Major League Baseball in 1947. Since then, opportunities for black athletes in a number of highly commercialized sports have substantially improved. In fact, racial progress in U.S. sport over the past several decades has been made largely by African-American males, although some women have been successful in major sports at the intercollegiate and Olympic levels. African-American female athletes have had to contend with the double challenge of being female as well as black.

Along with stacking, quotas, the limited number of "black" sports, discrimination in hiring for coaching and management positions, restricted commercial endorsement opportunities, and subtle and overt forms of prejudice and racist stereotyping, blacks have faced even more subtle or complex forms of racism in regard to marginal discrimination and elevated entry standards such as the NCAA's Proposition 48 and 42. The most researched type of racist practice in American sports is stacking, and there are a variety of explanations for its prevalence. Stacking can also be found in other nations in a variety of interesting forms. Canadian professional football shows a reversal of the American pattern of racial stacking, and NHL hockey in Canada feature stacking along linguistic-ethnic lines. Consequences of all types of racism in sport and society are the denial of opportunities and rewards and relegation to inferior status. Apartheid sport in South Africa provides dramatic examples of the extent that racism can diminish opportunities, rewards, and status for minority groups in sport and society.

CHAPTER 

Gender Relations and Sport

SPORT, GENDER, AND SOCIAL CHANGE

Stories can reveal much about human attitudes, expectations, and behavior. This one, involving one of the authors, reveals what it was like for American women in sport a quarter-century ago. A male and a female college student decided soon after they met that it would be fun to play tennis. The young man had played for several years as an adolescent but had hardly played at all for the previous four or five years. The young woman had played competitively since childhood and was at the time of this story the captain of her college team. In attempting to borrow a racket from friends, the young man was asked why he needed it. After indicating that he was going to hit the ball with his female friend, he received some good natured ribbing about playing against a "girl." He reassured them that he would "take it easy" on her and that he was playing only to be sociable and get some exercise. When he began hitting the tennis ball back and forth, he was surprised to find the ball returning as hard as he hit it. After being challenged to an informal set by his female friend, he found himself drawn into an increasingly serious competition. Facing a barrage of consistently solid and accurate returns of service, passing shots, and well-placed ground strokes, the young man saw the score mount against him. He did not win a game in the first or second set. Realizing the futility of further effort that day, he suggested that they end their game and do something else. More embarrassed than chastened by his experience, the young man was purposely vague about the outcome of his tennis game when he returned to the company of his male friends. With the passage of time, embarrassment turned to pride in the athletic ability of this young woman, who became his wife.

What makes this story unusual is not the athletic ability of the female, but the fact that she was willing to demonstrate it in competition with a male. We see in this story the typical male arrogance of the era. Not having played tennis for several years, the young man still believed that he could step on the court and dominate a woman who had been playing competitively for most of her life. In an era when it was uncommon for women to compete seriously at anything, he was surprised to find a woman willing to compete seriously against him—and beat him, even in one of the relatively few sports women played then. Along with sports such as gymnastics and swimming, tennis was one of the relatively few "ladylike" sports opportunities available to females twenty-five years ago. Beating a male surely was not considered ladylike at that time, however.

For men and women of younger generations, this story may seem very distant from their own experiences. Today, we see males and females of various ages jogging and competing in marathons alongside one another. Girls play Little League baseball. Women participate in an increasing array of high-level amateur and Olympic sports. Women compete in bodybuilding, wrestling, horse racing, boxing, auto racing, and a wide assortment of other sports that call for traditionally masculine qualities of strength, risk taking, and aggression. We sometimes see women competing against men in a number of these "macho" sports. Gender relations and opportunities for females in sport have changed a great deal over the past twenty-five years. Yet in many areas of sport, as in the larger society, change has been snail-like or seemingly nonexistent. This chapter will examine traditions and change in gender, gender relations, and sport.

POINT-COUNTERPOINT

DO FEMALES BENEFIT FROM PLAYING TRADITIONALLY MALE SPORTS?

YES

Females who stay in traditionally feminine roles will remain relegated to the inferior status accorded those roles. Females unwilling or afraid to participate in traditionally male sports, from basketball to bodybuilding, limit their chance to pursue activities they enjoy, to develop physical talents of which they are unaware, and to establish respect from males for doing what males (and some females) thought they could not do. Participating in the stereotypical masculine realm of sport teaches women that they can participate, compete, and succeed in other men's worlds as well. It teaches men these lessons, too. It also challenges a socially constructed division of labor that arbitrarily denies or minimizes opportunities and rewards for females merely because they are female.

NO

Females who play traditionally male sports risk physical danger to their more fragile bodies. They also risk a loss of their femininity. Masculinity and femininity, which are core aspects of identity formation in childhood and adolescent development, become confused when females participate in roles that have defined their participants as men. Sport has traditionally been a male realm in which young boys have developed their manhood, and many of the more aggressive or violent sports have been especially functional in creating male identities. The participation of females in these male realms dilutes their socialization value and makes it harder for males and females to find activities that clarify what it means to them to be feminine or masculine. A clear segregation of male and female sport avoids these problems.

How should this debate be resolved?

GENDER STRATIFICATION OF SOCIETY

One of the enduring historical facts about social relations and social structures in society is **gender stratification** involving structured inequalities of economic privilege, prestige, and power that favor men over women (Macionis, 1993: 362–371). Giddens asserted on the inequality of power between men and women that "although there are considerable variations in the respective roles of women and men in different cultures, there is no known instance of a society in which women are more powerful than men" (1991: 209). He observed that male dominance—or **patriarchy**—was not a result of male physical or intellectual superiority, but of the traditional mothering role of women, who have been charged with primary responsibility for child rearing. In the early stages of industrialization, the division of labor between men and women was sharply defined; women stayed at home to care for the household and children, and husbands were employed in the factories as breadwinner. Women depended on men to provide for their own and their children's material needs.

The dependence of women on men as breadwinners has lessened to some extent in industrialized countries in the twentieth century as increasing numbers of women have entered the paid work force, especially since the peak of the baby boom in 1955 (Hess, Markson & Stein, 1993: 215). Substantial gender inequalities remain, however, with gender stratification continuing to imprint the major social institutions, cultural beliefs, and cultural practices of entire societies (Brinton, 1988). What is most compelling about gender stratification is that it is produced and sustained by a set of beliefs, practices, and structural arrangements based on the premise that women are and ought to be inferior to men.

Since sport emphasizes qualities of physical strength and size, males are seen as naturally better than females at sport (Klein, 1990: 178–179). Birrell and Cole (1990: 18) have argued that because sport celebrates winning on the basis of physical superiority, "it is a major site for the *naturalization* [emphasis added] of sex and gender differences." They add that "sport's logic continually reproduces men as naturally superior to women" (1990: 18). The celebration of physicality and the assumption of natural male superiority contribute to the historic structuring of sport in many societies as a principally male preserve.

SPORT AS A MALE PRESERVE

Those who speak of "naturalizing" sex and gender differences in and through sport imply that it is natural for boys to want to play sports and for girls to want to avoid them. Also implied is that participation in sports is an important part of the establishment of male identity. Conversely, females who are involved in sports risk questions about their femininity. If sport is inappropriate for younger females, then by extension it also is inappropriate for adult females to aspire to coaching and other leadership positions in sport (Theberge, 1993). Much research and writing in sport sociology has focused on these ideas. While the increasing involvement of females in sport over the past two decades has diminished much of their discomfort with sports roles, sport has continued to provide males with opportunities to define or "prove" their masculinity.

In contemporary American society, universal formal masculine initiation rituals do not exist. A male preserve of sport serves informally as a lingering bastion of male identity and privilege as it separates and stratifies males and females. One particular bastion of male identity and fraternal bonding in sport is the locker room.

In his study of the locker room talk of two prominent male intercollegiate sports teams, Curry (1991) found that the talk often centered on sex and aggression. In trying to affirm their masculine identity and status, the male athletes conversed in ways that viewed women as objects, reaffirmed traditional and stereotypical ideas about masculinity, reflected attitudes demeaning women, and in the extreme, fostered thinking about rape. Such talk seemed to be intended to allay some of the anxieties or insecurities about identity and status that were produced by competition. Thus, male locker room talk of this sort contributed to sexism by desensitizing the athletes to women's rights and supporting ideas of male dominance. **Sexism** generally involves beliefs and practices that relegate females to inferior status in relation to men, demean them, deprive them of opportunities, and underreward them for their accomplishments.

Curry cited the example of the harassment of reporter Lisa Olson in the locker room of the New England Patriots as evidence of sexist attitudes promoted by the male locker room culture (see Montville, 1991). New England players harassed her by walking nude around her and making sexual gestures as she conducted an interview. Thus, the pro football locker room spawned narrow and distorted expressions of manhood and gender relations tied to a "macho" subculture antagonistic to women invading traditional male preserves. When training for manhood in sport has emphasized **macho**—that is, exaggerated stereotypical male-dominant and aggressive—conceptions of masculinity (Klein, 1990), the foundation of masculine identity often has been negative attitudes toward females.

In a study of the boyhood sports memories of male Finnish university students, Laitinen concluded that sport was a "cultural vehicle through which young boys become carriers of masculinity and patriarchy in our time" (Laitinen & Tiihonen, 1990: 192). She uncovered memories that reinforced ideas that girls were weaker and worth less than boys. In the friendship networks that evolved from sports, females were seldom accepted as equals and typically were excluded.

In the United States, sport has emphasized qualities associated with success and stereotypical masculinity such as competitiveness, aggression, ambition, toughness, risk taking, and dominance. The association of success with masculinity poses obvious problems for females: living in an achievement-oriented and success-oriented society, women are discouraged from trying to achieve and be successful. Zoble (1972: 219) called this the "woman's dilemma" in American society. The parallel, and often unrecognized, man's dilemma is that he is expected to achieve and be successful.

The *success imperative* in sport, as elsewhere in society, can be especially problematic and frustrating for men because success often is narrowly defined as win-

SPECIAL FOCUS

Sport as a Masculine Initiation Rite

Exploring questions of masculinity and sports through interviews with thirty male former athletes, Messner (1987, 1992, 1993) has argued not only that sport helps males to define what it means to be masculine, but also that sport plays an important role in determining how society defines masculinity. Sport, like fraternities and the military, can be seen as a masculine initiation rite, substituting for the ancient cultural rituals that bring boys into manhood in preindustrial societies (Raphael, 1988). According to Messner (1992), the role of sport in male socialization is not a reflection of a natural or biological propensity for males to be involved in sports and for females not to be involved. He argues instead that the foundations of sport are social.

Sport is a socially constructed institution of society, which varies across cultures, societies, and historical eras. As such, it embodies the dominant cultural ideas and practices and structural arrangements of the society; in particular, it tends to reflect and reinforce the established patterns of power, status, and class in the society. When those patterns change, sport is likely to change. When sport changes, the social construction of gender differences may change, too.

The traditional role of sport in gender socialization has depended largely on its status as a **male preserve**. That is, sport defines masculinity most clearly when it is predominantly or exclusively a realm for males. Zane Grey once wrote, "All boys love baseball. If they don't, they're not real boys" (cited in Messner, 1992: 24). Of course, there *are* boys who do not like baseball—or any other sport. As Messner (1992, 1993) has pointed out, though, boys are judged according to their athletic ability, whether or not they like sports. Insofar as sport is thought to be natural for males, a lack of sports interest and ability will seem unnatural and could cast doubt on one's masculinity.

ning and because so few are consistent winners. For boys growing into manhood who learn not only that they must be athletes but also that they must be winners, sport can be a source of significant insecurities (Messner, 1987). Laitinen (Laitinen & Tiihonen, 1990) found that the most positive sports memories held by the male university students she studied were about winning and that the most negative were about losing and getting injured. Losing, of course, is a common experience in sport at all levels. The fact that it can produce negative memories implies that winning in boyhood sports is both unrealistically and excessively emphasized. When winning is tied to masculinity, it can have an enduring impact on the development of masculine identity and esteem. Unrealistic or unattainable notions of masculinity and success bred by boyhood sports contribute to the pervasive feelings of inadequacy and failure among adult males in the United States that have been reported in the social science literature (Messner, 1987).

Females may be saved from the insecurities of having to prove themselves again and again as winners when they are excluded from sport, but this exclusion also creates a costly woman's dilemma for females who would like to be serious athletes. When sport is structured to be an arena where only males are supposed to be active participants, females realize that they enter at the peril of risking their femininity and only after overcoming some very high hurdles of discrimination, prejudice, and exclusion. Until relatively recently in American sports, the appropriate female role in sports was the passive supporter or cheerleader. This conception of women in sport began to change in the 1970s.

Evidence published at the end of the 1970s showed that an overwhelming majority of male and female students at two colleges rejected without reservation the idea that sport should be restricted mostly or entirely to males (Nixon, Maresca & Silverman, 1979). These beliefs were expressed several years after the groundbreaking Title IX legislation was passed in 1972 in the

United States, forbidding sex discrimination in educational programs or activities receiving federal financial assistance, including physical education and athletic programs (East, 1978; *CQ Researcher,* 1992b). The idea of sport as a male preserve may have been resoundingly rejected by the respondents to this questionnaire, but it is clear that the male—or in many cases, sexist—cast of sport has not disappeared. Male students in the survey just cited were substantially more likely than females to have reservations about the kinds and number of new sports opportunities that ought to be available to females (Nixon, Maresca & Silverman, 1979). This chapter will show that sport in the United States has become an arena that males have increasingly had to share with females and in which females have learned to feel much more comfortable. We also will see, however, that increased sports opportunities for females over the past twenty-five years have not eliminated perceptions of sport as a male world or eradicated all vestiges of prejudice and discrimination against females in sport. The Lisa Olson case is a dramatic example of the strong feelings against women in sport that continue to exist in some sports subcultures.

SPORT, GENDER ASSESSMENTS, AND HOMOPHOBIA

The structuring of sport as a male or macho preserve implies that male *non*participants have a flawed or suspect masculinity and that female participants have a flawed or suspect femininity. Furthermore, in the macho world of sport, unathletic males and athletic females have been derisively referred to as homosexuals and lesbians, respectively. In fact, in traditional male socialization, boys learn that becoming a man is avoiding stereotypically feminine *and* homosexual behavior (Herek, 1987; Messner, 1992: 35–36). Male and female athletes may engage in very visible and sometimes exaggerated displays of their masculinity and femininity, respectively, in order to guard against potentially negative and disapproving assessments of their gender identity. Males may assert macho qualities emphasizing toughness, aggression, competitiveness, and physical dominance; females may present an attractive appearance, with cosmetics and stylish "feminine" clothing.

Former tennis star Chris Evert was very conscious of her femininity on the tennis court and even rejected the idea that she was an athlete until the latter part of her career. As a baseline player, she was not risking her femininity. The men played the serve and volley power game, while women stayed back and hit graceful and well-placed ground strokes from the baseline. Speaking about her approach to tennis in the first half of her career, Evert said:

> I never felt at all like an athlete. I was just *someone who played tennis matches.* I still thought of women athletes as freaks, and I used to hate myself, thinking I must not be a whole woman. The nail polish, the ruffles on my bloomers, the hair ribbons, not wearing socks—all of that was very important to me, to compensate. I would not be the stereotyped jock. (Deford, 1981: 72)

Chris Evert ultimately established herself as both an athlete and a woman during a period when conceptions of women athletes were in transition. She shared the tennis stage during her career with Billie Jean King, who admitted a lesbian affair, and Martina Navratilova, who publicly accepted a lesbian identity (Axthelm, 1981; Kirshenbaum, 1981a; Albom, 1994). Sponsors, however, have emphasized heterosexual-type glamor in promoting women's sports and athletes (Axthelm, 1981), and they have been influenced by a **stigma**, or label of disgrace, that has discredited and devalued homosexuals and set them apart from "normal" people both in sport and in the larger society. Nevertheless, revelations of lesbianism have not necessarily tarnished the image of popular women's sports, such as tennis and golf, or their stars. According to recent comments by Martina Navratilova (Associated Press, 1993), her willingness to be more outspoken about lesbian and gay rights was encouraged by her sense that her sexual orientation did not adversely affect her endorsement deals. She believed that many companies were not concerned about an athlete's sexual preference, but she also acknowledged that others did not want to be stigmatized by having a lesbian or gay spokesperson. According to one commentator (Albom, 1994), Martina's "dirty little secret" is not that she is a lesbian because she has admitted and talked about it for years. It is that she keeps paying for it with less than full public and media recognition of her greatness as an athlete.

A certain amount of tolerance of lesbians in women's sports may be attributed to a common but not systematically documented perception that homosexuality is relatively widespread in a number of women's sports (Kirshenbaum, 1981a). Palzkill (1990) claimed that the presence of a disproportionately high number of lesbians in high-level women's sports was well-known to insiders in these sports. She interviewed nineteen top West German women athletes in handball, volleyball, basketball, soccer, athletics, badminton, rowing, and swimming who lived lesbian lives. She concluded that for these women the choice of a lesbian life style was related to a conscious decision to escape the usual male-defined dichotomy of masculinity versus femininity, which degraded women. She pointed out, though, that by choosing a lesbian life to resolve their role conflict as women athletes, these women paid the price of a great deal of discrimination. Incidents of sexual harassment and overt discrimination against lesbians have been documented in a number of different sports, physical education, and community recreation settings (Lenskyj, 1991).

Homophobia is an irrational fear and intolerance of homosexuality (Griffin, 1993). Claims of pervasive lesbianism in girl's and women's sports may be a means of scaring off possible participants, or they may demean the accomplishments of female athletes in general because these athletes suffer the stigma of lesbianism associated with women's sports, whether or not they are lesbians (*CQ Researcher,* 1992b: 199). According to sport sociologist Mary Jo Kane, sexual orientation is less important than intimidation and control in claims or accusations of lesbianism in women's sports: "Women are terrified of that label, so to that extent, it works.... It keeps us all fighting among ourselves and blaming lesbians for the problems" (quoted in *CQ Researcher,* 1992b: 199).

In a homophobic environment, both heterosexuals and lesbians may be afraid of being labeled gay. Heterosexuals are likely to make overt displays of their femininity to assert their heterosexuality, and lesbians are likely to try to disguise or conceal their sexual orientation. Whether or not lesbianism is widespread in women's sports, the assumption that it is widespread colors the image of women's sports.

In male sports, homosexuality and homophobia seem to have different meanings and implications than in female sports. The common assumption is that there are relatively few gay men in sports because sport is a realm where "real men" are made and compete. This assumption makes homophobia especially intense in male sports. Despite intense homophobia, gay men have participated at the highest levels of sport, with three prominent examples being tennis great Bill Tilden, former professional football player David Kopay (Kopay & Young, 1977), and Olympic diver Greg Louganis (McCallum & O'Brien, 1994). In addition, a 1977 survey of male athletes at three colleges indicated that 40 percent of the eighty-two respondents said they had engaged in homosexual acts at least twice in the previous two years (cited in Kirshenbaum, 1981a). Messner (1992: 34) more recently observed that "there is growing evidence that many (mostly closeted) gay males are competing in organized sport at all levels."

Homophobia prevails in male sports because perceptions of homosexuality threaten the stereotypical images of masculinity linked to the macho world of sport (Pronger, 1990). Although heterosexual women athletes may feel threatened by the presence of lesbian athletes, conforming to stereotypical conceptions of gender has seemed to be less important in the culture of women's sports than in the culture of men's sports. After all, sport traditionally has gained cultural significance as a *male* preserve where *masculine* identity is molded and *men* dominate. The erosion of male domination of sports helps explain the persistence of homophobia, as men try to hang on to stereotypical masculine images and traditions associated with the ascendancy of males in sport. Homophobia discourages public self-revelations of homosexuality among male athletes and makes it difficult to estimate the actual prevalence of homosexuals in sport.

Both heterosexual and homosexual women and men may face accusatory or derogatory reactions when they participate in "gender-inappropriate" sports. Women in traditionally masculine sports and events requiring power and strength, such as bodybuilding (Levin, 1980; Leerhsen & Abramson, 1985), and men in traditionally feminine sports and events requiring finesse, grace, and aesthetic performances,

POINT-COUNTERPOINT

DOES THE PUBLIC CARE MORE ABOUT THE SEXUAL ORIENTATION OF FEMALE THAN MALE ATHLETES?

YES

We read a great deal about lesbian athletes and suspicions of lesbianism among female athletes. The public is suspicious of lesbianism among female athletes because it still views women as out of place in the "man's world" of sport. If women are successful athletes, especially in more masculine sports, their status and identity as women are made even more dubious. The public feels more comfortable with males on the field and women on the sidelines because this gender segregation of sport makes it much easier to know what a man is and what a woman is. Thus, women who "invade" the traditionally male preserve of sport provoke more questions about their sexual orientation than men do when they play at something that is seen as naturally masculine. Lesbian athletes provoke more public concern than gay athletes do because there are relatively more lesbians in sport and lesbians are more open about their sexuality.

NO

The public may not warmly embrace lesbian athletes, but it cares much less about lesbians in sport than it does about gay men in sport. Despite the increasing involvement of women in sport, sport remains fundamentally a masculine enterprise. Its basic values of competitiveness, achievement, success, ambition, toughness, aggressiveness, risk taking, and dominance are still tied to masculinity. Thus, when gay men play sports, they challenge the basic identity of the sport institution itself in nations such as the United States. Because male athletes know that the public cares deeply about their masculinity and the masculine nature of sports, gay athletes are very reluctant to express their sexual orientation in mainstream sports. Lesbian athletes are more open about their sexual orientation than their male counterparts because they know the public cares less about them as athletes and women.

How should this debate be resolved?

such as figure skating or dance (Messner, 1992: 35), are likely to face degrading questions and accusations about their sexual orientation. These questions and accusations derive their sting from conceptions of gender and sport that emphasize traditional stereotypical dichotomies of masculinity and femininity and traditional ideas about the macho qualities of sport.

This potential stigma may keep females and males from pursuing their athletic interests in "gender-inappropriate" sports and roles (Messner, 1992: 35). On the other hand, those who choose to pursue such sports may have clearly nonstereotypical ideas about masculinity or femininity. Such ideas may precede their involvement in these sports, or they may result from involvement. Duff and Hong (1984) found in their study of self-images of competitive women bodybuilders that the bodybuilders held images of femininity that combined elements such as beauty from traditional definitions with dimensions of muscularity and body symmetry. These women believed that having muscles and being physically fit, strong, and healthy added to their femininity and increased their attractiveness and sex appeal to men. They did not see themselves as masculine, and few saw themselves as feminists or androgynists. In general, they viewed bodybuilding as a way of enhancing their femininity.

The culture of the male locker room studied by Curry (1991) encouraged homophobic displays to demonstrate genuine masculinity. By degrading gay men and mimicking stereotypical gay male mannerisms and behavior, male athletes fend off possible questions about their own masculinity. Demonstrative antigay displays may be especially necessary in sport because competition often forces opponents to touch each other and because men undress in close proximity to one another in the locker room. Athletes wanting to remain bonded to teammates try to maintain the appearance of conventional heterosexual male identity. A study of Little Leaguers (Fine, 1987: ch. 5) showed that homophobic talk is used even by eleven and twelve year olds to assert or protect an emerging masculine identity. Among the Little Leaguers, this type of talk combined with athletic prowess and sexist talk of females as sexual objects to gain status among male peers and to draw a boundary around their athletic world as a male preserve.

In the face of stigmatization and homophobia in high school, college, professional, and amateur sports, many lesbian and gay athletes participate in alternative sports. Lesbian and gay leagues have been established in a number of sports in many large cities around the United States. In 1982, the first Gay Games, modeled after the Olympics and founded by a former Olympic decathlete, Tom Waddell, were held in San Francisco. The Games have been a success, attracting 7,000 athletes from thirty countries to the third edition in Vancouver, Canada in 1990. The fourth Games were held in New York City in 1994 (Griffin, 1993).

BARRIERS AND PROGRESS FOR FEMALES IN SPORT

Historically, the image and structure of sport as a male preserve have discouraged many women from seriously pursuing sport and devalued the performances of women who have tried to be serious athletes. The cultural message has been that females do not belong in sports, and this message has been reinforced by a number of structural hurdles that have denied females access to sports opportunities, resources, and rewards. The existence of these hurdles did not prevent females from achieving in sport, however, as we see in Table 13.1.

Cultural Myths and Realities of Female Sports Participation

In a landmark series on women and sport, *Sports Illustrated* writers Bil Gilbert and Nancy Williamson (1973) identified three myths that were commonly used to justify discrimination against females in sport:

1. Athletics are physically harmful for women; competition may masculinize their appearance and adversely affect their sexual behavior.
2. Women do not play sports well enough to deserve athletic equality (or opportunity).
3. Women are not really interested in sports

The first of these myths is physiological, the second relates to performance, and the third concerns attitudes and interest. All three types have been expressed in a variety of specific ways. For example, the *physiological myths* have been expressed in terms of alleged risks and dangers regarding childbearing, damage to the reproductive organs or breasts, susceptibility to injury, menstrual problems, and the development of unappealing large muscles. Sport sociologists (e.g., Sabo, 1988; Coakley, 1990: 189–190; Eitzen & Sage, 1993: 354–358) have cited evidence and arguments to refute each one of these kinds of statements. In addition, they have pointed out that while women are typically warned about the dangers facing them in sport, men rarely hear such warnings. Indeed, the opportunity to confront and conquer risks and dangers is what enables males to prove their manhood in sport. By implication, these myths suggest that females are not capable of handling the challenges of sport or worthy of the chance to face them.

Performance myths are based on the notion that a person must demonstrate certain levels of ability or skill to be able to participate in sports. If females have been deficient in ability or skill, this deficiency can be attributed at least partially to the lack of opportunity to participate and develop athletically. Thus, they have been caught in a dilemma. They cannot participate because they are not good enough, but they have been denied the chance to participate that could enable them to become good enough to show that they deserve opportunities to participate. A parallel dilemma exists regarding the demonstration of interest. How is it possible to show interest in a sport if there are no opportunities to participate in that sport?

An underlying premise of the performance argument is that males are inherently better athletes than females because they are bigger and stronger. As Coakley (1990: 190) observed, this may be a reason to segregate males and females in certain sports where

TABLE 13.1 Time Line of American Women's Sports History[1]

Years	Significant Women's Sports Events, Figures, and Accomplishments
1831	Catherine Beecher's *Course of Calisthenics for Young Ladies* published
1837	Mt. Holyoke Seminary opened, with physical exercise part of curriculum
1865	First college women's physical education program established at Vassar College
1870	Middie Morgan first woman sportswriter
1887	First national women's singles tennis championships
1896	First women's intercollegiate competition is basketball game between University of California, Berkeley and Stanford University
1900	First women Olympians compete in golf and tennis
1916	The Amateur Athletic Union (AAU) holds its first national championships for women (swimming); Women's International Bowling Congress (WIBC) organized
1922	Federation Sportive Feminine Internationale holds first Women's World Games because Olympic Games do not include women's track and field
1925	Gertrude Ederle is first female to swim English Channel, breaking existing record by more than 2 hours
1926	Charlotte Shummel sets record in 150-mile swim from Albany to New York City
1927	Delphina Cromwell wins the President's Cup Power Boat Race on the Potomac River
1928	Females allowed to compete in Olympic track and field for the first time
1932	Babe Didrickson sets world records in three track and field events in Summer Olympics, but high jump record disallowed due to improper technique
1934	Fourth and final Women's World Games held in London, England
1938	Helen Wills wins eighth Wimbledon singles title
1940	Kathryn Dewey leads four-man bobsled team to National Championship against all-male competition, causing men to pass rule preventing women from future competition
1942	All American Girls Baseball League organized by P. K. Wrigley
1949	Ladies' Professional Golf Association (LPGA) formed
1951	Babe Didrikson Zaharias voted "Woman Athlete of First Half Century" by Associated Press
1953	Maureen Connolly is first woman to win Grand Slam in tennis
1957	Althea Gibson is first black woman to win Wimbledon and Forest Hills tennis titles
1960	Sprinter Wilma Rudolph wins three Olympic gold medals
1964	Volleyball is first team sport for women at Olympics
1967	Katherine Switzer (disguised as a male) is first woman to run Boston Marathon
1969	Diane Crump is first woman jockey on major track; Denise Long is first woman drafted by NBA
1971	Five-person full-court play officially adopted in women's basketball; tennis player Billie Jean King first woman athlete to earn $100,000
1972	Title IX of Education Amendments Act passed by U.S. Congress

(continued)

Table 13.1 *(continued)*

Years	Significant Women's Sports Events, Figures, and Accomplishments
1973	Billie Jean King defeats Bobby Riggs in $100,000 "Battle of the Sexes"; equal prize money for men and women at U.S. Open tennis tournament; Association for Intercollegiate Athletics for Women (AIAW) awards academic scholarships for female college athletes
1974	The Women's Sports Foundation established to foster development of women's sports; Little League Baseball allows females to play
1976	Janet Guthrie first female driver at Indianapolis 500
1977	Shirley Muldowney named top drag car racer of the year; for second time Sheila Young wins two world championships in two sports (speed skating and cycling) in same year
1978	LeAnne Schreiber becomes first woman sports editor of *New York Times*
1979	Ann Meyers (UCLA) first woman to sign NBA contract (Indiana); women reporters allowed to conduct postgame interviews with men
1981	Kathy Whitworth first female golfer to win $1 million in prize money; women gain membership on International Olympic Committee for first time
1982	AIAW dies
1984	Supreme Court ruling in *Grove City v. Bell* interpreted to restrict application of Title IX to sports; Joan Benoit wins first women's Olympic marathon; Lynette Woodward first woman to sign contract with Harlem Globetrotters
1988	Congress overturns Grove City decision with bill having broad interpretation of Title IX; Sarah Fulcher completes longest continuous solo run of 11,134 miles (approximately a marathon a day) around perimeter of U.S. in fourteen months
1989	Victoria Brucker (San Pedro, CA) first girl to play in Little League World Series at first base and pitcher; Paula Newby Fraser wins Ironman Triathalon second year in row with record-breaking performance and also wins World Biathalon Championship in same year
1990	Susan Butcher wins 4th Alaskan 1,049-mile Iditarod Trail Sled Dog Race; Martina Navratilova wins record ninth singles title at Wimbledon
1991	Judith Sweet first female president of NCAA; U.S. team wins first World Cup Championship in soccer for women

[1]Compiled from Cohen (1993) and Leonard (1993).

size and strength are important, but it is not a reason to deny females the chance to participate in sports. He also noted that females have been denied the opportunity to participate in sports before puberty, when they have been bigger than many of their male peers.

It is not clear how good is good enough to warrant serious opportunities or attention in sport. Women have demonstrated that with experience they can dramatically improve their performance to very high levels in a wide range of sports. For example, women were first permitted to compete in an Olympic marathon in 1984, and in that year Joan Benoit's winning time of 2:24:52 was better than the winning men's time in all but one of the men's Olympic marathons through 1956 (*Sports Illustrated,* 1991: 506, 510).

The performance argument is flawed in a number of ways. Most fundamentally, it is flawed by the notion that each sport has a universally understood standard of performance that justifies entry into the sport. With females excluded from sports, it was easy to maintain the fiction that they were not good enough because they were unable to develop athletically and

Source: AP/Wide World Photos

FIGURE 13.1 Breaking barriers: Joan Benoit wins the first women's Olympic Marathon in 1984.

demonstrate their capabilities. Like the other myths that have kept females out of many sports, the performance myth has been exposed as myth as females have increased their levels of sports performance with increasing participation.

The huge surge in female sports participation in the 1970s in the United States (*CQ Researcher,* 1992b: 198), spurred by the passage of Title IX, directly contradicted the *myth that females lack interest in sports.* This increase leveled off in high school sports in the late 1970s; the approximately 1.9 million girls—compared to approximately 3.5 million boys—who participated in 1990–91 is about the same number of females who participated during the peak years over a decade earlier.

A similar spurt in female sports participation occurred at the college level during the 1970s. Before Title IX, there were no women's athletic scholarships and virtually no intercollegiate athletic championships for women. By the mid-1980s, over 10,000 women held athletic scholarships, and women athletes competed for over thirty National Collegiate Athletic Association (NCAA) championships. From the early 1970s before Title IX to eight years after the passage of Title IX in 1980, the percentage of intercollegiate athletes who were female increased from 15 percent to 30 percent and the percentage of intramural athletes in colleges and universities who were female in 1980 was 44 percent (Leonard, 1993: 268–269). By 1990, over 150,000 women were in intercollegiate athletics (Carpenter & Acosta, 1991).

Limited enforcement of Title IX has allowed discrimination against females in high school and college sports to continue (Carpenter & Acosta, 1991; *CQ Researcher,* 1992b; Eitzen & Sage, 1993: 365–366). Given the chance, females have demonstrated that a substantial number of them are interested in sports participation, that they are capable of becoming quite accomplished as athletes in a variety of sports, and that their risks are no greater—and may be much less—than the risks males face in sport. The perpetuation of physiological, performance, and attitude and interest myths takes the chance away from females to participate and excel in sports.

Mass Media Images of Gender in Sport: From Invisibility to Ambivalence

The mass media of television, radio, newspapers, and magazines convey images of sport and athletes that may reflect, reinforce, or shape popular stereotypes in society. The virtual absence of images of female athletes in the major media has conveyed a very powerful message about the place of females in sport. The message is that women do not belong in sport. As female participation in sport has increased over the past two decades, we have seen somewhat more mass media coverage of female sports and athletes. This coverage, however, has been substantially less than the coverage of male sports and athletes, with one 1990 study showing that women received 5 percent of the television air time devoted to sports (Wilson, 1990: 2–3; Messner, 1992: 164–165). It also has reflected traditional stereotypes of females (Eitzen and Sage, 1993: 352).

In studies of the treatment of women and sports by magazines, a consistent pattern of traditional stereotypes of women and devalued female athletic accomplishments has been found (e.g., Poe, 1976; Boutilier

& San Giovanni, 1983; Hilliard, 1984; Rintala & Birrell, 1984; Kane, 1988, 1989; Duquin, 1989). In a recent study, Lumpkin and Williams (1991) examined *Sports Illustrated* feature articles between 1954 and 1987 and found that male authors often paid as much attention to the body dimensions and physical attractiveness of female athletes as to their sports accomplishments. They also found that nearly 91 percent of the articles focused on males, and articles about males were longer on average than articles about females. They concluded that the ways female athletes were described, the smaller number and shorter length of articles about females, and a predominant focus on females in the "gender-appropriate" sports of tennis, golf, and swimming reinforced traditional attitudes about females in sport.

A study by Leath and Lumpkin (1992) analyzed the treatment of sportswomen on the covers and in the feature stories of another magazine, *Women's Sports and Fitness*, between 1975 and 1989. This study revealed an interesting shift in the coverage of women, suggested by the change of the magazine's original title, *WomenSport*, in 1984. *WomenSport* initially focused on the sports achievements of outstanding women athletes. Then the magazine discovered that it could expand its target market by appealing to the growing number of women interested in physical fitness activities. As a result, nonathletes began to appear on covers, taking the place of star female athletes. The nonathletes typically were posed like fashion models rather than in action shots that displayed their athletic prowess. The effect of the shift in focus was to deemphasize the image of women as active athletes and to emphasize traditional stereotypical concerns among women about physical attractiveness.

Duncan (1990) analyzed 1984 and 1988 Olympic Games photographs appearing in popular North American magazines. She concluded that the content and context of these sports photographs conveyed an ideological message of sexual difference based on traditional stereotypes of masculinity and femininity. The focus on female athletes often emphasized physical attributes, such as hairdo, makeup, fingernails, poses, positions of the body, facial expressions, and emotional displays that reflected traditional femininity rather than athleticism. Thus, the photographs tended to reinforce textual messages in captions and stories that females were not to be taken as seriously as men in athletic roles. Their virtue was to be found in glamor and sex appeal and more passive roles. The many photographs of figure skater Katarina Witt and sprinter Florence Griffith Joyner, for example, typically emphasized their beauty and glamour. Their poses sometimes suggested soft-core pornography or pinup photography. Facial expression, body position, and emotional displays of female athletes often expressed availability, passivity, and vulnerability. Duncan believed that although the photographs legitimize patriarchal relations, more recent ones of athletes such as Florence Griffith Joyner convey the cultural message that women can be both beautiful and physically powerful.

Studies of newspaper coverage of the 1984 and 1988 Summer Olympics by the *Globe and Mail* in Canada and *The New York Times* (Lee, 1992) and of television coverage of men and women in basketball, surfing, and marathon running (Duncan & Hasbrook (1988) have revealed ambivalence toward women athletes. In Duncan and Hasbrook's (1988) study of television coverage, for example, there was a substantial amount of ambivalence in reports of women in these sports that was not present in reports of their male counterparts. Reports about the women combined positive messages with subtle negative suggestions that demeaned or trivialized the women's efforts. Along with comments about the women's strength or skill, other comments also implied that the women were to some extent weak, inferior, or otherwise unsuited for their sport or that their sport—basketball, for example—was a poor imitation of the men's game.

A study of television coverage of men's and women's college basketball showed that commentators used the men's game as the standard for viewing the women's game (Blinde, Greendorfer & Shanker, 1991). In addition, commentators generally spoke with inappropriate, sexist, or nonparallel language about the women's game. For example, they used terms such as "man-to-man defense," and they referred to the women players as "girls" but did not call the men players as "boys." They were also more likely to emphasize the athleticism and physicality of the men than women players.

SPECIAL FOCUS

Biases in the Newspaper Sports Department

Theberge and Cronk (1986) argued that the limited coverage of women in the sports section of newspapers is not necessarily a result of biases against women among sports writers. They proposed that it is affected by journalists' beliefs about the content of the news and their methods of gathering it. Their study of a U.S. newspaper revealed how certain aspects of the newspaper production process tended to keep women out of the sports news.

In trying to identify what is newsworthy, journalists look for subjects that are seen as newsworthy and capable of offering reliable and accessible information for reporting. In sport, men are viewed as playing sports that are followed more seriously and enthusiastically by the public, and news materials about these sports are more reliable and accessible than materials about less publicized and popular sports. The standard sources of sports news, such as wire services, media relations personnel of sports organizations, and press releases, provide reporters with a steady stream of information about the most commercialized sports, which are men's sports. Newspapers often rely heavily on these sources and, as a result, focus on the same relatively small set of highly commercialized and publicized men's sports.

With these biases in the normal work routines in newspaper sports departments, it is difficult for women—and men in less commercialized sports—to attract media coverage. This bias creates a self-perpetuating process wherein the most popular men's sports dominate the sports pages and women have to sneak into the sports news. Covering women's sports—and minor men's sports—requires more work for reporters and carries the risk of making readers disappointed or angry because space has been denied to the sports that they are more accustomed to seeing on the sports pages.

Evidence of sexist coverage of women in sport by writers, photographers, announcers, editors, publishers, and producers does not mean they are intentionally prejudiced against female sports participants. For example, both women and men may be treated as sex objects to generate more interest in a sport. Lumpkin and Williams (1991) identified an emphasis on the sex appeal of male athletes in swimming, soccer, and cycling in articles in *Sports Illustrated* over the past several years, which suggests that the rationale may be more commercialism than sexism. Thus, the coverage of females in sport may simply reflect the predominant media focus on the most commercialized professional and intercollegiate sports, which happen to be male. Of course, by covering women less, the media make it harder for women's sports to become more commercially successful.

Since bias against women in media sports coverage may result more from established structures of sport and reporters' work routines than from the conscious prejudices of individual reporters, hiring more women sports reporters does not produce more coverage of women in sport (Boutilier & San Giovanni, 1983; Theberge & Cronk, 1986). When women become sports reporters, they typically find that coverage of women's sports is not a high priority or a means of making a reputation in their profession. They also may find, like the much-publicized Lisa Olson (Montville, 1991), that male athletes, coaches, and management officials are cooler or more hostile toward them than they are toward their male counterparts. The adversarial relationship that exists between serious reporters and sports figures is exacerbated for women sports reporters because they are women. In fact, women often do not get a chance to get into the men's—or women's—locker rooms because they have faced barriers in trying to enter sports reporting, especially in major media markets (*CQ Researcher,* 1992c: 200).

Structural and Psychosocial Barriers and Changes

Giving access to the men's locker room to male but not female reporters puts women reporters at a significant disadvantage. Assigning women only to women's sports also disadvantages them because the limited amount of coverage of women's sports gives the reporters covering them less visibility. When Eberhard and Myers (1988) studied ninety women who worked in the sports departments of large American newspapers, they found that these women faced discrimination from male peers in their sports departments and from newspaper people outside sports. The main forms of discrimination the women sports reporters experienced in relations with coaches, managers, and athletes were denial of access to the locker room, sexual harassment, condescending treatment, and physical threats. They discovered, in general, that women sportswriters still lacked full acceptance in their profession.

Even though they face sexism, women in sports reporting have cracked a significant occupational barrier. Until recent years, sports journalism was an almost exclusively male (and white) occupation. A landmark event for women in sports journalism was the appointment of LeAnne Schreiber in November 1978 as the first woman to serve as sports editor of *The New York Times*. This appointment was preceded by a long struggle by women to reverse policies excluding them from stadium press boxes, locker rooms, and other sports facilities where sports news was gathered or made (Eitzen & Sage, 1993: 294–295). Although women have made progress by entering sports journalism, there were fewer than fifty women sportscasters at the more than 600 network affiliate stations in the United States, and even fewer in Canada in 1991. In addition, the top-paid women in these jobs earned less than 20 percent of the salaries of the best paid male sportscasters (Jenkins, 1991).

Women in Sports Leadership Positions

Within sports organizations, women have long faced substantial obstacles to securing leadership positions. Women have been very sparsely represented at the

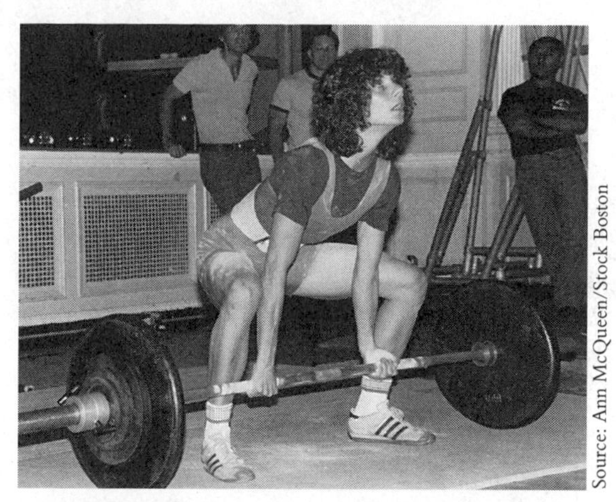

Source: Ann McQueen/Stock Boston

FIGURE 13.2 Challenging the male sports preserve.

highest levels of school, college, professional, and amateur sports organizations. For example, in the mid-1980s, women's professional tennis and golf, two of the most commercially successful women's sports, were run by men. In 1989, the eighty-six-member International Olympic Committee included six women. Before 1981, though, there were no female IOC members (Coakley, 1990: 186–187).

Women seem to have better chances for sports leadership positions in amateur sports below the international and national levels of sports administration. Research on provincial sports organizations in Canada showed that women held approximately 30 percent of the voluntary board positions in these organizations and did not seem to face higher standards than men in gaining access to these positions (Theberge, 1984; cited in McPherson, Curtis & Loy, 1989: 230). One sport, women's field hockey, was run entirely by women. The main difference between men and women board members seemed to be social status, with men more likely to have full-time jobs, to be in higher-status jobs, and to have more formal education.

It is ironic that after the passage of Title IX in the United States in 1972, a *decline* in the number and proportion of women in top athletic administration at many schools and colleges accompanied the dramatic *increases* in female participation in high school and college athletics. In addition, there was a decline in the number and proportion of women coaching many women's sports in schools and colleges after Title IX. Acosta and Carpenter (1992) reported that in 1972, over 90 percent of women's intercollegiate athletic programs were directed by women. By 1984, more than two years after the NCAA had taken over control of women's sports from the Association of Intercollegiate Athletics for Women (AIAW), only 17 percent of women's programs were run by women. The percentage of female head athletic directors of women's intercollegiate programs has changed little since 1984. In 1992, the figure was 16.8 percent. Less than one woman per college was in the administration of women's programs, even though over 75 percent of college athletic programs have more than one administrator. Female administrators were most likely to be at the NCAA Division I level, with an average of 1.10 women per 4.08 administrators. These women tend to have limited authority over policy, however.

According to Acosta and Carpenter (1992), the percentage of women coaching women's intercollegiate teams dropped from 90–100 percent in 1972 to 58.2 percent in 1977–78 and to a low of 47.3 percent in 1990. The percentage was up by one point in 1992 to 48.3. Also in 1992, at institutions with a male athletic director, 46.4 percent of the coaches of women's teams were female, while at institutions with a female athletic director 50.8 percent of the coaches of women's teams were female. Table 13.2 offers a graphic illustration of the changes from 1977 to 1984 to 1992 in the percentage of female coaches of the ten most popular women's sports. An increase in the percentage of female coaches occurred in only two cases, cross country and golf, between 1984 and 1992. The increase was more than 1 percent only in golf (+6%).

Women coaches generally exercise less power than men in mobilizing resources such as supplies, support, and information. In a study of 947 Division I intercollegiate coaches, Knoppers, Meyer, Ewing, and Forrest (1990) found that women coaching nonrevenue sports had the least access to crucial resources, whereas men coaching revenue sports had the most access and power. The high concentration of women coaches in nonrevenue women's sports means that they are generally less powerful than men coaches in college athletics. The biggest revenue producers are men's sports, and women do not coach these sports. Men, on the other hand, often coach women's revenue-producing sports. Since men are much more likely than women to be athletic directors and since athletic directors are most supportive of, and accessible to, coaches of the major revenue sports, we can see how both gender and type of sport conspire against women to limit their power in college athletics.

TABLE 13.2 Percentage of Women Coaches for the Ten Most Popular Women's Intercollegiate Sports, 1977, 1984, 1992[1]

SPORT	1977	1984	1992	Percentage (%) Change 1977–1992
Basketball	79.4	64.9	63.5	−15.9
Volleyball	86.6	75.5	68.7	−17.9
Tennis	72.9	59.7	48.0	−24.9
Cross country	35.2	19.7	20.1	−15.1
Softball	83.5	68.6	63.7	−19.8
Track	52.3	26.8	20.4	−31.9
Swim/dive	53.6	33.2	28.2	−25.4
Soccer	29.4	26.8	25.8	−3.6
Field hockey	99.1	98.2	97.0	−2.1
Golf	54.6	39.7	45.7	−8.9

[1]Based on Acosta and Carpenter (1992).

Although women coaches suffer because they are concentrated in nonrevenue sports, they also have less power and receive less compensation when they coach revenue sports. In one recent sex discrimination case, a court awarded $1.11 million in damages to the women's basketball coach at Howard University (Blum, 1993c). The coach had sued Howard because she believed that it had unfairly passed over her in hiring an athletic director. She also claimed that she had been paid less than her male counterparts and that she had received differential treatment because she was a woman. This court decision was followed by other gender discrimination lawsuits alleging violations of both the federal Equal Pay Act and Title IX (Blum, 1993c).

According to Knoppers, Meyer, Ewing, and Forrest (1991), NCAA Division I women coaches face discrimination based on both capitalistic biases concerning the emphasis on revenue generation and gender biases based on patriarchal ideas and structures. Women coaches experience constraints on their opportunities for access to better positions, for better pay, and for feedback from superiors. Thus, women coaches tend to have lower aspirations and less career satisfaction than their male counterparts, and they are more likely to leave coaching.

Even when women increase their numbers among the coaches of a woman's sport, they still face gender barriers and segregation. Women interact more frequently with female counterparts in women's sports as the percentage of women coaches in the sport increases. Men interact more frequently with other male coaches than women do, whether the ratio of female-to-male coaches in a women's sport is high, medium, or low (Knoppers, Meyer, Ewing & Forrest (1993). Thus, men always seem to seek out other male coaches in women's sports, but women seem to seek out female counterparts in such sports on a more frequent basis as their number increases. Women coaches believe it is necessary to interact with men until their number has become sufficient to permit the formation of their own social networks and subculture. The formation of gender-based networks and subcultures by women could offset male domination of their sport. The power of male coaches in a women's sport is enhanced by their tendency to form tight networks and strong coalitions with other men in their sport, even when their number is relatively small.

National statistics for trends in interscholastic sports are not as readily available as they have been for college athletics. One study of nine states across the United States demonstrated a clear trend, though. It showed that in 1971–72, 82 percent of the coaches of female high school sports teams were female and that in 1984–85, the percentage had dropped to 38 percent (True, 1983, as cited in Hasbrook, 1988). According to Stangl and Kane (1991), women had the best chance of becoming coaches under female athletic directors, especially at the interscholastic level. There were, however, far fewer female than male athletic directors in high school athletics. For example, Stangl and Kane (1991) found that in 1988–89, only 5.2 percent of all athletic directors in their sample of Ohio public schools were women.

Theberge (1993) interviewed forty-nine women who coached at the highest levels of thirteen Canadian sports. All worked mainly with females, but nearly two-thirds had coached males. These women were a tiny minority at their level of sport. In 1988, only 14 percent of the head and assistant coaches of Canadian national teams were women (*Sport Canada,* 1989). Women coaches were concentrated in the relatively few sports that historically have been organized separately for women and that have provided leadership opportunities for women. In Canada, primary examples of these sports have been synchronized swimming, field hockey, and figure skating.

In Kanter's (1977b) terms, less than 15 percent representation of a particular social group in an organization is considered skewed, giving the minority group members high visibility but token status. Theberge (1993) found that the women coaches with this kind of minority status had a clear sense that they were token representatives of their gender in their sport and had no trouble supplying examples of their isolation. These women were also very visible and confronted substantial pressure to demonstrate their competence. They responded to this pressure either by trying to fit in and become "one of the boys" or by trying to prove that they were good enough to "play with the boys." Women's respect and acceptance were limited, however, by ideological assumptions linking size and strength to coaching ability and by a reluctance of both men and women in integrated sports to be coached by a woman.

Residual Sexism in Sport

Female coaches and athletes today experience less role conflict than in the past (Eitzen & Sage, 1993: 358–359). Women in sport now often reject public attitudes about the incompatibility of femininity and sport (Del Rey, 1978), but cultural and social disapproval of their participation in a number of sports persists (Kane, 1987). Furthermore, despite progress in the United States, Canada, and many other countries, discrimination, prejudice, and minority status continue to mark the experience of females in sport (Riordan, 1985; Sfeir, 1985).

Public attitudes, media depictions and descriptions, and the structuring of opportunities often make females feel unwelcome, inferior, or uncomfortable in many sports roles and realms. A summary of the persisting forms of discrimination and prejudice includes lower budgets; fewer professional opportunities and less prize money in the professional sports they play; fewer and less lucrative commercial endorsements; inferior facilities and equipment; less access to facilities; shorter schedules; fewer athletic scholarships; fewer leadership opportunities in coaching and management and on governing boards; required sex-identity tests in international competition; less media coverage; less public and media respect; outright parental discouragement, especially in blue-collar families, or pressure during socialization to restrict participation to certain culturally and socially approved sports (McPherson, Curtis & Loy, 1989: 231–232).

TITLE IX AND THE GENDER EQUITY DEBATE

The continuing debate about gender (or sex) equity, which concerns the achievement of equal or equivalent opportunities and resources for women and men in athletics, captures the tensions between progress and regression, between change and tradition, for females in sport (*NCAA News,* 1993). This debate, which tends to be focused on the interpretation and implementation of Title IX for college athletics (e.g., Lederman, 1992a, 1992b, 1992c; Lopiano, 1992, 1993; Wolff, 1992), it has swung back and forth between the U.S. Congress and the courts. Congress passed Title IX of the Education Amendments in 1972, making it illegal to engage in gender discrimination in any educational institution receiving federal funds. In 1974, the Tower Amendment, which would have protected revenue sports from Title IX, died in a House-Senate conference committee. Title IX went into effect in July 1975. In 1984, the U.S. Supreme Court decided in the case of *Grove City College v. Bell* that Title IX applied only to programs that directly received federal aid. This ruling substantially narrowed the scope and departed from the original intent of Title IX. Congress passed the Civil Rights Restoration Act in 1988—over President Reagan's veto—to reverse the *Grove City* decision and outlaw gender discrimination throughout educational institutions that received federal funds. In 1992, the U.S. Supreme Court unanimously ruled that students could sue for monetary damages for sexual harassment and other forms of gender discrimination at schools and colleges. This ruling was viewed as an important boost for Title IX (*CQ Researcher,* 1992b: 203).

In the midst of the seesawing between Congress and the Supreme Court over Title IX have been a number of court battles over assorted allegations of discrimination against women at a number of institutions of higher education across the country. Women have fought underfunding, general preferential treatment of men's athletic teams, and efforts to cut athletic budgets by dropping women's sports. Men, too, have turned to federal gender-discrimination laws—as well as to the equal-protection clause of the U.S. Constitution—to charge that the dropping of their sports programs to make room for the women is a form of "reverse discrimination" (Blum, 1993d).

Recent court decisions have indicated that the *concept of proportionality* will be the standard for assessing compliance with Title IX (Blum, 1993d). Using this standard means that the number of men and women who participate in athletic programs at a particular institution should match the ratio of men and women in the student body of that institution. Critics have complained that the proportionality principle sets up gender quotas, but the basic issue for many observers of

SPECIAL FOCUS

Sexism in Names of College Sports Teams

One prominent symbolic indication of the persistence of sexist practices in sport is the naming of collegiate athletic teams. Eitzen and Zinn (1989, 1993) studied college and university nicknames, logos, and mascots for athletic teams. One study (Eitzen & Zinn, 1989) found that 451 (38%) of the 1,185 American institutions of higher education studied had sexist names for their athletic teams. When logos and mascots were examined in addition to names, over half (54.6%) of these colleges and universities had sexist elements in their symbolic representation of their women's sports teams. In the majority (55%) of cases of sexist names, the same male name, such as Cowboys, Knights, Rams, Gamecocks, Stags, and Steers, was used for male and female teams. In nearly one-third (32.2%) of the cases, "Lady" was used as a modifier, as in Lady Jets or Lady Eagles. In some cases, there was a double sexist gender marking, as in Lady Rams.While in others, feminine suffixes were added to the names of the male teams, as in Tigerettes.

According to Eitzen and Zinn, these naming practices and general symbolic representations of women's athletic teams demeaned women and promoted male supremacy and female subordination. In a case study of one university that used the name "Rams" (which refers to male sheep) for their male and female teams, Eitzen and Zinn (1993) found that there was substantial resistance to changing the name. The women's teams had previously been known as the "Lady Rams," but in the late 1980s, the university discontinued using the modifier "Lady." An effort on campus to give the women's teams a different and nonsexist name met substantial resistance from faculty, students, coaches of the women's teams, local journalists, the local community, and even the female athletes. These groups tended to reject the name change because they thought the issue was trivial or a nonissue or because they did not see the name as sexist. Eitzen and Zinn argued, however, that treating this matter as trivial failed to account for the powerful influence of symbols as reinforcers or conveyors of cultural and institutional patterns. The pattern they saw reinforced and represented by the naming practices at nearly half of the colleges and universities they studied was a sexism that made women seem inferior or subordinate to men.

sports law is underrepresentation of women or men in athletics. The problem colleges and universities face in trying to address gender equity is that efforts to balance athletic opportunities for women and men are being affected by general pressures and pressure within athletics to cut costs. Coaches of the major men's revenue sports, such as football and basketball, have protested the threatened cuts in their budgets.

Donna Lopiano (1992), executive director of the Women's Sports Foundation and former director of women's athletics at the University of Texas at Austin, has argued that recent pressure from women coaches and athletes to achieve gender equity has brought public attention to an issue that educational institutions have largely ignored. Failure by these institutions to implement the Title IX mandate to equalize athletic opportunities for male and female students left in place gross disparities in opportunity and funding between men's and women's programs (also see Lederman, 1992b, 1992c). Lopiano pointed out that Title IX did not require equal spending on men's and women's sports—because sports differ in their costs, but that it did require equal opportunity. The proportionality principle is an interpretation of equal opportunity. In her view, the key to achieving such equality is the restriction of exorbitant spending in big-time men's athletic programs. With the battle lines drawn, it is evident that the struggle over gender equity is likely to have serious implications for basic structure of athletics in colleges, universities, and schools in the United States.

If the courts continue to rule favorably for female complainants in antidiscrimination suits (Lopiano, 1992) and the financial squeeze continues in athletics

POINT-COUNTERPOINT

CAN COLLEGE ATHLETICS SURVIVE GENDER EQUITY?

YES

The aim of the gender equity movement is to create fairness in college athletics for women and men, not to destroy college athletics. If college athletics were destroyed, both women and men would suffer. Even with Title IX, women in college athletics experience underfunding, preferential treatment of men's athletic teams, and budget-cutting measures eliminating or curtailing a number of their programs. Women have a right to the same sports opportunities as men. Balancing the funding and opportunities for men and women will put financial pressure on excessively costly big-time men's athletic programs. Reducing the scale and aspirations of these programs will not destroy college athletics. It will result in athletic programs that are less commercially driven and more consistent with the educational values of higher education. By cutting or eliminating costly programs such as football on many campuses, a variety of more modestly funded programs for both women and men could be supported. The only losers in the implementation of genuine gender equity will be the most expensive programs.

NO

Gender equity will be very expensive, and college athletics is under great financial pressure now. Radical activists want separate and equally funded men's and women's athletic programs, and others want to give females more opportunities to administer and coach and equal pay, funding, and opportunities to participate. Taking resources away from men's athletic programs, especially the most commercialized ones in football and basketball, to create more resources and opportunities for women will bankrupt entire athletic programs. These programs depend on the big-time men's programs for media exposure, popular support,and revenue. Women's programs are incapable of generating the revenue produced by the big-time men's programs. Where will athletic directors get the money to fund the changes sought by equity advocates? Unless women want to give up their chance to participate in big-time programs of their own, women must modify their gender equity demands. Athletic programs with full gender equity will be on a very precarious financial footing, which colleges cannot afford.

How should this debate be resolved?

and education in general, providing more and more equivalent opportunities for females will force a restructuring of athletic programs. In particular, these forces will cause the most costly programs, which are men's programs in nearly all cases, to reduce their scale or find new funding sources. The most deficit-ridden of these programs may have to drop out of competition entirely.

CROSS-CULTURAL PATTERNS OF WOMEN IN SPORT

Stereotypical North American conceptions of femininity and women's roles have had a chilling effect on female participation in sport and other traditionally male-dominated roles and settings. In traditional Muslim culture, conceptions of women's roles and gender relations have been even more restricted. We picture Muslim women as veiled and secluded and relations between men and women in Muslim culture as highly segregated. Thus, the victory by a Moroccan woman, Nawal el Moutawakii, in the 1984 Olympic 400-meter hurdles represented a significant departure from such traditional ideas about Muslim cultural restrictions of women.

This athletic success reflected broader changes in Moroccan society in the status of women, brought about by women's liberation, increased educational opportunities, more employment opportunities, more involvement in politics and the economy, and the rise of secular ideological movements (Sfeir, 1985). Athletic participation by women could not happen in this society until traditional cultural practices and social patterns had dramatically changed. Once they did, prominent athletic success by women dramatically symbolized the changed status and opportunities for women in the society in general. Sfeir (1985) found in her study of the status of Muslim women in sport in the 1980s that there were variations across Islamic societies in attitudes about the status of women in general and about their role in sport.

These societal differences in attitude reflected differences in the complexity of society and in the ways that the cultures and social institutions of these societies expressed the Islamic religion. The result of these differences was a range of sports opportunities for women across societies, from very limited participation to participation at the highest level of international sport. Within and across the nations of the Muslim world, sports opportunities for women have tended to surge with the influence of nationalist movements and the diffusion of secular egalitarian beliefs through socialist ideologies and Western influences. Led by Turkey and Egypt, Muslim women have slowly responded to these influences of Westernization, modernization, and secularization, while still trying to hang on cultural traditions and identity (Sfeir, 1985).

In Egypt, for example, the nationalist movement brought about more sports opportunities for women. In particular, the Egyptian Revolution of 1952 increased female involvement in youth clubs, physical education, and sport. Increased ties with the Soviet Union under the leadership of Gamal Abdel Nasser during the 1960s further encouraged women's involvement in sport as an indication of their liberation from traditional roles. Under President Anwar Sadat, women were encouraged to assume leadership roles in sports clubs. Thus, the combination of nationalistic forces of change with increased contact with the non-Islamic cultures of other countries exposed Egyptian society and its women to new ideas about their roles and new options in sport (Sfeir, 1985).

Since sport is a visible and potent symbol and by-product of Western modernization, the increased sports participation of women can be seen as a threat to Islamic traditionalism. The return to veiling, seclusion, and segregation of women in some parts of the Muslim world is incompatible with the image of energetic, aggressive, and physically expressive women in sport. Thus, we should expect women's sports participation to be treated with suspicion or hostility in the revival of Islamic fundamentalism in the Muslim world. In general, the ways that sport is approached in Muslim countries will reflect broader resolutions of the tensions between Islamic traditionalism and secular modernism.

Women throughout the world have been influenced by the internationalization of sport (Riordan, 1985). As women from one country or part of the world achieve athletic prominence in certain sports, they may inspire women from other nations to seek more opportunities in those sports. In addition, international competition may challenge women to seek improvements in performance that they had not thought possible before seeing opponents achieve them. For example, the competition between Western women athletes and women from the former Soviet Union and Eastern Europe continually stimulated better performance on both sides in a variety of sports. In nations such as Nigeria, where sport was an exclusively male domain, cultural contact between traditional culture and that of the Western world has inspired women to participate in sport (Riordan, 1985: 119).

Sports participation for women is part of modern Western culture, and exposure to this culture for women of traditional cultures tends to create new images of women in nontraditional roles such as sport. As Riordan (1985: 119) pointed out, Soviet leaders intentionally introduced sport to the Muslim women of Soviet Central Asia (which bordered on Turkey, Iran, and Afghanistan) as a means of breaking the hold of traditional religion on the people in this region and of creating new roles for women.

The Olympic Games involve more nations in more sports than any other sports event in the world. For this reason, it provides a useful measuring stick for gauging women's progress in sport around the world. In general, male Olympic participants have greatly outnumbered females, but the percentage of females in the Olympics has substantially increased during this

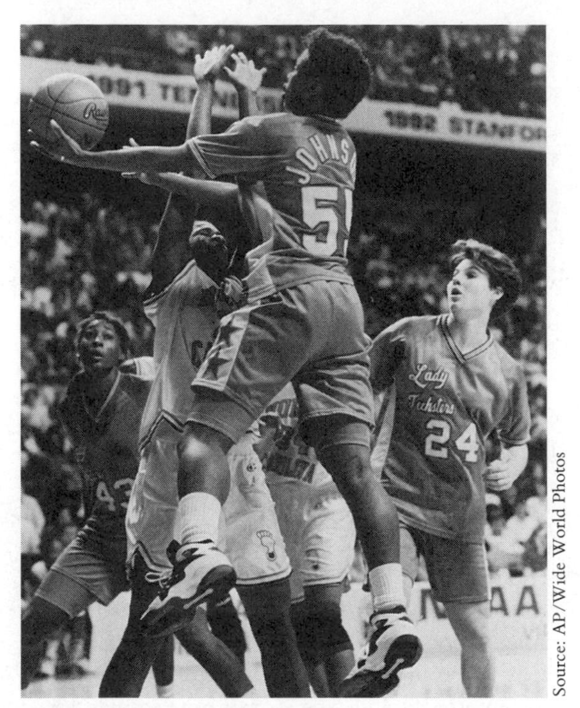
Source: AP/Wide World Photos

FIGURE 13.3 The name game in women's college athletics.

century. The percentage of females in the Summer Olympics increased from zero in the first Modern Olympic Games in 1896 to 25.8 percent in the Seoul Olympics of 1988 (Coakley, 1990: 186). Between 1924 and 1980, the percentage of women competing in the Winter Olympics increased from 4.4 percent to 21.1 percent (Riordan, 1985). Among women, participation rates have varied a great deal across different parts of the world. For example, in the 1976 Montreal Summer Games, the overall percentage of Olympians who were female was 20.65. Women made up 35 percent of the Soviet team and won 36 of their 125 medals (nearly 30%); women made up 40 percent of the German Democratic Republic (East German) team and won more than half of its gold and silver medals; U.S. women were slightly more than one quarter (26%) and British and West German women slightly over one fifth (20.6% and 21%, respectively) of their Olympic teams; and French women made up less than one-fifth (18.3%) of their nation's team. With the exception of Cuba—with a team that was 27.5 percent female—Latin American teams had virtually no female members (Riordan, 1985: 123). An Islamic nation first competed in the Olympics in 1908, and its one Olympic athlete was male. The first year in which women from Islamic nations competed in the Olympics was 1936, and at those Games, two of the 113 Muslim athletes from three Islamic countries were women. From the resumption of the Olympics after World War II in 1948 through the 1980 Games, thirty Muslim women competed. In 1984, twenty-two Muslim women participated in the Los Angeles Olympics. This number contrasted with the 565 Muslim men, and the 1,681 women and 5,914 men overall, who competed in the 1984 Games (Sfeir, 1985: 285).

CONCLUSION: AFTER THE REVOLUTION IN U.S. WOMEN'S SPORTS?

Dramatic changes in female sports opportunities and involvement have occurred in the United States since the early 1970s, despite persisting sexist beliefs and practices. Legislation such as Title IX has been a major factor in this virtual revolution. In addition, the women's movement, the health and fitness movement, and (moderately) increased mass media coverage and publicity for women athletes and sports have contributed to this revolution (Coakley, 1990: 177–179). As a result of these social forces, females have had more legal right to participate and to do so on an equal footing with males (legislation); have been encouraged to pursue new and nontraditional roles and experiences, expand their rights, and seek equality with men in sport (women's movement); have become more involved in physical activities to enhance their health and fitness (health and fitness movement); and have been exposed to more female sports role models (media coverage and publicity). Women have forged new conceptions of femininity in traditionally male sports such as bodybuilding and even have eclipsed the men in some major sports events, such as the 1992 Winter Olympics in Albertville, France (Callahan, 1992; Swift, 1992). In the 1992 Winter Games, American women won nine of the eleven medals and all five gold medals earned by the U.S. team. Research also suggests that high-level female college athletes are able

SPECIAL FOCUS

Eating Disorders and Female Athletes

The death of Christy Henrich on July 26, 1994 brought into sharp relief the issue of eating disorders among female athletes (Noden, 1994). At the height of her career as a world-class gymnast, the 4'10" Henrich weighed 95 pounds. When she died of multiple organ failure, eight days after her twenty-second birthday, she weighed 61 pounds. She had weighed only 47 pounds three weeks earlier, after her discharge from the St. Joseph's (Mo.) Medical Center. Henrich died from the effects of anorexia nervosa, which is an intense fear of gaining weight or becoming fat that causes victims to see themselves as fat even when they are grossly underweight and leads to obsessive dieting. A related eating disorder is bulimia, which involves repeated eating binges followed by purges of food by self-induced vomiting, laxatives, fasting, vigorous exercise, and other means. Like victims of anorexia, bulimia victims suffer from a distorted self-concept and a sense they cannot control their eating (Caldwell, 1993).

Eating disorders may be the most serious health problem faced by female athletes. The American College of Sports Medicine estimated that more than 60 percent of females competing in "appearance" sports (such as gymnastics and figure skating) and endurance sports suffer from an eating disorder. Eating disorders also occur in other women's sports, such as volleyball, tennis, and swimming. While eating disorders can affect males, it seems more closely linked to females. A 1992 NCAA study showed that 93 percent of the intercollegiate athletic programs reporting eating disorders were in women's sports (Noden, 1994: 54). Among female athletes and nonathletes, vulnerability to eating disorders is heightened by a desire for perfection. Not surprisingly, then, its victims tend to be young women who aspire to greatness. In pursuing their goals, these talented and ambitious young women struggle with identity and role conflicts revolving around being successful and being feminine (Nixon, 1989; Brownlee, 1992; Taub & Benson, 1992; Caldwell, 1993). In many sports, pressure on females to match a cultural ideal of a perfect (thin) body is exacerbated by the importance that sport places on weight for competitive reasons. This pressure is made worse by explicit or veiled messages from coaches and sports officials that chances of winning will be enhanced by losing weight and being thin. Liz Natale was a member of the 1986 NCAA champion cross-country team at the University of Texas. In reflecting on the anorexia from which she was recovering, Natale said that eating disorders for women were "like steroids ... for men. You'll get results, but you'll pay for them" (Noden, 1994: 60). The solution to eating disorders in sport and society is intimately tied to the emphases on winning in sport and on stereotypes of ideal body types for women in both sport and society. The solution is difficult because these emphases are so well entrenched.

to maintain a sense of personal empowerment despite participating in a largely male-dominated world (Blinde, Taub & Han, 1993). Whether sport actually has the capacity to *make* women feel empowered is a question still needing systematic study, though.

The landscape of sport today in the United States is much improved in many ways for women and girls over what it was just two decades ago. It is dotted with many more female fans (Clarke, 1993) as well as active participants (*CQ Researcher*, 1992b). This landscape still has many barriers and potholes for females, though. It is a landscape of traditions and change. The social forces bringing positive changes for females in sport have stirred forces of opposition that have slowed down or halted progress toward gender equality. Title IX has not been vigorously enforced, and court decisions have narrowed the application of it (Leonard, 1993: 269). The women's movement and federal antibias laws have created a backlash and produced reverse discrimination lawsuits (e.g., Blum, 1993b). The health and fitness movement has been as much about fad and glamor in physical exercise as it has been about competitive sport. Young female achievers still experience eating disorders or a desire for significant weight loss as they struggle with identity and role conflicts revolving around being successful and being feminine (Nixon, 1989; Brownlee, 1992; Taub & Benson, 1992; Caldwell, 1993). Women aspiring to leadership positions in sport have seen their opportunities erode after Title IX pushed the NCAA to take over women's intercollegiate sports. Mass media sports reporting, broadcasting, writing, and depictions of females in sport continue to be biased, albeit sometimes in subtle or ambivalent ways.

In addition, women's intercollegiate athletic programs have increasingly emulated the male model of college sports since Title IX and the merging of men's and women's programs (Blinde, 1989). The male sports model places more emphasis than the female sports model on athletic scholarships, recruitment, publicity, large and male-dominated coaching staffs, publicity, long seasons, and long hours each week in practice sessions. More emphasis on the male sports model in women's sports has been associated with more traditionally male sports values, such as strong competition, aggressiveness, seriousness, hard work, intimidation, being demanding and businesslike, autocratic control, commercialization, professionalization, and conformity (Blinde, 1989). The progress of women in sport mirrors the fluctuations in female status in society in general in countries around the world. Women have struggled, moved forward, and then faced resistance pushing them backward as they have sought to move forward once again. The battle over Title IX and gender equity provides a microcosm of this struggle. This battle indicates that the law—or the threat of legal action—can be a powerful tool for bringing about change, as women's participation in American sport surged after the passage of Title IX in 1972. The Title IX battle also shows, however, that gains in participation for women may not be paralleled by gains in other areas, especially when they involve authority and power. Furthermore, the quest for additional rights, resources, and opportunities and tangible progress in these areas create powerful forces of resistance and opposition. Women have found that powerful men have not willingly relinquished their advantage and control in sport.

SUMMARY

In many ways women and racial minorities have had parallel experiences in North American sports (Lumpkin, 1981). Both groups have been restricted from participating in sports because of formal and informal cultural and social barriers; both have participated in segregated competitions and leagues; both have struggled against discrimination as athletes, coaches, and sports officials; and both have been boosted by federal civil rights legislation in their efforts to fight discrimination. Female and black sports experiences have been fundamentally different in other respects. Whereas black males have been seen as naturally endowed to be athletes, females have been viewed as naturally unathletic in regard to most sports. Black male athletes have not had to contend with the ambivalent feelings, stigma, role conflict, or sense of cultural contradictions confronting female athletes trying to fit comfortably into their gender and athletic roles.

The existence of homophobia in men's and women's sports reflects persisting cultural ideas that sport remains a place for men where stereotypical masculinity is celebrated. Thus, although sport is no longer a strictly male preserve in the United States and many other nations, women still find their presence unwelcome or unappreciated in many sports. Female athletes are often presumed to be lesbians, and actual and presumed lesbian and gay athletes both feel the sting of stigma and homophobic reactions.

The progress of women in sport enables us to see the fluctuations in female status in society in general. Female athletes have had many opportunities since the passage of Title IX, but they also have had to struggle to achieve and retain their rights of access to sport. Women coaches and athletic administrators have seen their opportunities decline as more females have participated in sport over the past two decades. Gender equity is one of the major issues affecting the future of college athletics as women and men struggle over the allocation of resources and opportunities.

CHAPTER 14

Sport, Politics, and the Olympics

SPORT AND POLITICAL REALITIES

Avery Brundage, late president of the International Olympic Committee (1952 to 1972), always insisted that sport should operate without federal, state, or local government intervention and be free internally from factional political disputes. In his view sport, like music and art, presumably transcended the profane and mundane world of power plays and regulatory directives instigated by disagreeing political groups. That is, decisions about sport should be made by the participants, or by boards and commissions representing the participants, without interference from external regulators or officials.

The notion of separating sport from politics also suggests that struggles for power, domain, influence, priority, or resources among the participants and/or diverse factions within a sports community should never take precedence over or adversely influence the nature of participation. This position is guided by the premise that sport should be a simple diversion, with its outcomes having little significance beyond the boundaries of the specific competitive activity.

The reality is, however, that sport and politics are closely related, and many symbolic and dramatic incidents in modern Olympic history illustrate this point. In the opening ceremony for the 1908 Games, the United States intentionally refused to acknowledge the King of England; the infamous 1936 "Nazi Olympics" were used by Hitler for propaganda purposes; the symbolically powerful but silent black-gloved victory stand demonstration of Tommy Smith and John Carlos in the 1968 Mexico City Games was meant to draw attention to racism in the United States; Israeli athletes were kidnapped and killed by Arab terrorists in the 1972 Munich Olympic Village to dramatize the terrorists' dissatisfaction with Arab-Israeli relations; the 1976 Montreal Olympics were boycotted by thirty-two nations—with many from Africa—in protest of New Zealand's policy of maintaining sports relations with South Africa; and the United States and the Soviet Union boycotted each other's Olympic Games—in 1980 in Moscow and in 1984 in Los Angeles—as a result of Cold War politics.

The federal government and courts in the United States have had a long history dealing with regulatory issues concerning professional sports. Antitrust issues have arisen many times since the 1922 Supreme Court decision exempting Major League Baseball from the Sherman Antitrust Law of 1890. When the Supreme Court voted 5-3 against Curt Flood in 1972 in his challenge to the reserve clause, it turned to the Congress to take action to reverse the precedent of the 1922 decision (Nixon, 1984: 200). The U.S. Congress has also been concerned about college athletics in recent years, focusing on the academic progress of student-athletes and the integrity of university athletic programs. Federal, state, and local governments are playing an expanded role in sports in America.

SPORT AND GOVERNMENT IN AMERICA

The U.S. government has been playing a significant role in professional sports for a long time, but the intervention intensified in the 1950s and continues today. This involvement has expanded to reach into the amateur realm and the Olympics, in particular. The internal conflicts in sport and their potential implications for the public interest explain legislative and judicial interest in American sport. Sport has been

Source: AP/Wide World Photos

FIGURE 14.1 Politics and sport: Tommie Smith and John Carlos raising the black power salute at the 1968 Mexico City Olympics.

characterized by internal disputes between players and owners, cities and team management, coaches and administrators. There have been player strikes. Scandals about academic progress and financial shenanigans have been uncovered on college campuses. Cities have fought for franchises. Corporations and faceless investment syndicates have replaced sports-minded individuals as owners of many sports franchises. Furthermore, as we already have pointed out many times, television became a major player in the distribution of the athletic product and the financial health of leagues and franchises. In this context, it was almost inevitable that the government would expand its role in sport to the point that sport would be treated as a public policy domain in the same fashion as the areas of transportation, health, and education (Wilson, 1994).

Many aspects of American sport attracting government attention have involved monopolistic practices and related antitrust issues. For example, government agencies and officials have addressed reserve and option clauses, free agency, the pooling of television broadcast rights by leagues, television blackouts, and mergers between leagues (Weistart & Lowell, 1979; Closius, 1985; Wilson, 1994). In addition, legislation, government policy, and judicial decisions have affected discrimination in sport and discriminatory treatment of sports participants as members of society. For example, legislative and judicial pressure in recent years has pushed the National Collegiate Athletic Association (NCAA) to form a gender equity task force to resolve issues of gender discrimination in college athletics. Civil rights legislation has provided remedies for racial discrimination experienced by athletes and coaches as American citizens.

Violence is another aspect of sport provoking government attention. Members of Congress and state legislatures periodically express concern about violence at various levels of sport, and the courts have adjudicated cases involving charges of illegal sports violence. One proposal to control violence in sport, HR 2151, called for the formation of a Sports Court (Carlson & Walker, 1982). Introduced in the 99th Congress in 1985 by Representative Tom Daschle, this bill, called the Sports Violence Arbitration Act, established arbitration systems in professional sports league to resolve grievances regarding injuries or harm resulting from acts deemed to be excessively violent. HR 2151 never made it out of committee, but it raised the possibility of legal action if violence reached excessive levels in professional sports.

Although the United States government has been involved in sport in a variety of direct and indirect ways, it has not formulated or carried out a consistent or coherent sports policy. Government can have a significant indirect impact on sport, however, through its influence on the social, economic, and political conditions of the society in which sport exists (Wilson, 1994). In the United States, the corporatization and commercialization of sport have given rise to concerns about possible violations of the public trust in sport without more direct government regulation or oversight. In addition, as the economic stakes of sport have

SPECIAL FOCUS

College Athletics, Monopoly, and the Law

Like many of its professional sports counterparts, the NCAA has had increasing difficulty in the courtroom defending its monopolistic practices. In 1984, the U.S. Supreme Court ruled in *Board of Regents, University of Oklahoma v. NCAA* that the NCAA's practice of restricting the television broadcast of college games to those designated by that organization was monopolistic and in violation of antitrust regulations. The Universities of Oklahoma and Georgia, acting on behalf of the high-powered College Football Association (CFA), asserted that broadcast rights were the property of the institution unless relinquished to another entity such as the NCAA. They further contended that since they had not officially given the NCAA these rights, it was improper for the NCAA to prohibit individual institutions from negotiating their own television and radio contracts (Hochberg, 1985). The court decision meant that individual institutions had the right to establish their own networks and to negotiate individual television and radio packages. Notre Dame took advantage of this new legal environment to parlay its vast network of football fans into a lucrative television network for Fighting Irish games.

risen, so has a greater potential for conflict—for instance, between players and owners—and a concomitant need for a government arbitration.

Although many government agencies and officials have shown interest in sport in a variety of ways, the federal government in particular has generally been reluctant to exercise strong or centralized regulatory control over sport in the United States. An exception is the Amateur Sports Act of 1978. With encouragement from the President, Congress enacted this legislation in response to the declining performance of the United States in international competitions. A congressional commission asserted that the prime culprit in this problem was continual feuding and a general lack of cooperation among the major amateur sports organizations, including the NCAA, the U.S. Olympic Committee (USOC), the Amateur Athletic Union (AAU), and the assorted governing bodies of individual amateur sports. The provisions of this act made the USOC the primary coordinating body for amateur sports. The Athletics Congress (TAC) replaced the AAU as the coordinating body for track and field. With the exception of a one-time appropriation of $16 million for three Olympic Training Centers, the act did not provide funding. Thus, after giving some direction for the control of amateur sport and some money for training facilities, Congress displayed its traditional pattern of backing away from direct control over sport.

Governments have taken an active role in sports promotion at the local and municipal levels. For example, professional sports clubs have enjoyed public subsidies and tax breaks from municipalities, cities, and states, and both professional and college teams have been provided with public facilities that they have occupied at very low cost. As we noted earlier, the motivation for this sports involvement and support has been the desire to enhance the reputation of the city or region and to derive economic benefits from sports investment.

There are many reasons for government involvement in sport. Included among these reasons are the following ones: to promote a political ideology or set of values to maintain social order; to protest the actions of another country; to unify or increase a government's hold on its own people; to try to bring other nations into a government's sphere of influence; to arbitrate factional and jurisdictional disputes within or between sports governing bodies; to preserve equal public access to sport; to eliminate discrimination in a sport; to assure that sports practices conform to the laws of a society; to protect public investment in a sport; to improve physical fitness through sport; and to enhance the status or prestige of a community or nation through sport (Lever, 1983; Nixon, 1983; Johnson & Frey, 1985; Coakley, 1994). When governments use sport to increase popular support and

POINT-COUNTERPOINT

SHOULD GOVERNMENTS BE INVOLVED IN SPORT?

YES

Sport is now a major global or transnational economic entity with a high level of organizational complexity and bureaucratization. It is an integral part of every society and deserves the same governmental attention and regulation as other societal institutions, especially when the public interest or public trust is at stake. Without government intervention, racial and gender discrimination still would be prevalent in American sports; the American public would not be able to enjoy the performance of their favorite professional and college teams on television; and American professional athletes would not be able to control their careers today.

NO

Government intervention is not necessary because the market can dictate the organizational format and nature of sports participation and serve as a natural regulatory mechanism. The politics associated with government intervention in sport distracts from athletic performance. When governments get involved in sport, bureaucracy and red tape are introduced, making sports administration more cumbersome. Government regulation could prevent some athletes from competing because they fail to comply with a trivial bureaucratic rule. Government officials are not interested in sport for its own sake and would not regulate for the good of sport.

How should this debate be resolved?

elevate their stature in the nation or world, sport becomes a form of cultural capital that is traded along with other capital, such as technology, education, and information (Wilson, 1994).

It is not always clear that the public interest is served by government involvement in sport. Nor is it clear that the interests of the sporting public or athletes are well served by government involvement and regulation. When governments use sport for political purposes, the interests of sport and athletes become secondary to the interests of government and its representatives.

Government intervention raises questions about who is the major benefactor of this action, the general public or elite interests. The state-capitalist and liberal-pluralist perspectives offer different answers to this question. *State capitalism* assumes that sport reproduces existing class inequalities, serves the interests of the dominant class, and reinforces the dominant-class culture of state capitalism in general (Brohm, 1978; Cantelon & Gruneau, 1983; Johnson & Frey, 1985; Hargreaves, 1986). *Liberal pluralism* focuses on how sport serves the interests of society by performing social functions, such as the promotion of cohesion or unity, democracy, national or civic pride, and personal exhilaration (Johnson & Frey, 1985). In seeing class biases embedded in the structure of sport, government, and other societal institutions, state capitalists draw attention to the subversion of public interest by government involvement in sport. Liberal pluralists see government intervention in sport as a means to restore the benefits of sport to society when they are threatened.

THE ROLE OF SPORT IN INTERNATIONAL RELATIONS

Since the inception of the modern Olympic Games in 1896, sport has had a prominent and highly visible place in modern international relations. Strenk (1977) identified six ways that sport has been used in international relations, including:

1. To gain or deny diplomatic recognition
2. To spread nationalist propaganda or political ideology
3. To enhance a nation's prestige
4. To promote international understanding and peace
5. To make a protest

6. To engage in conflict without shooting or what has been called "war without weapons" (Goodhart and Chataway, 1968)

Diplomatic recognition or nonrecognition represents the use of sport as a way to acknowledge the acceptance or rejection a nation in the community of nations. When one nation is willing to enter into competition with another, this represents tacit diplomatic recognition of that country and its political system. On the other hand, recognition can be denied a country by refusing to compete against it in sport. According to Strenk (1977), the former German Democratic Republic (GDR) or East Germany, was the best at playing this diplomatic game. The GDR's "diplomats in track suits" worked feverishly to obtain membership in international sports organizations and to be able to participate in the Olympic Games. Participation as the GDR in international sports competition legitimized the country's separate status from West Germany, and its international sports success helped bring it recognition as a country to be reckoned with in its own right.

The case of South Africa represents another important example of the role of sport in diplomatic recognition. This country's racial segregation or apartheid practices, the official government policy from 1948 until 1994, resulted in its exclusion from the Olympic Movement and nearly all international sports competition for almost three decades. Until its recent reentry into the international sports world, South Africa was involved in the most dramatic, and perhaps longest running, political conflict in international sport (Lapchick, 1979). The oppressive segregation and discrimination against nonwhites in South Africa by the minority white regime included sport, and sport became an important arena where other nations pressured South Africa to dismantle the apartheid system. For example, the United Nations General Assembly adopted the International Declaration Against Apartheid in Sport in 1977 without a dissenting vote. The European Economic Community and the Commonwealth of African nations in the Gleneagles Agreement of 1978 called for governments across the world not to have sports contacts with South Africa. In 1981, the United Nations Centre Against Apartheid censured sports figures who had contacts with South Africa by naming them on a blacklist, called the *Register of Sports Contacts with South Africa.*

Sports contests can be interpreted as competition between political as well as athletic adversaries. As a result, a claim of athletic victory can be interpreted as a victory for a country's political and economic systems (Strenk, 1977; Frey, 1988a). In this way, sport is used to *serve propaganda and ideological purposes.*

The Soviet Union entered the Olympic Games at the end of World War II, after previously rejecting sport as a bourgeois practice. That nation's explicit aim was to promote its socialist system and spread its influence throughout the world. During the reign of communism and before the economic and social reforms of glasnost and perestroika initiated by Mikhail Gorbachev, sport was the Soviet Union's most important propaganda vehicle. International sports participation and achievement were intended to win support for the USSR and its system of government, especially in the regions of Africa, Latin America, and Asia. Immediately following World War II, the Soviet Union made a concerted effort to affiliate with every international sports organization, particularly those governing the major sports such as soccer and basketball. The USSR opened its doors to athletes and coaches from other countries, while at the same time allowing several hundred of its "sports ambassadors" the rare privilege of traveling abroad to represent their nation.

The 1936 Olympics in Berlin were viewed by Hitler's regime as an unprecedented propaganda opportunity to showcase the prowess of the German political state and the efficiency of their institutions (Mandell, 1971; Strenk, 1977). The Olympic Games also represented a diversion from Germany's military intrusion into neighboring countries and the genocidal atrocities it was performing on its own Jewish population and that of other countries. The United States has also used participation in international sport for political or ideological purposes, but its uses of sport for these purposes have generally been more subtle or indirect than the actions of the Nazis and socialist countries. The 1980 boycott of the Moscow Olympics was an exception to this pattern.

The use of sport to *gain prestige in the international community* has been especially characteristic of developing nations who "seek 'world class' standing in the

Source: UPI/Bettmann Newsphotos

FIGURE 14.2 Bringing the Olympic torch to the 1936 Nazi Olympics.

hierarchy of nations" (Frey, 1988a: 66). Sport represents a status opportunity. As countries develop, they are motivated to demonstrate to the world that their political, economic, and social institutions are governed by civil, not primordial, guidelines and that their citizens have achieved an acceptable standard of living. These nations also want to be recognized as responsible actors in the world arena (Geertz, 1963). In other words, "a country desires to be noticed, to be acknowledged as important, and to be accepted as a regular and equal member of the community of nations" (Frey, 1988a: 66).

International sports competition gives a country exposure to the world it might not otherwise be able to obtain. For example, the African nations of Kenya, Nigeria, and Ethiopia might receive less international attention if it were not for their participation in high-profile international sporting events. The same could be said for small island countries such as Jamaica, the Dominican Republic, and Trinidad. Sport can be a low-risk avenue to upward mobility in the hierarchy of nations. Participation generally involves little political, military, or economic risk. The political stability, defense stature, and economic security of a country are not compromised as the result of sports participation, but the rewards of recognition and status can be significant (Kyrolainen & Varis, 1981).

Sport can also be used as a vehicle to *promote international understanding and peace* by serving as a common basis that brings countries together (Strenk, 1979). Presumably as a result of this contact, countries can develop more sympathy for and understanding of each other; transnational networks are formed as a result of sporting contacts. Efforts at public diplomacy, which include sponsoring teams in international competition or sending sports ambassadors to other countries for exhibitions and training, are designed as peaceful tests of relations among countries. For example, before the normalization of relations between the United States and China in the early 1970s, the two nations competed in ping-pong exhibitions creating the now familiar term *ping-pong diplomacy* to be generally applied to all efforts to meld sport with diplomacy. Through this type of exchange, nations test the diplomatic waters by assessing whether they can cooperate sufficiently in less sensitive cultural activities in order

POINT-COUNTERPOINT

SHOULD SPORT BE USED FOR POLITICAL PURPOSES IN INTERNATIONAL RELATIONS?

YES

Sport is an excellent vehicle to test diplomatic relations between countries without the risks associated with military conflict or economic boycott. Athletic performers can serve as ambassadors of a country's cultural and ideological mission. They become spokespersons for a way of life and eradicate fears and correct misconceptions one country may have of another. The result can be international understanding and peaceful relations.

NO

We delude ourselves if we think that athletics has anything significant to contribute to international relations, especially where national interests such as economic survival are at stake. Athletics can have some public diplomacy value, but when nations exercise their political and economic clout sport falls into the background. To use sport for political purposes distracts from the basic nature of sport as athletic competition.

How should this debate be resolved?

to warrant the exploration of broader contacts and agreements.

Successful sporting contacts can set the stage for more serious negotiations about diplomatic matters. In this same spirit, Israel and the Palestinians competed against each other for the first time in an international table tennis tournament, following their historic agreement to establish more normal relations with each other. This was a test of the peace agreements they had signed, and the event took place without incident.

International sports competition often results in outcomes other than cooperation and understanding. Nations, groups, or individuals may be more *interested in protest* than cooperation or understanding when they compete in sport. Protest can be the conscious aim of a government, as when the United States boycotted the 1980 Olympics in Moscow and the USSR boycotted the 1984 Games in Los Angeles. Protests can be spontaneous or apparently unplanned, as when John Carlos and Tommy Smith raised their black-gloved hands at the 1968 Mexico City Olympics and athletes announce their defections from their country. Federations and teams also may organize a boycott, as the U.S. Tennis Federation did in 1976 when it decided not to participate in the Davis Cup because South Africa was involved.

The worldwide boycott of South Africa created pressure on that country to restructure its sports to provide desegregated opportunities and facilities for nonwhites. The boycott also focused attention on nations, such as New Zealand, that chose to disregard it, and such nations became targets of international protest themselves. In fact, a number of African nations refused to participate in the 1976 Montreal Summer Olympic Games because New Zealand was allowed to compete. The sports boycott helped to diminish South African national pride and prestige, and it set the stage for the powerful international economic sanctions that adversely affected the South African economy. South African leaders were finally ready to change their society by the beginning of the 1990s, and as a result of the promised changes their nation was allowed to return to the Olympic movement and international sport.

Sport can also be an *outlet for aggression or hostility*, a "war without weapons." Political conflict between nations is exacerbated when nations try to use their sports victories for an ideological advantage over their foes on the athletic field. Political tensions can turn into actual violence in the athletic arena when extremist or highly committed protest groups see opportunities to gain attention for their cause by dramatic actions in sport. The Black September Movement used the 1972 Summer Olympics in Munich to promote Palestinian recognition and display their hostility toward Israel; the soccer war between Honduras and El Salvador in 1969 was incited by a World Cup match between the two countries.

Although sport has been perceived as important by government leaders and the people of many nations and has been the stage for a number of prominent political actions, the political significance and implications of sport are sometimes exaggerated or overstated. We know that politics and sport are often closely linked, but we often lack evidence showing the precise impacts of the various political uses of sport. For example, there is only anecdotal evidence that most government leaders and citizens believe that international sports success is an important indicator of their nation's economic, political, or military prowess (Frey, 1984b). Furthermore, the amount of long-term intercultural understanding that results from international sports competition, even among participating athletes, seems to be very limited. Indeed, media coverage of international sports events, such as the Olympics, often reflects nationalistic biases. American coverage of the 1984 Los Angeles for the American audience received harsh criticism due to such bias. There seems to be more evidence that sport fosters ethnocentrism, nationalism, and hostility than international understanding, cooperation, or peace (Seppanen, 1982; Frey, 1984b).

Sport can play a role in international diplomacy and create pressure for change in international or domestic affairs, but it is less likely to have a direct impact on essential national interests or entrenched social patterns, such as economic development, political legitimacy, or social and economic inequalities. Sport is especially effective in providing a platform for the expression of national values or goals or for drawing attention to a particular political or social cause.

Neocolonialism through Sport

Sport can be used to reinforce colonial domination. The British imposed their sports, such as cricket and rugby, on nations making up the British empire, and Western nations have compelled the nations of Africa and Latin and Central America to compete at their sports to become part of the international sports world.In some cases, a country embraces a sport because it is consistent with internal developments or because the host country is interested in emulating the country from which the game has been imported. Japan was ready for the American version of baseball because it reflected modern values, such as order, perseverance, and equilibrium. Later, the Japanese took great pride in beating the Americans at their own game (Guttmann, 1993).

When sport is introduced to a colonized people, it may reinforce the colonial relationship, with the culture of the conqueror imposed on the culture of the colonized. Over time, however, as the colonized country becomes more proficient at the imported versions of sport, it can declare some measure of independence and superiority by defeating the powerful master at its own game. Thus, sport can play a role in development by *promoting a national identity separate from Western or European colonizing influence.* In this sense, sport can be a means of expressing anticolonialism. A country can resist the importation of another country's sports, as the United States did with cricket, in order to assert its unique national identity (Guttmann, 1993). Perhaps this desire at least partially explains the resistance of U.S. citizens to soccer, a sport that is defined as the most popular in the world but lacks American roots.

Critical observers of sport's role in international politics have portrayed it as an "imperialist phenomenon" and an "extension of monopoly capitalism" (Brohm, 1978; Hoberman, 1984; Kang, 1988). They see it as a tool to spread capitalism and impose a Western capitalist model of development on weaker and poorer Third World nations. The Peace Corps, church missionaries, formal sports exchanges and emissary programs, the training of native athletes in Western nations, and the media all may play a role in communicating the formal, rational, and corporate version of Western sport to nations around the world. In order for nations to participate in major international sports events, they must adopt a standardized version of the sport that is formally organized and regulated by international sports bodies. Thus, global sport is a homogenized version of physical activity, which eliminates unique cultural differences or variations in sport. The nature of this participation further reinforces the globalization and Westernization of a developing country's physical activities and erodes native physical culture. The result of these processes is a universal or standardized version of sport that bears the imprint of the rational, market-based, capitalistic model of modernity.

Developing countries wanting to make their mark in international sport but, lacking the resources to provide adequate training and facilities, can become culturally dependent on nations with more developed sports systems. Some developing nations, such as Cuba, have resisted this dependency by organizing their own high-level sports systems. Others have turned to more established nations in the world of sport for help. For example, many athletes from developing countries have turned to American colleges and universities for high-level sports training, facilities, and competition.

The recruitment and subsequent migration of athletes from their homelands to play modern sports in America and other countries is known as the **brawn drain** (Bale, 1991). For critics (e.g., Bale, 1991), this brawn drain is a form of American or Western cultural colonialism. Foreign athletes abroad may form negative images of their own country as they internalize and learn to express modern Western values of achievement, productivity, individuality, material progress, control, and the American Dream. Thus, these athletes become cultural emissaries for the United States and other host nations.

Developing nations encourage a brawn drain by default when they fail to provide the opportunities and resources needed by their most promising athletes for high-level sports participation. They do this more directly when they send their best athletes abroad to train and compete with official blessing. The expectation or hope of high-level officials in these countries is that these athletes will return home someday to represent their homeland in international sport. As a result, the country may be successful in international sport, but its population cannot take pride in the fact that the athlete trained and developed within its own sports system. There is no sign that the movement of athletic talent across international borders will decline; in fact, it may increase in intensity now that the borders of Eastern European countries no longer are the barriers they once were.

Kang (1988) pointed out contradictory roles of sport in international relations: on one hand, it is supposed to enhance nationalism and international understanding; on the other, it can be an instrument to create cultural dependency. In this process, indigenous mass physical activities are replaced by modern corporate and commercialized sports that involve only a relatively few elite participants.

The transition to modern sport has also meant a shift from active participation in sport to passive spectating. Thus, we see African dances involving over a hundred participants replaced by soccer matches with eleven players on a team that are viewed by thousands (Martin, 1983). The modern principles of spectatorship, competition, winning, and commercial return dominate traditional values of participation, enjoyment, and cooperation.

Kidd (1981) noted that the development of Canadian sports has followed the American pattern of formalization, rationalization, and commercialization. As in the United States, cartel-like associations, with the help of the Canadian federal government, have established control over competition. Gruneau (1983) noted that with the formation of the National Hockey League (NHL) in 1917, Canadian sport began to duplicate the corporate organization that was evolving in the United States. This pattern of sports development represented a movement from player-controlled club sport to corporate sport and was viewed by Gruneau (1983) as a form of "cultural betrayal" (p. 121).

As we suggested in Chapter 3, sport can be a form of cultural resistance to coloniallike domination by a larger and more powerful foreign nation. For example, the Dominican Republic infused its version of baseball with an emotional, reckless, yet leisurely style of play not found in the American game (Klein, 1991). In this way, Dominican national culture could be expressed through baseball, and the American effort to transform the Dominican game could be thwarted.

Arbena (1993) argued that sports introduced by foreign powers into less economically developed nations are changed over time by indigenous forces and often come to reflect values that are different from the culture of the foreigners. Along with Dominican baseball, bullfighting in Mexico, basketball in Trinidad and Tobago, polo in Argentina, and soccer in Peru are examples of sports that retain distinctive cultural features of the home country while operating in a context of modernization and industrialization. These sports have resisted complete enculturation.

Countries that have experienced colonial rule often have a difficult time adjusting to their newly acquired freedom and democratic political institutions. As a result, governments remain centralized until democracy is institutionalized. Mandle and Mandle (1988, 1990) observed that the extensive network of voluntary basketball organizations in the Caribbean "have opened a sphere of democratic choice and participation in everyday lives of the most economically deprived West Indians, a group for whom few opportunities exist" (1990: 68). The West Indian governments are centralized, have resisted the formation of autonomous institutions, and have continued to keep their populations in a state of dependency. Independent economic and political activities are very limited because people fear that they may duplicate the activities of the state or stimulate a hostile reaction from the central government (Mandle & Mandle, 1990).

Basketball in Caribbean nations such as Trinidad and Tobago has created an independent set of indigenous voluntary organizations that these nations so far have been unable to develop in economic, political, and social settings. The basketball leagues represent one setting where the general public can resist the controlling practices of the state. These conditions are likely to continue as long as elected authorities view basketball as an inconsequential political and cultural activity.

Arguments have been mounted against cultural imperialism hypotheses and the idea that sport has been an extension of neocolonialism in less economically developed or developing nations. It can be argued that dominant forms of sport were not simply imposed upon developing countries but instead were welcomed by these countries because these activities were consistent with their prevailing cultural traditions and values *or* with their aspirations. Japan welcomed baseball because it would help them emulate modern America. Asian and African nations participate in major sports, such as basketball and volleyball, even though they have no national tradition in these sports, because it gives them some national or international visibility and identity. Many developing, and one-time colonized, societies have been eager participants in the diffusion of major and minor nonnative physical activities within their country. In most cases, these sports have been adapted in form and content to cultural conditions and traditions in the recipient countries. In this way sport has been integrated into the societies and cultures of many developing nations through cultural cooperation rather than cultural imperialism.

THE ROLE OF SPORT IN NATIONAL DEVELOPMENT

Nations at all stages of development have used sport to increase the capability of various institutional spheres through enhancing human capital, attracting resources, and increasing the political and economic standing of that nation among nations, with the support of the public. The purpose of national development is to strengthen the internal workings of a society rather than concentrating on external relations. Sport has an important role in the development of a nation's infrastructure and resource availability.

The most commonly accepted role of sport in developing countries is *integrating otherwise diverse ethnic, class, political, regional, and cultural factions* (Lever, 1983; Frey, 1988a). When leaders are able to use sport to generate some cohesion among disparate groups, they create at least an appearance of political order and plant the roots for the establishment of national identity (Anthony, 1969; Shapiro, 1976). Sport has a particularly crucial role to play in establishing cohesion after a time of internal turmoil or even revolution. Nicaragua is a case in point. According to Wagner (1988), the ruling Sandinistas wanted to create a collective consciousness following a civil war by getting broad participation in an effort to build mass sports programs. The Nicaraguan leaders specifically used baseball to build national unity and legitimize their new regime. The aim was to maximize mass access to sport by creating local involvement at all levels of participation, including administration and organization.

Cuba, on the other hand, chose to emphasize high-level or elite sports competition to demonstrate to the world that the 1959 revolution did not totally disrupt political and social order. Castro also instituted extensive physical education programs in order to improve the physical fitness of the population, promote solidarity and discipline, and create popular support for the new regime (Arbena, 1993). Professional sports were abandoned, but elite amateur sport was emphasized so that Cuba could gain international respect as the result of athletic success, particularly in the Olympics.

Thus, sport helps a society to cohere by linking diverse groups to a common orientation in which symbols share meaning, overcoming political and social barriers. "The elite and the masses are brought together in the sports arena and status differences are neutralized and a semblance of democracy is established" (Frey, 1988a: 68). Solidarity is enhanced and the community is brought together despite opposition. Potential internal conflicts are avoided or neutralized by transferring aggression from the street to the stadium. This integration produces a national image of a cohesive and orderly political system.

A second role for sport in national development is *serving* as a *communication channel for prospective institutional change.* Sports events can be stages where representatives of local, regional, and national interests can inform people of impending economic, social, and political changes. Many African national leaders have used sports venues to communicate changes in agricultural policy, health care administration, and birth control practices (Uwechue, 1978).

The former Soviet Union saw sport as an important ingredient in their modernization efforts following World War II. Sport was used to socialize the Soviet people to accept the social changes needed to bring about a modern socialist society. Riordan (1980) noted that Soviet leaders gave sport important functions to perform in this modernization process. They associated sport with "health, hygiene, defense, patriotism, integration, productivity, international recognition, even nation-building" (Riordan, 1980: 3).

A third role for sport in national development is *establishing a model for coordinated planning and project cooperation among local, regional and national organizations.* That is, the planning and administration of major sports events in a nation can provide experience for people and organizations at different levels of government and society to work together on shared tasks and goals. The networks formed for this planning and cooperation in regard to sport may facilitate interagency and intergovernmental coordination and cooperation on other projects as a developing country builds its institutional infrastructure (Arbena, 1985). An additional outcome of cooperation within these networks is the test they provide for the national government of its effectiveness in articulating and implementing its policies and goals at the local and regional levels.

A fourth role for sport in national development is the opportunity it affords sports organizers *to acquire skills and experiences that could be used in the development process.* Experience with the organization and administration of sport can be a training ground for community and government leaders. For example, several countries in Latin America and the small African country of Cameroon have utilized their experience in coordinating soccer events to provide this type of training. As a result of this experience, leaders are identified and developed, the human capital of literacy and communication skills is enhanced, working in groups on collective tasks is made easier, and citizens learn how to operate in a structured environment according to standardized rules (Arbena, 1985; DeMatta, 1986). The development of indigenous human capital through sport prepares the people in a country to be effective in other institutional settings, and it enables them reduce their dependence on larger and more powerful nations. In Cuba, for example, the development of a national sports system created expertise and experience that enabled the people of this country to establish a broader infrastructure reducing their dependence on other countries.

A fifth important role for sport in development is *promoting a distinctively modern set of values and cultural practices* to the population. This role is part of the **civilizing process** (Elias & Dunning, 1986), and it is related to the second role cited here, sport as communication channel to facilitate social change. Soccer matches in Latin America, Olympic spectacles, Spartakiad competitions in the Soviet Union, and similar competitions serve as means by which a country's collective goals can be communicated to the larger population. The reduction of premodern, or primordial, sentiments, as Geertz (1963) called them, is a major factor in the modernization of a nation and its acceptance by the more powerful members of the world community. In emphasizing modern values, sport can clash with the traditional values of a developing nation. At the same time, the values expressed by sport can offer excitement and diversity that contrast with the routine cultural practices and beliefs tied to long-held traditions (Elias & Dunning, 1986). The civilizing role of sport is effective when a population perceives physical activity, particularly at the elite level, to be a proper setting for moral education. In his study of politics, spectator sport, and popular culture in the Soviet Union, Edelman (1993) observes that spectators are more likely to view elite sports as entertainment outlets rather than events where athletes can be showcased as normative and physical role models. The goals of the state for sports did not match those of the Soviet citizens. The government wanted sport to inspire the public to exercise in order to become better workers. Elite athletes would serve as heroic role models providing lessons in discipline, honesty, loyalty, and respect for authority. The state hoped to develop loyalty to its programs by associating itself with the values represented by athletics. Soviet spectators did not attend events to learn values, however; they attended to have fun and be entertained (Edelman, 1993).

The civilizing goals of the Soviet state were compromised by the entertainment orientation of sports fans. Sports events represented contested terrain where the purposes of fans stood in contrast to the aims of socialization or indoctrination and social control that the Communist party assigned to sport. Riots, fights, rowdiness, and other elements of unruly behavior characterized Soviet sport. Sports events represented opportunities to experience fun and excitement, to engage in raucous behavior, and to act in ways that reflected more diverse, and perhaps more deviant, behavior than was possible in normal Soviet life. For these reasons, Soviet sport was not an effective civilizing activity and therefore did not contribute to the general acceptance of the Communist party's authority in the Soviet Union.

A theme prevalent in the growing literature on the role of sport in national development is that indigenous sports are replaced by modern corporate sports imposed by powerful interests alien to these native cultures. Grand Prix races through the back roads of Kenya, golf tournaments in the homelands of South Africa, cricket in the Caribbean, and Major League Baseball in the Dominican Republic illustrate how local cultural and social conditions are ignored in favor of sports that reflect corporate and commercial values as well as interests foreign to the host country (Stoddart, 1989). At issue is the controversy over whether sports in developing nations should be an extension of indigenous traditions or a form of modernization guided or controlled by colonial or neocolonial forces (Riordan, 1986).

Stoddart (1989) and others have asserted that local sports must be treated on their own terms and not as exotic cultural deviations. Klein's (1991a, 1991b) work on baseball in the Dominican Republic and Lever's (1983) study of soccer in Brazil stand in contrast to the tendency to evaluate sport by Western standards. Both researchers acknowledge the importance of understanding the cultural context of these particular sports.

Stoddart (1989) noted that the Western or developed world view of sport may not be the most appropriate in developing countries that are economically depressed and in political upheaval. Sports as a vehicle of character development and value training and a symbol of peace is likely to have little relevance to the lives of people who are economically depressed and hungry. Furthermore, the International Olympic Committee displays little sensitivity to the status of poor nations when it adds new sports, such as baseball and tennis, that have long and rich histories in the developed world but little or no history in many nations of the developing world. The impoverished state of many of these Third World nations means that they are unlikely to have the resources to develop the training and facilities that will enable their athletes to compete successfully in the international arena. Thus, in making decisions to add modern sports, the ruling bodies of international sport bias the competition by favoring developed nations and disadvantaging many Third World nations. Indeed, even within more economically developed nations, many high-level amateur sports, such as polo, equestrian events, skiing, and kayaking, are accessible only to an elite group of performers because of their expense and class differences in access to early training and competition.

Despite these implied or actual criticisms of modern and Western sports, many developing nations have embraced these sports (Riordan, 1986). These countries are often poor and socialist and have committed themselves to modern sports development "to catch up with countries which have already taken a mighty leap forward in industry and commerce, in education, science and technology" (Riordan, 1986: 287). Western sports have not been viewed as vulgar or irrelevant by many non-Western nations in Asia, Latin America, or Africa that have enthusiastically participated in the Olympic Games and tried to emulate the success of the more powerful members of the sporting world.

THE ROLE OF SPORT IN THE RESTRUCTURING OF EASTERN EUROPE, THE USSR, AND SOUTH AFRICA

Yugoslavia, Czechoslovakia, Romania, and the former East Germany (now part of united Germany) have experienced considerable social and political upheaval since the demise of Soviet control over them and the subsequent relaxation of controls in these former satellite communist states. The Communist party no longer is the leader in this part of the world, and the Warsaw Pact, which bound together the nations of Eastern Europe under Soviet control, no longer exists. The Soviet Union has been replaced by a vaguely defined Commonwealth of Independent States, which have been striving to fill the power vacuum created by the collapse of the Soviet Union. In Germany, the 1989 collapse of the Berlin Wall led to the unification of East and West Germany and also a struggle to make the politically unified state socially unified and economically strong under capitalism (Allison, 1993). It is difficult to know what role sport will play in the future of these countries.

Sports officials in these countries seem committed to retaining continued involvement of elite programs in international competition (Hoberman, 1993). It is not clear how much government support sport will receive in the face of competing demands for resources in this period of change. National investment in sport is likely to depend on whether government officials perceive that participation in high-level international sport will serve national interests, foster national unity and pride, enhance or increase the visibility of their international stature, and help the nation accumulate hard currency.

Before the upheaval of 1989, the former East Germany (GDR) was able to translate its success in high-performance international sport into diplomatic uses. The East Germans were strongly committed to use their sports success for propaganda about the communist system and state socialism (Hoberman, 1993). The GDR quickly gained prominence in international sport after its first Olympic venture as an independent team in 1968. Modeling its sports system after the Soviet Union, it was especially adept at identifying and developing promising athletic talent across a variety of sports. With only 18 million citizens and a struggling economy, East Germany became one of the world's sports powers. This success was largely based on the skillful application of sport science for training, but we now know that it also included the controversial widespread use of performance-enhancing drugs (Hoberman, 1993). This system was beneficial for state purposes, but the reaction following the fall of the Berlin Wall demonstrated that the general population was not so enamored with elite sport.

The East Germans had neglected the Soviet slogan "Sport for All" in favor of an emphasis on high-performance elite sport to enhance the international political stature of the Communist regime. Many East Germans resented this emphasis because it symbolized the control the state had over their lives. Elite GDR sport was "identified in the popular consciousness with privilege, paramilitary coercion, distorted priorities, and . . . with an alien, Soviet-imposed institution" (Riordan, 1993a: 38). The people of Poland, Hungary, Romania, Bulgaria, and Germany associated participation in the Olympic Games with Soviet and communist dominance (Riordan, 1993a). An anti-elitist and pro-democratic spirit developed in these countries when they were freed from the grip of the old regime. As a result, popular sentiment resented commercial, professionalized high-performance sport in these countries in favor of sports embodying the principles of amateurism. State control and subsidies of sport will diminish, and self-regulation and self-support will increase, as popular sentiment against elite sports rises and public resources become more burdened by the costs of national reconstruction. However, declining state support will have the ironic effect of forcing sports organizations to seek independent funding. In other words, they will be pushed to adopt a commercial, Western model of sport. The population may be amenable to accepting this approach to sport if the goal is entertainment rather than political indoctrination and state enhancement, as was the previous case. Edelman's (1993) work on the Soviet Union suggests that this will be the case.

Russia and the Former Soviet Union

Soviet leadership had always publicly promoted the concept of *massovost* or "Sport for All"—as opposed to *masterstvo* or elite sport—with the goal of having the entire population embrace GTO (*Gotov k trudu i oborone*, "Prepared for Labor and Defense") (Riordan, 1990). In fact, only a small percentage of Soviet children met GTO standards and very few Russians engaged in sport regularly. The GTO was a program that catered to gifted athletes, rather than to the general population. One of the first outcomes of reform efforts under Mikhail Gorbachev was to abandon the GTO program. It had become increasingly unpopular with teachers and coaches, and its existence was shrouded in a history of inaccuracy and deceit.

In 1985, Mikhail Gorbachev initiated a massive effort to reform the Soviet Union. Gorbachev recognized the need to give new vitality to an economy that had become stagnant and outdated by virtue of "hyper-centralized" control by a rigid bureaucracy and an emphasis on heavy industry and military production (Edelman, 1993). It was necessary for the Soviets to participate in the developing global economy, but their money and products were not needed or wanted; they were isolated from global participation. The social and economic reforms undertaken by Gorbachev placed significant demands on the institutions of Soviet society, and the state no longer provided the support it once did. The highly bureaucratized and subsidized Soviet sports system, like other Soviet institutions, was forced to seek funding from nongovernmental sources. Sports had to consider commercialization and professionalization in order to survive in an emerging market economy. For the first time, commercial health clubs, charging user fees, were permitted to operate. Sports no longer were free to all (Riordan, 1990).

Russian sport has now begun to look increasingly like sport in the United States. The old Soviet sports figures were heroes, but they were heroes of the state; their status was never officially allowed to surpass the importance of the nation, community, group, or team for whom they competed. Today, however, top athletes in the states of the former Soviet Union have become stars. They are recognized as individuals and have their numbers retired and hung from the rafters of the arenas that were the sites of their greatest exploits (Lilley, 1994). Cheerleaders, commercial advertisements, merchandise sales, sports figures pitching various products, rock music during lulls in action, alcohol consumption at games, spotlights and light shows, mascots, and commercial logos for McDonald's, Coca-Cola, and other American companies emblazoned on uniforms are now common features of Soviet athletic events. American basketball players, such as Teresa Weatherspoon, Medina Dixon, and Tony Turner, have played for Russian teams and have been rewarded with five-figure salaries, housing, bodyguards, and free gasoline (Lilley, 1994). Their presence enhances the quality of local teams, resulting in greater attendance, commercial investment, and profit.

The influence of capitalism is prevalent now that teams and clubs no longer can depend on the state for financial support. Firms and individuals with a ready reserve of hard currency, usually American dollars or French francs, have the advantage in the contemporary climate of Russian sports. International sports competition represents one of the few ways hard foreign currency can be generated for the Russian economy. Edelman (1993) pointed out that this fact was not lost on the reformers who intensified the former Soviet athletes' involvement in international events. In order to keep the athletes' interest, however, a form of professionalism in sport, which was antithetical to the advertised Marxist-Leninist preference for amateurism, had to be implemented. In fact, the Soviet interpretation of amateurism involved government subsidies for athletes in the guise of pay for work in nonathletic settings, such as the military.

SPECIAL FOCUS

The Future of Sport in the Former Soviet States

The unified Soviet team of 1992 consisted of over 500 athletes from twelve provinces, and many of them either paid their own way to the 1992 Barcelona Summer Games or limited their participation in Barcelona. The Goskomsport—which was the state committee for sport in the Soviet Union—formerly financed Olympic competition with hard currency. The Goskomsport has virtually disappeared now, and teams and athletes are being sponsored by commercial firms such as Adidas and Smirnoff whose interest is primarily profit-driven—that is, to develop a market for their products in the nations of the former Soviet Union. Many of these athletes, such as world-class pole vaulter Sergei Bubka, have trained in other countries (Erlanger, 1988). Facility management, financial development, and administration are in a state of chaos and disarray. Athletes do not know whom they represent. Many are leaving their native country for better conditions in other countries. It is obvious that the condition of sport in the nations of the former Soviet Union will not stabilize until the political, economic, and social conditions in these nations stabilize. Such stability may take many years to attain.

Financial exigencies and the demand by the state that all institutions become self-sufficient or exhibit *khozraschet*—meaning "profit"—forced sports organizations to explore commercialism and professionalism (Edelman, 1993). Neither was unfamiliar to Soviet sports officials. Booster clubs, increased ticket prices, concessions, endorsements, memorabilia sales, incentive clauses in player contracts, a star system for marketing players, and other forms of professionalism appeared. Not everyone was happy with the changes.

Still in need of hard currency, Russian sports organizations have "sold" or "leased" their athletes to foreign teams (Edelman, 1993). The brawn drain, more commonly seen in European and Latin American countries, can now be seen in the former Soviet nations, with their basketball, hockey, and soccer stars competing in Finland, Austria, Germany, Greece, and Italy. The involvement of these athletes in American professional basketball and hockey leagues had begun before the collapse of the Soviet Union and continued afterward. Some athletes competing in former Soviet countries resisted the practice of giving a portion of their earnings to their sponsoring sports club at home; tennis player Natalia Zvereva refused to share any of her 1989 prize money with the USSR Tennis Federation (Riordan, 1990). The exodus of native talent reduced interest in domestic sport as it reduced the quality of the competing athletes, but the hard currency earned through the player contracts was sometimes the only basis on which many sports clubs could survive. Thus, the unwillingness of top athletes to return a portion of their earnings to their sponsoring club posed a serious financial threat to the survival of these clubs.

The reality of Russian sport under communism often contrasted with the images of it portrayed in Soviet sports ideology. Sport in the USSR was supposed to be free of the problematic and corrupting elements associated with capitalist sport, such as game fixing, elitism, illegal drug use, and under-the-table payments to athletes (Riordan, 1990). The new openness in Soviet and post-Soviet society revealed that Soviet sport had not lived up to the propagandists' conception. Despite the rhetoric of mass sport and physical culture, the Soviets—and more specifically, the Russians—were much more likely to be spectators rather than active participants in sport. Athletes in the former Soviet states were exposed as paid professionals, as Westerners always claimed, rather than amateurs, which was their status in the ideology of the Soviet state. At the same time that Soviet sports officials were condemning professional sport in the West, they were paying soccer players over 200 rubles a month (Riordan, 1990). As a result of this unveiling, athletes in the former Soviet Union now openly belong to professional associations and compete as professionals in many sports.

The environment of economic and political freedom, represented by the principles of *glasnost* and *perestroika,* has introduced a number of foreign sports to the people of these countries. Grand Prix auto racing, baseball, golf, darts, dog and horse racing, billiards, and even a professional football game have recently appeared in the nations of the former Soviet Union. The arrival of these sports raises questions about how much athletes and coaches in the former Soviet states will be influenced by foreign—typically Western—models of commercial, professional, entertainment-oriented corporate sport.

The Soviet sports system placed high expectations on coaches and athletes. At the same time, these coaches and athletes at the elite levels of this system could generally depend on the resources they needed to compete successfully and to live fairly comfortably. The unified team that was the last to represent the former Soviet Union in an Olympic Games competed in the 1992 Summer Games in Barcelona (see the Special Focus discussion). This team may have provided a glimpse at the future of sport in the former Soviet states, and it may have offered a hint of the chaos that was to characterize sport during the period of transition from communist consolidation and rule.

South Africa

South Africa was readmitted to the Olympic Games for the 1992 Summer Games after a thirty-two-year absence. Shortly afterward, the cities of Cape Town, Durban, and Johannesburg submitted bids to host the 2004 Summer Games. South Africa was allowed to reenter the Olympic arena because it had decided, in the face of massive and persisting world and internal pressure, to abandon its policy of apartheid.

On December 22, 1993, the white minority government voted itself out of existence and ended forty-five years of official apartheid and a much longer history of institutionalized racial segregation and discrimination. By a vote of 237–45 and despite strong resistance from white conservatives, South Africa's parliament adopted a new constitution giving all races equal rights. The multiracial elections in April 1994 resulted in nonwhite majority rule because nonwhites outnumbered whites by approximately 6.5 to 1.

Before these dramatic changes, there was some evidence that the racial divisions in South African sport were beginning to break down (Kidd, 1991a). In February 1991, the de Klerk government released African National Congress (ANC) leader Nelson Mandela and lifted the ban on liberation organizations such as the ANC. White and nonwhite organizations began to talk to each other about how the structure of government and other institutions, including sport, would change after apartheid. Several white sports organizations endorsed the ban on international competition, at the same time making moves to integrate sports competition and strengthen nonwhite sports organizations within the country (Kidd, 1991a). The liberation movement had gained such prominence within the country by 1990 that protests and demonstrations were held against white international sports competitors, something unheard of a few years earlier. Organizations such as the South African Council on Sport (SACOS) led the sports liberation movement with the slogan "No normal sport in an abnormal society," which called for a blanket boycott of all sports activity that included South African citizens, white or nonwhite, until apartheid was eliminated. This strategy was not popular among nonwhites, however, as it impeded their efforts to improve their sports experience.

At the same time that a boycott was being supported, efforts were being made to strengthen the infrastructure of nonwhite sports organizations and enhance the level of competition in anticipation of reentry into the international sports arena. The liberation movement advocated the nonracial character of sports organizations and competition, and it rejected the contradictory policy of multiculturalism that had limited access to sports resources and high-level competitive opportunities for nonwhite athletes. The movement even proposed sending a nonracial team to the Olympics under the sponsorship of the movement rather than under the South African flag to draw more world attention to continuing racial issues in the country (Kidd, 1991a).

An important force in the creation of nonracial sports in South Africa was the National Olympic and Sports Congress (NSC). The NSC organized massive protests against tours of international teams in cricket and rugby during the boycott period, and it has helped break the pattern of disproportionate (over 90%) corporate investment in white sports by increasing the amount of corporate sponsorship money for nonwhite training and competition (Kidd, 1991a: 43). The nonracial sports movement also forced the abandonment or modification of symbols, such as the national anthem and flag, associated with earlier times (Guelke, 1993).

Although sport did not independently cause an end to apartheid in South Africa, it played a significant role in pressuring the South African government and sports establishment to consider ending this policy in sport and the larger society. The pressure South Africa felt from the international sports boycott suggested, at least symbolically, that their membership in the community of nations was in jeopardy. Considerable pressure from other African nations, economic boycotts by the United States and many other nations, and an increase in sympathy for integration and racial tolerance among white South Africans also contributed to the demise of apartheid. Thus, we see in the case of South Africa a major example of the tight interweaving of politics and sport at the international level.

THE POLITICAL ECONOMY OF THE OLYMPICS

In 1894, Baron Pierre de Coubertin, a French nobleman who was disappointed by the performance of French soldiers in the 1870–1871 Franco-Prussian War, established an organization that later became known as the International Olympic Committee (IOC). De Coubertin believed that France's military prowess could be reestablished by means of a national fitness and sports program. He also believed that international sports competition could be an important vehicle for achieving international understanding and world peace. The Olympics would, at least for a short time, distract the nations of the world from their disagreements and compel them to deal with each other diplomatically (Lucas, 1980; Guttmann,1992).

The first modern Olympic Games in 1896 set an important precedent of recognizing national affiliations by allowing athletes and teams to display flags and other national symbols (Strenk, 1979). The opening ceremony is highlighted by a parade of nations, medal winners are recognized by a display of the flag and the playing of the national anthem of the country the athlete represents, and unofficial medal rankings by nation are kept. Efforts to denationalize the Games would, however, run the risk of lowering media interest, and consequently the commercial return, in the Olympics. There is a symbiotic relation between the Olympic Games' nationalistic tendencies and the commercial/corporate interest in this event. Thus, even though Olympic ideology stressed participation by individuals rather than nations and emphasized the goal of "sport for sport's sake," the stage was set for nationalistic and patriotic public displays. In this context, it easy to interpret Olympic competition in terms of its implications for national pride (Espy, 1979; Kanin, 1981; Nixon, 1988b; Guttmann, 1992; Hill, 1992).

The founding of the modern Olympic Games deepened the roots of politics that were planted in the ancient games. Political displays have occurred in every modern Games since their founding. In some cases, such as the Nazi Olympics of 1936, the Israeli massacre in 1972, the American boycott of the 1980 Games in Moscow, and the Soviet boycott of the 1984 Los Angeles Olympics, politics has overshadowed sport in the Olympic arena. There were many minor or ceremonial political displays in earlier Olympics, but the politicization of the Olympics reached a new level of sophistication and pervasiveness in 1936 with Hitler's Nazi Olympics.

The Cold War precipitated a number of prominent political controversies in the Olympic Games, especially after the Soviet Union entered the competition in 1952. Nationalistic concerns escalated at this time, since the Soviets and their East European satellites participated with the avowed purpose of promoting the virtues of socialist and communist systems. These problems reached their peak with the successive boycotts of the 1980 and 1984 Olympic Games.

The protest actions taken by the United States in Moscow in 1980 and the Soviet Union in Los Angeles in 1984 dramatize how the Olympic Games can become a political pawn in the chess game of international politics (Frey, 1984b). Moscow was the designated host of the 1980 Games. President Jimmy Carter asked the IOC to postpone, cancel, or move

SPECIAL FOCUS

The Nazi Olympics

Berlin was designated as the 1936 site of the Olympic Games before Hitler came to power. Hitler hoped to use the Games to mark Germany's return to respectability and to proclaim the prowess of the German system following the end of World War I (Mandell, 1971; Kanin, 1981). German officials also recognized that the Olympics provided a wonderful propaganda opportunity to unite that country's Aryan or white population with the existing political regime. The International Olympic Committee disregarded international concerns about Germany's intentions for the Games because the Germans assured the IOC that events would be held according to Olympic standards. Thus, the IOC adopted a stance of ignoring internal politics of members and host nations as long as their requirements for the Games were met (Leiper, 1981).

Jews were excluded from the 1936 Olympics, which prompted many protests from Jewish communities across the world, including the United States. A boycott proposed by the American Jewish Committee failed to materialize because of the resistance of Avery Brundage, newly elected president of the Association of American Universities (AAU), and the protests of German-Americans. Brundage felt that a boycott would improperly subvert the Games for political purposes, and he had been assured by German officials that Jews would be allowed to participate. Rudi Ball, a member of the German hockey team, and Helene Mayer, a fencer, were the only Jews permitted to compete in the Games. Brundage was misled and coopted by the Germans even as he praised the Germans for representing the true spirit of Olympism (Kanin, 1981). Only Ireland boycotted the Games. Despite prominent Jewish populations, other nations, including the United States, participated in the Hitler's Olympics. The Games were conducted in Berlin with only one major incident—the success of Jesse Owens and America's black athletes that countermanded the Aryan supremacy claims by German propaganda. Germany was declared the winner of the 1936 Games and Hitler was ecstatic.

the Games to another location that year if the Soviet Union did not withdraw its troops from Afghanistan. Since the Soviets did not yield and the IOC did not move the Games, the USOC withdrew America's team.

The boycott gave President Carter and the United States a chance to assert its human rights policy to the world. The White House felt that participation in the Moscow Games represented an endorsement of Soviet foreign policy in general and its invasion of Afghanistan in particular. The Carter administration clearly saw a direct connection between the Olympics and politics, and according to some poll results at the time, a majority of Americans seemed to favor the boycott (Leonard, 1993: 373–374). It is unclear whether the boycott had any political impact on the Soviet Union, since it did not seem to change the Soviet Union's foreign policy or force withdrawal from Afghanistan (Nafziger, 1980; Frey, 1984b). The major impact of the boycott was felt by the athletes of the several boycotting nations who lost their chance to be Olympians. American athletes filed a lawsuit to prevent the USOC from boycotting the Moscow Games, but without success (Leonard, 1993: 374).

Boycotts have drawn attention to world events and, in the case of South Africa, have contributed to significant international pressure for change within nations, but there is no evidence of a direct and immediate effect of an Olympic boycott on the internal or external policy of the impacted nation. In addition, none has caused the cancellation of an Olympics. The quality of the competition has suffered as the result of boycotts, but the stature and popularity of the event have not. Furthermore, in 1980, the boycott seemed to increase, rather than reduce or neutralize, cold war tensions. It strained relations among the United States and its allies, who participated in the boycott but were not entirely sympathetic to the U.S. position. The Soviet Union lost an opportunity for cultural exposure since worldwide television coverage was limited by the absence of the U.S. media and American athletes. Eighty-eight nations decided to compete in Moscow, and resistance to the American boycott by the European community and many other nations gave evidence of a changing reality in international relations. The era of the superpower dominance was coming to an end.

It is ironic that just a few years before the boycott, the President's Commission on Olympic Sports had decried abridgements of the rights of Olympic athletes by politics. Commission members wrote in the executive summary of their report that "the Commission deplores the actions of governments which deny an athlete the right to take part in international competition and calls upon world sports leaders to take whatever steps are necessary to eliminate the misuse of the Games" (President's Commission, 1977: l). It is obvious that this report had no impact on American foreign policy in 1980.

In retaliation for the 1980 boycott by the United States, the Soviet Union did not send a team to the 1984 Games held in Los Angeles. Retaliation for the 1980 American boycott is the generally accepted reason for the Soviet boycott, although the Soviets referred to difficulties with accreditation of officials, the inability to moor a Soviet ship in Long Beach harbor, fear that athletes would be mugged, and a concern that after the Soviets had shot down a Korean airliner in August 1983, protests and other actions would make the American scene too hostile for their athletes (Hill, 1992). In addition, some observers suggested that the Soviets did not want to contribute to the success of an Olympics that was organized for the first time by a private corporation rather than a city, that was spearheaded by a capitalist entrepreneur, Peter Ueberroth, and that was seen as an advertisement for capitalist ideology (Nixon, 1988b). A "Ban the Soviets" coalition had been formed, but even though this group was not significant it became an example of the hostility the Soviets feared. Cuba, East Germany, and all of Eastern Europe, except Romania joined the USSR in the protest. The boycott did damage the quality of competition but it did not diminish the commercial success or international media attractiveness of the 1984 Los Angeles games.

Other major political incidents have been associated with the Olympics, most of which reflect nationalistic concerns. The importance of official membership in the Olympics is exemplified by the diplomatic battle over whether the Republic of China (Taiwan) or the People's Republic of China (mainland China) should be the Chinese representative and whether the German team of the Cold War era should be unified or separated into East and West contingents. Eventually each of these four nations was granted the right to participate.

The South African boycott, the "Black Glove" salute protesting race relations in America at the 1968 Games in Mexico City, the ideological ploy of South Korea to offer to work toward unifying Korea if Seoul was to host the Games, boycotts or threatened boycotts of each Olympics by a variety of nations, and even tragic murders of athletes such as the killing of Israeli athletes by Palestinian terrorists in 1972 suggest that politics are embedded in the history of the Olympic Games. This association of the Olympics with politics has left a much deeper imprint on sport than the world order, though. Perhaps the most significant political consequences of the Olympics have been the diplomatic recognition and propaganda gains from hosting an Olympics or achieving athletic success in the Olympic spotlight. Of course, the gains for Olympic hosts have typically been offset by the huge financial costs incurred in staging an Olympics. The 1984 Los Angeles Olympics were an exception to this pattern (Nixon, 1988b).

The 1984 Capitalist Games

The total operating budget for the 1984 Games was approximately $500 million, and Los Angeles Olympic Organizing Committee (LAOOC) President Peter Ueberroth vowed that these Games would be profitable. The incentive to earn a profit was provided by the fact that previous host cities had to draw considerably from public funds, which resulted in huge public debts. Montreal, for example, was saddled with a $1 billion debt as a result of hosting the 1976 Summer Games.In the face of such potential debt, the citizens of Los Angeles had voted 3-1 not to provide public funds for the Games. Since no other city had offered

Source: AP/Wide World Photos

FIGURE 14.3 The 1984 (LA) Capitalist Games

to host the 1984 Summer Games, the IOC was faced with a dilemma. Its policy was to allow only cities to host and assume financial responsibility for Olympics. The Los Angeles bid, however, stipulated a private corporation as the host. To secure a host, the IOC accommodated Los Angeles and waived its rule for hosts. This move represented a major change in the way the Olympics were run, as the power shifted from the IOC to an independent and nongovernmental organizing committee (Hill, 1992). Thus, Ueberroth and the Los Angeles Olympic Organizing Committee (LAOOC) were able to operate on a free-enterprise basis and rely on private funding to stage the 1984 Olympic Games. For this reason, the 1984 Summer Games in Los Angeles were dubbed the "Capitalist Games," and they were the most financially successful modern Olympics (Nixon, 1988b).

In fact, the financial success of the Los Angeles Games and especially the LAOOC caused some controversy. Olympic dignitaries, visiting nations, LAOOC staff members and volunteers, and local communities were asked to make sacrifices or donations for the sake of austerity—these were to be the "Spartan Games"—and the financial solvency of the Games. Furthermore, taxpayers, who were not supposed to shoulder any of the costs of these Games, ultimately assumed the estimated $100 million cost of security. Many of these people felt misled or betrayed when they learned that the Games had generated an estimated $250 million in "surplus," which was ten times more than predicted. The surplus was used mainly to benefit the United States Olympic Committee and amateur sports organizations in Southern California and throughout the rest of the United States. Thus, among many individuals, organizations, and communities who had made the requested sacrifices to assure that the Los Angeles Games remained austere and avoided a deficit, there was "a surplus of ill will" (Creamer, 1984).

To be profitable, the LAOOC had to secure a lucrative television contract, which was achieved with a then-record $225 million from ABC, which earned the network a $300 million dollar profit after production and promotional costs had been paid (Seifart, 1984). The number of commercial sponsors was reduced from 381 in the 1980 Winter Games to just thirty-one so that the right of sponsorship could be made very exclusive—and very expensive. The price

POINT-COUNTERPOINT

WOULD THE OLYMPICS SUFFER FROM THE ELIMINATION OF ALL REFERENCES TO NATIONAL AFFILIATION?

YES

National affiliations are important to the popular and economic success of the Olympic Games, which now cost hosts approximately $1.5 billion to stage. The appeal of the Games to the mass media, corporate sponsors, and the international sports public is enhanced by the nation-versus-nation format. If national affiliations were not recognized as part of the structure of the Olympics, the mass media, sponsors, and the public would be much less interested, and the resources needed to stage such a large international sporting event would disappear.

NO

If the Games are to be a genuine international sports event focusing on the ideals espoused by Baron de Coubertin, they must eliminate all references to national affiliation. Nationalism politicizes and sometimes dominates the Games, taking precedence over the athletic accomplishments of individual athletes and teams. The elimination of medal counts, national anthems, and the parade of nations could reduce interest among those only wanting to exploit the Games for their economic or political purposes, but it would make them pure sport.

How should this debate be resolved?

of sponsorship was raised to a minimum of $4 million per corporation. The biggest corporations in America—IBM, Coca-Cola, Xerox, Budweiser, General Motors, and others—jumped at the chance to obtain global exposure of an estimated 2 billion viewers to its products without competition from any of its competitors. Their total contribution was estimated at just under $130 million.

Some criticized the 1984 Games as too commercial, but this criticism has to be assessed in the context of the pattern of increasing commercialization of the modern Games, especially since the infusion of large sums of television money began in 1960. In fact, broadcast revenue is projected at approximately $556 million for the 1996 Games in Atlanta or 34 percent of total revenues.

The role of corporate sponsorships was institutionalized during the 1984 Games, and now the Olympics could not survive without the millions provided by Coca-Cola, IBM, Kodak, and other companies seeking the visibility and revenue generated from an association with the Olympics. These firms will pay as much as $40 million to have the worldwide rights to be the exclusive auto, photocopy, soft drink, or bank card sponsor of the 1996 Games in Atlanta. The IOC, the host country Olympic Committee, and the city host share in the millions generated from sponsorships. For example, the USOC receives 30 percent of the fees paid for exclusive U.S. marketing rights. The 1994 Lillehammer Winter Games were seen by over 500 million viewers from 100 countries *each day.* The Olympics is the "most powerful marketing vehicle in the world." Sponsorships are necessary to offset the nearly $1.5 billion a city will expend to host the Games.

The financial success of the Los Angeles Games also prompted criticisms that the these Games had been a vehicle of American **jingoism**, or had been used by the American hosts as a means to promote capitalism and the American way of life. These messages came through in advertisements for American Express, IBM, and General Motors, which seemed to promote the American culture and standard of living in an apparently ethnocentric fashion. ABC and the print media were accused of flag waving and "us-ism" by focusing on American athletes in their broadcasts and stories (Deford, 1984). It was as if the only participants in the 1984 Games were Americans.

The history of the modern Olympics has been filled with contradictions between ideals and realities.

One of the great accomplishments of Olympic officials such as the late Avery Brundage is that despite their frustrations with the contradictory realities of commercialism, professionalism, and nationalism, they have been able to maintain the myth that the spirit of Olympism couched in the values of amateurism, sportsmanship, fair play, and individual achievement lives on in Olympic sports. In fact, Olympic history is a history of the interweaving of politics, economics, and sport. This history rests on the belief among nations that the Olympic arena is a highly visible transnational setting to pursue national interests and ideological goals without the risks associated with international commerce, conventional diplomacy, or armed aggression (Hargreaves, 1992). This history also involves the pursuit of profit and publicity by private investors.

CONCLUSION: THE NEW WORLD ORDER IN SPORT

Wagner (1990) noted several simultaneous trends that have had an impact on the world order of sport. One is the globalization trend that has exposed people in a number of countries to sports they had not seen before in their country. For example, basketball, volleyball, and auto racing increasingly have become global sports found all around the world. In recent years, we also have seen American football played in Great Britain and Russia. The National Basketball Association will open an office in Mexico City in 1995 joining existing offices in Hong Kong, Tokyo, and Madrid. The NBA will televise its games to over 400 million homes in 141 countries, including Iran, Ukraine, Armenia, Turkmenistan, and El Salvador, across the world.

Another aspect of this globalization trend has been the increasing tendency for athletes from one country to compete in leagues and teams based in other countries. For instance, we now see American athletes on basketball, volleyball, and soccer teams in Europe, Russia, and other countries of the world. Athletes from Eastern Europe, Africa, the Soviet Union, and Latin America routinely can be found on America's amateur and professional track, baseball, basketball, and football teams. Of course, we have long seen Americans compete in the Canadian Football League, and Canadians, Americans, and Europeans compete in the National Hockey League. In January 1994, the Los Angeles Dodgers signed Park Chan Ho, a pitcher on the South Korean national team, to a minor league contract. Ho is the first Korean player ever signed by a U.S. team.

The international migration of athletes is not just between the United States and the rest of the world. European soccer players are just as likely to be found on teams in Japan and Latin America as in the United States African track and field athletes compete for Great Britain, France, and other European countries. Athletes who formerly competed for the Soviet Union and East Germany can now be found on teams from France, the unified Germany, and the United States. All countries experience some of the brawn drain now that national boundaries seem to be blurred where athletic affiliations are concerned. The exception to this trend, of course, is the Olympics where athletes typically represent their country of national origin.

A second international sports trend is increasing participation by the smallest nations in major international sports events such as the Olympics, the Goodwill Games, the Commonwealth Games, and the Pan American Games. The IOC has instituted an Olympic Solidarity Program that is designed to increase investment in the sports of developing countries so that they can become more competitive at the international or regional level (Coakley, 1994).

A third trend is the increasing size of the television audience around the world for sports events of various kinds. This expanding television coverage has enabled many people to become familiar with a variety of sports with which they were previously unfamiliar. It has also enabled outstanding athletes and teams from small and poor countries, such as soccer players in Cameroon and Kenyan distance runners, to gain global exposure and respect.

A fourth trend is the growing awareness of the political importance of sport in international relations and in national development (Wagner, 1990: 401). In addition, there has been increasing recognition of the global economic or commercial potential of sports. For example, sports-related corporate sponsorships have extended their reach around the world. Adidas,

Nike, Coca-Cola, Pepsi, and other well-known corporations have been involved in supporting sporting events across the world in an effort to expand their markets and enhance their profits. The international exposure of consumer products through advertising at sports events is made even more possible by the formation of leagues that include teams representing cities of different countries (e.g., Montreal and Toronto in Major League Baseball).

While these trends have been described as representing the Americanization of sport (Maguire, 1990), Wagner has argued that the concept of Americanization does not accurately or fully capture what is happening in these trends. In fact, in many countries, there has been a blending of the old and the new, of American, European, Western, and traditional qualities. This blending is especially evident in the recent development of sports of Asia and Africa. Thus, the significant characteristic of globalization is not that it reproduces American culture around the world. The notion of Americanization oversimplifies the process. On the other hand, there is an imprint of American influence along with cultural influences from other nations that blend together to produce relatively homogenized or standardized forms of sport. Thus, globalization is part of a larger cultural process that represents the internationalization of sport.

The film *Rollerball* represents a vision of the future in which there is a world government of powerful corporations and war is banned. In this new world order, national affiliations in general and in sport are meaningless, and athletes instead compete for their corporate sponsors. The sport at which they compete is rollerball, a violent and often lethal competition. Thus, in the new world order of rollerball, sport is the primary stage for the expression of violent emotions and aggressive behavior that no longer can find their outlet in war. With intensifying rivalries and mortal combat to establish new national borders in the new world order of the 1990s, it is unlikely that the world will soon see the disappearance of national identities envisioned in *Rollerball*. On the other hand, the increasing power of multinational corporations and the expanding reach of the sports they sponsor could produce increasingly standardized forms of global sports. Whatever specific forms these sports take, it is likely that they will bear the clear imprint of the capitalist culture of large and powerful multinational corporations.

Thus, even as the world continues to erect political boundaries separating nations, sport may be blurring those boundaries by representing the global influence of transnational economic corporations. That is, the continuing globalization of sport may be helping the multinational corporations that sponsor them to expand their markets across more and more national boundaries. The corporate and commercial mass media are likely to be a major force behind the globalization of sport, both because they can increase worldwide exposure to sport and because sport is an effective vehicle for increasing their own audiences, markets, and profits around the world. In this changing world, governments may find that they need to pay increasing attention to sport because global sports cross national borders and test the scope of national control over global activities. Thus, it appears that sport may be at the intersection of powerful economic, political, and social forces of change as we move toward a new world order, and the globalization of sport may give us a hint of how that new world order may be formed.

SUMMARY

Despite the exhortations of athletes, athletic officials, and the public, politics and government influence are likely to remain embedded in sport, as they traditionally have been: it is the central political reality of sport. Sport, politics, and government have been intertwined in America in many different ways. Although the U.S. federal government has exercised less direct control over sport than other national governments have, on many occasions federal, state, and local government agencies have influenced American sport. A number of issues, such as restraint of trade and violence, have attracted substantial attention from government officials, politicians, the courts, and regulatory agencies. On the international level, governments and political groups have intentionally used sport to achieve political purposes, such as international understanding and peace, staging a protest, and promoting national unity. Efforts by colonial or neocolonial powers to dominate smaller nations have been expressed in part through efforts to control their sports, but these smaller nations have sometimes resisted such colonial domination through various forms of cultural resistance.

The history of the modern Olympic Games is a history of the political economy of international sport. In recent decades, the Olympics and international sport in general have reflected the major political currents that have changed nations and the world order. Political tensions and changes in South Africa, the Middle East, Eastern Europe, and the USSR all could be seen through the filter of sport. The so-called Capitalist Games in Los Angeles in 1984 reflected the ongoing tensions between the world's two superpowers at the time. Although sport may not change the world, we see in international and global sports the major social, cultural, political, and economic forces that are part of the changing world order.

APPENDIX

Science and Research Methods in Sport Sociology

THE SCIENTIFIC METHOD IN SPORT SOCIOLOGY

The scientific method, in general, involves the systematic or orderly collection of evidence through the objective or unbiased use of procedures that produce precise measurements (Hess, Markson & Stein, 1993: 26). *Precise measurements* are based on reliable instruments that produce the same results with repeated measurements and on valid instruments that measure what they were intended to measure. *Scientific objectivity* and *lack of bias* imply that researchers use procedures for the collection, analysis, and interpretation of data that conform to generally understood standards within their scientific community and that to the extent possible, personal attitudes, expectations, and values of individual researchers do not influence how data are collected, analyzed, or interpreted. As Giddens (1991: 21–22) has suggested, though, scientific objectivity in sociology is assured more by the public character of the discipline than by absence of bias among individual researchers.

When scientists make full public disclosure of their methods, peers in their discipline will be able to evaluate their methods and use them themselves to try to reproduce previously reported results. If scientists adhere to this standard of disclosure so that the nature and adequacy of their methods and results can be fully examined and debated, objectivity will be possible even when highly controversial issues are being studied. Scientific debate, which is a common feature of the scientific enterprise (Kuhn, 1970), is also more fruitful when scientists state the logic of their arguments in clear and systematic ways that are readily understood by their peers.

As social scientists, sport sociologists employ the scientific method to describe, explain, and predict social facts about sport. To be able to explain or predict, it is necessary to understand *cause-and-effect relationships*. In science, it is assumed that all events have causes and, under identical conditions, the same cause is assumed to produce the same effect repeatedly (Robertson, 1989). A world that did not meet these assumptions would be unpredictable and impossible to understand in a meaningful way. The challenge for sport sociologists is to identify and understand causes and effects in a social world that is complex and contains many possible causes and combinations of causes.

In trying to identify causes and effects, sport sociologists look for correlations between variables. *Variables* are factors with states or conditions that can change or differ. Among the most frequently studied variables in sociology are gender, social class, race, and age. It is easy to see how these factors differ or change. Other potentially important variables in sport sociology research are, for example, the rate of sports participation for particular groups or social categories, the amount of violence in different sports, and the amount of pressure to win that coaches feel over time or in different sports. The name given to variables that are seen as causes or factors influencing other variables is *independent variable*. A factor or condition that is seen as the effect of one or more independent variables is called a *dependent variable*.

Even though two variables may be *correlated* —that is, where change in one variable is regularly associated with a specific kind of change in the second one—it cannot be assumed that one variable is the cause of the other or that they are related at all. For example, playing on the high school varsity tennis or golf team may be highly correlated with earning high academic grades, but this correlation does not mean that either the athletic participation variable or the academic achievement variable is a direct or strong cause of the other variable. It may be that tennis and golf teams disproportionately have members from affluent families and that growing up in an upper middle class or wealthy family is the main cause of high academic achievement. Some aspect of participation on a tennis or golf team could contribute to higher academic motivation—for instance, if team members from more affluent backgrounds influence less affluent teammates to accept their values about academic achievement. Thus, one variable may have an indirect or complex causal influence on another.

A second example shows the importance of using caution in assuming *any* causation from correlation. It may be that watching Major League Baseball and drinking iced tea are highly correlated. One would not want to conclude that watching baseball causes iced tea drinking because the apparent relationship between these variables could be explained by a third variable, warm weather. Baseball is often played in warm weather, and people are very likely to drink more iced tea in warm weather. In this case, the apparent baseball watching–iced tea drinking correlation might largely disappear if baseball were played in a cool climate, where people might be more likely to drink hot chocolate or warm tea. This latter correlation is actually a *spurious correlation*—that is, a coincidental relationship that is not causal (Anderson & Zelditch, 1975: 184–186; Robertson, 1989: 17–18). Correlations are spurious when the relationship between variable *X* and variable *Y* can be entirely explained by a third variable, in this example, the weather.

The establishment of causality requires the application of experimental or statistical *controls* so that it is possible to eliminate the possibility that other factors may be influencing the correlation between two variables. This process could involve complicated procedures, which are discussed in books about research design and statistics for sociological research (e.g., Anderson & Zelditch, 1975; Babbie, 1995; Miller, 1991). If we are able to show that a correlation between two variables is not spurious and that other factors that could be causally related to the correlated variables are not, we then need to establish the time sequence of the relationship to be able show which of the variables is a cause and which is an effect. The logic of causality assumes that causes come before effects (Smith, 1987: 25). Of course, it is possible that the two variables are related by *reciprocal causation*. For example, successful athletic performances may cause athletes to spend more hours in training, and spending more hours in training may lead to more athletic success. We can see from this example, too, that a causal relationship can weaken, disappear, or change in direction beyond a certain state of development of one or both variables. In this case, when athletes reach a certain level of training, further training may be detrimental rather than beneficial; or, when a certain level of success of achieved, athletes could become so confident that they decide to reduce their training commitment.

In order to make generalizable statements about correlations and causes that are of enduring theoretical value, sport sociologists typically rely on *sampling*. Instead of looking at the entire population—for example, of people, of social relationships, or of groups or organizations—that is of interest, the researcher can study a portion of the population that is representative of the whole population. *Random sampling,* by which every member of the populution has an equal chance of being selected for inclusion in a study, allows social researchers to generalize about a population on the basis of evidence derived from relatively small segment of the population. For example, a researcher interested in learning how American sports fans in urban, suburban, and rural areas feel about further expansion of professional football will be able to generalize accurately, within several percentage points, on the basis of a sample of a few thousand Americans, as long as the sample is chosen by a random method. Randomization will assure representativeness, so that all the important features of the population, including place of residence, are represented in the sample. It is important to note that random selection, rather than sample size, assures representativeness. This fact often is misunderstood. If the researcher had surveyed 10 million New Yorkers, who were able to root for two teams in the New York City area and one in upstate Buffalo, the results could have produced a very unrepresentative picture about how urbanites, suburbanites, and people in rural areas across the nation felt about expansion.

RESEARCH METHODS IN SPORT SOCIOLOGY

Notions of variables, correlation, causation, control, generalizability, and sampling are basic ideas in social science. They guide the methods by which sport sociologists collect evidence. The most basic types of social research methods used in sport sociology are surveys, observational studies, experiments, and secondary data analysis (Robertson, 1989: 20–24; Miller, 1991).

Surveys

Social *surveys* usually rely on interviews or mailed questionnaires to gather standardized information in a systematic manner from a sample of respondents about their own attitudes, beliefs, or actions or those of groups, organizations, or communities to which they belong. One of the largest sports surveys undertaken in the United States was sponsored by the Miller Brewing Company (Miller Lite, 1983). With the assistance of an advisory panel that included prominent sport sociologists and psychologists, a research firm used fifty trained interviewers to conduct phone interviews of members of four population groups: the general public; coaches at the high school, college, amateur, and professional levels; sports writers and broadcasters; and sports physicians. The total number of respondents was over 1,300. The general public was the largest segment of the respondents, and over 60 percent of those contacted who were able to respond did so. During the 45 minutes to one hour of questioning, respondents gave answers to factual and attitudinal questions about many aspects of sport, including spectator interests and activities, forms of active sports participation, children's sports participation, athletes as role models, professional and Olympic sports, the sports media, violence, drugs, betting, sports programs in high school and college, and equal opportunity in sport for women and blacks.

An example of one type of finding in this study concerns the salaries earned by professional athletes (Miller Lite, 1983: 75–76, table 3.1). People were asked how much they agreed or disagreed (strongly agree, somewhat agree, somewhat disagree, strongly disagree, don't know) with the statement, "Professional athletes are overpaid." The researchers found that 76 percent of the 1,319 members of the general public, 72 percent of the 117 surveyed sports journalists, and 79 percent of the 230 surveyed coaches strongly or somewhat agreed with this statement. They also found that the youngest respondents (fourteen to seventeen years of age) were less likely to agree than those in older age groups; that blacks agreed less than whites; fans were less likely to agree than nonfans; and athletes were less likely to agree than nonathletes. Fan status and athlete status were measured on scales indicating degree of interest, commitment, or involvement in these roles. Thus, while fans generally believed that professional athletes were overpaid, age, race, fan status, and athlete status affected the likelihood and amount of agreement. A general theoretical explanation would focus on why these factors influence fans' perceptions of the payment of professional athletes as well as the general meaning of the results.

Surveys can be used in a variety of different ways and for a number of different purposes. Survey data gathered from mail, telephone, or face-to-face interviews may be collected to describe the characteristics (such as status attributes, attitudes, or behavioral patterns) of a population, to compare different populations at the same time or the same population at different times, to test hypotheses, or to analyze relationships between variables. Hypotheses are educated guesses by scientists about relationships between particular variables. More specifically, a survey researcher could try to determine the effects of a strike on fans' attitudes by surveying fans before an impending strike, during the strike, and immediately after it. The effects of general changes in the sports world over time could be studied by comparing responses to similar questions in different decades.

Common problems in survey research include possible untrue responses, responses from people who do not understand, care, or know anything about the questions to which they are responding; low response rates that limit the representativeness of respondents; costly preparation, collection, and analysis of data; and limited depth of responses to standardized questions. Despite these limitations, surveys can be very efficient means of collecting a substantial amount of data about large populations.

Observational Studies

Although surveys reveal what people are willing to report about what they think, feel, or do, observational studies reveal what researchers see as the actual behavior of people as individuals or as members of groups or larger collectivities. *Observational studies* generally involve systematic and intensive observation and provide rich and detailed descriptive detail about social behavior. The *case study* is the most common form of observational study and includes life histories of individual persons and field studies of the social patterns and processes of single groups, organizations, and communities. The method of collecting observational data may be *detached observation,* where the observer remains distant from the people and behavior being observed, or *participant observation,* where the researcher becomes a part of the group, organization, or community he or she is observing. Participant observers may reveal or disguise their identity as researchers, but the deception implied by a disguised identity can raise serious ethical problems.

An example of a participant observation study is the Adlers' (1991) investigation of a big-time college basketball team, which was based on five years of observation. Their research provides a detailed account, from an insider's perspective, of the socialization of big-time college athletes. It shows the ways in which college basketball players become immersed or engulfed in their athletic role and in the process, lose sight of their academic role.

An especially interesting aspect of the Adlers' report of their research is their prologue describing the nature of their roles in this study and how Peter's role changed as he increasingly became an insider. In discarding the role of detached observer, he embraced a subjectively immersed role from which he tried to understand the varied meanings of the experiences of the athletes, coaches, boosters, and others he observed and questioned. An important point they make is that a qualitative methodology can be a very rigorous methodology (Adler & Adler, 1991: 22). They carefully and systematically collected, evaluated, reported, and analyzed their evidence. On the basis of their research, they were able to produce new ways of understanding what it means to participate in big-time college basketball and how such participation changes the participants.

The Adlers' study provides a depth of understanding about its subject that is the hallmark of observational research. Such research has been criticized, though, for being too dependent on the skills and subjective interpretations of the researcher (Robertson, 1989: 23). This research can also be complicated by difficulties of gaining access and trust from those being studied or by the bias that arises from becoming too close to the subjects of the research, thereby compromising the role of researcher. In addition, depth of understanding from focusing on a single case or setting may be achieved at the expense of generalizability. One always must ask how much observational data can validly be generalized. Yet observational studies tend to provide considerably more insight into the perspectives and meanings of experiences of the people being studied than other forms of research provide.

Experiments

Experiments are a form of controlled observation used to test hypothesized relationships between variables. In a typical experiment, the researcher introduces an independent variable into a controlled situation and records its effects on a dependent variable. Experimenters are able to exercise a substantial amount of control over variables in laboratory settings, but experiments also are conducted in settings outside the laboratory. Sometimes researchers are able to observe natural experiments, which occur when changes are introduced as part of the normal functioning of a group, organization, or community or when one group of people is exposed to a new situation or treatment and a comparable group, called a *control group,* is not. Experimenters are able to infer causation with much precision and confidence.

An example of a field experiment in sport sociology is a study by Arms, Russell, and Sandilands (1987) of the effects of viewing aggressive sports on the hostility of spectators. Male and female subjects were exposed to one of three conditions of aggression in sports events: stylized (professional wrestling), realistic (ice hockey), or competitive nonaggression (swimming). The third condition was a control. Three paper-and-pencil measures of hostility were administered before and after the subjects were exposed to these three aggression conditions. The research was intended to test whether exposure to aggressive sports events has a cathartic effect of emotional release, which reduces a spectator's hostility level. The study replicated the results of a prior study, by Goldstein and Arms (1971), which refuted the catharsis hypothesis. In the Arms, Russell, and Sandilands experiment, it was found that hostility increased for wrestling and hockey spectators but not for swimming spectators.

The advantages of control in experiments may be offset by questions about the artificiality of experimental procedures and the generalizability of experimental results to less controlled or contrived settings. Experiments in natural settings, such as the one cited here, are likely to have more validity and, perhaps, generalizability, even though they sacrifice the control found in laboratory settings and open the results to questions of alternative explanations of the results. When the results of field or natural experiments agree with those of laboratory experiments, though, the validity and generalizability of the results would seem to be strengthened (Arms, Russell & Sandilands, 1987). Although experiments can be powerful tools for exploring causality, there are relatively few experimental studies in the sport sociology literature.

Secondary Data Analysis

It is not always necessary to collect new data to do research. Secondary data analysis involves the the coding, tabulation, and analysis of existing data sources, such as official records, historical documents, books, newspapers, and magazines. Books, newspapers, and magazines are often used for content analysis, which is a systematic and careful counting of how many times and in what ways specific ideas, images, or words appear in texts (Hess, Markson & Stein, 1991: 35). Content analysis can be used to infer trends, correlations, and even causes (Eckhardt & Ermann, 1977: 298–299).

Schafer's (1969) classic study of the relationship between interscholastic athletic participation and juvenile delinquency is an example of a study based on official records. Schafer used school and county juvenile court records to show that athletes from blue-collar homes were less likely than their nonathlete counterparts to have a court record for delinquent activities. While there has been much subsequent discussion about whether athletic participation actually deters delinquency (e.g., see Segrave & Hastad, 1984), Schafer's study shows how official records can be used to reveal relationships between variables.

A different kind of secondary data analysis is illustrated by Davis and Delano's (1992) study of the meanings of themes in antidrug campaigns in athletics. Their data sources were over forty antidrug campaign media texts aimed at athletes, which included posters, videotapes, and brochures. The themes they identified in their research led them to conclude that many of the antidrug media campaigns, especially involving steroids, tend to convey a problematic message that drugs are artificial substances that disrupt natural gender differences. They further argue that this message is problematic because in "naturalizing" physical gender differences, such media campaigns create influence on men and women in athletics to conform to traditional ideals of male and female physicality that limit opportunities for women (and men) to pursue nontraditional roles in sport and elsewhere. Thus, Davis and Delano used their content analysis to show how antidrug campaigns for athletes reinforce the traditional gender order in sport and society.

Another important form of secondary data analysis is the use of historical materials to trace the development of contemporary social patterns. An example of this type of research approach in sport sociology is Guttmann's (1988) analysis of the modernization of American sport. Guttmann looks at the development of contemporary patterns of American sport in terms of historical antecedents such as the rituals of pre-Columbian cultures, recreational and sports pursuits in colonial New England and the antebellum South, and manifestations of muscular Christianity in the late nineteenth and early twentieth centuries. In a more critical vein, Hoberman (1992) examines the historical roots of the scientific sport concept, which has created athletes absorbed in the values of industrial technology and willing to sacrifice their bodies to achieve higher and higher levels of performance. Those who use historical materials in their research remind us of Mills's exhortation to relate current social problems and issues to an historical context in trying to understand their larger sociological meaning.

Although secondary data analysis can be a relatively inexpensive way of pursuing a number of different types of research goals, it may limited because the secondary data sources were not produced by the research in which the data are being used. As a result, there may be questions about the validity, reliability, relevance, or completeness of the data. In addition, in content analysis, there may be the same kinds of problems of achieving validity and reliability associated with subjective interpretation in observational studies (Eckhardt & Ermann, 1977: 266–267).

MODEL OF THE RESEARCH PROCESS

Research in sport sociology and sociology in general rarely unfolds exactly as planned. It nevertheless should help clarify the nature of the social research process to provide an idealized model of how it generally is supposed to unfold. In Miller's (1991: 15–16) conception, a research design has ten basic elements:

1. State the nature and significance of a testable sociological problem.
2. Describe the relationship of the research problem to a theoretical framework and prior research.
3. Specify one or more hypotheses or questions to be addressed by the research and indicate their sociological significance, the nature of the variables, and how they are to be measured or "operationalized."
4. Describe the general methodological approach to the research, including controls to be used; the nature of the subjects, phenomena, and environment to be studied; the statistical procedures, if any, to be employed; and the desired confidence level for the results.
5. Specify sampling procedures; the nature of the population to be sampled or case to be selected; desired sample characteristics; and the method of randomization in the selection of subjects or respondents and, if relevant, the procedure for assignment to experimental and control conditions.
6. Describe the methods of data collection, including measures of quantitative variables or means of identifying qualitative variables or themes and the specific procedures that will be used to collect evidence (e.g., self-administered questionnaires, interviews, observation, and use of secondary data sources).
7. Prepare a working guide with time schedule, budget, and other cost and logistical estimates.
8. Specify how the results will be analyzed, including the types of quantitative or qualitative analytic techniques, statistics, and graphic and tabular presentations to be used.
9. Describe how the results will be interpreted in terms of the theoretical framework for the research.
10. Describe plans for reporting and publishing the research.

QUANTITATIVE AND QUALITATIVE RESEARCH

An important distinction implicit in the description of the research process is made between quantitative or qualitative research. This distinction determines whether the research will rely on numerical or statistical techniques to analyze data (*quantitative* research) or on interpretive or descriptive presentation of data in words rather than in numbers (*qualitative* research). The choice of a quantitative or qualitative research approach often reflects the researcher's theoretical orientation. For example, social structuralists of various sorts, social network analysts, and exchange theorists are likely to employ quantitative models (e.g., Burt & Minor, 1983; Cook, 1987; Wellman & Berkowitz, 1988; Burt, 1992) and statistical techniques. Symbolic interactionists, in contrast, tend to rely on qualitative approaches (e.g., Denzin, 1988).

In sport sociology, some interactionists, cultural theorists, feminists, and other critical or conflict theorists reject quantitative and statistical methods because they are seen as antihumanistic, representing a physical or natural science approach that is inappropriate for understanding the complex sources of how human behavior occurs and what such behavior means (see Nixon, 1991b). In this book, we assumed that sociological insights about sport come in a variety of forms and from a variety of perspectives and methodologies. Thus, our approach has been catholic and eclectic in the presentation of what sport sociologists know, but the assessment of the quality of existing data and sources of knowledge were based on the generally accepted canons of the scientific method that we have discussed in this appendix.

BASIC, APPLIED, AND EVALUATION RESEARCH

Sport sociologists may differ in their research orientations and commitments as well as in their theoretical and methodological approaches. A useful distinction among types of research orientations and commitments distinguishes basic, applied, and evaluation research (Miller, 1991: 3–12). *Basic* (or *pure*) *research* is meant to produce new knowledge about social facts and to establish general principles that will explain these social facts. In *applied* (or *policy-action*) *research,* the aim is to produce or obtain knowledge that can be immediately useful to policy makers or decision makers trying to deal with a concrete social problem or issue. Thus, knowledge in applied research is valuable not for its own sake, but because it can be used to formulate a policy or initiate action to solve a problem or resolve an issue. *Evaluation* (or *assessment-appraisal*) *research* has the goal of assessing the outcomes of a treatment, intervention, or program applied to a social problem. It is a means of accounting for and appraising the effects of social investment in a group, organization, community, or society.

More attention has been paid in recent years to applied research in sport sociology (e.g., Yiannakis & Greendorfer, 1992). This increasing attention has been accompanied by increasing debate, however (Sabo, Melnick & Vanfossen, 1988; Nixon, 1991). For instance, questions have been raised about whether sport sociologists have an obligation to pursue action-oriented and applied research and about who should be the primary beneficiaries of such research. On one side of this debate are sport sociologists such as Yiannakis (1989, 1990), who argue for applied approaches that are firmly rooted in the scientific method and value neutrality and that are used for solving practical problems, assisting in changing behavior, and improving the human condition. On the other side are sport sociologists such as Ingham and Donnelly (1990), who question the need for a special applied sport sociology because they believe that all sociological knowledge should be practical knowledge that could be used by powerless and disadvantaged people to improve their condition. Opponents of a special, or *programmatic,* orientation to applied sport sociology tend to reject the possibility of value neutrality and criticize applied researchers whose work seemed to be aimed at enhancing the power of the powerful over the powerless. Critics of applied researchers who assume a value-neutral scientific posture offer instead an agenda for a politicized and critical practical sport sociology meant to debunk, challenge the powerful, and empower the powerless.

The debate about applied research—and, perhaps implicitly, evaluation research as well—in sport sociology raises some very important questions about the nature of the research process, the value of different types of sociological knowledge about sport, the legitimacy of various uses of such knowledge, and the obligations of researchers. Such debates are a normal part of the scientific enterprise. They are useful and help advance a discipline precisely because they raise important questions. While sport sociologists will continue to debate and wrestle with the tough questions raised by their debates, they are unlikely to find simple or consensual resolutions or answers. Instead they will challenge each other to sharpen their grasp of what they know and become more systematic and self-conscious about the ways they produce, interpret, and use what they know. We drew from the diversity of these ways of understanding sport and society in presenting our sociology of sport. We recognize that in providing exposure to the diversity of sport sociology, we are likely to be seen as challenging conventional wisdom about many aspects of sport and society, thereby deflating a number of myths and making it more difficult to accept many prejudiced beliefs about the sports participation of women, minorities, and other disadvantaged groups (Eitzen & Hyatt-Hearn, 1992).

Glossary of Key Terms

AGEISM Prejudice and discrimination against people on the basis of their age. In sport, there is bias favoring youth; elderly people have often been viewed as incapable of vigorous physical activity and sport.

AGENT OF SOCIALIZATION People or social influences, such as the family, the school, the mass media, and the church, that affect the development of self and how a person learns to perform roles and participate in social relationships in various social settings.

AMERICAN DREAM A culture of materialism widely shared by Americans, which reflects the consumer-oriented economy of advanced industrial capitalism. It holds the promise that those who take advantage of opportunities and strive hardest will be rewarded with the "good life" of material comfort and perhaps social prominence and influence as well.

AMERICANIZATION The commercialization of sport in Canada and other countries outside the United States by American capitalists and state-subsidized American-based cartels, which have saturated foreign markets with American-focused events, images, and souvenirs. Canadization and Japanization are similar processes in which foreign investors acquire sports interests in the United States.

ANOMIE Conditions when social norms are weak, inconsistent, or vague; a state of normlessness or confusion.

ANTAGONISTIC COOPERATION The nature of association between males in many sports settings, in which men work together for the sake of teamwork while at the same time being in competition with each other for playing time, coaching attention, and status on the team.

ANTITRUST LAW Law that prohibits businesses from engaging in forms of contact with each other that restrain trade or commerce; a 1922 Supreme Court ruling exempted baseball from the Sherman Antitrust Law.

APARTHEID A racial caste system that places people in a stratum in society on the basis of their race—white, black, colored, Asian. Formerly the official racial policy of South Africa, apartheid placed nonwhites in lower castes; segregated them; and denied them access to privilege, power, prestige, and human rights in society and sport that were taken for granted by the dominant white caste; this system was dismantled in 1992 by constitutional change.

ATHLETICS A synonym for sport played in school or college settings.

AVERSIVE SOCIALIZATION Learning negative or deviant behaviors and attitudes from sports participation; an experience in sport that produces stress, anxiety, and other unpleasant feelings that can lead to withdrawal from sport.

BIRGing (basking-in-reflected-glory) The desire to increase identification with successful others. This process has an ego or self-esteem enhancement function.

BRAWN DRAIN The recruitment and training of athletes from one country by another country. Examples include the recruitment of Kenyan long-distance runners and Australian swimmers by U.S. colleges and the signing of soccer stars from various countries to compete for Italy.

BUREAUCRACY A type of formal organization emphasizing rational-legal authority linked to positions rather than specific individuals, specialized positions with a clear-cut division of labor, delegation of responsibilities within the hierarchy and division of labor, and rational calculation and efficiency in the pursuit of organizational tasks and goals. Bureaucracies are frequently large and relatively complex in administrative organization.

BURNOUT A condition that can cause people to reduce their role commitment or drop out of roles. Burnout occurs when role demands and stresses exceed a person's endurance and ability to cope, when people become highly frustrated or stressed by their roles, or when people experience a limited sense of control. It can take various forms, including emotional exhaustion, depersonalized treatment of other people, and a sense of diminished personal accomplishment.

CARTEL A self-regulating monopoly. Owners in a sports league form a cartel when they cooperate with each other to restrict access to and pricing of their product.

CIVIL RELIGION A shared public faith or belief system that elevates secular experiences, such as sport, almost to the level of the religious or sacred. (Similar to *secular religion*.)

CIVILIZING PROCESS The evolution of formal and informal mechanisms of social control in society and sport, which eventually leads to a control or minimization of violence and other aberrant behavior in a society.

CLASS AND STATUS MAINTENANCE ARGUMENT The argument (made for high school athletics, for example) that sport serves as a vehicle for maintaining the existing (and unequal) structures of social class and status in a community or society.

COMMERCIALIZATION OF SPORT The process whereby elements of a business orientation, such as profit, productivity, cost-benefit, and market demand, become increasingly prominent in the social organization of sport.

COMPETITION A form of interaction in which there are winners and losers. It is the predominant type of interaction in sports contests and emphasizes the use of physical skill to achieve victory.

CONDITIONAL SELF-WORTH In sport, self-worth that is defined and redefined by each successive performance. ("You are only as good as your last game").

CONFLICT THEORY A theoretical perspective inspired by Karl Marx that looks at society in terms of ongoing struggles or conflicts between social classes or groups that make societies and social order tenuous. Conflict theory emphasizes the coercion, exploitation, and oppression of subordinate classes or groups by dominant ones.

CONSPICUOUS CONSUMPTION (AND LEISURE) Public display of material goods, lifestyles, and behavior in a way that ostentatiously conveys privileged status to others for the purpose of gaining their approval or envy; public involvement in expensive or extravagant leisure pursuits to convey the appearance of privilege for approval or envy.

CORFing (cutting-off-reflected-failure) The effort to increase distance between oneself and unsuccessful others; this process has an ego or self-esteem protection function.

CORPORATE ATHLETICISM The organization and operation of sport in a manner that is modeled after large-scale commercial enterprises. Corporate athleticism is more like work than play, entertainment is more important than sport for its own sake, the outcome prevails over experience.

CORPORATE SPORT See corporate athleticism.

CORPORATE STRUCTURE A form of organization that is often formal and bureaucratic but is mainly commercial or businesslike in its orientation toward generating financial revenue or profits.

CULTURAL AND SOCIAL ASSIMILATION In the United States, becoming integrated into the mainstream of the community and society and gaining acceptance as "true Americans."

CULTURAL AND SOCIAL GATEKEEPING The functioning of activities such as sport to let in the accepted—that is, dominant class members—and keep out the marginal or unaccepted—that is, minority-group members and especially immigrant minorities.

CULTURAL CAPITAL Knowledge of ways of thinking, talking, and acting as well as cultural and social awareness needed to assume a position in the dominant stratum of society. It has been argued that participation in elite sports or sports participation at elite secondary schools is a means of acquiring such cultural capital.

CULTURAL DEPENDENCY A condition existing when the culture of a nation or people is shaped or imposed by the culture of another nation or people. In cases of cultural dependency in sport, a people borrow their sporting practices from other cultures and do not put their own distinctive cultural stamp on these practices.

CULTURAL DIFFUSION The spread of elements of one culture to another, as in the diffusion of American football to England and the diffusion of English rugby and soccer to the United States. As new elements are absorbed from other cultures, they are often modified by elements of the culture into which they have been introduced—thus, American football resulted from an amalgamation of English rugby and soccer.

CULTURAL IDEOLOGY An argument for the primacy of a particular set of cultural values, beliefs, and practices, which is supposed to be accepted without question, as a matter of faith.

CULTURAL RESISTANCE Efforts to display or assert indigenous cultural practices or forms in the face of foreign cultural influence, as in the case of the Dominican Republic response to the Americanization of baseball in their country.

CULTURAL UNIVERSAL Traits, behaviors, and other social patterns or arrangements thought to exist in every known culture; religious practices and play activities are examples.

CULTURE The shared ideas, ideals, and artifacts of a people that they create and sustain for their mutual survival and the perpetuation of their way of life; the distinctive way of life of the members of a society.

CULTURE OF RISK, PAIN, AND INJURY A set of cultural beliefs and practices in sport consistent with the Sport Ethic that emphasizes the value or necessity of accepting physical risks and playing with pain and injuries as a "normal part of the game"; this culture provides the beliefs and practices for the normalization of pain and injury in sport.

DEEP FAN Sports followers who engage in ritual identification with mass-mediated sports productions and portrayals of athletes and with related commercial advertising.

DEFINITION OF THE SITUATION A concept inspired by W. I. Thomas that points to the importance of how social actors subjectively interpret their social worlds. Thomas proposed that when people define situations as real, they will be real in their consequences.

DEGRADATION OF SPORT The corruption of athleticism that trivializes superior athletic performance, undermines the integrity and purity of athletic contests, and threatens the welfare of players and even serious fans.

DEMOCRATIC CITIZENSHIP The shared responsibility of participation in group, community, and societal decision making. In sport, it has meant participation in team decisions about who should start and what positions team members should play.

DEMOCRATIZATION The opening up of sports to people from varied social class and status backgrounds; the involvement of middle-class and working-class people in formerly elite sports.

DESOCIALIZATION The process of shedding and unlearning a role.

DOMINANT AMERICAN SPORTS CREED A term coined by Harry Edwards to refer to the dominant set of American cultural beliefs about the functions of sport. This belief system, viewed by Edwards as an ideology, is meant to convince people of the virtues of sports participation in emphasizing purported benefits of character building, discipline, preparation for life, physical and mental fitness, religiosity, and patriotism.

DOPING The use of substances alien or unnatural to the human body with the aim of obtaining an artificial or unfair increase of performance in competition.

EGALITARIANISM IN LEISURE SPORTS The ethos of mass sports in which maximizing the number of participants is more important than developing champions; it is the guiding philosophy of many public and community recreational programs and implies that everyone is welcome to participate.

ENTREPRENEURIAL CAPITALISM A competitive free enterprise economic system of private investors oriented toward maximizing their profits from the property they own. The stereotype of the entrepreneurial capitalist is the rugged individualistic risk taker.

ENTREPRENEURIAL CITY A city that aggressively pursues strategies, such as forging public-private partnerships and attracting professional sports franchises, to facilitate economic development and improve its public image.

EPHEMERAL ROLE A temporary role, such as a leisure sports role, chosen to meet social and/or psychological needs inadequately satisfied by the more central roles occupied in everyday life.

ETHNIC GROUP A category of people sharing a common cultural identity and cultural heritage.

FEMINISM A theoretical perspective and a social movement that emphasize how gender relations and structures oppress or discriminate against women and reinforce male domination in society.

FLOW An intense experience, as in a high-risk sport, that fuses self with activity and enables the participant to make decisions and act almost unconsciously. The body seems to take over the action and give the participant a sense of "pleasant total involvement" as he or she makes the right decisions and moves without thinking.

FOLK RELIGION Configuration of values upon which there is general agreement and that is a product of everyday living; a relatively diffuse and subtle religiouslike experience that has an enduring character for participants. Folk religion affirms the existing social arrangements and encourages people to accept prominent elements of popular culture, such as sport, as culturally important in an almost religious sense.

FORMAL SOCIAL STRUCTURE The organizational form of social relationships and social units that has explicit or official rules, enforcement by explicit or official regulatory bodies, and an explicit or official hierarchy of positions. Formal social structure often is bureaucratic.

FREE AGENCY An economic arrangement permitting athletes in professional team sports to sell themselves on the free market to the highest bidding team.

FUNCTIONAL EQUIVALENT Two social patterns, such as sport and religion, making the same or similar functional contributions to society.

GAME Relatively rule-bound activity that may be more or less institutionalized, formally organized, and serious. Games exist within sport as contests.

GENDER (OR SEX) EQUITY The achievement of equal or equivalent opportunities and resources for women and men in athletics. A proportionality principle stipulates that the number of men and women who participate in athletic programs at a given institution should match the ratio of men and women in the student body of that institution. The legal basis for gender equity in high school and college athletics in the United States is Title IX of the Education Amendments, passed by Congress in 1972, which forbids sex discrimination in public educational institutions.

GENDER STRATIFICATION Structured inequalities of economic privilege, prestige, and power that have historically favored men over women.

GLOBALIZATION A process that has created more extensive and complex links among different cultures and nations in the world.

GREAT WHITE HOPE A white athlete good enough to defeat a black champion or opponent. The term is linked to Jack Johnson, the black boxing champion who offended or outraged many whites with his bold, flamboyant, and assertive manner and his violations of race-related social mores of his time. Johnson's controversial behavior inspired a search for a "great white hope" to replace him.

HANDICAPISM Prejudice, discrimination, and stigmatization aimed at people with disabilities. Handicapism keeps athletes with permanent disabilities from competing in mainstream sports, even when they are able enough to do so.

HEGEMONY Imposition of the dominant group or class ideologies and structures on the values, beliefs, roles, and social relationships of other members of a society.

HIGH CULTURE Aesthetic practices, such as classical music, ballet, theatre, poetry, and the fine arts, often assumed to be the special province of the upper or highly educated strata, especially in Western nations.

HOMOPHOBIA An irrational fear and intolerance of homosexuality.

INDUSTRIALIZATION The development of technological innovations that produce changes in sectors of society. During industrialization, the application of science and technology, rather than tradition and personal relations, prevails in the distribution of goods and services; and a factory-based economy replaces one based on family farms.

INSTITUTIONALIZED Having an established structure of patterned and relatively persistent norms, statuses, roles, and social relationships.

JINGOISM Excessive nationalism or the extreme self-promotion of a country's values, ideas, and political/economic systems.

LEISURE SPORTS SUBCULTURE A network of people who are directly or indirectly tied to one another by the distinctive culture of a leisure sport in which they actively participate.

LITTLE LEAGUE SYNDROME Experience in youth sport that is marked by one of three major symptoms of excessive or dysfunctional behavior: *washout,* caused by insufficient success; *burnout,* caused by too much success too soon; and *superstar,* caused by exaggerated, antisocial competitiveness.

LITTLE LEAGUISM Adult domination of the organization and control of youth sports, which deprives children and youths of potentially valuable socialization experiences.

MACHO Describes exaggerated stereotypical male-dominant and aggressive conceptions of masculinity, which may be associated with sexist views and treatment of women.

MACRO LEVEL OF ANALYSIS A focus on whole social systems or societies and social networks that tend to be large and complex.

MALE PRESERVE A domain reserved exclusively for males or dominated by males. Sport in general and many specific sports have been male preserves, which emphasize distinctively and stereotypically male values and behavior and contribute to the development of masculine identity.

MARGINAL DISCRIMINATION A higher performance standard for members of a minority group, such as blacks, to make the team. As a result of this practice, teams only have room for marginally qualified players from the dominant racial or ethnic group; this type of discrimination is related to the need for black athletes to outperform white athletes to earn comparable pay.

MASS CULTURE The cultural practices and products brought to the masses in a society through the various media of mass communication. Televised sport is an example of a mass culture product. Related to popular culture.

McDONALDIZATION The incorporation of an increasing number of qualities of the fast-food industry—such as efficiency, quantification, calculation, predictability, control, and the substitution of nonhuman for human technology—into other parts of society, such as sport. McDonaldization in sport is associated with new technologies and social arrangements that are changing society.

MEDIATED SPORT Mass media interpretations of sports events. An event portrayed by the mass media is not shown as it naturally takes place but is described and interpreted through commentary and by the scene selection of the director before it is presented to the viewing public.

MERITOCRACY A system of dominance by the most talented; the term applies to more serious amateur sports programs and clubs in which there is an emphasis on competing for positions and rewards and moving to increasingly higher levels of competition. Meritocratic programs embody the middle-class ethos of the American Dream, which emphasizes achievement, success, and continual competitive striving for upward mobility.

MICRO LEVEL OF ANALYSIS A focus on relationships between individuals or on interaction in small groups.

MONOPOLY A restrictive economic arrangement with only one seller (which may be a group of firms organized as a cartel) providing a particular good, service, or form of employment.

MONOPOLY CAPITAL ENTERPRISE Large economic corporation that distorts the free play of the market by its size and control over one or more sectors of the market. In more advanced forms of monopoly capitalism, wealth and economic power are highly concentrated in a few very large corporations.

MONOPSONY A restrictive economic arrangement with only one buyer available to purchase a seller's goods, services, or employment.

MORTAL ENGINE An athlete treated as an experimental machine or physiological entity who is able to push his or her body beyond normal physical limitations as a result of the application of science and medicine.

MUSCULAR CHRISTIANITY The linking of sports participation to Christian character and development. The term also implies that sports success is evidence of Christian faith and witness.

MYTHIC SPECTACLE A cultural production of the electronic media, such as the Super Bowl, that functions in the manner of traditional mythic rituals in providing opportunities for collective participation and identification and a means of celebrating and reinforcing shared cultural meanings.

NEOCOLONIALISM One country's domination of another by gaining influence over key institutions such as the economy or polity without military conquest. Neocolonialism can also involve domination over one country's sport institution by another.

NORMALIZATION OF PAIN AND INJURIES Socialization process through which athletes learn to accept and minimize pain and injuries as a normal part of the game. This process draws from the culture of risk, pain, and injury in sport and makes it less likely that athletes will question the practice of playing hurt.

NORMS Formal and informal rules that govern or regulate social interaction.

ONE-PLUS-THREE MODEL OF COLLEGE ATHLETIC REFORM A model for reforming college athletics proposed by the Knight Commission in the 1990s and based on the principles of academic integrity, financial integrity, and independent certification of athletic programs, with presidential control over athletics.

ORDINARY DEVIANCE Violations of the rules of the game that are customary and are typically an expected part of sport; also, initially unusual practices that eventually become widely accepted.

ORGANIZATIONAL DEVIANCE Violations of regulations or the law by organizations or their subunits, such as athletic programs, as part of a deliberate strategy to gain an advantageous competitive position.

PARADOX OF PROFESSIONAL AND COMMERCIALIZED SPORTS Sport as simultaneous game and business, with the thrills and joys of the former at odds with the demands and constraints of the other. This paradox characterizes sport as a corporate enterprise in general.

PATRIARCHY Male dominance; technically, dominance by the father.

PERSONAL TROUBLES Individual matters that have to do with an individual's personality and the range of his or her immediate social relations, as in the case of the drug-related problems of an individual athlete.

PLAY Voluntary activity that is nonutilitarian and relatively spontaneous and unstructured.

POLITICAL ECONOMY PERSPECTIVE A way of looking at society and sport that emphasizes the powerful effects of the interrelationship of economics and politics, or government and business, or power and money.

POPULAR CULTURE Cultural practices and products designed for mass consumption, such as modern corporate sports and pseudosports.

POSITIVE DEVIANCE Conformity that is so intense, extensive, or extreme that it goes beyond the conventional boundaries of behavior. Positive deviance is *over*conformity rather than *counter*conformity but is deviant because of its extreme nature.

PRESTIGE The amount of social honor, respect, or deference accorded to positions, groups, or activities in society. Different positions on a sports team, such as starter and substitute, are accorded different levels of social honor just as different sports, such as golf and roller derby, have different levels of prestige associated with them.

PRIMARY GROUP A small cohesive group characterized by enduring and intimate relationships, which have the power to socialize their members.

PROCESS ORIENTATION TO SPORT Deriving pleasure from participation in sport or playing the game. Contrasted with a product or outcome orientation.

PRODUCT (OR OUTCOME) ORIENTATION TO SPORT Deriving pleasure from winning and the rewards it brings rather than playing the game. Contrasted with a process orientation.

PROFANE Quality assigned to things and activities related to everyday experience. Contrasted with sacred things and activities related to religious experience.

PROFESSIONAL MODEL OF SPORT A form of formal and bureaucratic sports organization that emphasizes outcomes and elitism over the intrinsic value of participation. The professional model can be a source of problems in children's and youth sports.

PUBLIC ISSUES Matters reflecting the larger social forces in a society, which are beyond the immediate experience, local environment, and personal awareness of individuals.

QUASI-RELIGION Established activities or institutions, such as sport, that seem to have the characteristics of a religion and serve as a functional equivalent of religion but lack essential elements of a genuine religion.

RACIAL DISCRIMINATION Unfair or arbitrary treatment of members of another racial group. Discrimination occurs when people are systematically underrewarded or denied opportunities solely because they are perceived to belong to a particular racial group.

RACIAL MINORITY GROUP A category of people who (1) are perceived to be genetically distinct and to possess distinctive physical characteristics; (2) are targets of prejudice and discrimination; (3) have an awareness of a common inferior status and identity that they share with other members of their race; and (4) think, feel, and act in terms of this shared sense of inferior status and identity.

RACIAL PREJUDICE Unfavorable feelings or attitudes toward members of another racial group based on their perceived differences.

RACIAL STRATIFICATION Persisting inequalities of economic privilege, prestige, and power that place certain races, such as blacks in the United States, beneath other races, such as whites in the United States.

RACISM The attribution of social significance to perceived racial differences and the resulting treatment of members of other racial groups with prejudice and discrimination.

RATIONALIZATION The process of decision making that emphasizes calculation, efficiency, and productivity in organizational or individual goal seeking. In sport, rationalization occurs in the application of science and medicine to produce highly efficient and productive "mortal engines."

RECREATION Mostly voluntary leisure activity meant to refresh the mind and/or body.

RELIGION A relatively coherent system of beliefs and practices centered on a spirituality, a deity, and related sacred activities and things; a sacred belief system that links a community of believers.

RESERVE CLAUSE The player reservation system, which legally binds players to a sports team in perpetuity or until an owner trades, sells, or releases the player.

RESOCIALIZATION A process that transforms the self and role conceptions and behavior into new forms. Resocialization occurs when people pursue new kinds of experiences and are exposed to new and different social influences.

RITUAL A rite or ceremony representing established practices and often having special shared meaning for those who engage in it. Rituals in secular activities such as sport can be perceived as having sacred or nearly sacred significance.

RITUAL OF INVERSION A cultural practice that reverses usual cultural roles and practices. In a "powder puff" football game, for example, females play the game and male players dress up as female cheerleaders.

ROLE The way people perform the behaviors expected of them in particular statuses.

ROLE CONFLICT Incompatible expectations or demands from different roles, such as coach and father or student and athlete, that can create confusion and stress for an individual trying to meet the competing expectations.

ROLE MODEL A person or group that serves as an example of personality qualities and behavior that ought to be emulated.

ROLE STRAIN Competing demands associated with a single role, such as coach—for example, when boosters to expect a coach to win, academic administrators and the athletic director expect the coach to stay within budget, and faculty expect the coach to reduce the time athletes spend in their sport and increase their time studying.

SACRED Quality assigned to objects or behavior that creates a sense of awe, reverence, respect, or even fear and sets them apart from everyday human existence; associated with objects or behavior having religious meaning.

SANITY CODE A set of regulations adopted by the NCAA before World War II to establish guidelines for recruiting and compensating athletes. The code was designed to assure that one member institution did not have an advantage, especially a financial one, over others in the ability to recruit prospective athletes.

SECULAR Associated with the material things and concerns of everyday life in this world.

SECULAR RELIGION Secular, or worldly, activity seen as comparable to religion in structural features, experiences, rituals, and fervor of its believers. Similar to civil religion.

SECULARIZATION Process in which religion loses its influence over everyday life and science takes precedence over religion in explaining life events; a social trend that often is believed to accompany urbanization and industrialization.

SERIOUS LEISURE Initially casual, then ephemeral, leisure roles that become central. The conversion from a casual, ephemeral role to a more serious pursuit occurs when leisure sports participants become increasingly involved in the subcultures of their leisure sports, and the satisfactions and rewards from their leisure sports roles become more important than the consequences of other roles.

SEXISM Beliefs and practices that relegate females to inferior status in relation to men, demean them, deprive them of opportunities, and underreward them for their accomplishments.

SIGNIFICANT OTHER A person or group whose views, judgments, opinions, and actions have substantial influence over the identity, roles, and behavior of another. Significant others serve as role models.

SOCCER HOOLIGANISM Violent behavior—including riotous invasions of the pitch, attacks on players and referees, property destruction, and fights with opposing fans—by soccer fans. The hooliganism of British soccer fans has been especially notorious.

SOCIAL ASCRIPTION Access to social position is based on qualities such as race, ethnicity, class background, and gender that are defined by (1) parental characteristics, (2) attributes possessed at birth, or (3) social class standing at birth.

SOCIAL CHANNELING Preparation of children to fit into roles consistent with their social origins. Social channeling has the effect of reproducing the social class structure of society.

SOCIAL DEVIANCE Behavior that breaks the rules or violates the norms of a group, organization, community, or society. Behavior becomes deviant because people with power or authority treat it as deviant; and these people generally treat certain behavior as deviant because they see it as threatening or disruptive to them, their interests, or the social order.

SOCIAL FUNCTION Positive effect or consequence of social structures or behaviors that contribute to the social stability, harmony, or well-being of social systems. A social function may be *manifest* and have obvious or intended consequences, or it may be *latent* and have consequences that are not recognized or intended. Social functions can be contrasted with *dysfunctions*, which are elements of society or social behavior having disruptive or negative consequences for social systems.

SOCIAL INSTITUTION A structured social arrangement in society that responds to enduring social needs of people, indicates the nature of problems and approved solutions in relation to the pursuit of enduring needs, and directs the actions of people in particular directions. Sport is a social institution.

SOCIAL NETWORK Web of social relationships that connects individuals, statuses, roles, groups, and larger social units in a community or society.

SOCIAL PROBLEM A perceived source of social disruption, tension, or difficulty in society. There are two major types of social problems: (1) a form of social deviance that powerful people or the mass media in society define as highly undesirable for the society, and (2) a condition of society that causes widespread or severe psychic or material suffering for some segment of society.

SOCIAL PROCESS A dynamic force in interaction that can reinforce the existing elements of social structure and culture or transform them into new forms.

SOCIAL RELATIONSHIP Enduring pattern of social interaction that links individuals, groups, and larger collectivities or social units.

SOCIAL RESEARCH The systematic means by which social scientists gather evidence and establish social facts to provide concrete answers to questions, hypotheses, and theoretical assumptions about society and social behavior.

SOCIAL STRATIFICATION A structural feature of all societies and groups that is based on an unequal distribution of valued resources among members. Sports and sports participation are stratified, reflecting the hierarchical arrangement of status and class in society. Certain sports are viewed as having higher or lower status than others; participants in different sports often are concentrated in a particular social class.

SOCIAL SYSTEM Set of social actors or social units, the patterns of interdependency linking them, and the boundaries that give them a distinct identity.

SOCIAL THEORY Abstract interpretation that makes general statements about the reasons for and meaning of a wide variety of specific and concrete facts about the social world.

SOCIALIZATION A lifelong process through which individuals develop and shape their identities, create a sense of self, and learn how to participate in social roles and relationships. It occurs in sport by processes of socialization *into* and *through* sport.

SOCIETAL REACTION TO LABELING INTERPRETATION A theoretical perspective that views social behavior as a social construction in which the meaning is determined by how people with status and power react to it and label it. For example, behavior becomes deviant when people with status and power, such as coaches and club owners, perceive and label it as deviant.

SOCIETY A collection of people who share a common territory or space and are at least loosely or indirectly tied together by interdependent networks of interaction and by possession of a common culture.

SOCIOLOGICAL IMAGINATION A form of consciousness or frame of mind that allows us to understand where individuals are located in society and in history, how their circumstances are like or unlike those of similar others in their society and historical period, and how human experiences are structured and made meaningful by the large social forces of their society, world, and period.

SOCIOLOGY The scientific study of human interaction and social organization. Its purpose as a social science is to produce knowledge to help us interpret, understand, and predict patterns of human social behavior and solve human social problems.

SOCIOLOGY OF SPORT An academic discipline that uses sociology to teach and learn about the social aspects of sport.

SOCIOLOGY THROUGH SPORT An approach to sociology that uses sport to teach the sociological perspective along with knowledge about society and social problems and issues.

SPORT Institutionalized physical competition occurring in a formally organized or corporate structure.

SPORT ETHIC A code of central values and norms in sport, including four basic beliefs about what being an athlete means or requires: (1) making sacrifices for one's sport or team; (2) striving for constant improvement and distinction (often reflected by winning); (3) accepting risks and playing through pain; and (4) refusing to accept limits in the pursuit of achievement and success.

SPORTIANITY A social movement of athletes and coaches who witness their (Christian) religious faith in the context of athletic events. Sportianity uses prominent sports figures such as athletes and coaches as evangelists.

SPORT SPECIALIZATION Restriction of athletic participation to a single sport, with training, practice, and/or competition throughout the year.

SPORTS HERO Symbolic representation and inspiring reminder of the dominant social myths and values of a society or subculture. Sports heroes are constructed by the mass media and public from idealized conceptions of real people.

STACKING Racial segregation by position in sport in which whites are disproportionately concentrated in certain positions and blacks are disproportionately concentrated in others.

STAR EFFECT An explanation for violent outbursts, including domestic abuse, by attractive male athletes. As stars accustomed to public adoration, these athletes or former athletes expect their spouse to be dutiful, submissive, and totally devoted; they react violently when their expectations are challenged or unfulfilled.

STATUS Position in the organizational structure of society that people occupy and that is defined by certain rights and duties. Statuses may be *achieved* by a person's own efforts or *ascribed* on the basis of characteristics possessed at birth—such as race and birth—or automatically assigned over the lifecycle—such as age.

STIGMA A deeply discrediting attribute or identity, such as homosexuality, which devalues people and sets them apart from "normal" people in society.

STRUCTURAL FUNCTIONALISM A theoretical perspective that focuses on the ways that constituent elements of societies or social systems function to contribute to the stability, harmony, and well-being of the whole system.

STRUCTURED UNCERTAINTY The lack of predictability built into the professional careers of individual-sport athletes. Structured uncertainty includes the variability and indeterminacy of tournament settings, uncertain performances and incomes, challenges trying to fraternize in the locker room with opponents, and itinerant lifestyles.

SYMBOLIC INTERACTIONISM A theoretical perspective that focuses on how people interpret their social world through the meanings they derive from interaction with others.

SYMBOLIC REFUGE An activity or experience with special symbolic significance that provides an escape from the dull or unfulfilling routines and the stresses of everyday life. The fantasized or romanticized world of sport can serve as a symbolic refuge.

THEORY Ideas, assumptions, and arguments proposed to clarify the underlying reasons or explanations for social facts. Involves abstract interpretations that make general statements about the reasons for and meaning of a wide variety of specific and concrete facts about the social world.

TIPPING POINT The proportion of minority group members needed to change the dynamics of interaction between races in a group or sports team.

TV SPORT SYSTEM A corporate commercial relationship between television and sport that exists primarily to generate profits for both television and sport.

VERSTEHEN The term coined by Max Weber for the ability to see the behavior and intentions of others through their eyes.

VERTICAL SOCIAL MOBILITY Movement up or down the stratification ladder, from lower to higher or from higher to lower social status or class standing. *Open* stratification systems permit and often encourage substantial vertical mobility; *closed* systems limit vertical mobility.

VICARIOUS TWO-PERSON SINGLE CAREER A marital situation in which both partners are committed to one partner's career. In the case of a professional baseball career, for example, both the employer of the husband and the husband himself expect the wife to conform happily to the stereotypical roles of supporter, comforter, helpmate, backstage manager, homemaker, and mother. In these roles, the baseball wife is supposed to be a "husband-oriented wife" devoted to helping her husband in his career.

VIOLENCE Physical contact that is intended to harm another person. Violence can vary in intensity, seriousness, and legal status in sport from body contact to borderline violence, quasicriminal violence, and criminal violence.

WORLD SPORT SYSTEM A network linking athletes and sports from around the world. Sports in different nations generally share the same formal structural characteristics in the modern world sports system.

Bibliography

Acosta, R. Vivian and Carpenter, Linda. 1990. "Women in Intercollegiate Athletics." Unpublished paper.

Acosta, R. Vivian and Carpenter, Linda J. 1992. "Women in Intercollegiate Sport: A Longitudinal Study—Fifteen Year Update 1977–1992." Brooklyn, NY: Brooklyn College. Unpublished report.

Adler, Peter and Adler, Patricia A. 1985. "From Idealism to Pragmatic Detachment: The Academic Performance of College Athletes." *Sociology of Education* 58: 241–250.

Adler, Patricia A. and Adler, Peter. 1991. *Backboards and Blackboards: College Athletes and Role Engulfment.* New York: Columbia University Press.

Agger, Ben. 1991. "Critical Theory, Poststructuralism, Postmodernism: Their Sociological Relevance." *Annual Review of Sociology* 17: 105–131.

Albanese, C. L. 1981. *American, Religions and Religion.* Belmont, CA: Wadsworth.

Albert, Edward. 1984. "Equipment as a Feature of Social Control in the Sport of Bicycle Racing." In N. Theberg and P. Donnelly (Eds.). *Sport and the Sociological Imagination* (pp. 318–333). Fort Worth, TX: Texas Christian University Press.

Albert, Edward. 1990. "Constructing the Order of Finish in the Sport of Bicycle Racing." *Journal of Popular Culture* 23: 145–154.

Albert, Edward. 1991. "Riding a Line: Competition and Cooperation in the Sport of Bicycle Racing." *Sociology of Sport Journal* 8: 341–361.

Albom, Mitch. 1994. "Navratilova the Greatest Win or Loss in Final Today." *Charlotte Observer* (July 2): 1B, 3B.

Aldrich, Howard E. 1979. *Organizations and Environments.* Englewood Cliffs, NJ: Prentice-Hall.

Alexander, Jeffrey C. 1988. "The New Theoretical Movement." In N. J. Smelser (Ed.). *Handbook of Sociology* (pp. 77–101). Newbury Park, CA: Sage.

Allen, George (with Joe Marshall). 1973. "A Hundred Percent Is Not Enough." *Sports Illustrated* (July 9): 74–86.

Allison, Lincoln. 1993. *The Changing Politics of Sport.* Manchester: Manchester University Press.

Alston, D. J. 1980. "Title IX and the Minority Women in Sport at Historically Black Institutions." Paper presented at the National Minority Women in Sport Conference, Washington, D.C.

Altheide, David L. and Snow, Robert P. 1978. "Sport Versus the Mass Media." *Urban Life* 7: 189–204.

Amdur, Neil. 1971. *The Fifth Down: Democracy and the Football Revolution.* New York: Delta.

American Council on Education (ACE). 1952. "Special Committee on Athletic Policy." *Educational Record* 33:246–255.

American Footwear Association. 1991. *American Youth and Sports Participation.* Palm Beach, FL: American Footwear Assoc.

An, Minseok and Sage, George H. 1992. "The Golf Boom in South Korea: Serving Hegemonic Interests." *Sociology of Sport Journal* 9: 272–384.

Anderson, Dave. 1992. "Baseball's Antitrust Exemption." *New York Times* (December 13): 27.

Anderson, Theodore R. and Zelditch, Morris, Jr. 1975. *A Basic Course in Statistics,* 3rd ed. New York: Holt, Rinehart & Winston.

Anson, Adrian C. 1900. *A Ball Player's Career.* Chicago: Era.

Anthony, Donald J. 1969. "The Role of Sport in Development." *International Development Review* 12: 10–11.

Apple, Michael W. 1983. "Curriculum in the Year 2000: Tensions and Possibilities." *Phi Delta Kappan* (January): 236–241.

Arbena, Joseph L. 1985. "Sport and the Study of Latin American History: An Overview." *Journal of Sport History* 13: 87–96.

Arbena, Joseph. 1989. "The Diffusion of Modern European Sport in Latin America: A Case Study of Cultural Imperialism?" Paper presented at the Southern Historical Association Annual Meeting, Lexington, Kentucky.

Arbena, Joseph L. 1993. "International Aspects of Sport in Latin America: Perceptions, Prospects, and Proposals." In E. G. Dunning, J. A. Maquire, and R. E. Pearton (Eds.). *The Sports Process: A Comparative and Developmental Approach* (pp. 151–167). Champaign, IL: Human Kinetics.

Arms, Robert L., Russell, Gordon W. and Sandilands, Mark L. 1987. "Effects on Hostility of Spectators of Viewing Aggressive Sports," In A. Yiannakis, T. D. McIntyre, M. Melnick and D. P. Hart (Eds.). *Sport Sociology: Contemporary Themes,* 3rd ed. (pp. 259–263). Dubuque, IA: Kendall/Hunt.

Armstrong, Christopher F. 1984. "The Lessons of Sports: Class Socialization in British and American Boarding Schools." *Sociology of Sport Journal* 1: 314–331.

Arnold, David O. 1976. "A Sociologist Looks at Sport Parachuting." In A. Yiannakis, T. D. McIntyre, M. J. Melnick, and D. P. Hart (Eds.). *Sport Sociology: Contemporary Themes* (pp. 143–145). Dubuque, IA: Kendall/Hunt.

Ashe, Arthur R., Jr. 1988. *A Hard Road to Glory: A History of the African-American Athlete.* 3 vols. New York: Warner Books.

Associated Press. 1987a. "Campanis Statement on Blacks Sets Off Furor." *Burlington Free Press* (April 8).

Associated Press. 1987b. "Campanis Forced to Resign." *Burlington Free Press* (April 9).

Associated Press. 1992. "Ten Years Ago, He Didn't Down the Ball." *Charlotte Observer* (December 15).

Associated Press. 1993a. "Controversy Doesn't Sell." *Charlotte Observer* (September 18): 2B.

Associated Press. 1993b. "Senate Considers Baseball Hearings." *New York Times* (October 1): B13.

Axthelm, Pete. 1981. "The Case of Billie Jean King." *Newsweek* (May 18): 133.

Baade, Robert A. and Dye, Richard F. 1988. "Sports Stadiums and Area Development: A Critical Review." *Economic Development Quarterly* 2: 265–275.

Baade, Robert A. and Tiehen, Laura J. 1990. "An Analysis of Major League Baseball Attendance, 1969–1987." *Journal of Sport & Social Issues* 14: 14–32.

Babbie, Earl. 1995. *The Practice of Social Research,* 7th ed. Belmont, CA: Wadsworth.

Bailey, Ian C. 1977. "Socialization in Play, Games and Sport." *Physical Education* 34: 183–187.

Bailey, Wilford S. and Littleton, Taylor D. 1991. *Athletics and Academe: An Anatomy of Abuses and a Prescription for Reform.* Washington, DC: American Council on Education.

Baker, James T. 1975. "Are You Blocking for Me, Jesus?" *The Christian Century* (November 5): 997–1001.

Baker, William J. 1982. *Sports in the Western World.* Totowa, NJ: Rowman and Littlefield.

Baldo, Anthony. 1993. "Secrets of the Front Office: What America's Pro Sports Teams Are Worth." In D. S. Eitzen (Ed.). *Sport in Contemporary Society: An Anthology* (pp. 187–194). New York: St. Martin's Press.

Bale, John. 1989. *Sports Geography.* London: E. & F. N. Spon.

Bale, John. 1991. *The Brawn Drain: Foreign Student-Athletes in American Universities.* Chicago: University of Illinois Press.

Ball, Donald W. 1976. "Failure in Sport." *American Sociological Review* 41: 726–739.

Bandura, Albert. 1969. "Social Learning Theory of Identificatory Processes." In D. A. Goslin (Ed.). *Handbook of Socialization Theory and Research* (pp. 213–262). Chicago: Rand McNally.

Bandura, Albert and Walters, R. H. 1963. *Social Learning and Personality Developments.* New York: Holt, Rinehart & Winston.

Barnett, Nancy P., Smoll, Frank L., and Smith, Ronald E. 1992. "Effects of Enhancing Coach-Athlete Relationship on Youth Sport Attrition." *The Sport Psychologist* 6: 111–127.

Bayles, Fred. 1993. "Study: Minority Hiring Improves Slightly." *Charlotte Observer* (July 9): 2B.

Bayless, Skip. 1990. *God's Coach: The Hymns, Hype, and Hypocrisy of Tom Landry's Cowboys.* New York: Simon & Schuster.

Beamish, Rob. 1985. "Sport Executives and Voluntary Associations: A Review of the Literature and Introduction to Some Theoretical Issues." *Sociology of Sport Journal* 2: 218–232.

Beamish, Rob B. 1991. "The Impact of Corporate Ownership on Labor-Management Relations in Hockey." In P. D. Staudohar and J. A. Mangan (Eds.). *The Business of Professional Sports* (pp. 202–221). Urbana and Chicago: University of Illinois Press.

Becker, Gary. 1987. "The NCAA: A Cartel in Sheepskin Clothing." *Business Week* (September 14): 24.

Becker, Howard. 1963. *Outsiders: Studies in the Sociology of Deviance.* New York: Free Press.

Becker, Howard. 1971. "Labeling Theory Reconsidered." Paper presented at the British Sociological Association Meetings. London, England.

Bellah, Robert N. 1967. "Civil Religion in America." *Daedalus* 96 (Winter): 1–21.

Bellamy, Robert V., Jr. 1989. "Professional Sports Organizations: Media Strategies." In L. A. Wenner (Ed.). *Media, Sports, & Society* (pp. 120–133). Newbury Park, CA: Sage.

Bend, Emil. 1971. "Some Potential Dysfunctional Effects of Sport Upon Socialization." Paper presented at the Third International Symposium on the Sociology of Sport. Waterloo, Ontario.

Berger, Peter. 1963. *Invitation to Sociology.* New York: Anchor Books.

Berghorn, Forrest J., Yetman, Norman R., and Hanna, William E. 1988. "Racial Participation and Integration in Men's and Women's Intercollegiate Basketball: Continuity and Change, 1958–1985." *Sociology of Sport Journal* 5: 107–124.

Berkowitz, S. D. 1982. *An Introduction to Structural Analysis: The Network Approach to Social Research.* Toronto: Butterworths.

Berler, Ron. 1993. "'Just Gimme One Chance.'" *Inside Sports* (May): 56–65.

Berryman, Jack W. and Loy, John W. 1976. "Secondary Schools and Ivy League Letters: A Comparative Replication of Eggleston's 'Oxbridge Blues.'" *British Journal of Sociology* 27: 61–77.

Best, Clayton. 1987. "Experience and Career Length in Professional Football: The Effects of Positional Segregation." *Sociology of Sport Journal* 4: 410–420.

Betts, John Rickards. 1974. *America's Sporting Heritage: 1850–1950.* Reading, MA: Addison-Wesley.

Birrell, Susan. 1988. "Discourse on the Gender/Sport Relationship: From Women in Sport to 'Gender Relations.'" *Exercise and Sport Science Review* 16: 459–502.

Birrell, Susan. 1989. " Racial Relations Theories and Sport: Suggestions for a More Critical Analysis." *Sociology of Sport Journal* 6: 212–227.

Birrell, Susan and Cole, Cheryl L. 1990. "Double Fault: Renee Richards and the Construction and Naturalization of Difference." *Sociology of Sport Journal* 7: 1–21.

Bissinger, H. G. 1990. *Friday Night Lights: A Town, a Team, and a Dream.* New York: HarperCollins.

Blalock, Hubert M., Jr. 1962. "Occupational Discrimination: Some Theoretical Propositions." *Social Problems* 9: 240–247.

Blalock, Hubert M., Jr. 1967. *Toward a Theory of Minority Group Relations.* New York: Wiley.

Blanchard, Kendall and Cheska, Alyce. 1985. *The Anthropology of Sport: An Introduction.* South Hadley, MA: Bergin & Garvey.

Bledsoe, Terry. 1973. "Black Dominance of Sports: Strictly from Hunger." *The Progressive* 37 (June): 16–19.

Blinde, Elaine M. 1989. "Participation in the Male Sport Model and the Value Alienation of Female Intercollegiate Athletes." *Sociology of Sport Journal* 6: 36–49.

Blinde, Elaine M., Greendorfer, Susan L., and Shanker, Rebecca J. 1991. "Differential Media Coverage of Men's and Women's Intercollegiate Basketball: Reflection of Gender Ideology." *Journal of Sport & Social Issues* 15: 98–114.

Blinde, Elaine M., Taub, Diane E., and Han, Lingling. 1993. "Sport Participation and Women's Personal Empowerment: Experiences of the College Athlete." *Journal of Sport & Social Issues* 17: 47–60.

Bloomfield, John, Fricker, Peter A., and Fitch, Kenneth D. (Eds.). 1992. *Textbook of Science and Medicine in Sport.* Champaign, IL: Human Kinetics.

Blum, Debra E. 1992. "Criminal Conduct by Some of Its Athletes Leaves Arizona State U. Wondering What's Gone Wrong." *Chronicle of Higher Education* (November 18): A27, A29.

Blum, Debra E. 1993a. "Graduation Rate of Scholarship Athletes Rose After Proposition 48 Was Adopted, NCAA Reports." *Chronicle of Higher Education* (July 7): A42, A44.

Blum, Debra E. 1993b. "Sharing the Football Pie." *Chronicle of Higher Education* (December 1): A39–A40.

Blum, Debra E. 1993c. "Two More Coaches of Women's Teams Go to Court to Press Claims of Sex Discrimination." *Chronicle of Higher Education* (September 1): A47–A48.

Blum, Debra E. 1993d. "Men Turn to Federal Anti-Bias Laws to Protect Teams From Chopping Block." *Chronicle of Higher Education* (August 11): A33–A34.

Blum, Debra E. 1993e. "Forum Examines Discrimination Against Black Women in College Sports." *Chronicle of Higher Education* (April 21): A39–A40.

Blumer, Herbert. 1969. *Symbolic Interactionism.* Englewood Cliffs, NJ: Prentice-Hall.

Bok, Derek. 1985. "Intercollegiate Athletics." In John W. Bennett and J. W. Peltason (Eds.). *Contemporary Issues in Higher Education: Self-Regulation and the Ethical Roles of the Academy* (pp. 123–146). New York: Macmillan.

Bourdieu, Pierre. 1977. *Outline of a Theory of Practice.* Cambridge: Cambridge University Press.

Bourdieu, Pierre. 1978. "Sport and Social Class." *Social Science Information* 17: 819–840.

Bourdieu, Pierre. 1982. "Leçon sur la Leçon." Paris: Editions de Minuit. Translated as "Lecture on the Lecture." In Pierre Bourdieu. 1990. *In Other Words: Essays Toward a Reflexive Sociology.* Stanford: Stanford University Press. (Cited in Bourdieu & Wacquant, 1992.)

Bourdieu, Pierre. 1984. *Distinction: A Social Critique of the Judgment of Taste.* Cambridge, MA: Harvard University Press.

Bourdieu, Pierre. 1988. "Program for a Sociology of Sport." *Sociology of Sport Journal* 5: 153–161.

Bourdieu, Pierre and Wacquant, Loïc J. D. 1992. *An Invitation to Reflexive Sociology.* Chicago: University of Chicago Press.

Boutilier, Mary A. and San Giovanni, Lucinda. 1983. *The Sporting Woman.* Champaign, IL: Human Kinetics.

Bouton, Bobbie and Marshall, Nancy. 1983. *Home Games.* New York: St. Martin's/Marek.

Bouton, Jim. 1971. *Ball Four: My Life and Hard Times Throwing the Knuckleball in the Big Leagues.* New York: Dell.

Braddock, J. H. II. 1980. "The Perpetuation of Segregation Across Levels of Education: A Behavioral Assessment of the Contact Hypothesis." *Sociology of Education* 53: 178–186.

Bradley, Bill. 1976. *Life on the Run.* New York: Quadrangle.

Brandmeyer, Gerard A. and Alexander, Luella K. 1983. "Private Life in the Public Domain: The Vicarious Career of the Baseball Wife." Paper presented at the Annual Conference of the North American Society for the Sociology of Sport, St. Louis.

Brandmeyer, Gerard A. and Alexander, Luella K. 1987. "Some Sociological Clues to Baseball as 'National Pastime.'" In A. Yiannakis, T. D. McIntyre, M. J. Melnick, and D. P. Hart (Eds.). *Sport Sociology: Contemporary Themes*, 3rd ed. (pp. 35–39). Dubuque, IA: Kendall/Hunt.

Brandmeyer, Gerard A. and McBee, G. Fred. 1986. "Social Status and Athletic Competition for the Disabled Athlete: The Case of Wheelchair Road-Racing." In C. Sherrill (Ed.). *Sport and Disabled Athletes* (pp. 181–187). Champaign, IL: Human Kinetics.

Brannigan, A. and McDougall, A. A. 1987. "Peril and Pleasure in the Maintenance of a High-Risk Sport: A Study of Hang-Gliding." In A. Yiannakis, T. D. McIntyre, M. J. Melnick, and D. P. Hart (Eds.). *Sport Sociology: Contemporary Themes*, 3rd ed. (pp. 284–292). Dubuque, IA: Kendall/Hunt.

Brinton, Mary C. 1988. "The Social-Institutional Bases of Gender Stratification: Japan as an Illustrative Case." *American Journal of Sociology* 94: 300–334.

Brohm, Jean-Marie. 1978. *Sport: A Prison of Measured Time.* London: Inks Links.

Brooks, Dana and Althouse, Ronald. 1993. "Racial Imbalance in Coaching and Managerial Positions." In Dana Brooks and Ronald Althouse (Eds.). *Racism in College Athletics* (pp. 101—142). Morgantown, WV: Fitness Information Technology.

Brooks, George A. 1981. *Perspective on the Academic Discipline of Physical Education.* Champaign, IL: Human Kinetics.

Broom, Eric F., Clumpner, Roy, Pendleton, Brian, and Pooley, Carol (Eds.). 1988. *Comparative Physical Education and Sport Volume 5.* Champaign, IL: Human Kinetics.

Broom, Leonard and Selznick, Philip. 1963. *Sociology.* New York: Harper & Row.

Brower, Jonathan J. 1978. "The Professionalization of Organized Youth Sports." *Annals of the American Academy of Political and Social Science* 445: 39–46.

Brown, Barbara. 1985. "Factors Influencing the Process of Withdrawal by Female Adolescents from the Role of Competitive Age Group Swimmer." *Sociology of Sport Journal* 2: 111–129.

Brown, Barbara A. and Curtis, James E. 1984. "Does Running Go Against the Family Grain? National Survey Results on Marital Status and Running." In N. Theberge and P. Donnelly (Eds.). *Sport and the Sociological Imagination* (pp. 352–367). Fort Worth, TX: Texas Christian University Press.

Brownlee, Shannon. 1992. "The Best of Times for American Women." *U.S. News & World Report* (January 13): 10.

Brubach, Holly. 1993. "Musclebound." *New Yorker* (January 11): 30–37.

Bruns, Bill and Tutko, Thomas. 1978. "Dealing with the Emotions of Childhood Sports." *Arena Review* 2: 3–13.

Bruscas, Angelo. 1994. "Athletes No Strangers to Domestic Violence." *Charlotte Observer* (June 25): 1B, 3B.

Brustad, Robert J. 1992. "Integrating Socialization Influences into the Study of Children's Motivation in Sport." *Journal of Sport and Exercise Psychology* 14: 59–77.

Buell, Charles. 1986. "Blind Athletes Successfully Compete Against Able-Bodied Opponents." In C. Sherrill (Ed.). *Sport and Disabled Athletes* (pp. 217–233). Champaign, IL: Human Kinetics.

Buhrmann, Hans G. 1972. "Scholarship and Athletics in Junior High School." *International Review of Sport Sociology* 7: 119–128.

Buhrmann, Hans G. and Bratton, Robert D. 1977. "Athletic Participation and the Status of Alberta High School Girls." *International Review of Sport Sociology* 12: 57–69.

Buhrmann, Hans G. and Jarvis, M. S. 1971. "Athletics and Status." *Canadian Association for Health, Physical Education and Recreation* 37: 14–17.

Buhrmann, Hans G. and Zaugg, Maxwell K. 1981. "Superstitions Among Basketball Players." *Journal of Sport Behavior* 4: 163–174.

Bunuel, Ana. 1991. "The Recreational Physical Activities of Spanish Women: A Sociological Study of Exercising for Fitness." *International Review for the Sociology of Sport* 26: 203–216.

Burt, Ronald S. and Minor, Michael. 1983. *Applied Network Analysis: A Methodological Introduction.* Beverly Hills, CA: Sage.

Cady, Edwin H. 1978. *The Big Game: College Sports and American Life.* Knoxville: University of Tennessee Press.

Caldwell, Marjorie. 1993. "Eating Disorders and Related Behavior Among Athletes." In G. L. Cohen (Ed.). *Women in Sport: Issues and Controversies* (pp. 158–167). Newbury Park, CA: Sage.

Callahan, Tom. 1992. "An Identity Crisis on Ice and Snow." *U.S. News & World Report* (March 2): 62.

Cantelon, Hart and Gruneau, Richard (Eds.). 1983. *Sport, Culture and the Modern State.* Toronto: University of Toronto.

Capel, Susan A., Sisley, Becky L., and Desertrain, Gloria S. 1987. "The Relationship of Role Conflict and Role Ambiguity to Burnout in High School Coaches," *Journal of Sport Psychology* 9: 106–117.

Capuzzo, Mike. 1992. "A Prisoner of Memory." *Sports Illustrated* (December 7): 80–92.

Carlson, Chris J. and Walker, Matthew S. 1982. "The Sports Court: A Private System to Deter Violence in Professional Sports." *Southern California Law Review* 55: 399–440.

Carpenter, Linda Jean and Acosta, R. Vivian. 1991. "Back to the Future: Reform with a Woman's Voice." *Academe* 77 (January/February): 23–27.

Carroll, John. 1986. "Sport: Virtue and Grace." *Theory Culture & Society* 3: 91–98.

Case, Bob, Greer, Scott, and Brown, James. 1987. "Academic Clustering in Athletics: Myth or Reality?" *Arena Review* 11: 48–56.

Cashmore, Ellis. 1990. *Making Sense of Sport.* London and New York: Routledge.

Center for the Study of Sport in Society. 1993. "The 1993 Racial Report Card." *CSSS Digest* 5 (Summer): 1, 4–8, 12–13.

Chalip, Lawrence, Csikszentmihalyi, M., Kleiber, Douglas, and Larson, R. 1984. "Variations of Experience in Formal and Informal Sport." *Research Quarterly for Exercise and Sport* 55: 109–116.

Chandler, Joan M. 1992. "Sport Is Not a Religion." In Shirl J. Hoffman (Ed.). *Sport and Religion* (pp. 55–61). Champaign, IL: Human Kinetics.

Chandler, Timothy J. L. 1988. "Secondary Schools and Oxbridge Blues: From the 1950s to the 1980s." In E. F. Broom, R. Clumpner, B. Pendleton, and C. A. Pooley (Eds.). *Comparative Physical Education and Sport,* Vol. 5 (pp. 111–122). Champaign, IL: Human Kinetics.

Charlotte Observer. 1990. "Other Arrangements." (March 29): 4B.

Charlotte Observer. 1993a. "True Value." Editorial. (November 10): 18A.

Charlotte Observer. 1993b. "How the Game Is Played." Compiled Reports. (July 12): 9A.

Chass, Murray and Goodwin, Michael. 1986. "Drug Abuse in Baseball." In R. E. Lapchick (Ed.). *Fractured Focus: Sport as a Reflection of Society* (pp. 277–309). Lexington, MA: Lexington Books.

Chu, Donald. 1982. *Dimensions of Sport Studies.* New York: John Wiley & Sons.

Chu, Donald and Griffey, David. 1985. "The Contact Theory of Racial Integration: The Case of Sport." *Sociology of Sport Journal* 3: 323–333.

Clarke, J. and Critcher, C. 1985. *The Devil Makes Work.* London: Macmillan.

Clarke, Liz. 1993. "Female Fans Root for Sports of All Sorts, Survey Reveals." *Charlotte Observer* (September 8): 1A, 6A.

Clarke, Liz and Greiff, James. 1993. "Pacts Vow Minority Jobs, Contracts." *Charlotte Observer* (July 2): 1A, 6A.

Clifton, Tony. 1987. "Soccer: Old Riot, New Riot." *Newsweek* (September 21): 8.

Closius, Phillip J. 1985. "Professional Sports and Antitrust Laws: The Ground Rules of Immunity, Exemption and Liability." In A. T. Johnson and J. H. Frey (Eds.). *Government and Sport: The Public Policy Issues* (pp. 140–161). Totowa, NJ: Rowman and Allenheld.

Clumpner, Roy A. and Pendleton, Brian B. 1978. "The People's Republic of China." In J. Riordan (Ed.). *Sport*

Under Communism (pp. 103–140). Montreal: McGill-Queen's University Press.

Coakley, Jay J. 1983. "Leaving Competitive Sport: Retirement or Rebirth?" *Quest* 35: 1–11.

Coakley, Jay J. 1986. "Socialization and Youth Sports." In C. R. Rees and Andrew W. Miracle (Eds.). *Sport and Social Theory* (pp. 135–144). Champaign, IL: Human Kinetics.

Coakley, Jay J. 1987a. "Sociology of Sport in the United States." *International Review for the Sociology of Sport* 22: 63–79.

Coakley, Jay J. 1987b. "Children and the Sport Socialization Process." In D. Gould and M. R. Weiss (Eds.). *Advances in Pediatric Sport Sciences, Vol II. Behavioral Issues* (pp. 43–60). Champaign, IL: Human Kinetics.

Coakley, Jay J. 1990. *Sport in Society: Issues and Controversies*, 4th ed. St. Louis: Times Mirror/Mosby.

Coakley, Jay J. 1992. "Burnout Among Adolescent Athletes: A Personal Failure or Social Problem?" *Sociology of Sport Journal* 9: 271–285.

Coakley, Jay J. 1994. *Sport in Society: Issues and Controversies*, 5th ed. St. Louis: Mosby.

Cohen, Greta L. (Ed.). 1993. *Women in Sport: Issues and Controversies*. Newbury Park, CA: Sage.

Coleman, James S. 1961. *The Adolescent Society.* New York: Free Press of Glencoe.

Contreras, Joseph. 1992. "Not Ready for Prime Time." *Newsweek* (November 23): 39.

Cook, Karen S. (Ed.). 1987. *Social Exchange Theory.* Beverly Hills, CA: Sage.

Cooley, Charles Horton. 1902. *Human Nature and the Social Order.* New York: Scribner's.

Coser, Lewis. 1973. *Greedy Institutions*. New York: Free Press.

Cox, Richard H. 1985. *Sport Psychology: Concepts and Applications*. Dubuque, IA: Wm. C. Brown.

CQ Researcher. 1992a. "Minority Women and Sports." In "Women and Sports" (March 6): 206. Washington, D.C.: Congressional Quarterly.

CQ Researcher. 1992b. "Women and Sports."(March 6): 193–216. Washington, DC: Congressional Quarterly.

CQ Researcher. 1992c. "Female Sports Reporters: Ready for Prime Time?" In "Women and Sports." (March 6): 200. Washington, DC: Congressional Quarterly.

Creamer, Robert. 1984. "A Surplus of Ill Will." *Sports Illustrated* (November 26): 21.

Csikszentmihalyi, M. 1975. *Beyond Boredom and Anxiety.* San Francisco: Jossey-Bass.

Curry, Timothy. 1991. "Fraternal Bonding in the Locker Room: A Profeminist Analysis of Talk About Competition and Women." *Sociology of Sport Journal* 8: 119–135.

Curry, Timothy and Jiobu, Robert M. 1984. *Sports: A Social Perspective.* Englewood Cliffs, NJ: Prentice-Hall.

Curry, Timothy and Strauss, R. 1988. "On the Normalization of Sport Injury: A Little Pain Never Hurt Anybody." Paper presented at the North American Society for the Sociology of Sport Annual Conference, Cincinnati.

Curtis, James E. and Ennis, R. 1988. "Negative Consequences of Leaving Competitive Sport? Comparative Findings for Former Elite-Level Hockey Players." *Sociology of Sport Journal* 5: 87–106.

Davidson, James D. 1982. "Social Differentiation and Sports Participation: The Case of Golf." In R. M. Pankin (Ed.). *Social Approaches to Sport* (pp. 181–224). Rutherford, NJ: Fairleigh Dickinson University Press.

Davis, Laurel R. and Delano, Linda C. 1992. "Fixing the Boundaries of Physical Gender: Side Effects of Anti-Drug Campaigns in Athletics." *Sociology of Sport Journal* 9: 1–19.

Davis, S. P. 1973. "A Study of Social Class and the Sport of Riding." Unpublished manuscript. University of Massachusetts, Department of Sport Studies, Amherst, MA.

Decker, June L. 1986. "Role Conflict of Teacher/Coaches in Small Colleges." *Sociology of Sport Journal* 3: 356—365.

Deem, Rosemary. 1988. "'Together We Stand, Divided We Fall': Social Criticism and the Sociology of Sport and Leisure." *Sociology of Sport Journal* 5: 341–354.

Deford, Frank. 1976. "Religion in Sport." *Sports Illustrated* (April 19, April 26, May 3): 88–100; 55–56, 68, 69; 43–44, 57–60, respectively.

Deford, Frank. 1979. "Religion in Sport." In D. S. Eitzen (Ed.). *Sport in Contemporary Society* (pp. 341–347). New York: St. Martin's Press.

Deford, Frank. 1981. "Love and Love." *Sports Illustrated* (April 27): 68–84.

Deford, Frank. 1984. "Cheer, Cheer, Cheer for the Home Team." *Sports Illustrated* 61 (June 13): 38, 40–41.

Deford, Frank. 1985. "No Longer a Cozy Corner." *Sports Illustrated* (December 23–30): 44–61.

Deford, Frank. 1993a. "Heavenly Game?" In D. S. Eitzen (Ed.). *Sport in Contemporary Society: An Anthology* (pp. 311–314). New York: St. Martin's Press.

Deford, Frank. 1993b. "Lessons from a Friend." *Newsweek* (February 22): 60–61.

DeFrantz, Anita. 1991. "We've Got to Be Strong." *Sports Illustrated* (August 12): 77.

Del Rey, Patricia. 1978. "The Apologetic and Women in Sport." In C. A. Oglesby (Ed.). *Women and Sport: From Myth to Reality* (pp. 107–111). Philadelphia: Lea and Febiger.

DeMatta, Roberto. 1986. "Soccer and Brazilian Nationalism." Paper presented at the Conference on Sport, Culture, and Society, Stanford University, April.

Denzin, Norman. 1978. *The Research Act,* 2nd ed. New York: McGraw-Hill.

Devereux, Edward C. 1976. "Backyard Versus Little League Baseball: The Impoverishment of Children's Games." In Daniel Landers (Ed.). *Social Problems in Athletics* (pp. 37–76). Champaign, IL: University of Illinois Press.

Dimaggio, Joe. 1946. *Lucky to be a Yankee.* New York: Randolph Field.

Dirix, A., Knuttgen, H. G., and Tittel, K. (Eds.). 1988. *The Olympic Book of Sports Medicine.* Champaign, IL: Human Kinetics.

Donnelly, Peter. 1981a. "Toward a Definition of Sport Sub-Cultures." In M. Hart and S. Birrell (Eds.). *Sport in the Sociocultural Process,* 3rd ed. (pp. 565–587). Dubuque, IA: Wm. C. Brown.

Donnelly, Peter. 1981b. "Athletes and Juvenile Delinquents: A Comparative Analysis Based on a Review of the Literature." *Adolescence* 16: 415–431.

Donnelly, Peter. 1982. "Social Climbing: A Case Study of the Changing Class Structure of Rock Climbing and Mountaineering in Britain." In A. O. Dunleavy, A. W. Miracle, and C. R. Rees (Eds.). *Studies in the Sociology of Sport* (pp. 13–28). Fort Worth, TX: Texas Christian University Press.

Donnelly, Peter. 1985. "Sport Subcultures." In R. Terjung (Ed.). *Exercise and Sport Sciences Review* (pp. 539–578). New York: Macmillan.

Donnelly, Peter and Young, Kevin. 1988. "The Construction and Confirmation of Identity in Sport Subcultures." *Sociology of Sport Journal* 5: 223–240.

Donohew, Lewis, Helm, David, and Haas, John. 1989. "Drugs and (Len) Bias on the Sports Page." In L. A. Wenner (Ed.). *Media, Sports & Society* (pp. 225–237). Newbury Park, CA: Sage.

DuBois, Paul E. 1986. "The Effects of Participation in Sport on the Value Orientations of Young Athletes." *Sociology of Sport Journal* 3: 29–42.

Duda, Joan L. 1987. "Toward a Developmental Theory of Children's Motivation in Sport." *Journal of Sport Psychology* 9:130–145.

Duff, Robert W. and Hong, Lawrence K. 1984. "Self-Images of Women Bodybuilders." *Sociology of Sport Journal* 1: 374–380.

Duncan, Margaret C. 1990. "Sports Photographs and Sexual Difference: Images of Women and Men in the 1984 and 1988 Olympic Games." *Sociology of Sport Journal* 7: 22–43.

Duncan, Margaret C. and Hasbrook, Cynthia A. 1988. "Denial of Power in Televised Women's Sports." *Sociology of Sport Journal* 5: 1–21.

Dunning, Eric. (Ed.). 1972a. *Sport: Readings from a Sociological Perspective.* Toronto: University of Toronto Press.

Dunning, Eric. 1972b. "The Development of Modern Football." In E. Dunning (Ed.). *Sport: Readings from a Sociological Perspective* (pp. 133–151). Toronto: University of Toronto Press.

Dunning, Eric. 1979. "The Figurational Dynamics of Modern Sport." *Sportwissenschaft* 9: 341–359.

Dunning, Eric. 1986a. "The Sociology of Sport in Europe and the United States: Critical Observations from an 'Eliasian' Perspective." In C. R. Rees and A. R. Miracle (Eds.). *Sport and Social Theory* (pp. 29–56). Champaign, IL: Human Kinetics.

Dunning, Eric. 1986b. "Sport as a Male Preserve: Notes on the Social Sources of Masculine Identity and its Transformations." *Theory, Culture & Society* 3: 79–90.

Dunning, Eric. 1990. "Sociological Reflections on Sport, Violence and Civilization." *International Review for the Sociology of Sport* 25: 65–82.

Dunning, Eric G., Maguire, Joseph A., and Pearton, Robert E. (Eds.). 1993. *The Sports Process: A Comparative and Developmental Approach.* Champaign, IL: Human Kinetics.

Duquin, Mary. 1978. "The Androgynous Advantage." In C. A. Oglesby (Ed.). *Women and Sport: From Myth to Reality* (pp. 89–106). Philadelphia: Lea and Febiger.

Duquin, Mary E. 1989. "Fashion and Fitness: Images in Women's Magazine Advertisements." *Arena Review* 13: 97–109.

Durkheim Emile. 1965. *The Elementary Forms of Religious Life.* New York: Free Press.

Dyreson, Mark. 1992. "America's Athletic Missionaries: Political Performance, Olympic Spectacle and the Quest for an American National Culture, 1896–1912." *Olympika* 1: 70–91.

East, Elizabeth R. 1978. "Federal Civil Rights Legislation and Sport." In C. A. Oglesby (Ed.). *Women and Sport: From Myth to Reality* (pp. 205–219). Philadelphia: Lea and Febiger.

Easterwood, Jim. 1979. "'Out-Coached' in Near Upset." *Honolulu Star-Bulletin*, March 2.

Eberhard, Wallace B. and Myers, Margaret Lee. 1988. "Beyond the Locker Room: Women in Sports on Major Daily Newspapers." *Journalism Quarterly* 65: 595–599.

Eckhardt, Kenneth W. and Ermann, M. David. 1977. *Research Methods: Perspective, Theory, and Analysis.* New York: Random House.

Edelman, Robert. 1993. *Serious Fun: A History of Spectator Sports in the USSR.* New York: Oxford University Press.

Edwards, Harry. 1973. *Sociology of Sport.* Homewood, IL: Dorsey.

Edwards, Harry. 1983. "Educating Black Athletes." *The Atlantic Monthly* vol: 31–38.

Edwards, Harry. 1986. "The Collegiate Athletic Arms Race: Origins and Implications of the 'Rule 48' Controversy." In R. E. Lapchick (Ed.). *Fractured Focus: Sport as a Reflection of Society* (pp. 21–43). Lexington, MA: Lexington Books.

Eggleston, John. 1965. "Secondary Schools and Oxbridge Blues." *British Journal of Sociology* 16: 232–242.

Eitzen, D. Stanley. 1976. "Sport and Social Status in American Public Secondary Education." *Review of Sport & Leisure* 1: 139–155.

Eitzen, D. Stanley. 1989. "Black Participation in American Sport since World War II." In D. S. Eitzen (Ed.). *Sport in Contemporary Society: An Anthology,* 3rd ed. (pp. 300–312). New York: St. Martin's Press.

Eitzen, D. Stanley and Furst, David. 1989. "Racial Bias in Women's Collegiate Volleyball." *Journal of Sport & Social Issues* 13: 46–51.

Eitzen, D. Stanley and Hyatt-Hearn, Susan. 1992. "Teaching 'Sport and Society': Problems and Consequences." *Sociology of Sport Journal* 9: 60–69.

Eitzen, D. Stanley and Sage, George H. 1986. *Sociology of North American Sport*, 3rd ed. Dubuque, IA: Wm. C. Brown.

Eitzen, D. Stanley and Sage, George H. 1993. *Sociology of North American Sport*, 5th ed. Madison, WI: Brown & Benchmark.

Eitzen, D. Stanley and Sanford, David C. 1975. "The Segregation of Blacks by Playing Position in Football: Accident or Design?" *Social Science Quarterly* 55: 948–959.

Eitzen, D. Stanley and Zinn, Maxine B. 1989. "The De-Athleticization of Women: The Naming and Gender Marking of College Sports Teams." *Sociology of Sport Journal* 7: 362–369.

Eitzen, D. Stanley and Zinn, Maxine B. 1993. "The Sexist Naming of Collegiate Athletic Teams and Resistance to Change." *Journal of Sport and Social Issues* 17: 34–41.

Eitzen, D. Stanley and Zinn, Maxine B. 1994. *Social Problems*, 6th ed. Boston: Allyn & Bacon.

Elias, Norbert. 1986. "The Genesis of Sport as a Sociological Problem." In E. Dunning (Ed.). *Sport: Readings from a Sociological Perspective* (pp. 88–115). Toronto: University of Toronto Press.

Elias, Norbert and Dunning, Eric. 1972. "Folk Football in Medieval and Early Modern Britain." In E. Dunning (Ed.). *Sport: Readings from a Sociological Perspective* (pp. 116–132). Toronto: University of Toronto Press.

Elias, Norbert and Dunning, Eric. 1986. *Quest for Excitement: Sport and Leisure in the Civilizing Process.* London: Basil Blackwell.

Erikson, Kai. 1964. "Notes on the Sociology of Deviance." In Howard Becker (Ed.). *The Other Side* (pp. 9–21). New York: Free Press.

Erlanger, Steven. 1988. "Unified Team Faces Splintered Future." *The New York Times.*

Espy, Richard. 1979. *The Politics of the Olympic Games.* Berkeley: University of California Press.

Ewald, Keith and Jiobu, Robert M. 1985. "Explaining Positive Deviance: Becker's Model and the Case of Runners and Bodybuilders." *Sociology of Sport Journal* 2: 144–156.

Farley, John E. 1994. *Sociology,* 3rd ed. Englewood Cliffs, NJ: Prentice-Hall.

Farr, Kathryn A. 1988. "Dominance Bonding Through the Good Old Boys Sociability Group." *Sex Roles* 18: 259–277.

Feltz, Deborah L. 1979. "Athletics in the Status System of Female Adolescents." *Review of Sport & Leisure* 4: 110–118.

Ferrante, Joan. 1992. *Sociology: A Global Perspective.* Belmont, CA: Wadsworth.

Figler, Stephen K. 1981. *Sport and Play in American Life.* Philadelphia: Saunders College Publishing.

Figler, Stephen K. and Whitaker, Gail. 1991. *Sport and Play in American Life: A Textbook in the Sociology of Sport*, 2nd ed. Dubuque, IA: Wm. C. Brown.

Fine, Gary A. 1981. "Preadolescent Socialization Through Organized Athletics: The Construction of Moral Meanings in Little League Baseball." In M. Hart and S. Birrell (Eds.). *Sport in the Sociocultural Process* (pp. 164–191). Dubuque, IA: Wm. C. Brown.

Fine, Gary Alan. 1987. *With the Boys: Little League Baseball and Preadolescent Culture.* Chicago: University of Chicago Press.

Fine, Gary Alan. 1989. "Mobilizing Fun: Provisioning Resources in Leisure Worlds." *Sociology of Sport Journal* 6: 319–334.

Finn, Robin. 1992. "An Assistant with a Different Image." *The New York Times* (December 30): B7–B8.

Finney, Henry C. and Lesieur, Henry R. 1982. "A Contingency Theory of Organizational Crime." In S. B. Bacharach (Ed.). *Research in the Sociology of Organizations.* (pp. 255–300). Greenwich, CT: JAI Press.

Fishwick, Lesley and Hayes, Diane. 1989. "Sport for Whom? Differential Participation Patterns of Recreation in Leisure-Time Physical Activities." *Sociology of Sport Journal* 6: 269–277.

Flake, Carol. 1992. "The Spirit of Winning: Sports and the Total Man." In S. J. Hoffman (Ed.) *Sport and Religion* (pp. 161–176). Champaign, IL: Human Kinetics.

Fleisher, Arthur A., Goff, Brian L., and Tollison, Robert D. 1992. *The National Collegiate Athletic Association: A Study in Cartel Behavior.* Chicago: University of Chicago Press.

Flint, William C. and Eitzen, D. Stanley. 1987. "Professional Sports Team Ownership and Entrepreneurial Capitalism." *Sociology of Sport Journal* 4: 17–27.

Foley, Douglas E. 1990. "The Great American Football Ritual: Reproducing Race, Class, and Gender Inequality." *Sociology of Sport Journal* 7: 111–135.

Freedman, Warren. 1987. *Professional Sports and Antitrust.* Westport, CT: Greenwood Press.

Frey, James H. 1980. "Youth Sports: Who Really Benefits?" *Journal of the Nevada Association for Health, Physical Education and Recreation* 1: 1–9.

Frey, James H. 1982. "Boosterism, Scarce Resources and Institutional Control: The Future of American Intercollegiate Athletics." *International Review of Sport Sociology* 17: 53–69.

Frey, James H. 1984a. "Gambling and College Sports: Views of Coaches and Athletic Directors." *Sociology of Sport Journal* 1: 36–45.

Frey, James H. 1984b. "The U.S. vs. Great Britain: Responses to the 1980 Boycott of the Olympic Games." *Comparative Physical Education and Sport* 6: 4–12.

Frey, James H. 1985. "Gambling, Sports, and Public Policy." In A. T. Johnson and J. H. Frey (Eds.). *Government and Sport: The Public Policy Issues* (pp. 189–218). Totowa, NJ: Rowman and Allanheld.

Frey, James H. 1986. "College Athletics: Problems of a Functional Analysis." In C. R. Rees and A. W. Miracle (Eds.). *Sport and Social Theory* (pp. 199–210). Champaign, IL: Human Kinetics.

Frey, James H. 1987–88. "Institutional Control of Athletics: An Analysis of the Role Played by Presidents, Faculty, Trustees, Alumni, and the NCAA." *Journal of Sport and Social Issues* 11: 49–60.

Frey, James H. 1988a. "The Internal and External Role of Sport in National Development." *Journal of National Development* 1: 65–82.

Frey, James H. 1988b. "The Meaning of Work to Participants in High-Risk Sport: The Case of Skydivers." Unpublished paper.

Frey, James H. 1991. "Social Risk and the Meaning of Sport." *Sociology of Sport Journal* 8: 136–145.

Frey, James H. 1994. "Deviance of Organizational Subunits: The Case of College Athletic Departments." *Journal of Sport & Social Issues* 18: 110–122.

Frey, James H. and Dickens, David R. 1991. "Leisure as a Primary Institution." *Sociological Inquiry* 60: 264–273.

Frey, James H. and Eitzen, D. Stanley. 1991. "Sport and Society." *Annual Review of Sociology* 17: 503–522.

Frey, James H. and Massengale, John D. 1988. "American School Sports: Enhancing Social Values Through Restructuring." *Journal of Physical Education, Recreation & Dance* 59: 40–44.

Friesen, David. 1976. "Academic-Athletic-Popularity Syndrome in the Canadian High School Society." In R. S. Gruneau and J. G. Albinson (Eds.). *Canadian Sport: Sociological Perspectives* (pp. 361–371). Don Mills, Ontario: Addison-Wesley.

Fuller, John R. and La Fountain, Marc J. 1987. "Performance-Enhancing Drugs in Sport." *Adolescence* 22: 969–976.

Furst, David M. and Heaps, J. E. 1987. "Stacking in Women's Intercollegiate Basketball." Paper presented at the North American Society for the Sociology of Sport Annual Conference, Edmonton, Alberta.

Gallmeier, Charles P. 1988. "Juicing, Burning, and Tooting: Observing Drug Use among Professional Hockey Players." *Arena Review* 12: 1–12.

Gallmeier, Charles P. 1989. "Traded, Waived, or Gassed: Failure in the Occupational World of Ice Hockey." *Journal of Sport & Social Issues* 13: 25–45.

Gallner, Sheldon. 1974. *Pro Sports: The Contract Game.* New York: Charles Scribner's Sons.

Gammon, Clive. 1983. "Swirling Shades of Gray." *Sports Illustrated* (May 16): 78–94.

Gammon, Clive. 1985. "A Day of Horror and Shame." *Sports Illustrated* (June 10): 20–35.

Gammons, Peter. 1987. "The Campanis Affair." *Sports Illustrated* (April 20): 31.

Garcia Ferrando, Manuel. 1986. *Hábitos Deportivos de los Españoles*. Madrid: Ministerio de Cultura.

Geertz, Clifford. 1963. "The Integrative Revolution: Primordial Sentiments and Civil Politics in New States." In C. Geertz (Ed.). *Old Societies and New States: The Quest for Modernity in Asia and Africa* (pp. 105–107). Glencoe, IL: Free Press.

Gent, Peter. 1973. *North Dallas Forty.* New York: William Morrow.

Giddens, Anthony. 1991. *Introduction to Sociology.* New York: Norton.

Gilbert, Bil. 1969. "Three-Part Series on Drugs in Sport." *Sports Illustrated* (June 23, 30; July 7).

Gilbert, Bil. 1988. "Competition." *Sports Illustrated* (March 16): 86–100.

Gilbert, Bil and Williamson, Nancy. 1973. "Three-Part Series on Women in Sport." *Sports Illustrated* (May 28; June 4, 11).

Gilley, J. Wade and Hickey, Anthony A. 1986. *Administration of University Athletic Programs: Internal Control and Excellence.* Washington, DC: American Council on Education.

Gilligan, C. 1982. *In a Different Voice: Psychological Theory and Women's Development.* Cambridge: Harvard University Press.

Givens, Ron. 1989. "Oklahoma Is Not OK." *Newsweek* (February 27): 80.

Glassner, Barry. 1989. "Men and Muscles." In M. S. Kimmel and M. A. Messner (Eds.). *Men's Lives* (pp. 310–320). New York: Macmillan.

Glock, Charles and Stark, Rodney. 1965. *Religion and Society in Tension.* Chicago: Rand McNally.

Gmelch, George. 1971. "Baseball Magic." *Transaction* 8: 54.

Goldstein, Warren. 1989. *Playing for Keeps: A History of Early Baseball.* Ithaca, NY: Cornell University Press.

Goode, Erich. 1990. *Deviant Behavior.* Englewood Cliffs, NJ: Prentice-Hall.

Goodhart, Phillip and Chataway, Christopher. 1968. *War Without Weapons.* London: W. H. Allen.

Gould, Daniel. 1987. "Understanding Attrition in Children's Sports." In D. Gould and M. R. Weiss (Eds.). *Advances in Pediatric Sports Sciences* (pp. 61–85). Champaign, IL: Human Kinetics.

Greendorfer, Susan L. 1977. "Role of Socializing Agents in Female Sports Involvement." *Research Quarterly* 48: 306–310.

Greendorfer, Susan L. 1990. "Motivation: The Neglected Factor in Children's Sport Socialization." Paper presented at the Annual Meetings of the American Alliance of Health, Physical Education, Recreation, and Dance. New Orleans, LA.

Greendorfer, Susan L. and Hasbrook, Cynthia A. 1991. *Learning Experiences in Sociology of Sport.* Champaign, IL: Human Kinetics.

Greendorfer, Susan L. and Lewko, John H. 1978. "Role of Family Members in Sport Socialization of Children." *Research Quarterly* 49: 146–152.

Gregory, C. J. and Petrie, P. M. 1975. "Superstitions of Canadian Intercollegiate Athletes: An Inter-Sport Comparison." *International Review of Sport Sociology* 10: 59–68.

Grey, Mark A. 1992. "Sports and Immigrant, Minority and Anglo Relations in Garden City (Kansas) High School." *Sociology of Sport Journal* 9: 255–270.

Griffin, Pat. 1993. "Homophobia in Women's Sports: The Fear That Divides Us." In G. L. Cohen (Ed.). *Women in Sport: Issues and Controversies* (pp. 193–203). Newbury Park, CA: Sage.

Gross, Edward. 1978. "Organizational Crime: A Theoretical Perspective." *Studies in Symbolic Interaction* 1: 55–85.

Grundman, Adolph H. 1986. "The Image of Intercollegiate Sports and the Civil Rights Movement: A Historian's View." In R. E. Lapchick (Ed.). *Fractured Focus: Sport as a Reflection of Society* (pp. 77–85). Lexington, MA: Lexington Books.

Gruneau, Richard S. 1975. "Sport, Social Differentiation and Social Inequality." In D. W. Ball and J. W. Loy (Eds.). *Sport and Social Order: Contributions to the Sociology of Sport* (pp. 117–184). Reading, MA: Addison-Wesley.

Gruneau, Richard S. 1976. "Class or Mass: Notes on the Democratization of Canadian Amateur Sport." In R. S. Gruneau and J. G. Albinson (Eds.). *Canadian Sport: Sociological Perspectives* (pp. 108–141). Don Mills, Ontario: Addison-Wesley.

Gruneau, Richard. 1983. *Class, Sports, and Social Development.* Amherst: University of Massachusetts Press.

Gruneau, Richard. 1989. "Making Spectacle: A Case Study in Television Sports Production." In L. A. Wenner (Ed.). *Media, Sports, & Society* (pp. 134–154). Newbury Park, CA: Sage.

Grusky, Oscar. 1963. "The Effects of Formal Structure on Managerial Recruitment: A Study of Baseball Organization." *Sociometry* 26: 345–353.

Guelke, Adrian. 1993. "Sport and the End of *Apartheid.*" In L. Allison (Ed.). *The Changing Politics of Sport* (pp. 151–170). Manchester: Manchester University Press.

Guttmann, Allen. 1978. *From Ritual to Record: The Nature of Modern Sports.* New York: Columbia University Press.

Guttmann, Allen. 1981. "Sports Spectacular from Antiquity to the Renaissance." *Journal of Sport History* 8: 5–27.

Guttmann, Allen. 1988. *A Whole New Ball Game: An Interpretation of American Sports.* Chapel Hill, NC: University of North Carolina Press.

Guttmann, Allen. 1991. "Sports Diffusion: A Response to Maguire and the Americanization Commentaries." *Sociology of Sport Journal* 8: 185–190.

Guttmann, Allen. 1992. *The Olympics: A History of the Modern Games.* Urbana, IL: University of Illinois Press.

Guttmann, Allen. 1993. "The Diffusion of Sports and the Problem of Cultural Imperialism." In E. G. Dunning, J. A. Maguire, and R. E. Pearton (Eds.). *The Sports Process: A Comparative and Developmental Approach* (pp. 125–138). Champaign, IL: Human Kinetics.

Haerle, Rudolf K. Jr. 1974. "The Athlete as 'Moral' Leader: Heroes, Success Themes and Basic Cultural Values in Selected Baseball Autobiographies, 1900–1970." *Journal of Popular Culture* 8: 392–401.

Hall, M. Ann. 1988. "The Discourse of Gender and Sport: From Femininity to Feminism." *Sociology of Sport Journal* 5: 330–340.

Hallinan, Christopher. 1991. "Aborigines and Positional Segregation in the Australian Rugby League." *International Review for the Sociology of Sport* 26: 69–81.

Hanford, George H. 1974. *An Inquiry into the Need for and Feasibility of a National Study of Intercollegiate Athletics.* Washington, DC: American Council on Education.

Hanford, George H. 1979. "Controversies in College Sports." *Annals of the American Academy of Political and Social Science* 445: 66–79.

Hardman, Ken, Krotee, March, and Chrissanthopoules, Andreas. 1988. "A Comparative Study of Interschool Competition in England, Greece, and the United States." In E. F. Broom, R. Clumpner, B. Pendleton, and C. Pooley (Eds.). *Comparative Physical Education and Sport,* Vol. 5 (pp. 91–102). Champaign, IL: Human Kinetics.

Hardy, Stephen H. and Berryman, Jack W. 1982. "A Historical View of the Governance Issue." In James H. Frey (Ed.). *The Governance of Intercollegiate Athletics.* (pp. 15–28). West Point: Leisure Press.

Hargreaves, John. 1986a. *Sport, Power and Culture.* Cambridge: Polity.

Hargreaves, John. 1986b. "The State and Sport: Programmed and Nonprogrammed Intervention in Contemporary Britain." In L. Allison (Ed.). *The Politics of Sport* (pp. 242–261). Manchester: Manchester University Press.

Hargreaves, John. 1992. "Olympism and Nationalism: Some Preliminary Considerations." *International Review for the Sociology of Sport* 27: 122–135.

Harootyan, Robert A. 1982. "The Participation of Older People in Sports." In R. M. Pankin (Ed.). *Social Approaches to Sport* (pp. 122–147). Rutherford, NJ: Fairleigh Dickinson University Press.

Harris, Stanley. 1925. *Playing the Game: From Mine Boy to Manager.* New York: Grosset & Dunlap.

Hart-Nibbrig, Nand. 1987. "Corporate Athleticism: An Inquiry into the Political Economy of College Sports." In A. Yiannakis, T. D. McIntyre, M. J. Melnick and D. P. Hart (Eds.). *Sport Sociology: Contemporary Themes* (pp. 156–161), 3rd ed. Dubuque, IA: Kendall/Hunt.

Hart-Nibbrig, Nand and Cottingham, Clement. 1986. *The Political Economy of College Sports.* Lexington: Lexington Books.

Harter, S. 1978. "Effectance Motivation Reconsidered: Toward a Developmental Model." *Human Development* 1: 34–64.

Harter, S. 1981. "A Model of Intrinsic Mastery Motivation in Children: Individual Differences and Developmental Change." In W. A. Collins (Ed.). *Minnesota Symposium on Child Psychology,* Vol. 14 (pp. 215–255). Hillsdale, NJ: Erlbaum.

Harvey, David. 1989. *The Condition of Post-Modernity.* Oxford: Blackwell.

Hasbrook, Cynthia A. 1986. "Reciprocity and Childhood Socialization into Sport." In J. Humphrey & L. Vander Velden (Eds.). *Current Selected Research in the Psychology and Sociology of Sport* (pp. 135–147). New York: AMS Press.

Hasbrook, Cynthia A. 1988. "Female Coaches—Why the Declining Numbers and Percentages?" *Journal of Physical Education, Recreation and Dance* 58 (2): 59–63.

Hastad, Douglas N., Segrave, Jeffrey O., Pangravi, Robert, and Petersen, Gene. 1984. "Youth Sport Participation and Deviant Behavior." *Sociology of Sport Journal* 1: 366–373.

Hastings, Donald W. 1987. *College Swimming Coach: Social Issues, Roles, and Worlds.* New York: University Press of America.

Hawthorne, Peter. 1976. "Cracks in the Racial Barrier." *Sports Illustrated* (November 22): 36–43.

Healey, Joseph. 1991. "An Exploration of the Relationships Between Memory and Sport." *Sociology of Sport Journal* 8: 213–227.

Heinemann, Klaus. 1993. "Sport in Developing Countries." In E. G. Dunning, J. A. Maguire, and R. E. Pearton (Eds.). *The Sports Process: A Comparative and Developmental Approach* (pp. 139–150). Champaign, IL: Human Kinetics.

Herek, G. M. 1987. "On Heterosexual Masculinity: Some Psychical Consequences of the Social Construction of Gender and Sexuality." In M. S. Kimmel (Ed.). *Changing Men: New Directions in Research on Men and Masculinity* (pp. 68–82). Beverly Hills, CA: Sage.

Hess, Beth B., Markson, Elizabeth W., and Stein, Peter J. 1991. *Sociology,* 4th ed. New York: Macmillan.

Hess, Beth B., Markson, Elizabeth W., and Stein, Peter J. 1993. *Sociology: 1993 Update.* New York: Macmillan.

Hewitt, Brian. 1993a. "A Game of Pain." *Chicago Sun-Times* (September 19): 17B–20B.

Hewitt, Brian. 1993b. "Playing at Any Cost." *Chicago Sun-Times* (September 20): 92–93.

Hewitt, Brian. 1993c. "Through the Tears." *Chicago Sun-Times* (September 21): 83–85.

Hewitt, Brian. 1993d. "Opposing Views." *Chicago Sun-Times* (September 22): 98–99.

Hewitt, Brian. 1993e. "Pension Benefits Good, But They Could Be Better." *Chicago Sun-Times* (September 22): 97, 99.

Higgs, R. J. 1983. "Muscular Christianity, Holy Play and Spiritual Exercises: Confusion About Christ in Sport and Religion." *Arete: Journal of Sport Literature* 1: 59–85.

Hill, Christopher R. 1992. *Olympic Politics.* Manchester: Manchester University Press.

Hill, Grant M. and Simons, Jeffrey. 1989. "A Study of the Sport Specialization on High School Athletics." *Journal of Sport & Social Issues* 13: 1–13.

Hilliard, Dan C. 1984. "Media Images of Male and Female Professional Athletes: An Interpretive Analysis of Magazine Articles." *Sociology of Sport Journal* 1: 251–262

Hilliard, Dan C. 1986. "Risk Taking and the Experience of Emotions in Sport and Recreation: Two Cases." Paper presented at the North American Society for the Sociology of Sport Annual Conference, Las Vegas, October 29–November 2.

Hoberman, John M. 1984. *Sport and Political Ideology.* Austin: University of Texas Press.

Hoberman, John M. 1992. *Mortal Engines: The Science of Performance and the Dehumanization of Sport.* New York: Free Press.

Hoberman, John M. 1993. "Sport and Ideology in the Post-Communist Age." In L. Allison (Ed.). *The Changing Politics of Sport* (pp. 15–36). Manchester: Manchester University Press.

Hochberg, Philip R. 1985. "Property Rights in Sports Broadcasting: The Fundamental Issue." In A. T. Johnson and J. H. Frey (Eds.). *Government and Sport: The Public Policy Issues* (pp. 162–170). Totowa, NJ: Rowman and Allenheld.

Hoffman, Shirl J. 1985. "Evangelism and the Revitalization of Religious Ritual in Sport." *Arete: Journal of Sport Literature* 3: 63–87.

Hoffman, Shirl J. 1992a. "Sport as Religion." In S. J. Hoffman (Ed.) *Sport and Religion* (pp. 1–12). Champaign, IL: Human Kinetics.

Hoffman, Shirl J. (Ed.). 1992b. *Sport and Religion.* Champaign, IL: Human Kinetics.

Hoffman, Shirl J. 1992c. "Evangelicalism and the Revitalization of Religious Ritual in Sport." In S. J. Hoffman (Ed.). *Sport and Religion* (pp. 111–125). Champaign, IL: Human Kinetics.

Hoffman, Shirl J. 1992d. "Sport, Religion, and Ethics." In S. J. Hoffman (Ed.). *Sport and Religion* (pp. 213–225). Champaign, IL: Human Kinetics.

Hoffman, Shirl J. 1992e. "Religion in Sport." In S. J. Hoffman (Ed.). *Sport and Religion* (pp. 127–141). Champaign, IL: Human Kinetics.

Hofmann, Dale and Greenberg, Martin J. 1989. *Sport$biz: An Irreverent Look at Big Business in Sports.* Champaign, IL: Human Kinetics.

Hogan, William R. 1967. "Sin and Sports." In Ralph Slovenko and James A. Knight (Eds.). *Motivation in Play, Games and Sports.* Springfield, IL: Charles C. Thomas.

Holland, Judith R. and Oglesby, Carole. 1979. "Women in Sport: The Synthesis Begins." *Annals of the American Academy of Political and Social Science* 445: 80–90.

Holland, T. Keating. 1993. "Field of Dreams." In D. S. Eitzen (Ed.). *Sport in Contemporary Society: An Anthology* (pp. 177–186). New York: St. Martin's Press.

Houlihan, Barrie. 1991. *The Government and Politics of Sport.* London: Routledge.

Hughes, Robert and Coakley, Jay. 1991. "Positive Deviance Among Athletes: The Implications of Overconformity to the Sport Ethic." *Sociology of Sport Journal* 8: 307–325.

Human Sciences Research Council. 1982. *Sport in the Republic of South Africa: Main Committee Report*. Pretoria.

Humber, L. 1988. "Vegas at Odds with Gretzky." *Toronto Globe and Mail* (May 7): A11.

Information Please. 1991. *The 1991 Information Please Almanac*. Boston: Houghton Mifflin.

Ingham, Alan and Donnelly, Peter. 1990. "Whose Knowledge Counts? The Production of Knowledge and Issues of Application in the Sociology of Sport." *Sociology of Sport Journal* 7:58–65.

Ingham, Alan G. and Hardy, Steven. 1984. "Sport, Structuration, and Hegemony." *Theory, Culture, and Society* 2: 85–103.

Irwin, John. 1973. "Surfing: The Natural History of an Urban Scene." *Urban Life and Culture* 2: 131–160.

Jacobs, Glenn. 1987. " Urban Samurai: The 'Karate Dojo.'" In A. Yiannakis, T. D. McIntyre, M. J. Melnick, and D. P. Hart (Eds.). *Sport Sociology: Contemporary Themes*, 3rd ed. (pp. 276–284). Dubuque, IA: Kendall/Hunt.

Jenkins, Dan. 1970. *Saturday's America*. Boston: Little, Brown.

Jenkins, Sally. 1991. "Who Let Them In?" *Sports Illustrated* (June 17): 78–90.

Jobling, Ian F. 1976. "Urbanization and Sport in Canada, 1867–1900." In R. S. Gruneau and J. G. Albinson (Eds.). *Canadian Sport: Sociological Perspectives* (pp. 64–77). Reading, MA: Addison-Wesley.

Johnson, Arthur T. 1993. *Minor League Baseball and Local Economic Development*. Urbana and Chicago: University of Illinois Press.

Johnson, Arthur T. and Frey, James H. (Eds.). 1985. *Government and Sport: The Public Policy Issues*. Totowa, NJ: Rowman and Allanheld.

Johnson, William Oscar. 1991a. "How Far Have We Come?" *Sports Illustrated* (August 5): 38–41.

Johnson, William Oscar. 1991b. "It Is Time, It Is Time." *Sports Illustrated* (April 29): 36–41.

Johnson, William Oscar. 1993. "The Agony of Victory." *Sports Illustrated* (July 5): 30–37.

Johnson, William Oscar and Moore, Kenny. 1988. "The Loser." *Sports Illustrated* (October 3): 20–27.

Kane, Martin. 1971. "An Assessment of Black Is Best." *Sports Illustrated* (January 18): 72–83.

Kane, Mary Jo. 1987. "The 'New' Female Athlete: Socially Sanctioned Image or Modern Role for Women?" *Medicine and Sport Science* 24: 101–111.

Kane, Mary Jo. 1988. "Media Coverage of the Female Athlete, Before, During and After Title IX: *Sports Illustrated* Revisited. *Journal of Sport Management* 2: 87–99.

Kane, Mary Jo. 1989. "The Post Title IX Female Athlete in the Media." *Journal of Physical Education, Recreation and Dance* 60 (March): 58–62.

Kane, Mary J. and Stangl, Jane M. 1991. "Employment Patterns of Female Coaches in Men's Athletics: Tokenism and Marginalization As Reflections of Occupational Sex-Segregation." *Journal of Sport & Social Issues* 15: 21–42.

Kang, Joon-Mann. 1988. "Sports, Media and Cultural Dependency." *Journal of Contemporary Asia* 18: 430–439.

Kanin, David B. 1981. *A Political History of the Olympic Games*. Boulder, CO: Westview Press.

Kanter, Rosabeth Moss. 1977a. *Men and Women of the Corporation*. New York: Basic.

Kanter, Rosabeth Moss. 1977b. "Some Effects of Group Life: Skewed Sex Ratios and Responses to Token Women." *American Journal of Sociology* 82: 965–1006.

Kaplan, H. Roy. 1986. "Sports, Gambling and Television: The Emerging Alliance." In R. E. Lapchick (Ed.). *Fractured Focus: Sport as a Reflection of Society* (pp. 247–257). Lexington, MA: Lexington Books.

Katz, Donald. 1993. "Triumph of the Swoosh." *Sports Illustrated* (August 16): 54–73.

Kearl, Michael C. and Gordon, Chad. 1992. *Social Psychology: Shaping Identity, Thought, and Conduct*. Boston: Allyn & Bacon.

Keefer, Robert, Goldstein, Jeffrey H. and Kasiarz, David. 1983. "Olympic Games Participation and Warfare." In J. H. Goldstein (Ed.). *Sports Violence* (pp. 183–193). New York: Springer-Verlag.

Kenyon, Gerald S. 1986. "The Significance of Social Theory in the Development of Sport Sociology." In C. R. Rees and A. W. Miracle (Eds.), *Sport and Social Theory* (pp. 3–22). Champaign, IL: Human Kinetics.

Kenyon, Gerald S. and Loy, John W. 1965. "Toward a Sociology of Sport." *Journal of Health, Physical Education, and Recreation* 36: 24–25, 68–69.

Kenyon, Gerald and McPherson, Barry D. 1973. "Becoming Involved in Physical Activity and Sport." In G. R. Rarick (Ed.). *Physical Activity* (pp. 304–332). New York: Academic Press.

Kerr, Clark. 1963. *The Uses of the University*. Cambridge, MA: Harvard University Press.

Kessler-Harris, Alice. 1992. "Multiculturalism Can Strengthen Our Common Culture." *Chronicle of Higher Education* (October 21): B3, B7.

Kidd, Bruce. 1981. "Sport, Dependency and the Canadian State." In M. Hart and S. Birrell (Eds.). *Sport in the Sociocultural Process* (pp. 707–721). Dubuque, IA: W. C. Brown.

Kidd, Bruce. 1991a. "How Do We Find Our Own Voices in the 'New World Order'? A Commentary on Americanization." *Sociology of Sport Journal* 8: 178–184.

Kidd, Bruce. 1991b. "From Quarantine to Cure: The New Phase of the Struggle Against Apartheid Sport." *Sociology of Sport Journal* 8: 33–46.

Kidd, Bruce. 1992. "The Culture Wars of the Montreal Olympics." *International Review for the Sociology of Sport* 27: 151–164.

Kiersh, Edward and Buchholz, Brad. 1993. "Annual Salary Survey." *Inside Sports* (April): 58–70.

King, Peter. 1991. "Two that Got Away." *Sports Illustrated 1992 Sports Almanac.* Boston: Little, Brown.

King, Peter. 1993. "The League of the Free." *Sports Illustrated* (July 26): 32–36.

Kirshenbaum, Jerry. 1981a. "Facing Up to Billie Jean's Revelations." *Sports Illustrated* (May 11): 13, 16.

Kirshenbaum, Jerry. 1981b. "Roger Wheeler and Jai Alai's None-Too-Watchful Watchdogs." *Sports Illustrated,* Scorecard Section (June 15): 13, 16.

Kirshenbaum, Jerry. 1989. "An American Disgrace." *Sports Illustrated* (February 27): 16–19.

Kirshnit, C. E., Ham, H. M., and Richards, M. H. 1989. "The Sporting Life: Athletic Activities During Early Adolescence." *Journal of Youth and Adolescence* 18: 601–615.

Klattell, David and Marcus, Norman. 1988. *Sports for Sale: Television, Money, and the Fans.* New York: Oxford University Press.

Kleiber, Douglas A. and Roberts, Glyn C. 1981. "The Effects of Sport Experience in the Development of Social Character: An Exploratory Investigation." *Journal of Sport Psychology* 3: 114–122.

Klein, Alan M. 1986. "Pumping Irony: Crisis and Contradiction in Bodybuilding." *Sociology of Sport Journal* 3: 112–133.

Klein, Michael. 1990. "The Macho World of Sport—A Forgotten Realm? Some Introductory Remarks." *International Review for the Sociology of Sport* 25: 175–184.

Klein, Alan. 1991a. "Sport and Culture as Contested Terrain: Americanization in the Caribbean." *Sociology of Sport Journal* 8: 79–85.

Klein, Alan M. 1991b. *Sugarball: The American Game, the Dominican Dream.* New Haven, CT: Yale University Press.

Knight Foundation. 1991. *Keeping Faith with the Student Athlete: A New Model for Intercollegiate Athletics.* Washington, DC: Commission on Intercollegiate Athletics.

Knoppers, Annelies, Meyer, Barbara B., Ewing, Martha, and Forrest, Linda. 1989. "Gender and the Salaries of Coaches." *Sociology of Sport Journal* 6: 348—361.

Knoppers, Annelies, Meyer, Barbara B., Ewing, Martha, and Forest, Linda. 1990. "Dimensions of Power: A Question of Sport or Gender?" *Sociology of Sport Journal* 7: 369–377.

Knoppers, Annelies, Meyer, Barbara B., Ewing, Martha, and Forrest, Linda. 1991. "Opportunity and Work Behavior in College Coaching." *Journal of Sport & Social Issues* 15: 1–20.

Knoppers, Annelies, Meyer, Barbara B., Ewing, Martha E., and Forrest, Linda. 1993. "Gender Ratio and Social Interaction Among College Coaches." *Sociology of Sport Journal* 10: 256–269.

Knuttgen, Howard G., Qiwei, Ma, and Zhongyuan, Wu. 1990. *Sport in China.* Champaign, IL: Human Kinetics.

Koch, James V. 1971. "The Economics of 'Big-Time' Intercollegiate Athletics." *Social Science Quarterly* 52: 248–260.

Koch, James V. 1973. "A Troubled Cartel: The NCAA." *Law and Contemporary Problems* 38: 39–69.

Kopay, David and Young, Perry Deane. 1977. *The Dave Kopay Story.* New York: Arbor House.

Korr, Charles P. 1991. "Marvin Miller and the New Unionism in Baseball." In P. D. Staudohar and J. A. Mangan (Eds.). *The Business of Professional Sports* (pp. 115–134). Urbana and Chicago: University of Illinois Press.

Kotarba, Joseph A. 1983. *Chronic Pain: Its Social Dimensions.* Beverly Hills, CA: Sage.

Kroll, Jack. 1989. "Race Becomes the Game." *Newsweek* (January 30): 56–59.

Kuhlman, Walter. 1975. "Violence in Professional Sports." *Wisconsin Law Review* III: 771–790.

Kuhn, Thomas. 1970. *The Structure of Scientific Revolutions.* Chicago: University of Chicago Press.

Kunesh, Monica A., Hasbrook, Cynthia A., and Lewthwaite, Rebecca. 1992. "Physical Activity Socialization: Peer Interactions and Affective Responses Among a Sample of Sixth Grade Girls." *Sociology of Sport Journal* 9: 385–396.

Kyrolainen, Hanna and Varis, Tapio. 1981. "Approaches to the Study of Sport in International Relations." *Current Research on Peace and Violence* 4: 55–58.

Laberge, Suzanne and Girardan, Yvan. 1992. "Questioning the Inference of Ethnic Differences in Achievement Values From Types of Sport Participation: A Commentary on White and Curtis." *Sociology of Sport Journal* 9: 295–306.

Laitinen, Arja and Tiihonen, Arto. 1990. "Narratives of Men's Experiences in Sport." *International Review for the Sociology of Sport* 25: 185–202.

Lance, Larry M. 1987. "Conceptualization of Role Relationships and Role Conflict Among Student Athletes." *Arena Review* 11: 12–18.

Lapchick, Richard E. 1979. "South Africa: Sport and Apartheid Politics." *Annals of the American Academy of Political and Social Science* 445: 155–165.

Lapchick, Richard E. 1987. *On the Mark: Putting the Student Back in Student-Athlete.* Lexington, MA: Lexington Books.

Lapchick, Richard E. 1989. "The High School Student-Athlete: Root of the Ethical Issues in College Sport." In R. E. Lapchick and J. B. Slaughter (Eds.). *The Rules of the Game: Ethics in College Sport* (pp. 17–28). New York: American Council on Education/Macmillan.

Lasch, Christopher. 1979. *The Culture of Narcissism: American Life in an Age of Diminishing Expectations.* New York: Warner Books.

Lavoie, Marc. 1989a. "Stacking, Performance Differentials, and Salary Discrimination in Professional Ice Hockey." *Sociology of Sport Journal* 6: 17–35.

Lavoie, Marc. 1989b. "The 'Economic' Hypothesis of Positional Segregation: Some Further Comments." *Sociology of Sport Journal* 6: 163–166.

Lavoie, Marc and Leonard, Wilbert M. II. 1990. "Salaries, Race/Ethnicity, and Pitchers in Major League Baseball: A Correction and Comment." *Sociology of Sport Journal* 7: 394–398.

Leahy, Michael. 1988. "Favorites Who Lost—How It Scarred Their Lives." *TV Guide* (September 17): 32–38.

Leath, Virginia M. and Lumpkin, Angela. 1992. "An Analysis of Sportswomen on the Covers and in the Feature Articles of *Women's Sports and Fitness* Magazine, 1975–1989." *Journal of Sport & Social Issues* 16: 121–126.

Lederman, Douglas. 1990. "In Glare of Public Spotlight, College Officials Struggle to Deal With Perceived Lawlessness of Their Athletes." *Chronicle of Higher Education* (November 7): A35–A36.

Lederman, Douglas. 1992a. "U.S. Drafts Memo on Sex Equity in College Sports." *Chronicle of Higher Education* (February 5): A1, A39–A40.

Lederman, Douglas. 1992b. "Men Get 70% of Money Available for Athletic Scholarships at Colleges That Play Big-Time Sports, New Study Finds." *Chronicle of Higher Education* (March 18): A1, A45–A46.

Lederman, Douglas. 1992c. "Men Outnumber Women and Get Most of Money in Big-Time Sports Programs." *Chronicle of Higher Education* (April 8): A1, A37–A40.

Lee, Judy. 1992. "Media Portrayals of Male and Female Olympic Athletes: Analyses of Newspaper Accounts of the 1984 and 1988 Summer Games." *International Review for the Sociology of Sport* 27: 197–222.

Leerhsen, Charles (with Abramson, Pamela). 1985. "The New Flex Appeal." *Newsweek* (May 6): 82–83.

Leerhsen, Charles. 1990. "Teen Queen of Tennis." *Newsweek* (May 14): 58–64.

Leiper, J. M. 1981. "Political Problems of the Olympic Games." In J. Segrave and D. Chu (Eds.). *Olympism* (pp. 104–117). Champaign, IL: Human Kinetics.

Leland, Elizabeth. 1990. "Owner Says Comments 'Misread.'" *Charlotte Observer* (March 29).

Lemert, Edwin. 1972. *Human Deviance, Social Problems, and Social Control*, 2nd ed. Englewood Cliffs, NJ: Prentice-Hall.

Lenk, Hans. 1986. "Notes Regarding the Relationship Between the Philosophy and the Sociology of Sport." *International Review for the Sociology of Sport* 21: 83–92.

Lenskyj, Helen. 1991. "Combating Homophobia in Sport and Physical Education." *Sociology of Sport Journal* 8: 61–69.

Leonard, Wilbert M. II. 1986. "Exploitation in Collegiate Sport: The Views of Basketball Players in NCAA Divisions I, II, and III." *Journal of Sport Behavior* 9: 11–30.

Leonard, Wilbert M. II. 1988a. *A Sociological Perspective of Sport*, 3rd ed. New York: Macmillan.

Leonard, Wilbert M. II. 1988b. "Salaries and Race in Professional Baseball: The Hispanic Component." *Sociology of Sport Journal* 5: 278–284.

Leonard, Wilbert M. II. 1989. "Salaries and Race/Ethnicity in Major League Baseball: The Pitching Component." *Sociology of Sport Journal* 6: 152–162.

Leonard, Wilbert M. II. 1993. *A Sociological Perspective of Sport*, 4th ed. New York: Macmillan.

Leonard, Wilbert M. II, Ostrosky, Anthony, and Huchendorf, Steve. 1990. "Centrality of Position and Managerial Recruitment: The Case of Major League Baseball." *Sociology of Sport Journal* 7: 294–301.

Leonard, Wilbert M. II and Reyman, Jonathan E. 1988. "The Odds of Attaining Professional Athlete Status: Refining the Computations." *Sociology of Sport Journal* 5: 162–169.

Leonard, Wilbert M. II and Schmitt, Raymond L. 1987. "The Meaning of Sport to the Subject." *Journal of Sport Behavior* 10: 103–118.

Lerch, Steve. 1984. "The Adjustment of Athletes to Career Ending Injuries." *Arena Review* 8: 54–67.

Lesieur, Henry R. 1977. *The Chase: Career of the Compulsive Gambler.* Garden City, NY: Anchor Books.

Lesieur, Henry R. 1987. "Deviance in Sport: The Case of Pathological Gambling." *Arena Review* 11: 5–14.

LeUnes, Arnold D. and Nation, Jack R. 1989. *Sport Psychology: An Introduction.* Chicago: Nelson-Hall.

Lever, Janet. 1976. "Sex Differences in the Games Children Play." *Social Problems* 23: 478–487.

Lever, Janet. 1978. "Sex Differences in the Complexity of Children's Play and Games." *American Sociological Review* 43: 471–483.

Lever, Janet. 1983. *Soccer Madness.* Chicago: University of Chicago Press.

Lever, Janet and Wheeler, Stanton. 1984. "The Chicago Tribune Sports Page, 1900–1975." *Sociology of Sport Journal* 1: 299–313.

Levin, Dan. 1980. "Here She Is, Miss, Well, What?" *Sports Illustrated* (March 17): 64–75.

Lewthwaite, Rebecca and Hasbrook, Cynthia. 1989. "Cognitive and Affective Predictors of Physical Activity Attraction and Involvement in Obese and Normal Weight Children." Paper presented at the Annual Conference of the Association for the Advancement of Applied Sport Psychology, Seattle, WA.

Lieber, Jill and Neff, Craig. 1989. "The Case Against Pete Rose." *Sports Illustrated* (July 3): 10–25.

Lilley, Jeffrey. 1994. "Russian Revolution." *Sports Illustrated* (January 10): 56–61.

Lindesmith, Alfred R., Strauss, Anselm L., and Denzin, Norman K. 1991. *Social Psychology,* 7th ed. Englewood Cliffs, NJ: Prentice-Hall.

Lipsitz, George. 1984. "Sports Stadia and Urban Development: A Tale of Three Cities." *Journal of Sport & Social Issues* 8 (2): 1–18.

Lipsyte, Robert. 1975. *SportsWorld: An American Dreamland.* New York: Quadrangle.

Lobmeyer, Hans and Weidinger, Ludwig. 1992. "Commercialization as a Dominant Factor in the American Sports Scene: Sources, Developments, Prospects." *International Review for the Sociology of Sport* 27: 309–327.

Locke, Lawrence F. and Massengale, John D. 1978. "Role Conflict in Teacher/Coaches." *Research Quarterly* 49: 162–174.

Longman, Jere. 1993. "Commentary: Cowboys' Johnson Cool for Playoffs." *Charlotte Observer* (January 6): 1B.

Lopata, Helen. 1971. *Occupation: Housewife.* New York: Oxford University Press.

Lopiano, Donna. 1992. "Colleges Can Achieve Equity in College Sports." *Chronicle of Higher Education* (December 2): B1–B2.

Lopiano, Donna A. 1993. "Political Analysis: Gender Equity Strategies for the Future." In G. L. Cohen (Ed.). *Women in Sport: Issues and Controversies* (pp. 104–116). Newbury Park, CA: Sage.

Lowe, Benjamin. 1977. *The Beauty of Sport: A Cross–Disciplinary Inquiry.* Englewood Cliffs, NJ: Prentice-Hall.

Loy, John W. 1972. "Social Origins and Occupational Mobility of a Selected Sample of American Athletes." *International Review of Sport Sociology* 7: 5–23.

Loy, John W., Kenyon, Gerald S., and McPherson, Barry D. 1987. "The Emergence and Development of the Sociology of Sport as an Academic Specialty." In A. Yiannakis, T. D. McIntyre, M. J. Melnick and D. P. Hart (Eds.), *Sport Sociology: Contemporary Themes* (pp. 2–12), 3rd ed. Dubuque, IA: Kendall/Hunt.

Loy, John W. and McElvogue, Joseph F. 1970. "Racial Segregation in American Sport." *International Review of Sport Sociology* 5: 5–24.

Loy, John W., McPherson, Barry D. and Kenyon, Gerald. 1978. *Sport and Social Systems.* Reading, MA: Addison-Wesley.

Loy, John W. and Sage, George W. 1978. "Athletic Personnel in the Marketplace." *Sociology of Work and Occupations* 5: 446–469.

Lucas, John A. 1980. *The Modern Olympic Games.* New York: A. S. Barnes and Company.

Lucas, John A. and Smith, Ronald A. 1978. *Saga of American Sport.* Philadelphia: Lea and Febiger.

Lumpkin, Angela. 1981. "Blacks and Females Striving for Athletic Acceptance." Paper presented at the Annual Convention of the North American Society for Sport History. Hamilton, Ontario.

Lumpkin, Angela and Williams, Linda D. 1991. "An Analysis of *Sports Illustrated* Feature Articles, 1954–1987." *Sociology of Sport Journal* 8: 16–32.

Lüschen, Günther. 1967. "The Interdependence of Sport and Culture." *International Review of Sport Sociology* 2: 27–41.

Lüschen, Günther. 1969. "Social Stratification and Social Mobility Among Young Sportsmen." In J. W. Loy and G. S. Kenyon (Eds.). *Sport, Culture and Society: A Reader on the Sociology of Sport* (pp. 258–276). Toronto: Macmillan.

Lüschen, Günther. 1980. "Sociology of Sport: Development, Present State, and Prospects." *Annual Review of Sociology* 6: 315–347.

Lüschen, Günther. 1986. "On Small Groups in Sport: Methodological Reflections with Reference to Structural-Functional Approaches." In C. R. Rees and A. W. Miracle (Eds.). *Sport and Social Theory* (pp. 149–157). Champaign, IL: Human Kinetics.

Lutz, Ronald. 1991. "Careers in Running: Individual Needs and Social Organization." *International Review for the Sociology of Sport* 26: 173.

MacAloon, John J. 1987. "An Observer's View of Sport Sociology." *Sociology of Sport Journal* 4: 103–115.

MacAloon, John and Czikzentmihalyi, Mihaly. 1983. "Deep Play and the Flow Experience in Rock Climbing." In J. C. Harris and R. J. Park (Eds.). *Play, Games, and Sports in Cultural Contexts* (pp. 361–384). Champaign, IL: Human Kinetics.

MacDonald, William W. 1981. "The Black Athlete in American Sports." In W. J. Baker and J. M. Carroll (Eds.). *Sports in Modern America* (pp. 88–100). St. Louis: River City Publishers.

Macionis, John J. 1993. *Sociology*, 4th ed. Englewood Cliffs, NJ: Prentice-Hall.

MacKillop, Allyson and Snyder, Eldon E. 1987. "An Analysis of Jocks and Other Subgroups Within the High School Status Structure." Paper presented at the Fifth Canadian Congress on Leisure Research, Halifax, Nova Scotia.

Maguire, Joe A. 1986. "The Emergence of Football Spectating as a Social Problem 1880–1985: A Figurational and Developmental Perspective." *Sociology of Sport Journal* 3: 217–244.

Maguire, Joe. 1988a. "The Commercialization of English Elite Basketball 1972–1988: A Figurational Perspective." *International Review for the Sociology of Sport* 23: 305–323.

Maguire, Joe A. 1988b. "Race and Position Assignment in English Soccer." *Sociology of Sport Journal* 5: 257–269.

Maguire, Joe A. 1990. "More than a Sporting Touchdown: The Making of American Football in England 1982–1990." *Sociology of Sport Journal* 7: 213–237.

Malinowski, Bronislaw. 1948. *Magic, Science, and Religion and Other Essays*. Glencoe, IL: Free Press.

Malnic, Eric. 1993. "'Rainbow Man' Guilty in Airport Hotel Siege." *Los Angeles Times* (June 12): B3.

Maloney, T. R. and Petri, B. 1972. "Professionalization of Attitude Toward Play Among Canadian School Pupils as a Function of Sex, Grade and Athletic Participation." *Journal of Leisure Research* 4: 184–195.

Mandell, Richard D. 1971. *The Nazi Olympics*. New York: Macmillan.

Mandle, Jay R. and Mandle, Joan D. 1988. *Grass Roots Commitment: Basketball and Society in Trinidad and Tobago*. Parkersburg, IA: Caribbean Books.

Mandle, Jay R. and Mandle, Joan D. 1990. "Basketball, Civil Society and the Post Colonial State in the Commonwealth Caribbean." *Journal of Sport & Social Issues* 14: 59–75.

Mantel, Richard C. and Vander Velden, Lee. 1974. "The Relationship Between the Professionalization of Attitude Toward Play of Preadolescent Boys and Participation in Organized Sport." In G. H. Sage (Ed.). *Sport and American Society: Selected Readings* (pp. 172–178). Reading, MA: Addison-Wesley.

Marbeto, J. A. 1967. *The Incidence of Prayer in Athletics as Indicated by Selected California Collegiate Athletes and Coaches*. Master's Thesis. University of California, Santa Barbara.

Marcum, John P. and Greenstein, Theodore N. 1985. "Factors Affecting Attendance of Major League Baseball: II. A Within-Season Analysis." *Sociology of Sport Journal* 2: 314–322.

Margolis, Jon. 1990. "When It Comes to Sports, Sociologists Have All the Answers." *Chicago Tribune* (May 4): sect. 4, pp. 1–2.

Marsh, Herbert W. 1993. "The Effects of Participation in Sport During the Last Two Years of High School." *Sociology of Sport Journal* 10: 18–43.

Martin, L. J. 1983. "Africa." In J. C. Merrille (Ed.). *Global Journalism: A Survey of the World's Mass Media* (pp. 190–248). New York: Longman.

Martin, Thomas W. and Berry, Kenneth J. 1987. "Competitive Sport in Post-Industrial Society: The Case of the Motocross Racer." In A. Yiannakis, T. D. McIntyre, M. J. Melnick, and D. P. Hart (Eds.). *Sport Sociology: Contemporary Themes*, 3rd ed. (pp. 269–284). Dubuque, IA: Kendall/Hunt.

Martinez del Castillo, Jesus. 1968. "Activitats Físiques de Recreació. Noves Necessitats, Noves Polítiques." In *Review Apunts, Educació Física* 4 (June): 9–17. Barcelona.

Marx, Karl. 1977 (1864). *Capital: A Critique of Political Economy*, vol. 1. New York: Random House.

Massengale, John D. 1974. "Coaching as an Occupational Subculture." *Phi Delta Kappan* 56: 140—142.

Mathisen, James A. 1992. "From Civil Religion to Folk Religion: The Case of American Sport." In Shirl J. Hoffman (Ed.). *Sport and Religion*, (pp. 17–34). Champaign, IL: Human Kinetics.

Maxa, Rudy and Elliott, Laura. 1984. "The Perils of Celebrity." *The Washingtonian* (September): 142–149.

Maxwell, Joe. 1992. "Evangelism by the Hook." *Christianity Today.* 36 (October 26): 12.

McAll, Christopher. 1992. "English/French Canadian Differences in Sport Participation: Comment on White and Curtis." *Sociology of Sport Journal* 9: 307–313.

McCallum, Jack and O'Brien, Richard. 1994. "Graceful Voice." *Sports Illustrated,* Scorecard section (July 18): 11–12.

McChesney, Robert W. 1989. "Media Made Sport: A History of Sports Coverage in the United States." In L. A. Wenner (Ed.). *Media, Sports & Society* (pp. 49–69). Newbury Park, CA: Sage.

McCormick, Robert E. and Meiners, Roger. 1987. "Bust the College Sport Cartel." *Fortune* (October 12): 235–236.

McIntosh, Peter C. 1966. "Mental Ability and Success in School Sport." *Research in Physical Education* 1: 1.

McKay, Jim and Miller, Toby. 1991. "From Old Boys to Men and Women of the Corporation: The Americanization and Commodification of Australian Sport." *Sociology of Sport Journal* 8: 86–94.

McLuhan, Marshall. 1964. *Understanding Media: The Extensions of Man.* New York: McGraw-Hill.

McMillan, Tom. 1992. *Out of Bounds: How the American Sports Establishment Is Being Driven by Greed and Hypocrisy-and What Needs to Be Done About It.* New York: Simon & Schuster.

McPherson, Barry D. 1975. "The Segregation by Playing Position Hypothesis in Sport: An Alternative Explanation." *Social Science Quarterly* 55: 960–966.

McPherson, Barry D. 1978. "The Child in Competitive Sport: Influence of the Social Milieu." In R. Magill, M. J. Ash, and F. L. Smoll (Eds.). *Children in Sport: Contemporary Anthology.* (pp. 219–250). Champaign, IL: Human Kinetics.

McPherson, Barry D. 1981. "Socialization Into and Through Sport Involvement." In G. F. Luschen and G. H. Sage (Eds.). *Handbook of Social Science of Sport.* (pp. 246–273). Champaign, IL: Stipes.

McPherson, Barry D. 1986. "Socialization Theory and Research: Toward a 'New Wave' of Scholarly Inquiry in a Sport Context." In C. R. Rees and A. W. Miracle (Eds.). *Sport and Social Theory.* (pp. 111–134). Champaign, IL: Human Kinetics.

McPherson, Barry D., Curtis, James E., and Loy, John W. 1989. *The Social Significance of Sport.* Champaign, IL: Human Kinetics Publishers.

Mead, George Herbert. 1934. *Mind, Self, and Society.* Chicago: University of Chicago Press.

Medoff, Marshall H. 1977. "Positional Segregation and Professional Baseball." *International Review of Sport Sociology* 12: 49–56.

Medoff, Marshall H. 1986. "Positional Segregation and the Economic Hypothesis." *Sociology of Sport Journal* 3: 297–304.

Medoff, Marshall H. 1987. "A Reply to Yetman: Toward an Understanding of the Economic Hypothesis." *Sociology of Sport Journal* 4: 278–279.

Meggyesy, David. 1970. *Out of Their League.* Berkeley, CA: Ramparts.

Meggyesy, David. 1992. "Agents and Agency: A Player's View." *Journal of Sport & Social Issues* 16: 111–112.

Melnick, Merrill J. 1988. "Racial Segregation by Playing Position in the English Football League." *Journal of Sport & Social Issues* 12: 122–130.

Melnick, Merrill J., Vanfossen, Beth E., and Sabo, Donald F. 1988. "Developmental Effects of Athletic Participation Among High School Girls." *Sociology of Sport Journal* 5: 22–36.

Melnick, Merrill J., Vanfossen, Beth E., and Sabo, Donald F. 1992. "Educational Effects of Interscholastic Athletic Participation on African-American and Hispanic Youth." *Adolescence* 27: 295–308.

Menn, Joseph. 1993. "Lawyers Call Denny's Fair Share Pact Weak, Unenforceable." *Charlotte Observer* (October 5): 1A, 4A.

Merton, Robert K. 1938. "Social Structure and Anomie." *American Sociological Review* 3: 672–682.

Merton, Robert K. 1957. *Social Theory and Social Structure.* Glencoe, IL: Free Press.

Merton, Robert K. 1968. *Social Theory and Social Structure,* 2nd ed. New York: Free Press.

Messner, Michael A. 1987. "The Meaning of Success: The Athletic Experience and the Development of Male Identity." In H. Brod (Ed.). *The Making of Masculinities: The New Men's Studies* (pp. 193–209). Winchester, MA: Allen & Unwin.

Messner, Michael A. 1988. "Sport and Male Domination: The Female Athlete as Contested Ideological Terrain." *Sociology of Sport Journal* 5: 197–211.

Messner, Michael A. 1990. "When Bodies Are Weapons: Masculinity and Violence in Sport." *International Review for the Sociology of Sport* 25: 203–220.

Messner, Michael A. 1992. *Power at Play: Sports and the Problem of Masculinity.* Boston: Beacon Press.

Messner, Michael A. 1993. "Boyhood, Organized Sports, and the Construction of Masculinities." In A. Yiannakis, T. D. McIntyre, and M. J. Melnick (Eds.). *Sport*

Sociology: Contemporary Themes (259–272). Dubuque, IA: Kendall/Hunt.

Messner, Michael A. and Sabo, Don. (Eds.). 1990. *Sport, Men, and the Gender Order: Critical Feminist Perspectives.* Champaign, IL: Human Kinetics.

Meyer, Barbara B. 1990. "From Idealism to Actualization: The Academic Performance of Female Collegiate Athletes." *Sociology of Sport Journal* 7: 44–57.

Miller, Delbert. 1991. *Handbook of Research Design and Measurement.* Newbury Park, CA: Sage.

Miller, Lori K., Fielding, Lawrence W., and Pitts, Brenda G. 1992. "A Uniform Code to Regulate Athlete Agents." *Journal of Sport & Social Issues* 16: 93–102.

Miller, James Edward. 1990. *The Baseball Business: Pursuing Pennants and Profits in Baltimore.* Chapel Hill, NC: University of North Carolina Press.

Miller Lite. 1983. *Miller Lite Report on American Attitudes Toward Sports.* Milwaukee, WI: Miller Brewing Company.

Mills, C. Wright. 1959. *The Sociological Imagination.* New York: Oxford University Press.

Miracle, Andrew W. and Rees, C. Roger. 1994. *Lessons of the Locker Room: The Myth of School Sports.* Amherst, NY: Prometheus Books.

Mizruchi, Ephraim. 1964. *Success and Opportunity.* New York: Free Press.

Mizruchi, Mark. 1991. "Urgency, Motivation, and Group Performance: The Effect of Prior Success on Current Success among Professional Basketball Teams." *Social Psychological Quarterly* 54: 181–189.

Monaghan, Peter. 1992. "U. of Colorado Football Coach Accused of Using His Position to Promote His Religious Views." *Chronicle of Higher Education* (November 11): A35, A37.

Montville, Leigh. 1991. "Season of Torment." *Sports Illustrated* (May 13): 60–65.

Moore, John M. 1967. "Football's Ugly Decades, 1893–1913." *The Smithsonian Journal of History* 2: 49–68.

Morgan, William J. and Meier, Klaus V. (Eds.). 1988. *Philosophic Inquiry in Sport.* Champaign, IL: Human Kinetics.

Morris, David B. 1991. *The Culture of Pain.* Berkeley, CA: University of California Press.

Morrison, L. Leotus. 1993. "The AIAW: Governance by Women for Women." In G. L. Cohen (Ed.) *Women in Sport: Issues and Controversies* (pp. 59–68). Newbury Park, CA: Sage.

Mortimer, J. T. and Simmons, R. G. 1978. "Adult Socialization." *Annual Review of Sociology* 4: 421–452.

Mottram, D. R. (Ed.). 1988. *Drugs in Sport.* Champaign, IL: Human Kinetics.

Muir, Donal E. 1991. "Club Tennis: A Case Study in Taking Leisure Very Seriously." *Sociology of Sport Journal* 8: 70–78.

Murphy, Austin. 1989. "The Rose Probe." *Sports Illustrated,* Scorecard section. (March 27): 13.

Murphy, Austin. 1991. "Unsportsmanlike Conduct." *Sports Illustrated* (July 1): 22–27.

Nafziger, James A. R. 1980. "Diplomatic Fun and Games: A Commentary on the United States Boycott of the 1980 Summer Olympics." *Willamette Law Review* 17: 67–81.

Nash, Jeffrey E. 1976. "The Short and the Long of It: Legitimizing Motives for Running." In J. E. Nash and J. P. Spradley (Eds.). *Sociology: A Descriptive Approach* (pp. 161–181). Chicago: Rand McNally.

Nash, Jeffrey E. 1979. "Weekend Racing as an Eventful Experience: Understanding the Accomplishment of Well-Being." *Urban Life* 8: 199–217.

National Center for Educational Statistics. 1986. *High School and Beyond, 1980: Sophomore Cohort Second Follow-up (1984). Data File Users Manual.* Ann Arbor, MI: Inter-university Consortium for Political and Social Research.

National Federation Handbook 1984–85. 1984. Kansas City, MO: National Federation of State High School Associations.

NCAA News. 1993. "Gender-Equity Task Force Final Report." (August 4): 14–16.

Neff, Craig. 1985. "Can It Happen in the U.S.?" *Sports Illustrated* (June 10): 27.

Neff, Craig. 1989. "Rose Probe (Cont.)." *Sports Illustrated,* Scorecard section. (May 8): 11.

Neil, Graham L., Anderson, Bill, and Sheppard, Wendy. 1981. "Superstitions Among Male and Female Athletes at Various Levels of Involvement." *Journal of Sport Behavior* 4:139–140.

Neimark, Jill. 1991. "Out of Bounds: The Truth About Athletes and Rape." *Mademoiselle* (May): 196–199, 244–248.

Newman, Bruce. 1989. "Japan: Coming on Strong." *Sports Illustrated* (August 21): 48–65.

NIAAA. 1991. "NIAAA Round Table: Six High School Athletic Directors Discuss Key National Issues." *Interscholastic Athletic Administration* 17 (Summer): 6–9.

Nisbett, Robert E. 1968. "Birth Order and Participation in Dangerous Sports." *Journal of Personality and Social Psychology* 8: 351–353.

Nixon, Howard L. II. 1976. *Sport and Social Organization.* Indianapolis, IN: Bobbs-Merrill.

Nixon, Howard L. II. 1979. *The Small Group.* Englewood Cliffs, NJ: Prentice-Hall.

Nixon, Howard L. II. 1984. *Sport and the American Dream.* Champaign, IL: Human Kinetics/Leisure Press Imprint.

Nixon, Howard L. II. 1985. "The Future of Sport in America: A Simulation Game." In W. C. Whit (Ed.). *Syllabi and Instructional Materials for Courses on Sociology of Sport* (pp. 171–173). Washington, DC: American Sociological Association Resources Center.

Nixon, Howard L. II. 1986. "Social Order in a Leisure Setting: The Case of Recreational Swimmers in a Pool." *Sociology of Sport Journal* 3: 320–332.

Nixon, Howard L. II. 1988a. "Sport as Refuge or Reality? Arguments and Evidence about the 'Degradation' of American Sports." Paper presented at the North American Society for the Sociology of Sport Annual Conference, Cincinnati, OH.

Nixon, Howard L. II. 1988b. "The Background, Nature, and Implications of the Organization of the 'Capitalist Olympics.'" In J. O. Segrave and D. Chu (Eds.). *The Olympic Games in Transition* (pp. 237–251). Champaign, IL: Human Kinetics.

Nixon, Howard L. II. 1989. "Reconsidering Obligatory Running and Anorexia Nervosa as Gender-Related Problems of Identity and Role Adjustment." *Journal of Sport & Social Issues* 13: 14–24.

Nixon, Howard L. II. 1990a. "Voice of the People: Don't Underestimate Sport Sociology." *Chicago Tribune* (August 17): sect. 1, p. 22.

Nixon, Howard L. II. 1990b. "Rethinking Socialization and Sport." *Journal of Sport & Social Issues* 14: 33–47.

Nixon, Howard L. II. 1991a. "Sport Sociology that Matters: Imperatives and Challenges for the 1990s." *Sociology of Sport Journal* 8: 281–294.

Nixon, Howard L. II. 1991b. *Mainstreaming and the American Dream: Sociological Perspectives on Parental Coping with Blind and Visually Impaired Children.* New York: American Foundation for the Blind.

Nixon, Howard L. II. 1992a. "A Social Network Analysis of Influences on Athletes to Play with Pain and Injuries." *Journal of Sport & Social Issues* 16: 127–135.

Nixon, Howard L. II. 1992b. "Coaches' Views of Risk, Pain, and Injury in Sport." Paper presented at the North American Society for the Sociology of Sport Annual Conference, Toledo, OH.

Nixon, Howard L. II. 1993a. "Social Network Analysis of Sport: Emphasizing Social Structure in Sport Sociology." *Sociology of Sport Journal* 10: 315–321.

Nixon, Howard L. II. 1993b. "Accepting the Risks of Pain and Injury in Sport: Mediated Cultural Influences on Playing Hurt." *Sociology of Sport Journal* 10: 183–196.

Nixon, Howard L. II. 1993c. "The Effects of Status, Injury Experience, and Social Relations on Pain and Injury Tolerance Levels, Talk, Help Seeking, and Avoidance Behavior of Athletes in a College Sportsnet." Paper presented at the International Sunbelt Social Network Conference, Tampa, FL.

Nixon, Howard L. II. 1993d. "Cultural Beliefs, Status Factors, and Vulnerability to Pain and Injuries in Sport." Paper presented at the American Sociological Association Annual Conference, Miami, FL.

Nixon, Howard L. II, Maresca, Philip J., and Silverman, Marcy A. 1979. "Sex Differences in College Students' Acceptance of Females in Sport." *Adolescence* 14: 755–764.

Noden, Merrill. 1993. "Shot Down." *Sports Illustrated* (March 15): 42–46.

Noden, Merrell. 1994. "Special Report: Dying to Win." *Sports Illustrated* (August 8): 52–60.

Noll, Roger G. (Ed.). 1974. *Government and the Sports Business.* Washington, DC: The Brookings Institution.

Noll, Roger G. 1991. "Professional Basketball: Economic and Business Perspectives." In P. D. Staudohar and J. A. Mangan (Eds.). *The Business of Professional Sports* (pp. 18–47). Urbana and Chicago, IL: University of Illinois Press.

Novak, Michael. 1976. *The Joy of Sports.* New York: Basic Books.

Noverr, Douglas A. and Ziewacz, Lawrence E. 1983. *The Games They Played: Sports in American History, 1865–1980.* Chicago: Nelson-Hall.

O'Brien, Richard. 1993. "Fighting Words." *Sports Illustrated,* Scorecard section (March 15): 9–10.

Observer News Services. 1993. "Jackson Leads Protest." *Charlotte Observer* (July 14): 4B.

Observer Staff and News Services. 1993. "Dwindling Opportunities." *Charlotte Observer* (July 12): 2B.

Ogilvie, Bruce C. and Tutko, Thomas A. 1971. "Sport: If You Want to Build Character, Try Something Else." *Psychology Today* 5: 61–63.

Okner, Benjamin A. 1971. "Direct and Indirect Subsidies to Professional Sports." Paper presented at the Brookings Conference on Government and Sport. Washington, DC.

Oliver, Chip. 1971. *High for the Game*. New York: William Morrow.

Olsen, Jack. 1968. *The Black Athlete*. New York: Time-Life Books.

Orelove, Fred P. and Moon, M. Sherril. 1984. "The Special Olympics Program: Effects on Retarded Persons and Society." *Arena Review* 8: 41–45.

Orelove, Fred P., Wehman, Paul, and Wood, Judy. 1982. "An Evaluative Review of Special Olympics: Implications for Community Integration." *Education and Training of the Mentally Retarded* 17: 325–329.

Orlick, Terry D. 1973. "Children's Sports: A Revolution Is Coming." *CAHPER Journal* 39: 12–14.

Orlick, Terry D. 1974. "The Athletic Dropout: A High Price of Inefficiency."*Canadian Association for Health, Physical Eduction and Recreation Journal* 41: 24–27.

Orlick, Terry D. and Botterill, Cal. 1975. *Every Kid Can Win*. Chicago: Nelson-Hall.

Ozanian, Michael K. 1994. "The $11 Billion Pastime." *Financial World* 163 (May 10): 50–59.

"P. K." 1992. "White Guys Can't Run." *Sports Illustrated* (September 7): 28–29.

Palmer, Denise & Howell, Maxwell L. 1973. "Sport and Games in Early Civilization." In E. F. Zeigler (Ed). *A History of Sport and Physical Education* (pp. 21–34). Champaign, IL: Stipes.

Palzkill, Birgit. 1990. "Between Gymshoes and High Heels—The Development of a Lesbian Identity and Existence in Top Class Sport." *International Review for the Sociology of Sport* 25: 221–234.

Papanek, Hannah. 1973. "Men, Women, and Work: Reflections on the Two-Person Career." *American Journal of Sociology* 78: 852–872.

Parsons, Talcott. 1951. *The Social System*. Glencoe, IL: Free Press.

Peterson, Robert W. 1970. *Only the Ball Was White*. Englewood Cliffs, NJ: Prentice-Hall.

Phillips, D. 1987. "Socialization of Perceived Academic Competence Among Highly Competent Children." *Child Development* 58: 1308–1320.

Phillips, John C. 1993. *Sociology of Sport*. Boston: Allyn & Bacon.

Phillips, John C. and Boelter, J. 1985. "Diffusion of the 'Flop' Technique in the High Jump Along Racial Lines." Paper presented at the Olympic Scientific Congress, Eugene, OR.

Picou, J. Stephen and Curry, E. W. 1974. "Residence and the Athletic-Aspiration Hypothesis." *Social Science Quarterly* 55: 768–776.

Pitter, Robert. 1990. "Power and Control in An Amateur Sport Organization." *International Review for the Sociology of Sport* 25: 309–322.

Plagens, Peter. 1992. "No More Show Time." *Newsweek* (February 24): 66, 69.

Poe, A. 1976. "Active Women in Ads." *Journal of Communication* 26: 185–192.

Porter, H. V. *H.V.'s Athletic Anthology*. Chicago: H. V. Porter, 1939.

Prebish, Charles S. 1992. "'Heavenly Father, Divine Goalie': Sport and Religion." In S. J. Hoffman (Ed.). *Sport and Religion* (pp. 43–53). Champaign, IL: Human Kinetics.

President's Commission on Olympic Sports. 1977. *The Final Report of the President's Commission on Olympic Sports: 1975–1977* Washington, DC: Government Printing Office.

Price, Joseph L. 1984. "The Super Bowl as Religious Festival." *The Christian Century* (February 22): 190–191.

Pronger, Brian. 1990. *The Arena of Masculinity: Sport, Homosexuality, and the Meaning of Sex*. New York: St. Martin's Press.

Purdy, Dean A., Eitzen, D. Stanley, and Hufnagel, Rick. 1982. "Are Athletes Also Students? The Educational Attainment of College Athletes." *Social Problems* 29: 439–447.

Putnam, Pat. 1986. "Another View of Gambling: It's Good for You." *Sports Illustrated* (March 10): 56.

Rader, Benjamin G. 1983. *American Sports*. Englewood Cliffs, NJ: Prentice-Hall.

Rader, Benjamin G. 1990. *American Sports*, 2nd ed. Englewood Cliffs: Prentice-Hall.

Randall, Aaron J., Hall, Stephen F., and Rogers, Mary F. 1992. "Masculinity on Stage: Competitive Male Bodybuilders." *Studies in Popular Culture* 14: 57–69.

Raphael, Ray. 1988. *The Men from the Boys: Rites of Passage in Male America*. Lincoln: University of Nebraska Press.

Real, Michael R. 1975. "Super Bowl: Mythic Spectacle." *Journal of Communication* 25: 31–43.

Real, Michael R. 1976. "Super Bowl: Mythic Spectacle." In A. Yiannakis, T. D. McIntyre, M. J. Melnick, and D. P. Hart (Eds.). *Sport Sociology: Contemporary Themes*, 1st ed. (pp. 22–28). Dubuque, IA: Kendall/Hunt.

Real, Michael R. 1989a. "Super Bowl Football Versus World Cup Soccer: A Cultural-Structural Comparison." In L. A. Wenner (Ed.). *Media, Sports, & Society* (pp. 180–203). Newbury Park, CA: Sage.

Real, Michael R. 1989b. *Super Media: A Cultural Studies Approach*. Newbury Park, CA: Sage.

Real, Michael R. and Mechikoff, Robert A. 1992. "Deep Fan: Mythic Identification, Technology, and Advertising in Spectator Sports." *Sociology of Sport Journal* 9: 323–339.

Rees, C. Roger, Howell, Frank M., and Miracle, Andrew W. 1990. "Do High School Sports Build Character? A Quasi-Experiment on a National Sample." *The Social Science Journal* 27: 303–315.

Rees, C. Roger and Miracle, Andrew W. 1984. "Participation in Sport and the Reduction of Racial Prejudices: Contact Theory, Super Ordinate Goals, Hypothesis or Wishful Thinking." In N. Theberge and P. Donnelly (Eds.). *Sport and the Sociological Imagination* (pp. 140–152). Fort Worth, TX: Texas Christian University Press.

Rees, C. Roger and Miracle, Andrew W. (Eds.). 1986. *Sport and Social Theory.* Champaign, IL: Human Kinetics.

Rees, C. Roger and Miracle, Andrew. 1988. "The Crisis in High School Sport: Changing Times, Changing Myths." Paper presented at the Annual Meeting of the North American Society for the Sociology of Sport. Cincinnati.

Rehberg, Richard and Schafer, Walter E. 1968. "Participation in Interscholastic Athletics and College Expectations." *American Journal of Sociology* 63: 732–740.

Reilly, Rick. 1989. "What Price Glory?" *Sports Illustrated* (February 27): 32–34.

Renson, R. 1976. Social Status Symbolism of Sport Stratification." *Hermes* 10: 433–443.

Rhee, Foon. 1990. "Shinn Suggested Original Lease." *Charlotte Observer* (March 29).

Riesman, David and Denney, Reuel. 1951. "Football in America: A Study in Cultural Diffusion." *American Quarterly* 3: 309–319.

Riess, Steven A. 1991. "A Social Profile of the Professional Football, 1920–1982." In P. D. Staudohar and J. A. Mangan (Ed.). *The Business of Professional Sports* (pp. 222–246). Urbana and Chicago: University of Illinois Press.

Rintala, Jan and Birrell, Susan. 1984. "Fair Treatment for the Active Female: A Content Analysis of *Young Athlete* Magazine." *Sociology of Sport Journal* 1: 231–250.

Riordan, James (Ed.). 1978. *Sport Under Communism. Sport in China, Cuba, Czechoslovakia, German Democratic Republic and USSR.* London: C. Hurst.

Riordan, James. 1980. *Soviet Sport.* New York: New York University Press.

Riordan, James. 1985. "Some Comparisons of Women's Sport in East and West." *International Review for the Sociology of Sport* 20: 117–126.

Riordan, James. 1986. "State and Sport in Developing Societies." *International Review for the Sociology of Sport* 21: 287–299.

Riordan, James. 1990. "Playing to New Rules: Soviet Sport and Perestroika." *Soviet Studies* 42: 133–145.

Riordan, James. 1993a. "Sport in Capitalist and Socialist Countries: A Western Perspective." In E. G. Dunning, J. A. Maguire, and R. E. Pearton (Eds.). *The Sports Process: A Comparative and Developmental Approach* (pp. 245–264). Champaign, IL: Human Kinetics.

Riordan, James. 1993b. "Soviet-Style Sport in Eastern Europe: The End of an Era." In L. Allison (Ed.). *The Changing Politics of Sport* (pp. 37–57). Manchester: Manchester University Press.

Ritchie, O. and Koller, M. R. 1964. *Sociology of Childhood.* New York: Appleton.

Ritzer, George. 1993. *The McDonaldization of Society.* Thousand Oaks, CA: Pine Forge Press.

Robbins, James M. and Joseph, Paul. 1980. "Commitment to Running: Implications for the Family and Work." *Sociological Symposium* 30: 87–108.

Roberts, Gary A. 1991. "Professional Sports and the Antitrust Laws." In P. D. Staudohar and J. A. Mangan (Eds.). *The Business of Professional Sports* (pp. 135–151). Urbana and Chicago, IL: University of Illinois Press.

Roberts, Randy and Olson, James. 1989. *Winning Is the Only Thing: Sports in America Since 1945.* Baltimore, MD: The Johns Hopkins Press.

Robertson, Ian. 1989. *Society: A Brief Introduction.* New York: Worth.

Rokeach, Milton. 1973. *The Nature of Human Values.* New York: Free Press.

Rooney, John F., Jr. 1974. *A Geography of American Sport: From Cabin Creek to Anaheim.* Reading, MA: Addison-Wesley.

Rooney, John F., Jr. and Pillsbury, Richard. 1992. "Sports Regions of America." *American Demographics* 14: 30–44.

Rosenblatt, Aaron. 1967. "Negroes in Baseball: The Failure of Success." *Transaction* 4 (September): 51–53.

Ross, Stephen F. 1991. "Break Up the Sports League Monopolies." In P. D. Staudohar and J. A. Mangan (Eds.). *The Business of Professional Sports* (pp. 152–174). Urbana and Chicago, IL: University of Illinois Press.

Rotenberk, Lori. 1992. "Pray Ball." In Shirl J. Hoffman (Ed.). *Sport and Religion* (pp. 177–181). Champaign, IL: Human Kinetics.

Rothstein, Lawrence E. 1986. *Plant Closing: Power, Politics, and Workers*. Dover, MA: Auburn House.

Roversi, Antonio. 1991. "Football Violence in Italy." *International Review for the Sociology of Sport* 26: 311–332.

Rudman, William J. 1986. "The Sport Mystique in Black Culture." *Sociology of Sport Journal* 3: 305–319.

Rudman, William J. 1989. "Age and Involvement in Sport and Physical Activity." *Sociology of Sport Journal* 6: 228–246.

Rushin, Steve. 1993. "On the Road Again." *Sports Illustrated* (August 16): 18–27.

Ryckman, Richard M. and Hamel, Jane. 1992. "Female Adolescents' Motives Related to Involvement in Organized Team Sports." *International Journal of Sport Psychology* 23: 147–160.

Sabo, Donald F. 1985. "Sport, Patriarchy, and Male Identity: New Questions About Men and Sport." *Arena Review* 9: 1–30.

Sabo, Donald F. 1986. "Pigskin, Patriarchy and Pain." *Changing Men: Issues in Gender, Sex, and Politics* 16: 24–25.

Sabo, Donald F. 1988. "Psychosocial Impacts on Athletic Participation on American Women: Facts and Fables." *Journal of Sport & Social Issues* 12: 83–96.

Sabo, Donald F., Melnick, Merrill J., and Vanfossen, Beth E. 1989. *The Women's Sports Foundation Report: Minorities in Sport*. New York: Women's Sports Foundation.

Sabo, Donald F., Melnick, Merrill J., and Vanfossen, Beth E. 1993. "High School Athletic Participation and Post-secondary Educational and Occupational Mobility: A Focus on Race and Gender." *Sociology of Sport Journal* 10: 44–56.

Sack, Allen L. 1991. "The Underground Economy in College Football." *Sociology of Sport Journal* 8: 1–15.

Sack, Allen L. and Thiel, Robert. 1985. "College Basketball and Role Conflict: A National Survey." *Sociology of Sport Journal* 2: 195–209.

Sagarin, Edward. 1985. "Positive Deviance: An Oxymoron." *Deviant Behavior* 6: 169–185.

Sage, George H. 1974. "Machiavellianism among College and High School Coaches." In G. H. Sage (Ed.). *Sport and American Society*, 2nd ed. (pp. 187–207). Reading, MA: Addison-Wesley.

Sage, George H. 1980. "Sport and American Society: The Quest for Success." In G. H. Sage (Ed.). *Sport and American Society*, 3rd ed. (pp. 112–121). Reading, MA: Addison-Wesley.

Sage, George H. 1981. "Sport and Religion." In G. R. F. Lüschen and G. H. Sage (Eds.). *Handbook of Social Science of Sport* (pp. 147–159). Champaign, IL: Stipes.

Sage, George H. 1982. "The Intercollegiate Sport Cartel and Its Consequences for Athletes." In James H. Frey (Ed.). *The Governance of Intercollegiate Athletics* (pp. 131–143). West Point: Leisure Press.

Sage, George H. 1987. "The Social World of High School Athletic Coaches: Multiple Role Demands and Their Consequences." *Sociology of Sport Journal* 4: 213–228.

Sage, George H. 1990a. *Power and Ideology in American Sport: A Critical Perspective.* Champaign, IL: Human Kinetics.

Sage, George H. 1990b. "High School and College Sports in the United States." *Journal of Physical Education, Recreation & Dance* 81: 59–63.

Sage, George H. 1993. "Stealing Home: Political, Economic, and Media Power and a Publicly-Funded Baseball Stadium in Denver." *Journal of Sport & Social Issues* 17: 110–124.

Sage, George H. and Loudermilk, Sheryl. 1979. "The Female Athlete and Role Conflict." *Research Quarterly* 50: 88–96.

Sailes, Gary. 1985. "Sport Socialization Comparisons Among Black and White Male Athletes and Non-athletes." Paper presented at the Annual Meetings of the North American Society for the Sociology of Sport, Boston, MA.

Santomeier, James. 1979. "Myth, Legitimation, and Stress in Formal Sport Organizations." *Journal of Sport & Social Issues* 3(2): 11–16.

Savage, Howard J. 1929. *American College Athletics*, bulletin 23. New York: Carnegie Foundation for the Advancement of Teaching.

Scanlan, Tara K. and Lewthwaite, Rebecca. 1986. "Social Psychological Aspects of Competition for Male Youth Sport Participants: IV Predictors of Enjoyment." *Journal of Sport Psychology* 8: 25–35.

Schafer, Walter E. 1969. "Some Social Sources and Consequences of Interscholastic Athletics: The Case of Participation and Delinquency." *International Review of Sport Sociology* 4: 63–81.

Schafer, Walter E. 1975. "Sport and Male Sex Role Socialization." *Sport Sociology Bulletin* 4: 224–233.

Schafer, Walter E. and Armer, J. Michael. 1968. "Athletes Are Not Inferior Students." *Trans-Action* (November): 21–26, 61–62.

Schafer, Walter E. and Armer, J. Michael. 1972. "On Scholarship and Interscholastic Athletics." In E. Dunning (Ed.). *Sport: Readings from a Sociological Perspective* (pp. 198–229). Toronto: University of Toronto Press.

Schafer, Walter E. and Rehberg, Richard. 1970. "Athletic Participation, College Aspirations, and College Encouragement." *Pacific Sociological Review* 13: 182–186.

Schlozman, Kay Lehman and Verba, Sidney. 1979. *Insult to Injury.* Cambridge: Harvard University Press.

Schultz, Richard D. 1989. "The Role of the NCAA." In Richard E. Lapchick and John B. Slaughter (Eds.). *The Rules of the Game* (pp. 161–169). New York: ACE/Macmillan.

Scully, Gerald W. 1973. "Economic Discrimination in Professional Sports." *Law and Contemporary Problems* 39: 67–84.

Scully, Gerald W. 1989. *The Business of Major League Baseball.* Chicago: University of Chicago Press.

Seefeldt, Vern, Ewing, M., Hylka, T., Trevor, C. and Walk, S. 1989. *Participation and Attrition in Youth Sports, With Implications for Soccer.* Paper presented at the Annual Conference of the United States Soccer Association, Seattle, WA.

Seefeldt, Vern and Gould, Daniel. 1980. *Physical and Psychological Effects of Athletic Competition on Children and Youth.* Washington, DC: ERIC Clearinghouse on Teacher Education.

Seff, M. A., Gecas, V. and Frey, J. H. 1993. "Birth Order, Self Concept, and Participation in Dangerous Sports." *The Journal of Psychology* 127: 221–232.

Segrave, Jeffrey O. 1981. "Role Preferences Among Prospective Physical Education Teacher/Coaches: Its Relevance to Education." In V. Crafts (Ed.). *NAPEHE Proceedings*, vol. 2 (pp. 53–61). Champaign, IL: Human Kinetics.

Segrave, Jeffrey O. 1983. "Sport and Juvenile Delinquency." In R. Terjung (Ed.). *Exercise and Sport Sciences Review,* vol. 2 (pp. 161–209). Philadelphia: Franklin Institute Press.

Segrave, Jeffrey O. and Hastad, Douglas. 1984. "Interscholastic Participation and Delinquent Behavior: An Empirical Assessment of Relevant Variables." *Sociology of Sport Journal* 1: 117–137.

Seifart, H. 1984. "Sport and Economy: The Commercialization of Olympic Sport by the Media." *International Review for the Sociology of Sport* 19: 305–316.

Selcraig, Bruce. 1988. "The Deal Went Sour." *Sports Illustrated* (September 5): 32–33.

Sellers, Robert M. 1992. "Racial Differences in the Predictors for Academic Achievement of Student-Athletes in Division I Revenue Producing Sports." *Sociology of Sport Journal* 9: 48–59.

Semyonov, Moshe and Farbstein, Mira. 1989. "Ecology of Sports Violence: The Case of Israeli Soccer." *Sociology of Sport Journal* 6: 50–59.

Seppanen, Paavo. 1982. "The Idealistic and Factual Role of Sport in International Understanding." *Current Research on Peace and Violence* 5: 113–121.

Serwer, Andrew E. 1993. "How High?" *Sports Illustrated* (November 8): 88–99.

Sewart, John J. 1981. "The Rationalization of Modern Sport: The Case of Professional Football." *Arena Review* 5 (1): 45–53.

Sfeir, Leila. 1985. "The Status of Muslim Women in Sport: Conflict between Cultural Tradition and Modernization." *International Journal for the Sociology of Sport* 20: 283–306.

Shapiro, Richard L. 1976. "The Lombardian Ethic: Sport and Socialization in the United States and Implication for Africa." Paper presented at the Conference of the African Studies Association, Boston.

Shecter, Leonard. 1969. *The Jocks.* Indianapolis, IN: Bobbs-Merrill.

Sheehan, George A. 1978. *Running and Being.* New York: Simon & Schuster.

Sherman, Lawrence W. 1982. "Deviant Organizations." In David Ermann and Richard J. Lundman (Eds.). *Corporate and Governmental Deviance* (pp. 63–72). New York: Oxford.

Sherrill, Claudine (Ed.). 1986. *Sport and Disabled Athletes.* Champaign, IL: Human Kinetics.

Shirley, Bill. 1984. "There Isn't Any Turning the Other Cheek."*Los Angeles Times* (November 1): 1, 16, 17.

Shuster, Rachel. 1991. "Pressure for Wins, Making Money Create Turnover." *USA Today* (March 28): 10C.

Siegel, Donald. 1994. "Higher Education and the Plight of the Black Male Athlete." *Journal of Sport & Social Issues* 18: 207–223.

Simon, Robert L. 1985. *Sport and Social Values.* Englewood Cliffs, NJ: Prentice-Hall.

Simpson, Kevin. 1985. "Religion, Football Don't Mix, ACLU Says." *Los Angeles Times* (September 22): 10.

Sipes, Richard G. 1973. "War, Sports and Aggression: An Empirical Test of Two Rival Theories." *American Anthropologist* 75: 64–86.

Slowikowski, Synthia S. 1991. "Burning Desire: Nostalgia, Ritual, and the Sport-Flame Ceremony." *Sociology of Sport Journal* 8: 239–257.

Smith, Claire. 1992. "At Capitol Hill, It's Batters Up." *The New York Times* (December 10): B8.

Smith, Garry J. 1988. "The Noble Sports Fan." *Journal of Sport & Social Issues* 12: 54–65.

Smith, Garry J. 1990. "Pools, Parlays, and Point Spreads: A Sociological Consideration of the Legalization of Sports Gambling." *Sociology of Sport Journal* 7: 271–286.

Smith, H. W. 1987. *Introduction to Social Psychology.* Englewood Cliffs, NJ: Prentice-Hall.

Smith, Kevin B. and Stone, Lorene H. 1989. "Rags, Riches, and Bootstraps: Beliefs about the Causes of Wealth and Poverty." *Sociological Quarterly* 30: 93–107.

Smith, Michael D. 1979. "Getting Involved in Sport: Sex Differences." *International Review of Sport Sociology* 14: 93–99.

Smith, Michael D. 1986. "Sports Violence: A Definition." In R. E. Lapchick (Ed.). *Fractured Focus: Sport as a Reflection of Society* (pp. 221–227). Lexington, MA: Lexington Books.

Snyder, Eldon E. 1972. "High School Athletes and Their Coaches: Educational Plans and Advice." *Sociology of Education* 45: 313–325.

Snyder, Eldon E. 1975. "Athletic Team Involvement, Educational Plans, and the Coach-Player Relationship." *Adolescence* 10: 191–200.

Snyder, Eldon E. 1991. "Sociology of Nostalgia: Sport Halls of Fame and Museums in America." *Sociology of Sport Journal* 8: 228–238.

Snyder, Eldon E. and Purdy, Dean A. 1982. "Socialization into Sport: Parent and Child Reverse and Reciprocal Socialization." *Research Quarterly for Exercise and Sport* 53: 263–266.

Snyder, Eldon E. and Spreitzer, Elmer. 1977. "Participation in Sport as Related to Educational Expectations Among High School Girls." *Sociology of Education* 50: 47–55.

Snyder, Eldon E. and Spreitzer, Elmer. 1978. *Social Aspects of Sport.* Englewood Cliffs, NJ: Prentice-Hall.

Snyder, Eldon E. and Spreitzer, Elmer A. 1989a. *Social Aspects of Sport,* 3rd ed. Englewood Cliffs: Prentice-Hall.

Snyder, Eldon E. and Spreitzer, Elmer A. 1989b. "Baseball in Japan." In D. S. Eitzen (Ed.). *Sport in Contemporary Society,* 3rd ed. (pp. 46–49). New York: St. Martin's Press.

Sorensen, Tom. 1990. "Shinn's Comments Weren't Misread, As Public Knows." *Charlotte Observer* (March 29).

Sperber, Murray. 1990. *College Sports Inc.: The Athletic Department versus the University.* New York: Henry Holt.

Sport Canada. 1989. "Women in Leadership: An Issue for Sport. Summary of a National Level Survey." Ottawa: Fitness and Amateur Sport.

Sport. 1987. "Police Blotter." (December): 101–102.

Sport. 1993. "Ball Figures, Then and Now." (March): 37.

Sports Illustrated. 1986. "The Biggest Game in Town: Special Report on Gambling." (March 10): 31–57.

Sports Illustrated. 1988. "Special Report: Sports in China." (August 15).

Sports Illustrated. 1989. "The Big Story." Scorecard section (August 28): 15.

Sports Illustrated. 1991. *Sports Illustrated 1992 Sports Almanac.* Boston: Little, Brown.

Spreitzer, Elmer. 1992. "Does Participation in Interscholastic Athletics Affect Adult Development: A Longitudinal Analysis of an 18–24 Age Cohort." Paper presented at the Annual Meeting of the American Sociological Association. Pittsburgh, PA.

Spreitzer, Elmer and Pugh, Meredith. 1973. " Interscholastic Athletics and Educational Expectations." *Sociology of Education* 46: 171–182.

Spreitzer, Elmer and Snyder, Eldon E. 1976. "Socialization into Sport: An Exploratory Analysis." *Research Quarterly* 47: 238–245.

Spreitzer, Elmer and Snyder, Eldon E. 1990. "Sports Within the Black Subculture: A Matter of Social Class or a Distinctive Subculture." *Journal of Sport & Social Issues* 14: 48–58.

Stangl, Jane M. and Kane, Mary Jo. 1991. "Structural Variables that Offer Explanatory Power for the Underrepresentation of Women Coaches Since Title IX: The Case of Homologous Reproduction." *Sociology of Sport Journal* 8: 47–60.

Starr, Mark, Barrett, Todd, and Smith, Vern E. 1993. "Baseball's Black Problem." *Newsweek* (July 19): 56–57.

Start, K. B. 1966. "The Substitution of Games Performance for Academic Achievement as a Means of Achieving Status among Secondary School Children." *British Journal of Sociology* 17: 300–305.

Start, K. B. 1967. "Sporting and Intellectual Success among English Secondary School Children." *International Review of Sport Sociology* 2: 47–53.

Staudohar, Paul D. 1992. "McNeil and Football's Antitrust Quagmire." *Journal of Sport & Social Issues* 16: 103–110.

Staudohar, Paul D. and Mangan, James A. 1991. "Introduction." In P. D. Staudohar and J. A. Mangan (Eds.). *The Business of Professional Sports* (pp. 1–17). Urbana and Chicago: University of Illinois Press.

Steadward, Robert and Walsh, Catherine. 1986. "Training and Fitness Programs for Disabled Athletes: Past, Present, and Future." In C. Sherrill (Ed.). *Sport and*

Disabled Athletes (pp. 3–19). Champaign, IL: Human Kinetics.

Stebbins, Robert A. 1987. *Canadian Football: The View from the Helmet.* London, Ontario: Centre for Social and Humanistic Studies of the University of Western Ontario.

Stebbins, Robert A. 1993. "Stacking in Professional American Football: Implications for the Canadian Game." *International Review for the Sociology of Sport* 28: 65–73.

Steele, Paul D. and Zurcher, Louis A., Jr. 1973. "Leisure Sports as 'Ephemeral Roles.'" *Pacific Sociological Review* 16: 345–356.

Stein, Michael. 1977. "Cult and Sport: The Case of Big Red." *Mid-American Review of Sociology* 2: 29–42.

Steinberg, Leigh. 1991. "The Role of Sports Agents." In P. D. Staudohar and J. A. Mangan (Eds.). *The Business of Professional Sports* (pp. 247–262). Urbana and Chicago: University of Illinois Press.

Steinberg, Leigh. 1992. "Agents and Agency: A Sports Agent's View." *Journal of Sport & Social Issues* 16: 113–115.

Steinbreder, John. 1993. "The Owners." *Sports Illustrated* (September 13): 64–86.

Stevenson, Christopher L. 1986. "The Culture of the Weight Room." Paper presented at the North American Society for the Sociology of Sport Annual Conference, Las Vegas.

Stevenson, Christopher L. 1990. "The Early Careers of International Athletes." *Sociology of Sport Journal* 7: 238–253.

Stevenson, Christopher L. and Nixon, John E. 1972. "A Conceptual Scheme of the Social Functions of Sport." *Sportswissenschaft* 2: 119–132.

Stoddart, Brian. 1989. "Sport in the Social Construct of the Lesser Developed World: A Commentary." *Sociology of Sport Journal* 6: 125–135.

Stoddart, Brian. 1990. "Wide World of Golf: A Research Note on the Interdependence of Sport, Culture, and Economy." *Sociology of Sport Journal* 7: 378–388.

Stone, Gregory. 1955. "American Sport: Play and Display." *Chicago Reiew* 9: 83–100.

Straw, Philip. 1986. "Point Spreads and Journalistic Ethics." In R. E. Lapchick (Ed.). *Fractured Focus: Sport as a Reflection of Society* (pp. 259–274). Lexington, MA: Lexington Books.

Strenk, Andrew. 1977. "Sport as an International Political and Diplomatic Tool." *Arena Newsletter* 1: 3–10.

Strenk, Andrew. 1979. "What Price Victory? The World of International Sports and Politics." *Annals of the American Academy of Political and Social Science* 445: 128–140.

Sullivan, Robert. 1986. "Sermon on the Mound." *Sports Illustrated* (November 10): 11.

Swift, E. M. 1991a. "Why Johnny Can't Play." *Sports Illustrated* (September 23): 60–72.

Swift, E. M. 1991b. "Reach Out and Touch Someone." *Sports Illustrated* (August 5): 54–58.

Swift, E. M. 1992. "Women of Mettle." *Sports Illustrated* (March 2): 38–39.

Tatum, Jack. 1972. *They Call Me Assassin.* New York: Doubleday.

Taub, Diane E. and Benson, Rose Ann. 1992. "Weight Concerns, Weight Control Techniques, and Eating Disorders Among Adolescent Competitive Swimmers: The Effect of Gender." *Sociology of Sport Journal* 9: 76–86.

Taylor, Ian. 1972. "'Football Mad': A Speculative Study of Football Hooliganism." In E. Dunning (Ed.). *Sport: Readings from a Sociological Perspective* (pp. 352–377). Toronto: University of Toronto Press.

Tedeschi, Bob. 1991. "Will a Reunified Germany Continue to Dominate the International Sports Scene?" *Women's Sports and Fitness* (January/February): 45–50.

Telander, Rick. 1988. "Sports Behind the Walls." *Sports Illustrated* (October 17): 82–96.

Telander, Rick and Sullivan, Robert. 1989. "You Reap What You Sow." *Sports Illustrated* (February 27): 20–31.

The Final Report of the President's Commission on Olympic Sports. 1977. Washington, DC: Superintendent of Documents.

Theberge, Nancy. 1977. *An Occupational Analysis of Women's Professional Golf.* Unpublished doctoral dissertation, University of Massachusetts, Amherst.

Theberge, Nancy. 1978. "The World of Women's Professional Golf: Responses to Structured Uncertainty." In M. Salter (Ed.). *Play: Anthropological Perspectives.* West Point, NY: Leisure Press.

Theberge, Nancy. 1984. "Some Evidence on the Existence of a Sexual Double Standard in Mobility to Leadership Positions in Sport." *International Review for the Sociology of Sport* 19: 185–195.

Theberge, Nancy. 1985. "Toward a Feminist Alternative to Sport as a Male Preserve." *Quest* 37: 193–202.

Theberge, Nancy. 1993. "The Construction of Gender in Sport: Women, Coaching, and the Naturalization of Difference." *Social Problems* 40: 301–313.

Theberge, Nancy and Cronk, Alan. 1986. "Work Routines in Newspaper Sports Departments and the Coverage of Women's Sports." *Sociology of Sport Journal* 3: 195–203.

Thirer, Joel and Wright, Stephen. 1985. "Sport and Social Status for Adolescent Males and Females." *Sociology of Sport Journal* 2: 164–171.

Thomas, William I. 1927. *The Unadjusted Girl.* Boston: Little, Brown.

Tillich, Paul. 1948. *The Shaking of the Foundations.* New York: Scribner's.

Todd, Terry. 1987. "Anabolic Steroids: The Gremlins of Sport." *Journal of Sport History* 14: 87–107.

Torrez, Danielle Gagnon and Lizotte, Ken. 1983. *High Inside.* New York: G. P. Putnam's Sons.

Tow, Ted C. 1982. "The Governance Role of the NCAA." In James H. Frey (Ed.). *The Governance of Intercollegiate Athletics.* (pp. 108–116). West Point, NY: Leisure Press.

True, S. 1983. *Data on the Percentage of Girls' High School Athletic Teams Coached by Women.* Kansas City, MO: National Federation of State High School Associations.

Turner, Ralph. 1978. "The Role and the Person." *American Journal of Sociology* 84: 1–23.

Turner, Ralph H. 1988. "Personality in Society: Social Psychology's Contribution to Sociology." *Social Psychology Quarterly* 51: 1–10.

Tygiel, Jules. 1983. *Baseball's Great Experiment: Jackie Robinson and His Legacy.* New York: Oxford University Press.

U.S. Bureau of the Census. 1991. *Statistical Abstract of the United States 1990.* Washington, DC: U.S. Government Printing Office.

Uehling, Barbara S. 1983. "Athletics and Academe: Creative Divorce or Reconciliation." *Educational Record* 64: 13–15.

USA Today. 1991. "How Colleges Compare in Minority Hiring." (March 19): 11A.

Uwechue, Ralph C. 1978. "Nation Building and Sport in Africa." In B. Lowe, D. B. Kanin, and A. Strenk (Eds.). *Sport and International Relations.* Champaign, IL: Stipes.

Van Bottenburg, Maarten. 1992. "The Differential Popularization of Sports in Continental Europe." *The Netherlands' Journal of Social Science* 28: 3–30.

Vande Berg, Leah R. and Trujillo, Nick. 1989. "The Rhetoric of Winning and Losing: The American Dream and America's Team." In L. A. Wenner (Ed.). *Media, Sports, & Society* (pp. 204–224). Newbury Park, CA: Sage.

Vanreusel, B. and Renson, R. 1982. "The Social Stigma of High-Risk Sport Subcultures." In A. O. Dunleavy, A. W. Miracle, and C. R. Rees (Eds.). *Studies in the Sociology of Sport* (pp. 183–202). Fort Worth, TX: Texas Christian University Press.

Vaughan, Diane. 1983. *Controlling Unlawful Organizational Behavior: Social Structure and Corporate Misconduct.* Chicago: University of Chicago Press.

Vaz, Edmund W. 1972. "The Culture of Young Hockey Players: Some Initial Observations." In A. W. Taylor and M. L. Howell (Eds.). *Training: Scientific Basis and Application* (pp. 222–234). Springfield, IL: Charles C Thomas.

Vaz, Edmund W. 1982. *The Professionalization of Young Hockey Players.* Lincoln, NE: University of Nebraska Press.

Veblen, Thorstein. 1899. *Theory of the Leisure Class.* New York: Macmillan.

Verducci, Tom. 1993. "Sign of the Times." *Sports Illustrated* (May 3): 14–21.

Voigt, David Q. 1976. *America Through Baseball.* Chicago: Nelson-Hall.

Voigt, David Q. 1991. "Serfs Versus Magnates: A Century of Labor Strife in Major League Baseball." In P. D. Staudohar and J. A. Mangan (Eds.). *The Business of Professional Sports* (pp. 95–114). Urbana and Chicago: University of Illinois Press.

Voy, Robert with Deeter, Kirk D. 1991. *Drugs, Sport, and Politics.* Champaign, IL: Leisure Press.

Wacquant, Loïc J. D. 1992. "The Social Logic of Boxing in Black Chicago: Toward a Sociology of Pugilism." *Sociology of Sport Journal* 9: 221–254.

Wadler, Gary I. and Hainline, Brian. 1989. *Drugs and the Athlete.* Philadelphia: F. A. Davis.

Wagner, Eric A. 1988. "Sport in Revolutionary Societies: Cuba and Nicaragua." Unpublished paper.

Wagner, Eric A. 1990. "Sport in Asia and Africa: Americanization or Mundialization?" *Sociology of Sport Journal* 7: 399–402.

Wankel, Leonard M. and Sefton, Judy M. 1989. "A Season-Long Investigation of Fun in Youth Sports." *Journal of Sport and Exercise Psychology* 11: 355–366.

Wann, Daniel L. and Branscombe, Nyla R. 1990. "Die-Hard and Fair-Weather Fans: Effects of Identification on BIRGing and CORFing Tendencies." *Journal of Sport & Social Issues* 14: 103–117.

Watson, Geoffrey G. 1975. "Sex Role Socialization and the Competitive Process in Little Athletics." *The Australian Journal of Health, Physical Education and Recreation* 70: 10–21.

Watson, Geoffrey G. 1976. "Reward Systems in Children's Games: The Attraction of Game Interaction in

Little League Baseball." *Review of Sport and Leisure* 1: 93–117.

Watson, Geoffrey G. 1977. "Games, Socialization and Parental Values: Social Class Differences in Parental Evaluation of Little League Baseball." *International Review of Sport Sociology* 13: 17–48.

Watson, Geoffrey G. and Kando, Thomas M. 1976. "The Meaning of Rules and Rituals in Little League Baseball." *Pacific Sociological Review* 19: 291–316.

Webb, Harry. 1969. "Professionalization of Attitudes Toward Play Among Adolescents." In G. S. Kenyon (Ed.). *Proceedings of C.I.C. Symposium on the Sociology of Sport* (pp. 161–178). Chicago: The Athletic Institute.

Weber, Max. 1922. *Economy and Society.* New York: Bedminster Press. Excerpted in H. H. Gerth and C. Wright Mills (trans. and eds.), *From Max Weber: Essays in Sociology.* New York: Oxford University Press, 1958.

Weber, Max. 1946. *From Max Weber: Essays in Sociology.* H. H. Gerth and C. Wright Mills (Trans./Eds.). New York: Oxford University Press.

Weber, Max. 1958 (1904). *The Protestant Ethic and the Spirit of Capitalism.* New York: Scribners.

Weiss, Paul. 1969. *Sport: A Philosophic Inquiry.* Carbondale, IL: Southern Illinois University Press.

Weistart, John. 1987. "College Sports Reform: Where Are the Faculty?" *Academe* 73: 12–17.

Weistart, John C. and Lowell, Cym H. 1979. *The Law of Sports.* Indianapolis, IN: Bobbs-Merrill.

Weitzer, J. E. 1989. *Childhood Socialization into Physical Activity: Parental Perceptions of Competence and Goal Orientations.* Unpublished master's thesis. University of Wisconsin, Milwaukee.

Wenner, Lawrence A. (Ed.). 1989. *Media, Sports & Society.* Newbury Park, CA: Sage.

White, Philip and Curtis, James. 1990a. "Participation in Competitive Sport among Anglophones and Francophones in Canada: Testing Competing Hypotheses." *International Review for the Sociology of Sport* 25: 125–141.

White, Philip and Curtis, James E. 1990b. "English/French Canadian Differences in Types of Sport Participation: Testing the School Socialization Explanation." *Sociology of Sport Journal* 7: 347–368.

Whitson, David and Macintosh, Donald. 1993. "Becoming a World-Class City: Hallmark Events and Sport Franchises in the Growth Strategies of Western Canadian Cities." *Sociology of Sport Journal* 10: 221–240.

Widmeyer, W. Neil and Birch, J. S. 1984. "Aggression in Professional Ice Hockey: A Strategy for Success or a Reaction to Failure?" *Journal of Psychology* 117: 77–84.

Wilkinson, Rupert. 1984. *American Tough: The Tough-Guy Tradition and American Character.* Westport, CT: Greenwood.

Will, George. 1994. "Strike 8?" *Charlotte Observer* (July 10): 3C.

Willey, Fay. 1988. "Traveling Hooligans." *Newsweek* (June 27): 37.

Williams, Robin M., Jr. 1970. *American Society: A Sociological Interpretation* 3rd ed. New York: Alfred A. Knopf.

Williams, Roger L. and Youssef, Zakhour I. 1975. "Division of Labor in College Football Along Racial Lines." *International Journal of Sport Psychology* 6: 3–13.

Wilson, John. 1991. "Efficiency and Power in Professional Baseball Players' Employment Contracts." *Sociology of Sport Journal* 8: 326–340.

Wilson, John. 1994. *Playing By the Rules: Sport, Society and the State.* Detroit: Wayne State University Press.

Wilson, Wayne. (Ed.). 1990. *Gender Stereotypes in Televised Sports.* Los Angeles: Amateur Athletic Foundation.

Wilstein, Steve. 1993. "Illegal Gambling An Obsession? You Can Bet On It." *Charlotte Observer* (January 24): 1C, 8C.

Winbush, Raymond A. 1987. "The Furious Passage of the African-American Intercollegiate Athlete." *Journal of Sport & Social Issues* 11: 97–103.

Wolff, Alexander. 1987. "Playing By Her Own Rules." *Sports Illustrated* (July 6): 38.

Wolff, Alexander. 1992. "The Slow Track." *Sports Illustrated* (September 28): 52–64.

Womack, Mari. 1979. "Why Athletes Need Ritual: A Study of Magic Among Professional Athletes." In W. J. Morgan (Ed.). *Sport and the Humanities: A Collection of Original Essays* (pp. 27–38). Knoxville: University of Tennessee.

Wrigley, J. R. 1970. "Magic in Sport." Paper presented at the Seventh World Congress of the International Sociological Association. Varna, Bulgaria.

Wulf, Steve. 1989. "No Saints." *Sports Illustrated* (April 24): Scorecard section.

Yablonsky, Lewis and Jonathan Brower. 1979. *The Little League Game: How Kids, Coaches, and Players Really Play It.* New York: Times Books.

Yamaguchi, Yasuo. 1984. "A comparative Study of Adolescent Socialization into Sport: The Case of Japan and Canada." *International Review for the Sociology of Sport* 19:63–82.

Yamaguchi, Yasuo. 1987. "A Cross-National Study of Socialization Into Physical Activity in Corporate Settings: The Case of Japan and Canada." *Sociology of Sport Journal* 4: 61–77.

Yates, A., Leehey, K., and Shisslak, C. M. 1983. "Running—An Analogue of Anorexia?" *New England Journal of Medicine* 308: 251–255.

Yesalis, Charles E. (Ed.). 1993. *Anabolic Steroids in Sport and Exercise.* Champaign, IL: Human Kinetics.

Yetman, Norman R. 1987. "'Positional Segregation and the Economic Hypothesis': A Critique." *Sociology of Sport Journal* 4: 274–277.

Yetman, Norman R. and Eitzen, D. Stanley. 1984. "Racial Dynamics in American Sport: Continuity and Change." In D. S. Eitzen (Ed.). *Sport in Contemporary Society,* 2nd ed. (pp. 324–343). New York: St. Martin's Press.

Yiannakia, Andrew. 1989. "Toward an Applied Sociology of Sport: The Next Generation." *Sociology of Sport Journal* 6: 1–16.

Yiannakis, Andrew. 1990. "Some Additional Thoughts on Developing an Applied Sociology of Sport: A Rejoinder to Ingham and Donnelly." *Sociology of Sport Journal* 7: 66–71.

Young, Kevin. 1991. "Violence in the Workplace of Professional Sport from Victimological and Cultural Studies Perspectives." *International Review for the Sociology of Sport* 26: 3–14.

Zelman, Walter A. 1976. "The Sports People Play." *Parks and Recreation Magazine* 11 (February): 27–39.

Zoble, Judith E. 1972. "Femininity and Achievement in Sports." In D. V. Harris (Ed.). *Women and Sport: A National Research Conference* (pp. 203–223). Penn State HPER Series, no. 2.

Index

continued

O

P

Q

R

T